Neotropical Social Wasps

Fabio Prezoto
Fabio Santos Nascimento
Bruno Corrêa Barbosa • Alexandre Somavilla
Editors

Neotropical Social Wasps

Basic and Applied Aspects

Springer

Editors
Fabio Prezoto
Universidade Federal de Juiz de Fora
Juiz de Fora, Minas Gerais, Brazil

Bruno Corrêa Barbosa
Universidade Federal de Juiz de Fora
Juiz de Fora, Minas Gerais, Brazil

Fabio Santos Nascimento
Universidade De Sao Paulo
Ribeirão Preto, São Paulo, Brazil

Alexandre Somavilla
Instituto Nacional de Pesquisas da Amazônia
Manaus, Amazonas, Brazil

ISBN 978-3-030-53509-4 ISBN 978-3-030-53510-0 (eBook)
https://doi.org/10.1007/978-3-030-53510-0

This Springer imprint is published by the registered company Springer Nature Switzerland AG
The registered company address is: Gewerbestrasse 11, 6330 Cham, Switzerland

Foreword

Vespidology in Brazil: An Appreciation

This is a landmark volume for anyone interested in social wasps and social insects in general. It contains authoritative chapters on a very wide range of topics, from keys for the identification of genera to specific aspects of morphology, ecology, behavior, and evolution. Most of the chapters feature the social wasps of Brazil through the eyes of Brazilian researchers and others who have worked in Brazil or jointly with Brazilian colleagues. The first chapter of the volume describes the lives of some important figures in the history of Brazilian studies of social wasps. Some of them are mentioned in this preface, where I take advantage of the chance to write a personal appreciation of that history.

It would be difficult to overstate the importance of research in Brazil for the history of social-wasp science, especially for studies of Neotropical species. I first began to appreciate this when preparing for fieldwork in Colombia on tropical wasps in the middle of the twentieth century (1964–65). Among the few publications I took along as guides were copies of monographs by Adolpho Ducke (1876–1959) (Fig. 1), the Austro-Hungarian Trieste-born Brazilian biologist, who, before turning to botany (see Egler 1963), was an entomologist at the Museu Paraense (now Goeldi Museum) in Belem. Even though Ducke's 21 papers with information on social wasps were published many years earlier (during the period 1904–1918), and in three different languages (German, French and Portuguese), they were clearly written and well illustrated. Some contained photographs and drawings of nests, for Ducke considered nests important evidence for studies of vespid phylogeny. His monographs also contained information on natural history and behavior not available elsewhere. Richards and Richards (1951, p. 2) paid tribute to "the great Adolfo Ducke, whose abilities as a naturalist, collector and taxonomist were alike unrivalled" and whose 1910 paper "is still the foundation of all work on S. American POLYBIINAE." Largely because of Ducke's papers I still feel a sentimental attachment to now obsolete names of Neotropical social wasp genera such as *Nectarina*, *Tatua*, and *Gymnopolybia* (now *Brachygastra*, *Epipona*, and

Fig. 1 Adolpho Ducke (1876–1959). (From Egler 1963)

Agelaia, respectively). And I regret the loss of the sonorous name *Marimbonda* (now *Leipomeles*) assigned by Richards (1978) to two Brazilian species, one discovered by Richards in the Mato Grosso and the other named by Ducke.

In addition to Ducke's publications, I used early publications by other Brazilians on Brazilian social wasps. Some of those with information on castes and reproduction were by Hermann von Ihering, a German-born naturalized Brazilian, (1850–1930) and his son Rodolpho (1883–1939). (For references to their work on social wasps see Richards and Richards (1951) and Chap. 1 of the present volume.) Like Ducke, the von Iherings published on many biological topics other than entomology and were primarily based in museums, not in teaching institutions. Similarly, R. L. Araujo, who published several important taxonomic papers on Brazilian social wasps, including notes on nests in the 1940s, was based at the Institute of Biology of São Paulo, dedicated primarily to agricultural research and outreach rather than teaching. Jose Francisco Zikán (1881–1949), who published monographs on the taxonomy of *Mischocyttarus* and other social wasps beginning in 1935 (see references in Ross and Matthews 1991), was employed as a naturalist of the Parque Nacional do Itatiaia (from introduction to Zikan 1949 by W. Duarte de Barros). In a paper published posthumously (Zikan 1949) he described several species of *Mischocyttarus* that he believed to be parasitic. To my knowledge this idea has not subsequently been investigated.

The Brazilian authors of the present volume are in large part the second and third intellectual descendants of a later generation of university-based social-wasp researchers who began their work in the 1960s, including Professors Ronaldo Zucchi, Vilma Maule Rodriques, Vera Lígia Letízio-Machado, and Carminda do Cruz-Landim. At about the same time there was an influx of important social-wasp researchers from outside. At least some of this surge of activity was due to the promotion of interest in social insects by the charismatic and energetic Brazilian honeybee geneticist Warwick Kerr. It was Kerr who encouraged and advised V.M. Rodriquez, for example, beginning when she was an undergraduate and also during her doctoral work; Kerr was also host to a 1964 visit to Brazil by William D. Hamilton, who then met Rodriquez while he was beginning work on social wasps. Hamilton shared information on the natural history of wasps and supplied

Fig. 2 O. W. Richards (1901–1984) at the campo-cerrado boundary near the Mato Grosso base camp of the Xavantina-Cachimbo expedition of the Royal Society of London and the Royal Geographical Society, March 1968. Richards, a noted naturalist and ecologist, is contemplating a plant, which he identified as *Paepalanthus* (Eriocaulaceae). (Photograph by Maude J. Richards; sent by O.W. Richards to the author in 1971)

difficult-to-obtain reprints of articles (Rodriquez, 1968). Hamilton was in Brazil when he did the final revisions on his influential 1964 papers (Hamilton 1964a, b) on the evolution of social behavior by what was later called "kin selection." While in Brazil he divided the original manuscript into two parts and added his own observations on Brazilian social wasps (see Hamilton 1996, p.29), including especially *Polistes*. Hamilton's papers appeared in print while he was still in Brazil and, coincidentally, while I was in Colombia studying *Polistes*. Hamilton's mention of Brazilian *Polistes* attracted my attention and enabled me to immediately attempt the first field tests of Hamilton's ideas (see West, 1966, 1967).

O. W. Richards, a leader in taxonomic and biological studies of social wasps (see especially Richards and Richards 1951, Richards 1978) made three trips to Brazil (Richards, 1978, p. 1). The first and longest was in 1968 when he spent 14 weeks at a base camp in the central Mato Grosso maintained by the Royal Society of London-Royal Geographical Society Xavantina-Cachimbo expedition (Fig. 2). That research, and collections made by W.D. Hamilton at the same site, led to the decision by Richards to write a monumental treatise on the social wasps of the Americas (Richards, 1978). He returned to Brazil in 1970 to study type specimens of *Mischocyttarus* species named by Zikán at the Universidade Federal do Paraná, where his host was Padre J. S. Moure. Then, in 1971, Richards travelled to São Paulo to study the historic collections at the Museu de Zoologia da Universidade de São Paulo, hosted by its director Paolo E. Vanzolini. Richards was 77 years old when the book was finally published. He had typed the entire manuscript himself, as I learned from conversations with Richards.

The success of such visitors always depends on the experience and hospitality of their hosts, another notable contribution of Brazilian scientists. The hospitality of rural landowners has also facilitated fieldwork on the social wasps. Robert Jeanne, for example, spent more than a year in Brazil as a graduate student. His doctoral thesis is now well known as a classic study of social wasp *Mischocyttarus drewseni*. In his monograph on that species he acknowledged not only the help of Brazilian scientific colleagues but also the indispensable role of the Hagmann family at Fazenda Taperinha, located on the Rio Ayayá, a branch of the Amazon east of Santarém, Pará. Jeanne wrote that their "assistance, patience, understanding, sympathy, encouragement and hospitality were such as I have never experienced and shall not soon forget" (Jeanne 1972, p.141).

This volume features the special contribution of modern research on social wasps in Brazil. It shows how Brazilian scientists have capitalized on the diversity of species available for study, the opportunity for in-depth studies by people actually living where the wasps live, their special open-ness to international collaborations, and their own long and rich tradition of research on social wasps.

Mary Jane West-Eberhard
Smithsonian Tropical Research Institute
c/o Escuela de Biología, Universidad de Costa Rica
San Pedro, Costa Rica
e-mail: mjwe@sent.com

References

Egler W (1963) Adolpho Ducke – traços biográficos, viagens e trabalhos. Boletin do Museu Paraense Emílio Goeldi. Nova Série. Botánica No. 18:15–130

Hamilton WD (1964a, b) The genetical evolution of social behavior. I, II. J Theor Biol 7:1–16, 17–52

Hamilton WD (1996) Narrow roads of gene land, Volume 1. W.H. Freeman and Company, New York

Jeanne RL (1972) Social biology of the Neotropical wasp *Mischocyttarus drewseni*. Bull Museum Comparative Zool 144(3):63–150

Richards OW (1978) The Social wasps of the Americas excluding the Vespinae. British Museum, London

RichardsOW, Richards MJ (1951) Observations on social wasps of South America (Hymenoptera, Vespidae). Trans R Entomol Soc, London 102:1–107

Rodriguez VM (1968) Estudo sobre vespas sociais do Brasil (Hymenoptera, Vespidae). Doctoral Thesis, Universidade de Campinas, Rio Claro

Ross KG, Matthews RW (eds) (1991) The social biology of wasps. Comstock Publishing Associates, London

West MJ (1966) The nature and determination of castes in *Polistes* wasps. Am Zool 6(4):Abstract 346.

West MJ (1967) Foundress associations in polistine wasps: dominance hierarchies and the evolution of social behavior. Science 157:1584–1585

Zikán JF (1949) O genero *Mischocyttarus* Saussure (Hym. Vespidae), com a descricao de 82 espeies novas. Boletin Parque Nacional do Itatiaia 1:1–251

Prologue

The Neotropical region has the greatest diversity of social wasps in the world, and this is the reason that, throughout the years, attracted the attention of specialist researchers, who investigated different biological, ecological, behavioral, and evolutionary aspects using the species of these social insects as a model.

In recent decades, studies with Neotropical social wasps have increased significantly, as this group offers fertile ground for many researchers; however, the results of these works are fragmented in thousands of scientific articles, thus hindering an insight into the state of art of the taxon.

In this way, the aim of this book is to share the knowledge obtained by several research groups around different countries (like Belgium, Brazil, Colombia, Costa Rica, Japan, New Zealand, Portugal, the UK, and the USA) whose efforts resulted in a huge amount of information on social wasps. This book provides an updated overview of different aspects of Neotropical social wasps and also pays a tribute to pioneering work, rescuing the natural history of these wasps and pointing out trends in different areas of research.

We hope that the experience shared in this book gives readers an exciting glimpse into the fascinating history of Neotropical social wasps, thereby inspiring a new generation of motivated vespidologists.

Best regards,
The Editors

Contents

Chapter 1
A Brief Review of Studies on Social Wasps in Brazil

Sergio R. Andena, Luis Filipe Lopes, James M. Carpenter, and Orlando T. Silveira

Abstract Social wasps in Brazil are common elements being important components of interactions with other organisms, acting as predators as well pollinators. In this chapter, we review the study of this group since the discovery of Brazil, in 1500. In the first two centuries (1500–1700), references to this group are rare, being most of the references to Hymenopteran being about ants and bees. During this period, most of the animal descriptions were done by naturalists who focused on their practical use, derived products, and damage caused by them. Later, with the development of natural sciences and onset of taxonomic systems, studies became more systematic. Many foreign researchers and expeditions visited the country to study its wildlife and particularly social wasps. Furthermore, with the establishment of scientific societies and universities in Brazil, many researchers have pursued the study of this group.

Keywords Wasp researchers · Expeditions · Natural history · Naturalists

---

S. R. Andena (✉)
Universidade Estadual de Feira de Santana, Museu de Zoologia, Divisão de Entomologia, Feira de Santana, Bahia, Brazil

L. F. Lopes
Global Health and Tropical Medicine, GHTM, Instituto de Higiene e Medicina Tropical, IHMT, Universidade Nova de Lisboa, UNL and Centre for Ecology, Evolution and Environmental Changes (cE3c), Faculdade de Ciências da Universidade de Lisboa, Lisbon, Portugal

J. M. Carpenter
American Museum of Natural History, New York, NY, USA

O. T. Silveira
Museu Paraense Emilio Goeldi, Belém, PA, Brazil

F. Prezoto et al. (eds.), *Neotropical Social Wasps*,
https://doi.org/10.1007/978-3-030-53510-0_1

Social wasps are widely distributed in the Neotropical Region and are common elements in tropics (Pickett and Carpenter 2010). Also, they are important components of complex interactions with other organisms in neotropical ecosystems (Silveira et al. 2012) as predators – acting in biological control – as well as pollinators (Carpenter and Marques 2001). Research on this important group of insects, sometimes misunderstood by the general public, because some a small fraction of species have huge nests or are very aggressive, has been the key to important theories such as natural selection, where caste differentiation challenged Charles Darwin (1859), inclusive fitness by Willian Hamilton (Hamilton 1964a, b), and sociobiology by Edward Wilson (1975).

Knowledge about social wasps has been reported since the ancient Greeks, Egyptians, Romans, Native Indians in South and Central America, and the Orient. Wasps, in some places, have played an important role in rituals, legends and symbolism of ancient civilizations. A more detailed account of this part of the history can be found in Edwards (1980).

In this chapter we trace the records of social wasps in Brazil, since its discovery, in 1500, until recently, building from the early reports on Brazilian nature, through the first naturalists, where specific reports on social wasps are scarce, to a period where extensive research work is focused on species of this group.

1.1 Wasp Records in Brazil: Since 1500

Brazil was discovered on April 22 of 1500 by Pedro Álvares Cabral, as part of Kingdom of Portugal marine expansion (Fausto 1996:9). When the Portuguese arrived in Brazil for the first time, they were surprised by the exuberance of the nature in the "new land." Pero Vaz de Caminha (1450–1501), one of the crew members of the ship, wrote a letter addressed to King Don Manuel I (1469–1521), describing the discovery of the "new land", including records of the nature, the animals, and the first contact with the culture of Native Indians. In an excerpt from the letter, Pero Vaz describes "the presentation" of some animals to natives and vice-versa:

> Mostraram-lhes um papagaio pardo que o Capitão traz consigo; tomaram-no logo na mão e acenaram para a terra, como se os houvesse ali.
>
> Mostraram-lhes um carneiro; não fizeram caso dele.
>
> Mostraram-lhes uma galinha; quase tiveram medo dela, e não lhe queriam pôr a mão. Depois lhe pegaram, mas como espantados.

> They showed them a brown parrot that the Captain brings with him, they took him in his hand and beckoned to the earth, as if they were there.
>
> They showed them a ram; they ignored him.
>
> They showed them a chicken; they were almost afraid of her, and they did not want to lay hands on her. Then they took him, but they were astonished.

This first and friendly contact with the native Indians was indispensable for success in the colonization of the Brazil, including the knowledge of nature, although, unfortunately, there is no record of social wasps in this early stage of Brazil.

Records of hymenopterans were rare, and most of the references found are to bees and ants, not specifically to social wasps. However, an anecdote without date, which we thought may be from the time of the early "Brazil Colony," makes some reference to wasps (Lenko and Pappavero 1996: 169–170).

> Dizem que…. Durante o período do Brasil Colônia, um dos colonizadores esbarrou descuidadamente em um ninho de vespas. Estas saíram furiosas, em bando, para atacá-lo. Pondo-se em defesa, com a espada desembainhada, o pobre colonizador pôs-se a dar vãs cutiladas no ar, tentando desesperadamente defender-se dos terríveis ferrões. Mas sem resultado – ficou todo disforme e inchado. Bastante vexado, curtindo as picadas, voltou a embainhar a espada, ferido física e moralmente, dizendo: –Ah! Covardes.... Tivessem vindo uma a uma....

> They say... During the period of the Brazil Colony, one of the settlers bumped carelessly into a wasp nest. They left, furious, to attack him. Turning in defense, with his drawn sword, the poor colonizer began to whip in the air, trying desperately to defend himself from the terrible stings. But without result – it was all disfigured and swollen. Quite vexed, resenting the stings, he again sheathed the sword, wounded physically and morally, saying: –Ah! Cowards …. They had come one by one…

The first hymenopteran records, as with those of other species, are mostly based on their practical use, either as a food source (honey) or derived goods (wax), or, as in the previous example, on the threat that they may pose. Some of the early visitors who reported on Brazilian animal species mentioned them.

The German Hans Staden (1525–1576), who was prisoner of the Tupinambás – a tribe of cannibals (Staden 2006) – depicted, in his diary, some landscapes, and records of bees, probably of the species *Tetragonisca angustula* (Latreille, 1825), *Melipona quadrifasciata* (Lepeletier 1836), and *Scaptotrigona postica* (Latreille, 1807) (Almaça 2002). Some species of bees were widely employed by natives for harvesting honey (Staden 2006) (Fig. 1.1).

Jean de Léry was sent by J. Calvin to "Antarctic France," as was called an island in Guanabara Bay, which was controlled at the time by the French. Later, he refers to bees unlike those in Europe that resemble little black flies and that deposit honey and wax in tree trunks (Almaça 2002). The indigenous people would eat the honey and use the wax to insulate wooden chests. He produced excellent illustrations of the fauna and flora of Brazil in his *Histoire d'un voyage fait en la terre du Bresil, dite Ameriquein*, in 1578.

The priest José de Anchieta (1534–1597) is considered the first major naturalist in Brazil, producing quality descriptions of many animal species (Almaça 2002; Paiva 2000). In May of 1560, in a letter addressed to the general priest of the Companhia de Jesus, he describes about 70 species, among which are some hymenopteran insects, especially ants, considered one of the major destructive forces of the region, as well as a reference to *Atta sexdens* (Linnaeus, 1758), whose winged forms were collected by the indigenous people. A wasp of the Pompilidae

Fig. 1.1 Drawing of a routine of Tupinambás natives' life by Hans Staden (Chapter XLVI)

family is also mentioned, probably *Pepsis elevata* Fabricius, 1804, which is deemed to kill spiders and transport them to the nest (Almaça 2002).

Yves d'Évreux (1577–1632) was sent to Maranhão State, northeastern of Brazil, in 1612, and mentions a wide range of flies, as well as wasps, among those a black wasp that builds complex pottery nests in the tops of trees. Seeking to better observe these nests, he was stung by one and reported feeling excruciating pain and being sick for several days (Almaça 2002).

The publication *Historia Naturalis Brasiliae* must be referred to for this period, as it is considered the first scientific work to be published about Brazilian natural history (Santos 2014; Vanzolini 1996), being published in Leiden and Amsterdam in 1648 (Marcgrave and Piso 1648). The first volume of this publication is of the authorship of Willem Piso on medical subjects and the second of George Marcgrave (1610–1643) about the natural history of Brazil, divided into eight books, entitled *Historiae rerum naturalium Brasiliae libri octo*. Four out of eight chapters are dedicated to zoology and became a work of reference for the Brazilian fauna (Almaça 2002). In Chap. IX (*Infectum volans*), page 255, he pictured the *Paipai guacu*, a Crabronidae).

The little and scattered information about the Brazilian fauna from the 1500s to 1700s is due mainly to the study of natural history in Brazil being hindered by the Portuguese inquisition, which caused a stagnation in science and culture, and by the limitations imposed by the crown on the publication of knowledge obtained in all areas of the natural history of Brazil, treating it as "a state secret to avoid foreign coveting" (Almaça 2002; Paiva 2000; Santos 2014). Furthermore, Brazil was strictly forbidden from having its own press (Nomura 1998). Therefore, most of the Portuguese texts and descriptions on the natural resources of Brazil were only published many years after they were written. In this period, most of the reports of the Brazilian natural history were published by foreigners, especially from areas that were temporarily controlled by France and the Netherlands. Furthermore, since 1604 foreigners were forbidden to enter Brazil to study its natural resources, as was the case with the famous naturalist Alexander von Humboldt, which further hampered the study of Brazilian zoology (Nomura 1998). Only, much later, in 1801, Count Johann von Hoffmannsegg obtained authorization to send his collector Friedrich Wilhelm Sibber to Brazil to collect insect specimens (Papavero 1971).

Nomura (1998) analyzed the work of 93 naturalists who worked on Brazilian zoology in the eighteenth century, describing animals that occur in Brazil, listing the animals referred to by them and when possible trying to identify its current scientific name. Of those 93 naturalists, only some refer to insects and even fewer species of the order Hymenoptera. Despite the few known expeditions, wasp species, collected in Brazil, were described by Europeans during the eighteenth century, as follows:

Carl Linnaeus (1707–1778) named many South American insect species, among them several Hymenoptera, of the families Formicidae, Vespidae, and Tenthredinidae. The Vespidae species were *Synoeca surinama* (Linnaeus, 1767) and *Polistes canadensis* (Linnaeus, 1758).

Francisco Antônio de Sampaio (unknown birth and death, 1700) referred to several insects, among them two hymenopteran species of the family Apidae, which should correspond to *Melipona scutellaris* (Latreille, 1811) and *Tetragonisca angustula* (Latreille, 1825).

Guillaume Antoine Olivier (1756–1814) referred mainly to species of the order Coleoptera in which he specialized, but included some hymenopterans, of the families Apidae, Formicidae, and Vespidae, namely, *Polistes versicolor, Agelaia pallipes, Polybia dimidiata, Polybia occidentalis, Polybia sericea,* etc. (Olivier, 1792).

Johann Christian Fabricius (1745–1808) was responsible for the description of the largest number of insect species of Brazil from the eighteenth century, including some hymenopterans of the families Apidae, Formicidae, and Vespidae, namely, from this last family, *Synoeca cyanea* (Fabricius, 1775).

João Daniel (1722–1776) referred to many insect species and among them several hymenopterans of the families Apidae, Formicidae, Sphecidae, and Vespidae. The Vespidae referred to would be *Polistes* sp. and *Synoeca cyanea*.

Still in the eighteenth century, under the influence of Sebastião José de Carvalho e Melo (1699–1782), Portuguese prime minister, the influence of the inquisition was limited and public instruction was reformed, including the education at University of Coimbra (Paiva 2000; Simon 1983). This opened the way for the creation, in Lisbon, of the Academia Real das Ciencias in 1779. That same year, Joaquim Veloso de Miranda was sent to Brazil with instructions to collect materials and information pertaining to the natural history of Brazil, to be sent to the Real Gabinete de História Natural da Ajuda (Paiva 2000).

A expedition was prepared to the Brazilian region of Pará, organized by Domingos Vandelli (1730–1816), who proposed Alexandre Rodrigues Ferreira (1756–1815), a Brazilian who studied under his direction in the University of Coimbra, to conduct this expedition (Simon 1983). In his "Phylosophycal Voyage", Alexandre Rodrigues Ferreira described and collected many new animals and amassed an incredible collection of the Brazilian fauna, unique at its time. However, he never got to publish any results, and with the occupation of Portugal by French troops (see below), most of these collections were taken by Etienne Geoffroy Saint-Hilaire to France, including over 500 specimens of insects (Paiva 2000; Simon 1983). Most of the work is focused on other groups, but some references to insects and a picture of a wasp nest were included in the illustrations produced in that expedition (Fig. 1.2). Joaquim José Codina (unknown birth date, 1790) and José Joaquim Freire (1760–1847) were artists accompanying the expeditions headed by Alexandre Rodrigues Ferreira.

In the second half of the eighteenth to early nineteenth centuries, some new scientific academies were formed in Brazil, such as the Academia Brasílica dos Renascidos (1759), in Salvador, and the Academia Científica (1772) do Rio de Janeiro. Later, the Casa de História Natural was created, known as Casa dos Pássaros

Fig. 1.2 A nest of wasp depicted by José Joaquim Freire (1760–1847), in Desenhos de gentios, animais quadrúpedes, aves, anfíbios, peixes e insetos da Expedição Filosófica do Pará, Rio Negro, Mato Grosso e Cuiabá. Originais. Volume 1. Deposited at Arquivo Histórico dos Museus da Universidade de Lisboa - Museu Nacional de História Natural e da Ciência, Universidade de Lisboa | PRISC INV. MUHNACMUL-AH, PT-MUL-RMJBA-TC-02-0006

(1784), in Rio de Janeiro, with the objective to collect, prepare, store, and send to the Real Gabinete de História Natural da Ajuda in Portugal. This was the precursor of the current Museu Nacional of Rio de Janeiro, created in 1818 (Paiva 2000) and burned recently, in 2018. In 1810, finally the Royal Press was installed in Brazil, which allowed the publishing of scientific works in Brazil.

All these institutions had a great impact in the Brazil colony, but it was after the move of the Portuguese Court to Brazil that other institutions were created, and science was free to proliferate in the country.

1.2 The Kingdom of Portugal Moves to Brazil in 1808

The Napoleonic Wars, in the nineteenth century, brought serious consequences to Europe. In November 1807 French troops marched into Lisbon, and Prince Dom João, in a few days, decided to move the Portuguese Court to Brazil (Fausto 1996: 75). The transfer of the Court to Brazil involved around 10,000–15,000 people, including ministers, counselors, judges, clergy, armed forces, etc. Furthermore, documents, manuscripts, royal treasure, and libraries were transferred, which later became the National Library in Rio de Janeiro (Fausto 1996:75).

The transfer of the Royal Family to Brazil changed, definitively, the administrative center of the colony to Rio de Janeiro (Fausto1996: 78). In September 1808, the first newspaper in Brazil was published; theaters, libraries, and literary and scientific academies were also opened to meet the requirements of the royal family and an urban population in rapid expansion (Fausto 1996: 78–79).

After Napoleon's defeat Dom João raised Brazil to the United Kingdom of Portugal and Algarves. He left his son Dom Pedro I (named Pedro IV in Portugal) in Brazil, who assumed the throne. Dom Pedro I later went back to Portugal and left his son, Dom Pedro II, an enthusiast of arts and sciences, as the Emperor of Brazil.

1.3 Dom Pedro II and Science in Brazil

No doubt that the political and intellectual effervescence that characterized the coming of the Royal Family to Rio de Janeiro brought great benefits to Brazil. Here we highlight Dom Pedro II – the most learned emperor of Brazil (from 1831 through 1889). Pedro II's interests included anthropology, geography, medicine, law, religion, philosophy, sculpture, music, theater, poetry, technology, arts, language, and natural science (Lyra 1977). He had an extensive correspondence with the most prominent scientists, artists, and intellectuals of the period, just to cite a few: Richard Wagner, Alexander Graham Bell, Louis Pasteur, and Friedrich Nietzsche (Calmon 1975; Lyra 1977; Gray 2006). Some institutions created under Pedro II were: Colegio Pedro II (Pedro II School), Ópera Nacional (National Opera), Instituto Histórico e Geografico Brasileiro (Brazilian Institute of Geography and History), Escola Imperial de Belas Artes (Imperial School of Fine Arts), Instituto Pasteur (Pasteur Institute), and Museu Nacional (National Museum). Another important contribution of Pedro II was to provide means for Brazilian students to attend universities in Europe, fostering the acquisition of new knowledge in the country. These actions had a deep influence on the science of Brazil until the present day.

In the period of the late Empire and early Republic of Brazil (installed in 1889), many naturalists interested in the fauna and flora of Brazil developed their research in the country. Here we emphasize some of those who worked on wasps.

1.4 Amédée Louis Michel Lepeletier of Saint-Fargeau (1770–1845)

Amédée Louis Michel Lepeletier of Saint-Fargeau (1770–1845) was an entomologist in France, dedicating most of his life to the study of Hymenoptera. Also, he was president of the Entomological Society in France, from 1833 until his death, in 1845.

In 1836 Lepeletier published the "Histoire naturelle des insectes – Hyménoptères," as part of the series "Suites à Buffon," edited and published by Roret. This work, comprising four volumes, about 2500 pages and 48 plates – drawn by Priest and having excellent coloring for that time – is still, despite its age, an excellent reference for those interested in studying Hymenoptera. Lepeletier (1836) employed the social behavior as the main feature for separating the species into two groups: solitary and social. He named the social vespids the Polistides, with eight genera, including the new *Polybia, Agelaia, Apoica,* and *Chartergus,* which presently comprise only neotropical species.

1.5 Karl August Möbius (1825–1908)

Möbius was born in Eilenburg in Saxony. In 1844 he passed the exams with distinction and began working as teacher in Seesen, on northwest edge of the Harz mountain range. In 1849 he began studying natural science and philosophy at Humboldt University in Berlin. After he graduated, he taught zoology, botany, mineralogy, geography, physics, and chemistry at the Johanneum High School in Hamburg (https://peoplepill.com/people/karl-moebius-1/ accessed on December, 01 2019)

Most of his works concerned study of oysters of coastal regions of northern Germany, publishing important titles on this subject, as the *Über Austern- und Miesmuschelzucht und Hebung derselben an der norddeutschen Küste* [About oysters and mussel farming and uplifting on the north German coast], published in 1870, which had the object of the production of oysters commercially.

Despite his dedication to oysters, Möbius also published articles on wasps. His article "Die Nester der geselligen Wespen. Beschreibungen neuer Nester - und einiger neuen Wespen-Arten des naturhistorischen Museums zu Hamburg, nebst Betrachtungen über den Nesterbau im Allgemeinen" [The nests of the social wasps. Descriptions of new nests and some new wasps' species of the Museum of Natural History in Hamburg, as well as considerations about nest building in general], published in 1856, comprised 51 pages and 19 colored plates devoted to nest architecture (Fig. 1.3).

In 1888, Möbius became curator of the Zoological Collection at Museum für Naturkunde in Berlin, and professor of Systematic and Geographical Zoology at the Kaiser Wilhelm University, also in Berlin, where he taught until he retired at age of 80, in 1905 (https://peoplepill.com/people/karl-moebius-1/ accessed on December, 01 2019).

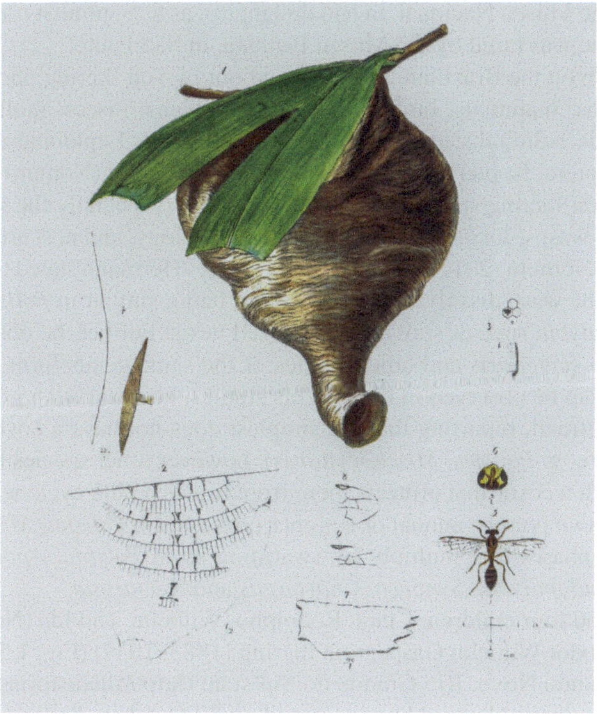

Fig. 1.3 Nest of *Angiopolybia pallens*, figured by Möebius as *Polybia ampullaria*

1.6 Hermann (1850–1930) and Rodolpho von Ihering (1883–1939): Father and Son

Hermann Friedrich Albrecht von Ihering was born in Germany and graduated in Medicine (1886–1873) in Berlin and Göttingen. Just after his graduation, he became a zoological assistant of Carl Claus, a crustacean specialist, however von Ihering worked on the ontogeny of *Cyclas* (a freshwater Mollusca), in his PhD thesis, defended at the University of Göttingen. According to Lopes and Podgorny (2014) von Ihering was pressed to publish his articles and get a job, between 1872 and 1880. During this time, he published at least 30 articles in several zoological and anthropological journals. Also, these authors stated that, if several factors of a personal and professional nature must have influenced von Ihering in coming to Brazil, there is no doubt that the competitive environment among German zoologists and the lack of jobs was also an influencing factor. At the age of 30, in 1880, he moved to Rio de Janeiro. From there he settled in Rio Grande do Sul state, southern Brazil, where he started collecting specimens and sending them to museums in Germany and England in exchange for monetary compensation (Nomura 2012). In 1883, he was hired by the Museu Nacional, in Rio de Janeiro, as a "naturalist traveler; 4 years later, in 1887 he was hired by the Museu Paulista, in São Paulo.

The research on the Brazilian fauna, carried out by von Ihering, included a wide range of groups: mammals, birds, reptiles, amphibians, pisces, mollusks, crustaceans, arachnids, helminths, and several groups of insects – Lepidoptera, Coleoptera, Hemiptera, Diptera, Isoptera, Orthoptera, and Hymenoptera (Nomura 2012).

Hermann von Ihering studied ants (Fig. 1.4), bees (especially the stingless bees of Brazil), and wasps, focusing on their behavior, biology, and nest architecture. As pointed out by Nomura (2012), in an article of 1896, "Hermann stated that in Europe the colony of the wasp dissolves in early winter, being similar in *Polistes* in Brazil, and that the *Polybia* species stay into the closed nests. Further, he observed that in Brazil, *Polybia scutellaris* and other species of the same genus form new nests by swarming, as can be observed in bees. In his studies, he observed a great variety of wasp nests in Brazil, reporting that the simplest does not have a cover, but only a layer of cells, (e. g. *Polistes, Mischocyttarus*), however other species build complicated nests with a cover that protects them from rain. The life cycle was divided by Hermann into two phases: annual or summer phases as in *Polistes, Mischocyttarus,* and perennial phases that multiply by swarming as in *Polybia, Apoica, Tatua* (= *Epipona*), *Pseudopolybia, Synoeca, Chartergus* and *Nectarinia.*"

Hermann had four children: Clara, Rodolpho, Wilhelm, and Ida (Nomura 2012). Rodolpho Theodor Wilhelm Gaspar von Ihering (1883–1939) (Fig. 1.5) was born in Taquara do Mundo Novo, Rio Grande do Sul state (http://dichistoriasaude.coc.fiocruz.br/iah/pt/verbetes/ihenrod.htm, accessed on 30 October, 2019), and was a disciple of his father.

In 1901 Rodolpho graduated in Science and Modern Language, in São Paulo, and, in 1902, he was nominated by his father as Assistant Director of Finances of the

Fig. 1.4 Hermann von Ihering with his wife, Meta Buff von Ihering, possibly in the Reserva Florestal do Alto da Serra, São Paulo State. Available at http://www.kb.dk/images/billed/2010/okt/billeder/object147776/en/)

Fig. 1.5 Picture of Rodolpho Theodor Wilhelm Gaspar von Ihering (1883–1939). (Modified from Paiva and Mesquita (2013))

Museum Paulista (http://dichistoriasaude.coc.fiocruz.br/iah/pt/verbetes/ihenrod.htm, accessed on 30 october, 2019; Nomura 2012). In 1905 he went to Europe to study natural sciences; however, after 6 months he had to come back to Brazil after the death of his brother and his mother (Nomura 2012).

Rodolpho, since 1904, dedicated his studies to the freshwater fishes of Brazil including taxonomy, reproduction, behavior, systematics, etc. (http://dichistoriasaude.coc.fiocruz.br/iah/pt/verbetes/ihenrod.htm, accessed on 30 October, 2019).

Despite his dedication to fishes, he also worked on invertebrates and published several articles and books on zoology and biogeography. He traveled to several parts of Brazil for more than 30 years collecting specimens and records about rituals and stories regarding the Brazilian fauna. One of his most prominent works was the *Dicionário dos animais do Brasil* [Dictionary of animals of Brazil], in 1940.

In 1904, Rodolpho published the article "As vespas sociais do Brazil" [The social wasps of Brazil]. In this revision, he pointed out that "the present review of the Brazilian species shows 11 genera and 130 species." Furthermore, he provided an identification key to the social wasps of South America.

1.7 William J. Fox (Unknown Birth and Death)

Fox was not a researcher on wasps; he was hired at the Academy of Natural Sciences, in Philadelphia, as an assistant librarian. "Nothing is recorded of Fox's early life, but the fact that his first paper, published in 1891, dealt with a trip to Jamaica made in company with C. W. Johnson suggests that his interest in insects may have been first aroused, or at least encouraged, by that beloved dipterist who, as is well known, initiated and nurtured an interest in insects in several boys who later became well known entomologists" (Bradley 1959).

Despite Fox having worked most on the fauna of Hymenoptera of North America, he also published a report on the Hymenoptera collected by Donaldson Smith on an expedition in Africa and a very important series of papers on the Hymenoptera collected by Herbert H. Smith during some years' residence in Brazil (Bradley 1959). His knowledge of the neotropical fauna of social wasps was mainly based on the collection made by Herbert H. Smith, of which he described the genus and species *Charterginus fulvus* (Fox 1902), and 12 more species from other polistine genera. In Vespidae, Fox described nearly 80 species, mostly in Eumeninae. Bradley (1959) stated that Fox's work on Hymenoptera was discontinued in 1898, not because of loss of interest but, he was told, because he was officially discouraged by his superior officer from continuing it; however he, later, published an article about the Eumenidae of Brazil in 1902.

1.8 Henri de Saussure (1829–1905)

Born in Switzerland, Saussure graduated from University of Geneva, where he was introduced to entomology by François Jules Pictet de la Rive and got his PhD at the University of Giessen (Yung 1905; de Claplarède 1905).

"Beside entomology, de Saussure surprised those who approached him by the extent and variety of his knowledge; a naturalist educated in all the fields of zoology, he was also a geologist, archaeologist, historian, and geographer; he stood at the heart of the progress of agronomy as well as of physics or alpinism, and very few were the questions to which he could not immediately give a precise and thorough answer" (Yung 1905).

Under the direction of Dr. Pictet, he began his great monograph of *Guêpes Sociales*, which he pursued in Paris, where he spent several years, following courses at the Sorbonne, often in the laboratories of the museum, making friends with the professors of this establishment (de Claparède, 1905). In 1854 he and his friend Henri Peyrot set out on a long voyage of exploration in the Antilles and Mexico, a difficult journey through the Mexican provinces, then in permanent revolution, during which time he amassed considerable numbers of objects (Yung 1905).

This voyage was an important event in the career of de Saussure; he engaged in research on hygrology, volcanoes, insects, and myriapods; and he made several first-rate discoveries. "Among these were his notes on the Volcan de Jurullo et le pic d'Orizba, his Description des ruines d'une ancienne ville mexicaine, his Observations sur les mammifères et sur divers aiseaux du Mexique, his Recherches sur les orthoptères de l'Amérique moyenne, his Divers Crustacés nouveaux du Mexique et des Antilles, his Essai d'une faune des myriapodes du Mexique and letters which he addressed to our journal, describing in a sober and captivating style the events of which he was the witness and the remarkable men he met during his perilous expedition. His letters on Mexico were for all those who read them a true revelation." The interest was still exceeded, if possible, by the tales de Saussure loved to tell to his friends about this decisive period of his life (Yung 1905; Léjéallien 1906).

He returned from Mexico in 1856, after visiting the United States, where he established friendly relations with Louis Agassiz, Professor Henry, head of the Smithsonian Institute, and other eminent scholars of the New World. De Saussure became a great authority, not only on the Hymenoptera, a group on which he wrote thousands of pages and hundreds of cleverly drawn figures, but also on the Orthoptera, of which he had formed an unrivaled collection. He was sent insects from all the parts of the globe to determine, and his correspondence was immense. While engaged in an investigation of the anatomical forms, he did not lose sight of the living insect, and his observations on the nesting of the *guêpes* will remain a model of the study of the manners of the animals (Yung 1905).

The travels of de Saussure resulted in important collections, which, together with the material deposited in European collections, were the base of his great monograph on the vespids, describing around 50 genera (Van der Vecht and Carpenter 1990). His most important work *Études sur La famille des Vespides*, published in three volumes from 1852 to 1858, divided the Vespidae into three tribes – Masariens, Euméniens, and Vespiens – based on the level of sociality, which, according to the author, would be enough to separate the groups. Besides taxonomic work, de Saussure presented extensive biological data, as well as detailed studies on nest architecture. Some terms created by him to describe general architectural features like "stelocyttarous" and "phragmocyttarous" were widely employed by Richards and Richards (1951) and Richards (1978) and are still in use.

1.9 Adolpho Ducke (1876–1959)

The naturalist Adolpho Ducke (born in Trieste, Italy) became one of the most prominent botanists/entomologists working in the Amazon region in the early 1900s.

As part of the scientific development of the Museu Paraense – in Belém, Pará State, northern Brazil – in 1899, Dr. Emilio Goeldi, head of the Museum, hired several technicians, among them, Adolpho Ducke, at that time at the age of 23. Ducke initially worked in the Zoology Department, as collector and curator of the entomological collection (Miranda 1999), being the first Hymenoptera taxonomist resident in the Brazilian Amazon.

In several articles published in the first two decades of the nineteenth century, Ducke (1904, 1905, 1907, 1908a, b, 1910a, b, 1914, 1918) carried out important work on the neotropical fauna, creating nine new genera of social wasps – most used until very recently and three of them still being valid nowadays. With these new taxa, Ducke promoted the recognition of considerable diversity of form, social organization, and architecture in natural groups traditionally treated as parts of *Polybia*. Ducke's last work on wasps was a catalog of the Brazilian fauna (Ducke 1918). Despite his works on wasps, Ducke was also a botanist. Jacques Huber (1909) *apud* Miranda (1999) says that "Mr. Ducke, entomologist at the Museu Goeldi, collect in his travels to country of the state, in behalf of the Museu Goeldi, to collect, besides insect, a great number of dried plants." In 1914 he started the collection of botany, and, in 1918, he became the head of the Botanical Sector of the Jardim Botânico do Rio de Janeiro (Botanical Garden of Rio de Janeiro), where he produced his last work about wasps. From Rio de Janeiro, Ducke, in 1933, moved back to Northern Brazil, to Manaus, to study the region called "Alto Amazonas."

According to Miranda (1999), in the 1950s, Ducke proposed, along with other botanists, the preservation of the area surrounding Manaus, due to the great floristic diversity. In 1956 that area became the Center of INPA (Instituto Nacional de Pesquisas na Amazônia, National Institute for Amazon Research) Forest Surveys. In 1962, the 10,000 hectare Center of INPA was donated to INPA, receiving the name of the Adolpho Ducke Forest Reserve.

1.10 Joseph Francisco Zikán (1881–1949)

Joseph Francisco Zikán was one of the most important taxonomists of the neotropical social wasps during the twentieth century. Zikán was born in Teplitz-Schönau, present-day Czech Republic, and completed his basic studies in his native country. Migrating at a young age to Brazil, in 1902, he initially settled in São Paulo. He lived for some time in Minas Gerais, where he worked as a teacher and an insect collector, carrying out expeditions there and in Espirito Santo State, where he lived from 1911 until 1923 when he finally settled with his family in Itatiaia, Rio de Janeiro. During the years 1927 and 1928, he made collections in the Amazon, invited by the Rio Negro Prelature. He was then hired as a technical assistant to the current Itatiaia National Park, where he was later promoted to naturalist, having devoted 26 years of his life to the study of the insects in that region (Fundação Oswaldo Cruz – Adolfo Lutz Virtual Library; www.bvsalutz.coc.fiocruz.br; acessed on 06.viii.2019).

Zikán published about 60 works on, besides social wasps, insects of the orders Coleoptera, Lepidoptera, and Diptera. He published two works of major impact on the taxonomy of wasps of the genus *Mischocyttarus* de Saussure, with emphasis on the fauna of the Itatiaia region and other locations in southeastern and southern Brazil. The first of these studies was published in 1935, where he described 27 new species. Besides, the study contained detailed descriptions of nest architecture and other aspects of the biology and distribution of these wasps. The second article was published in 1949, the year of his death, where he described 82 new species of *Mischocyttarus*. The mere consideration of the number of new taxa described by Zikán leaves no doubt about the importance of his work in this field, but it is further reinforced by the notable insect collection he produced, currently deposited in the collection of the Instituto Oswaldo Cruz, in Manguinhos, Rio de Janeiro. It contains close to 150,000 specimens, mainly from the fauna of Itatiaia National Park, in which we highlight the collections of Lepidoptera (57,329 specimens), Coleoptera (56,744 specimens), and Hymenoptera (32,785 specimens). It was acquired by the Instituto Oswaldo Cruz in 1952 (Instituto Oswaldo Cruz Entomological Collection; http://ceioc.fiocruz.br/; assessed on 17.x.2019; Academia Itatiaiense de História; https://acidhisoficial.blogspot.com/P/acidhis patrons.html; assessed on 17.x.2019) (see also Richards 1978).

1.11 Renato Lion de Araujo (1912–1978)

Renato Lion de Araujo was born in Lambarí, Minas Gerais State, and moved to São Paulo to pursue his studies. In 1931 he graduated as an accountant and was hired by the Instituto Biológico, Secretaria da Agricultura, as a supervisor. Three years later, in 1934, he started working in the Museu Paulista, where he developed most of his studies. Initially, Araujo dedicated most of his time working on Isoptera; however, he also worked on wasps and birds. He traveled to most parts of Brazil collecting, especially Isoptera, and compiling a vast collection, presently deposited in the Museu Paulista, São Paulo.

In 1937, Silvestri, visiting Brazil, suggested the creation of a Brazilian society to serve entomologists. Araujo followed through to create the Sociedade Brasileira de Entomologia, of which he was the head from 1947 to 1951.

Regarding Araujo's work on Hymenoptera, one of his most prominent projects was the revision of the genus *Metapolybia,* describing two new species (Araújo 1945). Moreover, he proposed a new name, *Angiopolybia*, for the new concept of *Stelopolybia* (Ducke 1914) (see Araújo 1946) and other important works about Vespidae, such as his work of 1951, in which he expands the knowledge of *Clypearia angustior*, about Polybiinae (=Epiponini) (Araújo 1960), and also the knowledge of *Synoecoides* (Araújo 1944).

1.12 Owain Westmacott Richards (1901–1984)

Richards was born in Croydon, United Kingdom, in 1901, had a long interest in natural history, and, according to his own record, had from "about the age of 10" been determined to take up entomology (Southwood,1987). In 1912, he moved with his family to Cardiff and was sent as a boarder to Hereford Cathedral School, but on his weekends at home, he roamed the countryside on his bike collecting butterflies and plant specimens for his youngest brother Paul Westmacott Richards (Southwood 1987; http://herbariaunited.org/wiki/Owain_Westmacott_Richards).

Richards graduated from Oxford (Southwood 1987). After leaving Oxford he worded at Imperial College where he remained until 1967 when he retired from the Headship of the Department of Zoology and Applied Entomology and from the Directorship of the College Field Station at Silwood Park (Waloff 1986). In 1959 he was elected to the Fellowship of the Royal Society, and it was his achievements, rather than any other person's, that sustained the international reputation of this department of entomology (Waloff 1986).

Richard approached fieldwork as one of the milestones in the development of ecology and he contributed much to the understanding of quantitative life budgets (Waloff 1986). He firmly believed that a good grounding in basic taxonomy is essential for an ecologist and liked to point out that even closely related species differ in their ecological behavior and physiological requirements (Waloff 1986).

Despite most Hymenopterists knowing Richards as a taxonomist of wasps, he also worked on other groups of insects, such as Diptera and Orthoptera. His wide biological interests lay in ecology, taxonomy, and the theory of evolution (Waloff 1986). He was a world authority on the taxonomy of Diptera, Sphaeroceridae in particular, and, of course, on aculeate Hymenoptra, which were his life-long interest and on which his long list of papers and monographs culminated in 1978 in the publication of his *opus magnum, The social wasps of the Americas, excluding the Vespinae* (Waloff 1986).

Between 1950 and the mid-1970s, Richards travelled widely, always collecting, identifying, observing and often shaming local biologists with the extent of his knowledge of their flora and fauna (Southwood 1987). International congresses and meetings were one cause, perhaps one might say excuse, for such trips (Southwood 1987). More extensive journeys included a visit to the Rukwa Valley in Tanzania in 1952, to advise the International Red locusts Control Service on methods of estimating red locusts (*Nomadacris septemfasciata*) (Southwood 1987). Also, his travels included two expeditions to Guyana, in the first of which his ecological and youngest brother – Professor Paul W. Richards – also participated. The two brothers derived much pleasure and inspiration from discussions of their common interests (Waloff 1986). In 1968, he and his first wife Maud, a well-known entomologist, joined the Royal Society Mato Grosso Expedition to Brazil. Later he visited there alone and paid numerous visits to North and South America in his pursuit of the aculeate Hymenoptera (Waloff 1986).

Most of this work was undertaken in the British Museum (Natural History), where he was an Honorary Associate (Southwood 1987). When he retired from his chair at Imperial College, he was offered and accepted space in the Hymenoptera Section at the Museum and worked there full-time, except for the day or so a week he spent in Silwood (Southwood 1987). He was a prodigious worker and even long after his retirement confessed to feeling shame-faced when he spend time away from his studies on insects (Waloff 1986). He donated to the Museum his collection of 69,000 Hymenoptera and 14,000 Diptera and incorporated much of this material to the museum collection (Southwood 1987).

O.W. Richards was the author or joint author of more than 180 papers and six books, the first of which, *The Variations of Animals in Nature,* was written with J.C. Robson and published in 1936. In 1957 and again in 1977, together with R.G. Davies, he revised the entomologists' bible, *Imm's Textbook of Entomology* (Waloff 1986; Southwood 1987). With his wife, Maud Jessie Richards, he published in 1951 a study about social wasps of South America, especially based on his Guiana work. In 1978 he published *The Social Wasps of the Americas, excluding the Vespinae* (see above), his most important publication on social wasps. In his publications, he described 15 new Vespidae genera (Van der Vecht and Carpenter 1990) and around 150 species and subspecies of Neotropical Polistinae, the taxonomic revision of *Mischocyttarus* de Saussure, and others.

Despite all the advances in the field since its publication, *The social wasps of the Americas, excluding the Vespinae* still remains a fundamental reference for those studying the New World's social wasps and certainly has been inspiring generations since then.

1.13 Universidade de São Paulo (USP, Ribeirão Preto) and Universidade Estadual Paulista (UNESP, Rio Claro): Two Groups Involved in the Study of Wasps in Brazil

These two universities employed important hymenopterists that have been involved in the most recent studies on Brazilian wasps. Most of these researchers were deeply influenced by or were disciples of Warwick Estevam Kerr (1922–2018), who worked in both universities. Kerr was born in 1922 in Santana de Parnaíba, São Paulo State, and started his academic career at a time when there was an extraordinary development of this field in São Paulo, thanks to the presence of Carlos Arnaldo Krug, Friedrich Gustav Brieger, André Dreyfus, and Theodosius Dobzhansky (Coelho and Kerr 2005).

In 1955 Kerr became head of the Rio Claro Department of Biology at the UNESP, when this University was created. In 1965, he became head of the Department of Genetics at the Faculty of Medicine of USP – Ribeirão Preto, where he became a full professor in 1971.

In these two universities, Kerr began the study of Hymenoptera, and although he worked mainly on stingless bees, some of his students were stimulated to work on social wasps.

In the 1960s, Ronaldo Zucchi, one of Kerr's students, started his studies on the behavior of social wasps in Rio Claro and later moved to Ribeirão Preto, having published more than 130 articles on bees and wasps. Zucchi was also influenced by Shôichi F. Sakagami (1927–1996), Emeritus Professor of Hokkaido University, who was a visiting Professor during 1971–1977 and studied stingless bees and social wasps. Zucchi was the advisor of Dércio Simões, Nivar Gobbi, and Sulene Noriko Shima, who were hired by UNESP Rio Claro and all have worked with social wasps. Presently all of them are retired.

Also, in the 1960s Vilma Maule Rodrigues (1938–2019) and Vera Lígia Letízio Machado were precursors of a generation of wasp researchers at UNESP Rio Claro. Their former students José Roque Raposo Filho and Edilberto Giannotti (all retired) supervised many of the current generation of researchers presently working in Brazil.

1.14 Final Remarks

This brief review attempts to compile the historic research on wasps carried out in Brazil for the last 520 years, from the early naturalists, whose reference to this group was scarce and focused on eminently practical faunistic features, to the most recent academic research. Even with an increase of researchers and publications devoted to the group in the last decades, there is still much to be done in this field, considering the country's vast territory and the environmental variability it holds, comprising six main biomes (Atlantic Rain Forest, Amazon Rain Forest, Cerrado, Caatinga, Pampa, and Pantanal). We believe that the work done in this field has fostered the formation of a new generation of Brazilian researchers that we hope may continue in the front line of creating new knowledge on this important group of Hymenoptera.

All these researchers, cited above, left deep roots on the wasps' studies in Brazil, and this book is a tribute to those who have, direct or indirectly, contributed to this knowledge.

Along the chapters of this book is presented the "state-of-art" of this fascinating group of insects.

Acknowledgment Lopes LF was funded by FCT through funds to GHTM – UID/Multi/04413/2013 and CE3C, UID/BIA/00329/2013. We acknowledge the Museu Nacional de História Natural e da Ciência, Universidade de Lisboa, and the Portuguese Infrastructure of Scientific Collections (PRISC) for the access to the illustrations in the book *Desenhos de gentios, animais quadrúpedes, aves, anfíbios, peixes e insetos da Expedição Filosófica do Pará, Rio Negro, Mato Grosso e Cuiabá. Originais. Volume 1*.We dedicate this chapter to all researchers along the history who worked on wasps and contributed to the scientific knowledge of the Brazil, specially Ronaldo Zucchi, Sulene Shima, and Nivar Gobbi, recently retired, who left many students for keeping the wasp's studies alive in Brazil. Also, this chapter is dedicated to Edilberto Giannotti for teaching about the wasp's history for decades at UNESP (Rio Claro), as well his suggestions and friendship.

References

Academia Itatiaiense de História; https://acidhisoficial.blogspot.com/p/patronos-da-acidhis.html; assessed on 17.x.2019

Almaça C (2002) A zoologia pré-lineana no Brasil. Museu Bocage – Museu Nacional de História Natural, Lisbon

Araújo RL (1944) Contribuição para o conhecimento do gênero *Synoecoides* Ducke, 1905 (Hym., Vespidae). Brazil J Biol 5:29–35

Araújo RL (1945) Contribuição para o conhecimento do gênero *Metapolybia* Ducke, 1905 (Hymenoptera, Vespidae). Arq Instit Biol 16:65–82

Araújo RL (1946) *Angiopolybia* nom, n. para o conceito revalidado de "*Stelopolybia* Ducke, 1914" (Hymenoptera, Vespidae-Polybiinae). Pap Avul Zool 7:165–170

Araújo RL (1951) Contribuição para o conhecimento de *Clypearia angustior* Ducke, 1906. (Hym. Vespidae). Arq MusNac 42:49–56

Araújo RL (1960) Insecta Amapaensia (Hymenoptera: Vespidae, Polybiinae). Studia Entomol 3:251–253

Bradley JC (1959). The influence of the American Entomological Society upon the study of Hymenoptera. Trans Amer Entomol Soc (1890) 85:4. The Centennial Number (Dec., 1959) pp 277–301

Calmon P (1975) História de D. Pedro II. 1–5. J. Olympio, Rio de Janeiro

Carpenter JM, Marques OM (2001) Contribuição ao Estudo dos Vespídeos do Brasil. Universidade Federal da Bahia, Departamento de Fitotecnia, Bahia, CD-ROM, 147p

Coelho MA, Kerr WE (2005) A Amazônia, os Índios e as Abelhas. Rev Estud Avan 19 (53):51–69. Entrevista de Warwick E. Kerr concedida a Marco A. Coelho

Darwin C (1859) The origin of species (by means natural selection). John Murray, London

de Saussure H (1853–58) Monographie des Guêpes Sociales ou de la Tribu des Vespiens. Masson, Paris, and J. Cherbuliez, Genève. [1–96, 1853; 97–256, 1854]
Ducke A (1904) Sobre as Vespidas sociaes do Pará. Bol Mus Goeldi 4:317–374
Ducke A (1905) Sobre as Vespidas sociaes do Pará. Bol Mus Goeldi 4:652–698
Ducke A (1907) Connaissance de la faune Hyménoptérologique du nord-est du Brésil. Revue D'entomologique 23:73–96
Ducke A (1908a) Contribution à la connaissance de la faune hyménoptèrologique du nord-est du Brésil. II. Hyménoptères récoltés dans l'Etat de Ceará en 1908. Rev Ent 27:57–87
Ducke A (1908b) Contribution à la connaissance des hyménoptères des deux Amériques. Rev Ent 27:28–55
Ducke A (1910a) Contribution á la connaissance de la faune hymènoptérologique du nord-est du Brésil. Rev Ent 29:78–122
Ducke A (1910b) Contribution à la connaissance des scoliides de l'Amérique du Sud. III. Rev Ent 29:73–77
Ducke A (1914) Über Phylogenie und Klassifikation der sozialen Vespiden. Zool Jb Abt Syst Geogr Biol Tiere 36:303–330
Ducke A (1918) Catálogo das vespas sociaes do Brazil. Rev Mus Paulista 10:314–374
Edwards R (1980) Social wasps, their biology and control. Rentokil Lib
Fausto B (1996). História do Brasil, Edusp . 89pp
Fox WJ (1902) Contributions to a knowledge of the Hymenopera of Brazil, N 8 – Eumenidaae, continued (genus *Odynerus*). Proc Acad Sci Phil 1902:44–69
Fundação Oswaldo Cruz – Adolfo Lutz Virtual Library; www.bvsalutz.coc.fiocruz.br; assessed on 06.viii.2019
Gray C (2006) Reluctant genius: the passionate life and inventive mind of Alexander Graham Bell. Harper Collins, Toronto
Hamilton W (1964a) The genetical evolution of social behaviour. I. J Theor Biol 7(1):1–16. https://doi.org/10.1016/0022-5193(64)90038-4
Hamilton W (1964b) The genetical evolution of social behaviour. II. J Theor Biol 7(1):17–52. https://doi.org/10.1016/0022-5193(64)90039-6
http://herbariaunited.org/wiki/Owain_Westmacott_Richards, accessed on 30/x/ 2019
http://dichistoriasaude.coc.fiocruz.br/iah/pt/verbetes/ihenrod.htm, accessed on 30/x/ 2019
https://peoplepill.com/people/karl-moebius-1/ accessed on 01 Dec 2019
Léjéallien L (1906) Henri de Saussure. J Soc Am 3(1):97–99
Lenko K, Pappavero N (1996) Insetos no Folclore. Ed. Pleiade/FAPESP, São Paulo
Lepeletier A (1836) Histoire naturelle des insectes – Hyménoptères. Tome 1. Librairie Encyclopedique de Roret, Paris
Lery J (1578) Histoire d'un voyage fait en la terre du Bresil, dite Ameriquein. La Rochelle
Lopes MM, Podgorny I (2014) Entre mares e continentes: aspectos da trajetória científica de Hermann von Ihering, 1850-1930. História, Ciências, Saúde – Manguinhos, Rio de Janeiro 21(3):809–826
Lyra H (1977) História de Dom Pedro II (1825–1891): Declínio (1880–1891). 3. Itatiaia, Belo Horizonte
Marcgrave G, Piso W (1648) Historia Naturalis Brasiliae, Leiden, Amsterdam
Miranda IPA (1999) Adolpho Ducke, Uma visão de conservação da Amazônia. Acta Amaz 29(3):501
Nomura H (1998) História da Zoologia no Brasil: Século XVIII. Museu Bocage – Museu Nacional de História Natural, Lisbon
Nomura H (2012) Hermann von Ihering, o Naturalista. Card Hist Ciênc 8(1):9–60
Paiva MP (2000) Contribuição Portuguesa para o estudo das Ciências Naturais no Brasil colonial (1500–1822). Museu Bocage – Museu Nacional de História Natural, Lisbon
Papavero N (1971) Essays on the history of neotropical Dipterology. Museu de Zoologia. Universidade de São Paulo

Pickett KM, Carpenter JM (2010) Simultaneous analysis and the origin of eusociality in the Vespidae (Insecta: Hymenoptera). Arthr Syst Phyl 68:3–33
Richards OW (1978) The social wasps of the Americas excluding the Vespinae. British Museum (Natural History, London, p 580
Richards OW, Richards MJ (1951) Observations on the social wasps of South America (Hymenoptera, Vespidae). Trans Ent Soc Lond 102:1–170
Santos MS (2014) Naturalists in nineteenth-century Brazil. Archiv Weltmuseum Wien 63–64:38–59
Silveira OT, Silva SS, Pereira JLG, Tavares IS (2012) Local-scale spatial variation in diversity of social wasps in an Amazonia rain forest in Caxiuanã, Pará, Brazil (Hymenoptera, Vespidae, Polistinae). Rev Brasil Entomol 56:329–346
Simon WJ (1983) Scientific expeditions in the Portuguese overseas territories (1783–1808) and the role of Lisbon in the intellectual-scientific community of the late eighteenth-century. Instituto de Investigação Científica Tropical, Lisbon
Southwood (1987). Sir Richard Westmacott Richards, 31 December 1901–10 November 1984. Biogr Mems Fell R Soc 33
Staden H (2006) Duas Viagens ao Brasil. [trad.] Guiomar de Carvalho Franco, 1st edn. Martins Claret, São Paulo
Van der Vecht J, Carpenter JM (1990) A catalogue of the genera of Vespidae. Zoolo Verhande 260:62
Vanzolini PE (1996) A contribuição zoológica dos primeiros naturalistas viajantes do Brasil. Rev USP São Paulo 30:190–238
Von Ihering H (1896) Zur Biologie der socialen wespen brasiliens. Zoolog 19:449–453
Von Ihering R (1904) As vespas sociaes do Brasil. Rev Mus Paul (6):9–309
Von Ihering R (1940) Dicionário dos Animais do Brasil. Secretaria de Agricultura, Indústira e Comércio, Diretoria de Publicidade AgrícolaVon, São Paulo
Waloff N (1986) Orbituary O. W. Richards, F.R.S. J Anim Ecol 55(1):392–394
Wilson EO (1975) Sociobiology. Belknap/Harvard University Press, Cambridge, MA
Yung E (1905) "Henri de Saussure". Journal de Genève
Zikán JF (1935) Die sozialen Wespen der Gattung *Mischocyttarus* Saussure, nebst Beschreiburg 27 neuen Arten (Hym., Vespidae). Arq Inst Biol Veget, Rio de Janeiro 1:143–203.
Zikán JF (1949) O gênero *Mischocyttarus* Saussure (Hym, Vespidae), com a descrição de 82 espécies novas. Parque Nacional do Itatiaia Bol Parq Nac Itat 1:125

Chapter 2
The Evolution of Swarm Founding in the Wasps: Possible Scenarios

Robert L. Jeanne

Abstract Independent- and swarm-founding wasps represent two discrete social syndromes, differing from each other in a number of traits that include colony size, level of social complexity, queen number, division of labor among workers, nest architecture, body size, and ecological dominance. Swarm founding evolved independently at least four times in the Vespidae. While much attention in recent decades has been paid to unraveling the steps leading to eusociality in the vespids, virtually none has been devoted to understanding how swarm founding evolved from its ancestral independent-founding state. Here I suggest possible scenarios by which the transition could have occurred. I argue that the key initial step was the evolution of pheromonal queen signaling, which enabled the evolution of larger colonies. Larger colonies in turn led to the decentralization of colony control away from a dominant queen and onto the workers. Other traits of the swarm founders, including polygyny, nest envelopes, nocturnality (in *Apoica* and *Provespa*), and small body size, probably evolved later. Swarm founding appears to be an inevitable outcome of the evolution of larger colony size among tropical vespids.

Keywords Social complexity · Decentralization · Colony size · Major transitions · Queen signaling

2.1 Introduction

The eusocial wasps comprise some 1000 species worldwide and encompass a tremendous range of colony size and degrees of social complexity (Jeanne 1991). This diversity is especially evident in the subfamily Polistinae, by far the largest and most widely distributed of the three subfamilies of social vespids. Species of

R. L. Jeanne (✉)
Department of Entomology, University of Wisconsin–Madison, Madison, WI, USA
e-mail: rljeanne@wisc.edu

F. Prezoto et al. (eds.), *Neotropical Social Wasps*,
https://doi.org/10.1007/978-3-030-53510-0_2

Vespidae fall into two conspicuously different behavioral groups, based on their mode of colony founding (Hölldobler and Wilson 1977; Jeanne 1980). In the *independent founders (IF)*, new colonies are initiated by one or a few inseminated females, without the aid of workers. In contrast, *swarm founders (SF)* initiate new colonies by means of a group comprising many workers and a smaller number of inseminated females (queens). Although the independent-founding species (including the Polistinae and the Vespinae) outnumber the swarm founders by roughly 3 to 1 (J.M. Carpenter, pers. comm.), the SF are the more ecologically dominant (Wilson 1990) group in the tropics, especially in the Neotropics (Jeanne 1991; Kojima and Van Achterberg 1997; Carpenter and Wenzel 1999).

IF and SF appear to comprise largely discrete behavioral syndromes, with remarkably little overlap between them (Jeanne 2003) (see also (Bourke 1999). The most discrete trait appears to be the mode of founding itself. We know of no intermediates between independent founding and swarm founding, and in fact it is hard to imagine what such an intermediate would look like. In addition to mode of founding, each group is characterized by a set of associated traits (Table 2.1). In column 3 of the table, I have given most of these traits an estimated degree-of-overlap score. A few of these associated traits appear to be almost as discrete as the mode of founding itself—most notably worker control and queen number—although deeper research on little-studied SF genera may challenge this assessment. Most of the other traits characterizing the two groups show greater or lesser degrees of overlap.

The two groups also represent two discrete levels of social complexity (Jeanne 2003). IF species can be characterized as having simple societies. Colonies in this group are small, comprising less than 100 adults and often many fewer. Colony members engage in conflict over access to direct reproduction. Colony control—regulation of the colony's activity level (Jeanne 2003)—is centralized in the queen, who maintains her position as top reproductive via dominance interactions with her subordinate co-foundresses and worker offspring (Reeve 1991). Division of labor among workers is minimal or non-existent. Selection at the individual level appears to play the dominant role in maintaining sociality in this group.

In contrast, SF species form complex societies. Colonies are typically large enough that workers have little chance of direct reproduction, leading to the convergence of their self-interests with those of the queens (Bourke 1999; Jeanne 2003). Consequently, colony control is decentralized onto the workers. Interactions among colony members give rise to self-organizing processes that in turn lead to adaptive patterns of behavior at the colony level. Age polyethism among workers is strongly developed, and some species have evolved queen-worker dimorphism. Selection on these colony-level traits predominates over selection on individual-level traits (Jeanne 2003). Drawing an analogy between these two levels of social organization and solitary animals, we could say that the IF are roughly analogous to the Parazoa (sponges) and the SF to the Eumetazoa.

Table 2.1 Comparison of traits of IF vs. SF social wasps

IF traits	SF traits	IF-SF overlap
Colony founding		
Propagules are individual gynes	Propagules are swarms comprising workers and gynes or queens	None
Founding female/queen searches for and selects nest site	Workers search for and select nest site	None
Associate foundresses ("helpers") join founding female hours or days later [Apparent exception: *Parapolybia varia* (Yamane 1985)]	Scout workers recruit swarm (queens + workers) to new nest site via mechanical(?) and chemical signals and queens and workers emigrate together to the new site	None?
Queen signaling		
Queen physically dominates subordinates, inducing them to take up worker roles. Evidence for queen pheromone in some species (Landolt et al. 1998), but apparently absent in others (Gadagkar 2001)	Queens are passive; hypothesized queen pheromone signals reproductive status	Much
Queen- vs. *worker-control of colony activity*		
Queen domination of workers stimulates colony activity. Exceptions in some *Polistes* (Jha et al. 2006), *Ropalidia* (Bruyndonckx et al. 2006)	Workers regulate colony activity, stimulate activity via biting attacks	Some
Division of labor		
Founding female/queen initiates nest	Workers initiate nest	None
Founding female/queen initiates new nest cells	Workers initiate new nest cells. Exception: *Chartergellus golfitensis*: (Chavarria-Pizarro and West-Eberhard 2010)	Some
Founding female/queen forages for nest material for cell initiation. Exceptions in some species: *Mischocyttarus drewseni* (Jeanne 1972)	Workers forage for all nest material and do all nest construction	Some
Colony size		
Colonies are small, rarely exceeding 100 adults (range: 10^0–10^2)	Colony sizes large, exceeding one million workers (range: 10^1–10^6)	Some
Queen number		
Short-term monogyny: typically one egg-laying queen. Exceptions: *Ropalidia rufoplagiata* (Gadagkar 2001); *R. revolutionalis* (Henshaw et al. 2004, 2015)	Long-term polygyny: multiple egg-laying queens cycling down to several or even one during colony development. Exception: *Provespa* is monogynous (Matsuura 1999)	Some

(continued)

Table 2.1 (continued)

IF traits	SF traits	IF-SF overlap
Queen competition		
Founding females engage in contest competition among themselves for the right to be the sole egg-layer, establishing a dominance hierarchy with the egg-layer in the alpha position. Exception: *Ropalidia marginata* (Gadagkar 2001)	Queens do not act aggressively toward one another; by constructing cells faster than queens can fill them with eggs (in newly founded nests), workers create the condition in which queens engage in scramble competition via egg-laying	None
Nest architecture and construction		
Nest growth gradual and continuous, responds to demand for oviposition sites; controlled by queen	Nest construction/expansion explosive/episodic, construction of new cells exceeds demand for oviposition sites; controlled by workers Abundance of empty cells may create context for queen-queen competition. Exceptions exist (Jeanne and Bouwma 2004)	Some
No nest envelopes (except in Vespinae)	Nest envelopes in most species. Exceptions: *Apoica*; most *Agelaia* spp.	Some
Nests of single combs, or if multiple combs, separately attached to the substrate	Nests of single or multiple combs	Some
Behavioral specialization		
Little or no specialization among workers	Age polyethism. Task partitioning: foraging and nest work performed by different individuals Exception: *Chartergellus golfitensis* (Chavarria-Pizarro and West-Eberhard 2010)	Some
Life history traits		
Egg-to-adult development times longer: 38–87 days	Egg-to-adult development times shorter: 28–30 days	None?
Worker lifespans longer	Worker lifespans shorter	?
Colony survival		
Low rates of colony survival	High rates of colony survival	None?
Defense		
Chemical defense of nest against scouting-and-recruiting ants	Active defense of nest against scouting-and-recruiting ants. Exceptions: *Nectarinella* and *Leipomeles* employ sticky traps	Some

Traits are generalized and are focused primarily on the Polistinae. Examples of exceptions, if known, are indicated. The column headed "IF-SF overlap" provides a subjective estimate of the degree of overlap between the two groups for each trait: none, some, and much. Note that some of these are tentative and with more research will likely need to be revised

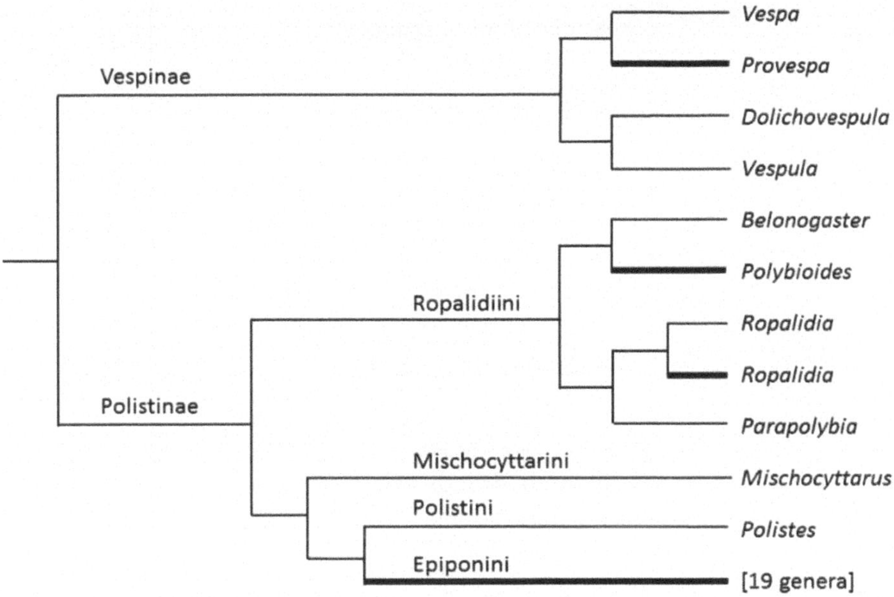

Fig. 2.1 Phylogenetic tree of the social vespids (*Stenogastrinae not shown*), showing the subfamily Vespinae and the four tribes of Polistinae. Heavy lines indicate lineages in which swarm-founding evolved. Note that *Ropalidia* contains both IF and SF species. Phylogeny based on Pickett and Carpenter (2010) and J. M. Carpenter, pers. comm.

Swarm founding evolved independently at least four times in the Vespidae (Carpenter 1991), in each case from an independent-founding ancestor (Fig. 2.1). Compared with the origin of eusociality itself from solitary ancestors, the transition from IF to SF has been little-addressed (but see West-Eberhard (1982) and Henshaw et al. (2004)). Yet it can be argued that the IF-to-SF transition was the more difficult of the two, because it represents a more significant shift from individual-level selection to selection at the level of the group, the colony (Szathmáry and Maynard Smith 1995). My aim here is to explore how this transition may have happened by examining the steps required to evolve swarm founding. My main focus will be on the Polistinae and its three origins of SF, although references to the Vespinae will be made when informative.

Howard Evans was able to identify a series of steps from solitary to eusocial wasps, with most steps represented by extant species (Evans 1958). In contrast, species representing intermediate steps along the path from IF to SF appear to be non-existent. That is, the two groups appear to occupy discrete adaptive peaks separated by a fairly deep valley. This suggests that intermediate conditions are not evolutionarily stable, i.e., that the whole suite of SF traits had to evolve more or less together. Nevertheless, it may be possible to identify a key trait, or traits, that may have been a first step enabling the shift from IF to SF (Fig. 2.2). What clues do we have as to what that critical step might have been? I emphasize that the ideas expressed below are speculative, and others may argue for different evolutionary pathways.

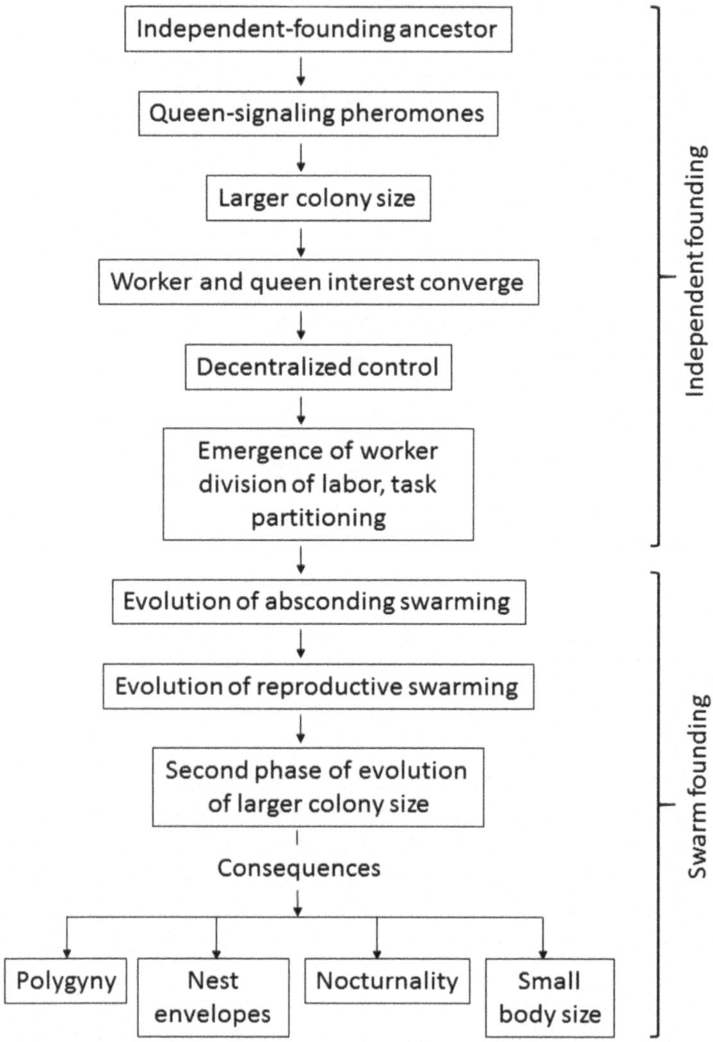

Fig. 2.2 Diagram of hypothesized steps in evolving swarm founding from an independent-founding ancestor

2.2 Decentralized Control of Colony Activity

I have argued elsewhere that the most essential underlying difference between the two groups is decentralized (worker) control over colony activity in the SF vs. centralized (queen) control in the IF (Jeanne 2003). In the typical IF species, the queen is the pacemaker of the colony, stimulating worker activity via physical dominance (Reeve and Gamboa 1983; Bruyndonckx et al. 2006). Furthermore, queens engage in certain non-reproductive tasks, such as foraging for nest material and the initiation of new cells. In contrast, in the SF the workers take over all or most non-reproductive tasks, while the queens are largely limited to oviposition and are generally behaviorally passive. This derives, at least in part, from large colony size, wherein any single individual has a small chance of becoming the egg layer (Bourke 1999). Thus, workers are shut out of direct reproduction, leading to the convergence of their self-interests with those of the queen(s) (Bourke 1999; Jeanne 2003).

How does this decentralized control come about? As just argued, the shift from queen to worker control requires large colony size. Colony size in the IF appears to be constrained to a few dozen adults by the limited reach of the behavioral means used by the queen to signal her status to the workers. In this group, queen signaling is primarily via physical dominance attacks on subordinates and workers (Spradbery 1991) and vibrational signaling to developing larvae (Jeanne and Suryanarayanan 2011; Suryanarayanan et al. 2011). In *Mischocyttarus drewseni*, for example, virtually every female offspring is physically dominated by the queen during its first few days as an adult (Jeanne 1972). This requires repeated, close physical contact, which limits the number of offspring over which this kind of signaling can be effective. Two factors lead to the breakdown of this form of colony control. First, the rate of emergence of young females increases during the ergonomic stage, and second, the absolute rate of domination by the queen decreases as she ages. Both lead to an increasing likelihood that some young females do not receive enough of this signal during their first few days as adults to shift them into worker roles. Some of these become non-working potential gynes and some may go on to challenge and supersede the reigning queen, leading to serial polygyny (Jeanne 1972). This puts the colony into the reproductive stage, producing not only these non-working females, but males, at the expense of additional workers and continued growth of the size of the colony. The resulting increase in mouths to feed, coupled with the decreasing number of workers to feed them, leads to the decline of the colony after 5–8 months from founding. Colonies of *Polistes*, *Ropalidia*, *Belonogaster*, and other IF polistines appear to be size-limited for similar reasons (Owen 1962; West-Eberhard 1996).

2.3 Queen-Signaling Pheromone

This suggests that the necessary first step enabling the escape from the limits of behavioral queen signaling on colony size was very likely the evolution of effective queen pheromones. In the SF there is little or no direct physical aggression by queens toward workers, strongly suggesting that queen signaling is by chemical means, i.e., via queen pheromones (West-Eberhard 1978b; Spradbery 1991; Kelstrup et al. 2014). On the other hand, in some SF aggression and displays by workers toward females appear to be a means of biasing the development of emerging females into becoming queens (Platt et al. 2004; Kelstrup et al. 2014). Effective queen pheromones enable the evolution of larger colony size, which in turn is essential for the convergence of worker and queen interests. Queen pheromones have not as yet been chemically characterized for any SF species, nor is it known how queen pheromones are delivered to all the members of the colony. Possibilities include direct contact with the queens, airborne dispersal of volatile compounds, substrate-borne contact pheromones via the nest carton, and surface hydrocarbons on queen-laid eggs (Endler et al. 2004; Oi et al. 2015).

Intermediate stages in the transition from physical to chemical signaling are apparently stable, as suggested by the variation in intensity of domination behavior among species of IF. The assumption here is that decreasing intensity of dominance behavior across taxa reflects an increasing role for chemical signaling. The degree of reliance on physical signals ranges from the despotic and violent domination seen in *Polistes canadensis* and *Belonogaster grisea* to mild or non-existent interactions, as in *P. instabilis, P. versicolor, Mischocyttarus angulatus*, and *Ropalidia revolutionalis* (Spradbery 1991; Ito 1993; de Souza and Prezoto 2012)). *R. marginata*, an IF species, establishes linear dominance hierarchies, but they are not headed by the queen (Gadagkar 2001). Instead, the queen is docile, rarely engaging in physical attacks on nestmates. This suggests that she maintains her monopoly on egg-laying via pheromones, and indeed there is evidence that one source is the Dufour's gland and that she applies the pheromone to the nest surface (Mitra and Gadagkar 2011; Mitra et al. 2011; Saha et al. 2012). If the effectiveness of the pheromone is experimentally reduced, queens resort to aggression toward workers to maintain their status (Saha et al. 2012). *Polistes gallicus* has also been shown to produce a queen-signaling pheromone, in this case via the van der Vecht's gland (Dapporto et al. 2007).

A similar variability is seen in the Vespinae. In small-colony vespines, both physical dominance and presumed queen pheromones are utilized, whereas in species forming larger colonies (*Vespula vulgaris* group), there is apparently complete reliance on chemical signaling, with no sign of residual physical signals (Spradbery 1991). Two compounds, n-C29 and 3-MeC29, have been shown to be queen pheromone components in both *Vespula vulgaris* and *Dolichovespula saxonica* (Oi et al. 2016).

Among SF species we also see variation in the degree of queen-worker interactions. Aggressive dominance displays in which queens bend the gaster laterally (the "bending display") toward approaching workers or other queens was first described for *Metapolybia aztecoides* by West-Eberhard (1978b). Similar displays and even aggression by queens have since been reported for *Chartergellus* spp., *Synoeca* sp., *Protopolybia acutiscutis*, *Leipomeles*, *Asteloeca ujhelyii*, and *Parachartergus fraternus*, but are apparently absent in *Protopolybia fuscatus*, *Parachartergus colobopterus*, and *Nectarinella championi* (reviewed in Chavarria-Pizarro and West-Eberhard 2010). Older workers of *Parachartergus colobopterus* are aggressive toward younger ones, but queens are not involved in these encounters (Platt et al. 2004). These displays and acts of aggression by queens and workers suggest an incomplete reliance on queen pheromones to signal queen status.

Both physical and chemical queen control over reproduction by workers appear to be reliable ("honest") signals (Mitra and Gadagkar 2012). Although in the proximate sense the signals have physiological effects on the receivers, at the level of ultimate causation, the receivers extract honest information about the sender that they respond to so as to obtain a fitness benefit for themselves (Peso et al. 2015). This view is supported by the recent finding that cuticular hydrocarbons, specifically saturated hydrocarbons, act as a conserved class of queen pheromones, inhibiting worker reproduction across all three groups of eusocial Hymenoptera (Mitra and Gadagkar 2012; Van Oystaeyen et al. 2014). If queen pheromones were detrimental to workers' fitness— that is, if queens were manipulating workers against their personal interests—selection would act on workers to evolve resistance and in turn on queens to overcome that resistance. The path of the resulting evolutionary arms race would result in rapid change in the composition of queen pheromones in each lineage, leading to species-specific differences among them (Peso et al. 2015).

2.4 Larger Colony Size

An IF ancestor that evolved an effective queen pheromone would be able to evolve larger colony size. The simplest route would be through the lengthening of the ergonomic phase, i.e., by enabling the continued increase in worker numbers before switching to the reproductive phase. Such colonies could still look much like those of *Polistes* or *Mischocyttarus*, i.e., they would retain independent founding, except that the colony cycle would be longer and colonies would grow larger. The combination of relatively large colony size and strong and effective queen-signaling pheromones would move the colony well along the road to the convergence of workers' interests with those of the queen and thus toward worker control of colony activities. Among existing species, this stage may be most closely represented by the Vespinae, for which there is good evidence for queen-signaling pheromones (Oi et al. 2016). However, no known IF polistine has these characteristics, suggesting that ancestral tropical IF that evolved along this path went on to evolve swarm founding.

2.5 Evolution of Swarm Founding Itself

Among tropical polistines, the next step was probably the evolution of swarm founding itself. Here I characterize a true "swarm" as worker-controlled, to distinguish it from associations of independent-founding females. In some temperate-zone *Polistes*, satellite nests are sometimes founded by groups of workers in mid-summer (Strassmann 1981; Page et al. 1989) in a process that has been likened to swarming (Rau 1941), but there is little evidence that it is any closer to true swarming than is ordinary springtime independent founding in these species.

It seems certain that worker control had to be in place before swarming could evolve, as it is the workers that scout out a suitable nest site and coordinate the eventual move to it by the rest of the swarm (Forsyth 1981). How the transition could have occurred is less evident than the steps enabling the evolution of larger colony size. One possibility is that the swarm as the colony-founding unit first evolved in the context of loss of the brood to a predator; that is, the absconding swarm (Jeanne 1991) was the first step. In the Neotropics, mass-foraging ants (e.g., *Eciton* spp.) are important predators of social wasps, and most wasp species have no effective defense against them. Most of the adults usually escape and can renest, starting the colony cycle over. In many parts of the tropics, nesting is possible year-round, so there is no seasonal constraint on this. Colonies ought to be at an advantage if the adults can stay together and renest as a single, large group, rather than dispersing to form small independent-founding groups of one or a few individuals. One advantage is seen in today's SF wasps: the ability to quickly build a nest and stock it with a large number of eggs. Another is a reduction in the risk of failure during the founding stage. A founding swarm of many individuals not only vastly reduces the risk of failure by attrition due to predation on individual group members (West-Eberhard 1982), but it increases the likelihood that it will successfully defend the brood against natural enemies. This step may not have been hard to take. It would require that the workers scout out and reach consensus on a nearby site for renesting. One hurdle to achieving this step is the need to evolve some mechanism by which the numerous scouts arrive at a consensus on a single site. Recruiting the swarm members to the new site may require only a modest ability to communicate location, e.g., by chemical marking of the chosen site itself. If the distance between the old and new sites is short, members of the group should be able to arrive at the new site by detecting the airborne volatiles emitted from it. Several SF species have been observed to scent-mark the new site, but omit the chemical trail when the distance to the new site is less than 20 m or so (West-Eberhard 1982). This suggests that the laying of a chemical trail could have been a secondary step, enabling emigration over greater distances. On the other hand, behavior similar to that seen in the epiponines sometimes occurs among the IF. Queens of *Mischocyttarus labiatus* scent-mark (drag the gaster) on leaves between the old and new nest sites shortly after nest initiation (Litte 1981), suggesting that scent trails could have preceded the evolution of swarming. But in this case, it is presumably inseminated foundress females, not workers, that lay down the scent marks.

Once such absconding swarming was in place, the swarm as the reproductive propagule would have to secondarily evolve. This may have been a more difficult step: absconding swarming is simple—the entire adult population (except males in some species (Bouwma et al. 2000)) makes the move, whereas reproductive swarming requires decisions about when to emit a swarm, what proportion of the colony to bud off into each swarm, and which workers will go with which queens. This added complexity is another reason that absconding swarming may have led the way.

The move to swarm founding appears to have been an all-or-none transition. As far as known, existing polistine species unambiguously fall into either the IF or the SF category. Indeed, it is hard to imagine what an intermediate IF-SF stage might look like. The absence of large colonies (>100 adults) among the ~750 species of IF polistines suggests that the transition to swarm founding was an inevitable endpoint of the sequence "queen pheromone → large colony size → worker control →

swarm founding," at least in the tropics. Alternatively, there may be other, possibly ecological, limits on colony size for tropical IF. There are no known reversals of the swarming habit within the Epiponini, despite that numerous species have secondarily evolved colony sizes that are well within the range of those of independent founders (Pickett and Wenzel 2007), suggesting that swarming is a "point of no-return" (Hölldobler and Wilson 2009) and that the swarm-founding wasps occupy a high peak in the adaptive landscape.

There are variants of the paradigm exemplified by *Polybia* of rapid regrouping following absconding, with scouts reaching consensus on a new site and guiding swarm members to it via a chemical trail (Jeanne 1981; Sonnentag and Jeanne 2009). *Apoica,* for example, apparently does not deploy scouts and does not lay scent trails. Instead, volatiles are wafted into the air after the swarm is airborne (Howard et al. 2002). Some species of *Agelaia* do not emigrate to a new nest site in a single event, but spread the move from the old to the new nest over several days (Jeanne 1975b). Exploring species in these and other genera in more detail may shed some light on the evolutionary steps taken.

Once swarm founding had evolved, the evolution of much larger colony size could have followed relatively quickly. There is no known theoretical limit to the size of a reproductive swarm. Colonies of some species of *Agelaia* attain sizes of 10^4–10^6 adults (Jeanne 1991; Zucchi et al. 1995) and doubtless send out proportionately large swarms. Smaller colony size could also have evolved. A number of extant epiponine species have mature colony sizes of well under 100 adults (Jeanne 1991; Jeanne 2003; Pickett and Wenzel 2007).

2.6 Consequences and Correlates of Swarm Founding

Among the SF polistines are several other traits that seem clearly to have evolved after swarm founding was achieved. Because they are not known in the IF polistines, I call them consequences, or correlates, of swarm founding. For the same reason, they are worth calling attention to, in that they may help provide insight into why the swarm founders have become so ecologically dominant.

2.6.1 Polygyny

As far as is known, long-term polygyny characterizes all species in the three groups of polistine swarm founders, although this remains unconfirmed for many species and even for entire genera. West-Eberhard postulated that polygyny was either directly derived in the wasps from the polygyny of presocial groups or was secondarily derived from the short-term monogyny characterizing the IF (West-Eberhard 1978a). A phylogenetic test between these two hypotheses comes down in favor of the latter (Carpenter 1991). The same path was evidently taken in the ants, where polygyny is strongly correlated with large colony size (Boulay et al. 2014). That colonies of *Provespa* are monogynous (Matsuura 1999) indicates that polygyny is not a necessary concomitant of swarm founding and suggests instead that we look for extrinsic causes. A source of selection favoring polygyny may come from natural enemies. One advantage of spreading the egg-laying function among multiple queens instead of a single physogastric individual is the retention of mobility in the event of an attack on the colony (Richards and Richards 1951). The type of predator may also matter. In the Neotropics, army ants, particularly those in the genus *Eciton*, are major predators of social wasps. Their primary prey are the larvae and pupae in the nest, but they also will take any adult that fails to evacuate the nest quickly. In 2013 at the Reserva Florestal Adolpho Ducke north of Manaus, I witnessed an attack by *E. hamatum* on a nest of *Chartergellus jeannei*. Although most of the adults managed to flee, less than a minute after the first ant reached the nest, ants blocked the entrance, trapping significant numbers of wasps inside the nest and thereby subjecting them to capture (Fig. 2.3). Since in many wasp species the queens are the last to leave the nest in event of threat (Richards and Richards 1951), some of those killed may well have been queens. By spreading egg-laying capacity among many reproductives, the absconding swarm would likely still have enough queens to quickly stock its new nest with enough eggs for rapid growth.

In the Old World tropics, on the other hand, driver ants are apparently less important predators on social wasps than are army ants in the Neotropics (Yamane 1996). Instead, certain *Vespa* species appear to be the more important threat (Matsuura and

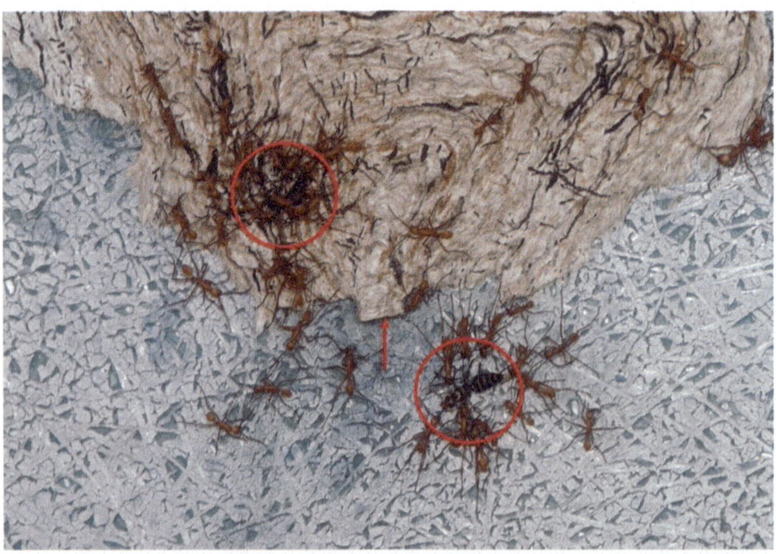

Fig. 2.3 Raid of a nest of *Chartergellus jeannei* by *Eciton hamatum* at Reserva Florestal Adolpho Ducke, Manaus, Amazonas, Brazil. Two adult wasps are shown being captured (red circles). Nest entrance is shown by the arrow. Photo was taken barely 2 minutes after the first ant reached the nest. Substrate is the outer wall of a fiberglass water tank

Yamane 1990). These hornets present a different kind of predation on colonies. *Vespa* workers are much larger than are those of most of their social wasp prey, and their raids involve just one or a few workers acting with impunity to carry off the brood over several days (Jeanne and Hunt 1992). In the face of such a predator, the defending adults are at comparatively little risk, and a single queen would have a high chance of surviving these attacks and eventually emigrating with the workers to start a new nest. This may in part explain the monogyny of *Provespa* in that region. On the other hand, swarm-founding *Ropalidia* in the same habitats are polygynous (Kojima and Jeanne 1986; Spradbery and Kojima 1989; Kojima 1996); R. L. Jeanne, unpublished data), so intrinsic factors may have played some role in the evolution of polygyny in the *Ropalidia*.

As far as we know, the rule in SF polistines is long-term polygyny with periodic reduction to one or a few queens (cyclical oligogyny) (Ross and Carpenter 1991). In contrast, colonies of most IF polistines are monogynous. Seeming exceptions include *Ropalidia revolutionalis*, an IF species in which some colonies have multiple co-occurring egg-layers (Henshaw et al. 2004). There are also several reports of polygynous colonies of *Polistes* spp., especially when old nests are reused (Liebert et al. 2008). Based on such observations, Henshaw and colleagues have suggested that the evolution of polygyny may have preceded swarm founding in some lineages and that the evolution of cyclical oligogyny was preceded and enabled by worker manipulations that helped preserve high relatedness under multiple queens (Henshaw et al. 2004).

2.6.2 Nest Envelopes

Nest envelopes are another trait of interest in this context. The IF polistines are noteworthy for their universal lack of nest envelopes. Why this is so is an interesting question in itself, but is beyond the scope of this chapter. In contrast, the vespines universally have them, as do the great majority of SF polistines. The fact that all *Apoica* and most *Agelaia*, both basal epiponine genera, lack envelopes suggests that the first species to cross the threshold to swarm founding also lacked an envelope. Similarly, the common ancestors of the clade *Belonogaster* (no envelopes) + *Polybioides* (envelopes) and of the clade *Parapolybia* (no envelopes) + *Ropalidia* (envelopes in some SF species) very likely also built naked nests. Envelopes appear to have evolved at least six times independently in the polistine and vespine wasps: once each in *Polybioides*, *Ropalidia*, and *Provespa*, at least twice in *Agelaia,* and once (or more) in the clade representing the remaining epiponine genera (Wenzel 1991). *Agelaia* is the most basal epiponine genus to have evolved envelopes. In the two species that build them (*A. areata, A. flavipennis*), the envelope is an extension of the petiolate comb, and so is little-removed from the IF nest type, whereas envelopes in the other genera differ fundamentally from this and may well have evolved independently of *Agelaia* and even of each other. In fact, given the tremendous diversity of epiponine nest architecture (Jeanne 1975a; Wenzel

1991), it seems plausible to hypothesize that envelopes had multiple origins in that clade. In other words, even if swarm founding evolved only once in the Epiponini, there could have been an adaptive radiation of envelope design, with many independent attempts, most successful enough to have been universally adopted. It has been argued that envelopes were secondarily lost in the SF *Ropalidia* that moved to cavity nesting (Kojima and Jeanne 1986); the same argument could be raised in the case of *Agelaia* (Wenzel 1991). However, that line of reasoning seems weak in light of the fact that vespine species that have made the same move retain at least rudimentary envelopes. Nevertheless, the question of which is the apomorphic state—naked nests or envelopes—in these lineages is an interesting and unresolved one.

The thick, air-filled envelope of the vespines is almost certainly an adaptation that insulates the nest against low temperatures, enhancing homeostatic control of the temperature around the brood (Himmer 1931; Schmolz et al. 2000). But why have the SF polistines, in their tropical habitats, evolved envelopes several times? I suggest that it ultimately has to do with the novel selection pressures faced by larger colonies and the larger nests they require. Nests of many SF species comprise more than one comb, arranged in a large diversity of architectural forms (Jeanne 1975a; Kojima and Jeanne 1986; Wenzel 1991) to accommodate the added weight while maintaining expandability. With the exception of *Apoica*, SF either enclose their nests in an envelope (typically carton, but in some cases cemented-together leaves) or nest in cavities Richards 1978a) (Wenzel 1991). The basal epiponine genus *Apoica* (Pickett and Wenzel 2007) builds a single naked comb [see (Pickett et al. 2009)], and the multi-combed nests of most species of *Agelaia*, sister group to the remaining Epiponini, are naked, but all remaining genera of epiponine wasps construct nests with envelopes. The selective advantages of covered nests probably involve both homeostatic control of physical conditions and protection from rain and natural enemies, as well as escape from the limitations of finding suitable cavities to nest in. Whereas the small, single-combed nests of IF polistines are often built under leaves of understory plants, which provide at least some shelter from the elements, larger nests are heavier and require sturdier substrates such as trunks, branches, and twigs, which provide little shelter. Envelopes effectively prevent rain from reaching the brood in the combs.

Perhaps an equally important function is to restrict access to the brood by arthropod enemies. Envelopes limit access by scouting-and-recruiting ants to the narrow entrance and increase the likelihood that the workers can detect approaching scout ants on the envelope and remove them (Jeanne 1975a). The universal occurrence among the IF polistines of tough nest pedicels made up mostly of chitin-like oral secretion and their coating of ant-repellent glandular secretion stands in stark contrast to the apparent complete absence of these adaptations among the SF species (London and Jeanne 2000; Makino 2010) and supports the notion that envelopes evolved as a more effective means of defense against ants than is a chemically defended pedicel.

Envelopes may be even more effective in preventing parasitoids from reaching the brood. The brood of IF species are subject to attack by four to five times as many species as are swarm founders (Makino 1985; Yamane 1996). In contrast to the large diversity of lepidopteran species reported to attack brood of IF, none have as yet been recorded from SF and are rarely found in vespine nests (Yamane 1996). It has been experimentally shown for *Polybia occidentalis* that the envelope significantly reduces access to the brood by phorid flies (London and Jeanne 1998). This wasp also responds facultatively to high numbers of phorids flying near the nest by reducing the size of the entrance (London and Jeanne 1998). In addition, species lacking envelopes appear to be more likely to suffer colony failure from heavy infestation than are SF species (Schremmer 1972; Jeanne 1979; Litte 1981; Schmid-Hempel 1998) (but see (Simões et al. 1996)). Envelopes have also been cited for reducing the infestation of vespine nests by parasitoid moths (Matsuura and Yamane 1990).

Wasps in the genus *Apoica* are unique among epiponines in that they neither construct an envelope nor nest in cavities. In nine of the ten species (exception, *A. arborea* de Saussure; see below), the single comb is provided with a thick (25 mm in *A. flavissima* (Yamane et al. 2009) felt-like mat of plant hairs above the brood cells. Two explanations of the function of this feature have been proposed (Yamane et al. 2009). One is that it serves to strengthen the attachment of the combs to the substrate. This seems unlikely, given that the thickening tapers to the outer margins of the comb and does not occur just at the point of attachment. The second explanation is that it insulates the brood cells from fluctuations in temperature. This also seems unlikely, in that temperature fluctuations in a central cell in an active nest of *A. flavissima* have been shown to be much less than those measured in an empty nest (Yamane et al. 2009), indicating that the adults and brood have a much greater effect on moderating temperature than does the nest structure. A third possibility is that the thick felt functions to prevent parasitoids such as ichneumonids from reaching the backs of the brood cells with their ovipositors, as does *Pachysomoides*, for example, in parasitizing *Polistes* (West-Eberhard 1969). It is noteworthy that the cell openings on the face of the comb are protected during the day by the multiple layers of closely packed adults, which are likely very effective in preventing access to the brood from that direction. No such clustering occurs on top of the comb. Although records are few, the rate of parasitization suffered by *Apoica* appears to be exceptionally low. Other than the unusual case of heavy infestation by phorids in a nest of *A. pallens* (Schremmer 1972) [cited by Schremmer as *A. pallida* (Richards 1978b)], the only brood parasitoid ever recorded for the genus (two species) is the hymenopteran *Seminota marginata* (Westwood) (Bertoni 1911; Trindade et al. 2012; Santos and Noll 2013). This trigonalid may get around the defensive curtain of adults by making its way into the nest indirectly, either via a primary host (caterpillar) brought to the nest as food or by laying its eggs on plants utilized as nesting material by this wasp (Santos and Noll 2013). Interestingly, the adult/brood-cell ratio in the genus is much higher than in other swarm founders—close to 1.0, compared to well under 1.0 for other epiponines (Richards 1978b) (Jeanne, unpublished data). Such a high ratio of workers to nest cells may be an adaptation to maintaining the large number of adults required by this form of defense. An alternative nest architecture is seen in *A. arborea*, whose nest is a long, narrow comb comprising entirely sessile cells built under a branch (Pickett et al. 2009). The supporting branch may provide the same barrier to probing parasitoids as does the felt-like thickening of the comb-back of its congeners.

2.6.3 Nocturnality

Nocturnality has evolved twice in swarm-founding wasps—in *Apoica* and *Provespa*—but apparently never in the IF. We can only speculate as to the selective forces favoring this form of specialization, but escape from natural enemies may have played a role here as well. Having the entire adult population forming an inactive, tightly packed mass covering the face of the comb maximizes their effectiveness in physically blocking parasitoids from reaching the brood during the day, when most parasitoids presumably are active. While escape from parasitoids has also been proposed for nocturnality in certain bees, escape from competition for food resources is an alternative explanation (Wcislo et al. 2004; Warrant 2008). An interesting difference between the two genera is that while *Provespa* swarms emigrate at night (Matsuura 1999), *Apoica* emigrates during the day (Hunt et al. 1995).

2.6.4 Small Body Size

The adults of most New World IF wasps are medium (~1 cm) to large (2+ cm) in size. Most SF species tend to be smaller, with a number of species very small indeed. Analysis (Fig. 2.4) shows that *Polistes* spp. are on average the largest polistine wasps and are significantly larger than *Mischocyttarus* (t = 14.4; $p < 0.001$). *Mischocyttarus* spp. in turn are significantly larger than the epiponines ($t = -7.35$; $p < 0.001$). In fact, 98% of *Mischocyttarus* species are smaller than the average-sized *Polistes*, and 78% of the epiponine species are smaller than the average *Mischocyttarus*. Thus it is clear that in the New World the SF polistines are significantly smaller than are the IF species. Karsai and Wenzel (1998) came to the same conclusion based on head widths and suggested that small body size may have evolved repeatedly in different epiponine lineages.

What factors might favor smaller body size in the SF than in the IF? Several possible explanations come to mind.

1. *More workers per unit of resource*. Smaller worker size allows more workers to be produced from a given amount of resources, enabling in turn a greater increase in colony size (Karsai and Wenzel 1998; Bourke 1999).
2. *Escape from competition for food*. Another hypothesis is that small body size evolved in response to competition with larger-bodied independent founders by enabling the exploitation of smaller prey. If this is true, it begs the question of how the move to swarm founding enabled this.

   Neither of these hypotheses explain why IF species have not also evolved smaller body size.

   The following explanations avoid this shortcoming.
3. *Mode of defense against natural enemies*. It may be that for the modest-sized colonies of the IF, large body size is maintained by the need for an effective stinging defense against vertebrates. In contrast, the larger colonies of the SF can

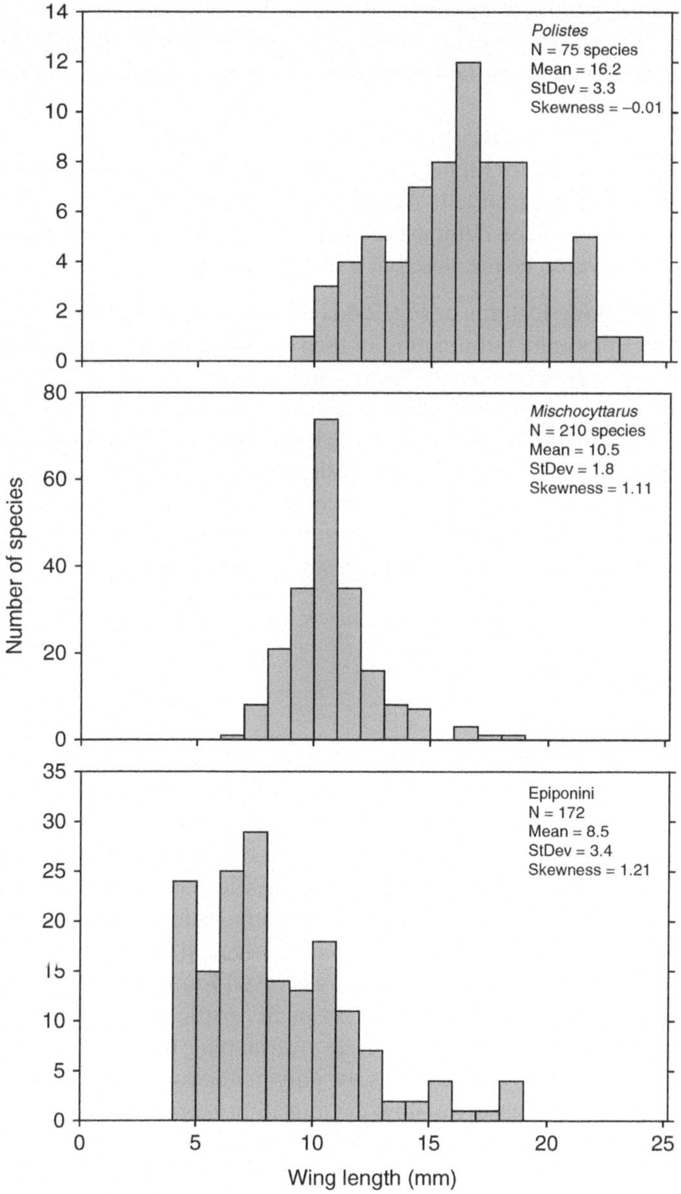

Fig. 2.4 Wing lengths of New World polistine wasps. Shown are the frequency distributions of wing length for each of the three taxa. Where data were available for two or more subspecies, only one entry was made for the species. Where a range was given, the middle value was used. (Data from: (Zavattari 1906, Bequaert 1938, Bequaert 1943, Bequaert 1944, Araujo 1945, Richards 1945, Araujo 1946, Araujo 1949, Richards and Richards 1951, Willink 1959, Naumann 1968, Richards 1978b, Cooper 1993, Silveira and Carpenter 1995, Mateus and Noll 1997, Cooper 1999, Raw 1999, Cooper 2000, Cooper 2001, Carpenter and Kojima 2002, Pickett 2003, Carpenter et al. 2004, Pickett and Wenzel 2007, West-Eberhard et al. 2010, dos Santos et al. 2015, Grandinete et al. 2015))

sting in larger numbers, making up for the reduced deterrent effect of individual stings by their smaller workers. Arguing against this scenario, on the other hand, is the fact that many small-bodied SF have quite small colonies (e.g., *Leipomeles*, some *Protopolybia*).

4. *Co-evolution with increased task complexity*. Small body size may have evolved in SF under selection for higher tempo of interaction and greater task complexity in larger colonies (Karsai and Wenzel 1998). Greater task complexity in large-colony species has been demonstrated (Karsai and Wenzel 1998; Jeanne 2003), but tempo has yet to be measured in a social wasp species.

One of the striking features of Fig. 2.4 is the increase in skewness in going from *Polistes* to the Epiponini, culminating in the strongly truncated size-frequency distribution shown in the latter group. Twenty-four species have wing lengths of 4 mm, but no species in the dataset is smaller than that. This suggests, first, strong selection for small size and, second, that 4 mm is a strict lower limit. What might impose such a limit? One possibility is that wasps smaller than this lack the size and mandible strength to collect and manipulate vegetable fibers into a nest. Whatever it may be, the sphecid genus *Microstigmus* appears to have escaped it. *Microstigmus comes*, for example, has a wing length of just 2.4 mm (Matthews 1968).

2.7 Conclusions

It can be argued that the IF and SF represent two discrete social syndromes, each characterized by a set of mutually stable traits (Bourke 1999; Jeanne 2003). This is especially the case for the mode of founding itself; no intermediates are known. Yet several of the other traits do not cleanly sort into one or the other of the groups, as the examples in Table 2.1 show. As more species in both groups are investigated with regard to such traits, we will begin to discern patterns in how they correlate with colony size, life-history traits, and phylogeny, and these in turn will lead to better explanations of how the SF may have evolved from their IF ancestors. There is a need for such studies of more species of SF, particularly those in the lesser-known genera and those with colony sizes overlapping those of the IF. *Ropalidia* deserves special attention, because it is the only genus containing members of both groups, suggesting that the key to how the transition was made may be more apparent than in the Epiponini (Henshaw et al. 2004; Henshaw et al. 2015).

In addition to colony size and body size, the IF and SF differ in certain life history traits. One example is the significant negative correlation across all social wasps between mean worker lifespan and colony size (Toth et al. 2016). Another is the ratio of worker lifespan to egg-to-adult development time for workers, a strong determinant of the rate of colony growth (Richards and Richards 1951). Both appear to be significantly lower in the SF than in the IF. As an aside, it is striking that in the few species we have data on, development time exceeds the mean worker longevity, refuting the Richardses' assertion that the length of life of workers must exceed the average developmental period, or true social life is not possible (Richards and

Richards 1951) (p. 120). This dictum may apply to the crossing of the threshold from solitary to eusocial life, but not to workers in eusocial colonies, where the more relevant requirement is that a worker's contribution to the colony during her lifetime must be sufficient to rear the equivalent of her replacement; anything above that, and the colony will grow in size. Unfortunately, we have data on these and other life-history traits for too few species to know how they correlate with colony size, body size, or the swarm-founding habit. Data on these and other aspects of life history for both IF and SF species would be very useful, as would comparative analyses of rates of work and per-worker lifetime contributions to colony productivity for both groups.

The solution to such multi-faceted puzzles will depend on the creative thinking of future vespidologists to refine and expand these ideas and others into testable hypotheses.

References

Araujo RL (1945) Contribuição para o conhecimentio de "*Polybia minarum*" Ducke, 1906 (Hymenoptera, Vespidae). Rev Bras Biol 5(2):157–163

Araujo RL (1946) Contribuição para o conhecimento do gênero "*Protonectarina* Ducke, 1910" (Hym., Vespidae). Arquivos do Instituto Biologico 17:229–238

Araujo RL (1949) Contribuição para o conhecimento de "*Polybia punctata* Buysson, 1908" (Hym. Vespidae). Arquivos do Instituto Biologico 19:269–274

Bequaert JC (1938) A new *Charterginus* from Costa Rica, with notes on *Charterginus, Pseudochartergus, Chartergus, Pseudopolybia, Epipona* and *Tatua* (Hymenoptera, Vespdae). Revista de Entomologia (Rio de Janeiro) 9(1–2):99–117

Bequaert JC (1943) A new species of *Polybia* from Panama. Ann Entomol Soc Am 36(3):394–396

Bequaert JC (1944) A revision of *Protopolybia* Ducke, a genus of neotropical social wasps (Hymenoptera, Vespidae). Revista de Entomologia (Rio de Janeiro) 15(1–2):97–134

Bertoni A (1911) Contribucion a la biologia de las avispas y abejas del Paraguay (Hymenoptera). Anales del Museo Nacional de Buenos Aires 15:97–146

Boulay R, Arnan X, Cerda X, Retana J (2014) The ecological benefits of larger colony size may promote polygyny in ants. J Evol Biol 27(12):2856–2863

Bourke AFG (1999) Colony size, social complexity and reproductive conflict in social insects. J Evol Biol 12(2):245–257

Bouwma PE, Bouwma AM, Jeanne RL (2000) Social wasp swarm emigration: males stay behind. Ethol Ecol Evol 12:35–42

Bruyndonckx N, Kardile SP, Gadagkar R (2006) Dominance behaviour and regulation of foraging in the primitively eusocial wasp *Ropalidia marginata* (Lep.) (Hymenoptera: Vespidae). Behav Process 72(1):100–103

Carpenter JM (1991) Phylogenetic relationships and the origin of social behavior in the Vespidae. In: Ross KG, Matthews RW (eds) The social biology of wasps. Cornell University Press, Ithaca, pp 7–32

Carpenter JM, Kojima JI (2002) A new species of paper wasp from Costa Rica (Hymenoptera: Vespidae; Polistinae, Epiponini). J New York Entomol Soc 110(2):212–223

Carpenter JM, Wenzel JW (1999) The relative abundance of swarm-founding social wasps in the Congo Basin (Insecta: Hymenoptera; Vespidae, Polistinae). Nat History Bull Ibaraki Univ 3:9–14

Carpenter JM, Do Nascimento FS, Mateus S, Noll FB, Kojima JI (2004) A revision of the genus *Asteloeca* (Hymenoptera: Vespidae; Polistinae). Am Museum Novitates 3427:1–12

Chavarria-Pizarro L, West-Eberhard MJ (2010) The behavior and natural history of *Chartergellus*, a little-known genus of neotropical social wasps (Vespidae Polistinae Epiponini). Ethol Ecol Evol 22(4):317–343

Cooper M (1993) A new species of *Chartergellus* (Hym., Vespidae, Polistinae, Polybiini) from Bolivia. Entomologist's Monthly Magazine 129:165–166

Cooper M (1999) New species of *Metapolybia* Ducke (Hym., Vespidae, Polistinae). Entomologist's Monthly Magazine 135:107–110

Cooper M (2000) Five new species of *Agelaia* Lepeletier (Hym., Vespidae, Polistinae) with a key to members of the genus, new synonymy and notes. Entomologist's Monthly Magazine 136:177–198

Cooper M (2001) Two new species of *Agelaia* Lepeletier (Hym., Vespidae, Polistinae). Entomologist's Monthly Magazine 137:233–235

Dapporto L, Santini A, Dani FR, Turillazzi S (2007) Workers of a *Polistes* paper wasp detect the presence of their queen by chemical cues. Chem Senses 32(8):795–802

de Souza AR, Prezoto F (2012) Aggressive interactions for a decentralized regulation of foraging activity in the social wasp *Polistes versicolor*. Insect Soc 59(4):463–467

dos Santos JNA, Silveira OT, Carpenter JM (2015) Phylogeny of *Protopolybia* Ducke, 1905 and taxonomic revision of the *Protopolybia exigua* species-group (Hymenoptera: Vespidae, Polistinae), with description of four new species. Zootaxa 3956(2):151–182

Endler A, Liebig J, Schmitt T, Parker JE, Jones GR, Schreier P, Hölldobler B (2004) Surface hydrocarbons of queen eggs regulate worker reproduction in a social insect. Proc Natl Acad Sci U S A 101(9):2945–2950

Evans HE (1958) The evolution of social life in wasps. Proceedings of the 10th international congress of entomology Montreal, 1956, 2:449–457

Forsyth AB (1981) Swarming activity of polybiine social wasps (Hymenoptera: Vespidae: Polybiini). Biotropica 13(2):93–99

Gadagkar R (2001) The social biology of *Ropalidia marginata*. Toward understanding the evolution of eusociality. Harvard University Press, Cambridge, MA

Grandinete YC, Andena SR, Noll FB (2015) *Chartergellus jeannei,* a new species of the swarming social wasps from the Amazon forest (Hymenoptera: Vespidae: Epiponini). Sociobiology 62(1):120–123

Henshaw MT, Robson SKA, Crozier RH (2004) Queen number, queen cycling and queen loss: the evolution of complex multiple queen societies in the social wasp genus *Ropalidia*. Behav Ecol Sociobiol 55(5):469–476

Henshaw MT, Woller-Skar MM, Pence AN (2015) Multiple queens and biased sex ratios in the independent-founding *Ropalidia* Guerin-Meneville, 1831 (Hymenoptera: Vespidae) wasps. Austral Entomol 54(1):100–109

Himmer A (1931) Über die Wärme im Hornissennest (*Vespa crabro* L.). Z Vgl Physiol 13:748–761

Hölldobler B, Wilson EO (1977) The number of queens: an important trait in ant evolution. Naturwissenschaften 64:8–15

Hölldobler B, Wilson EO (2009) The superorganism: the beauty, elegance, and strangeness of insect societies. W. W. Norton and Company, New York

Howard KJ, Smith AR, O'Donnell S, Jeanne RL (2002) Novel method of swarm emigration by the epiponine wasp, *Apoica pallens* (Hymenoptera Vespidae). Ethol Ecol Evol 14(4):365–371

Hunt JH, Jeanne RL, Keeping MG (1995) Observations on *Apoica pallens*, a nocturnal neotropical social wasp (Hymenoptera, Vespidae, Polistinae, Epiponini). Insect Soc 42(3):223–236

Ito Y (1993) Behaviour and social evolution of wasps: the communal aggregation hypothesis. Oxford University Press, Oxford, UK

Jeanne RL (1972) Social biology of the Neotropical wasp *Mischocyttarus drewseni*. Bull Mus Comp Zool Harvard Univ 144:63–150

Jeanne RL (1975a) The adaptiveness of social wasp nest architecture. Q Rev Biol 50:267–287

Jeanne RL (1975b) Behavior during swarm movement in *Stelopolybia areata* (Hymenoptera: Vespidae). Psyche 82:259–264

Jeanne RL (1979) Construction and utilization of multiple combs in *Polistes canadensis* in relation to the biology of a predaceous moth. Behav Ecol Sociobiol 4:293–310
Jeanne RL (1980) Evolution of social behavior in the Vespidae. Annu Rev Entomol 25:371–396
Jeanne RL (1981) Chemical communication during swarm emigration in the social wasp *Polybia sericea* (Olivier). Anim Behav 29:102–113
Jeanne RL (1991) The swarm-founding Polistinae. In: Ross KG, Matthews RW (eds) The social biology of wasps. Cornell University Press, Ithaca, pp 191–231
Jeanne RL (2003) Social complexity in the Hymenoptera, with special attention to the wasps. In: Kikuchi T, Azuma N, Higashi S (eds) Genes, behaviors and evolution of social insects. Japan Hokkaido University Press, Sapporo, pp 81–130
Jeanne RL, Bouwma AM (2004) Divergent patterns of nest construction in eusocial wasps. J Kansas Entomol Soc 77:429–447
Jeanne RL, Hunt JH (1992) Observations on the social wasp *Ropalidia montana* from peninsular India. J Biosci 17(1):1–14
Jeanne RL, Suryanarayanan S (2011) A new model for caste development in social wasps. Commun Integr Biol 4(4):373
Jha S, Casey-Ford RG, Pedersen JS, Platt TG, Cervo R, Queller DC, Strassmann JE (2006) The queen is not a pacemaker in the small-colony wasps *Polistes instabilis* and *P. dominulus*. Anim Behav 71(5):1197–1203
Karsai I, Wenzel JW (1998) Productivity, individual-level and colony-level flexibility, and organization of work as consequences of colony size. Proc Natl Acad Sci U S A 95(15):8665–8669
Kelstrup HC, Hartfelder K, Nascimento FS, Riddiford LM (2014) The role of juvenile hormone in dominance behavior, reproduction and cuticular pheromone signaling in the caste-flexible epiponine wasp, *Synoeca surinama*. Front Zool 11:78
Kojima J (1996) Colony cycle of an Australian swarm-founding paper wasp, *Ropalidia romandi* (Hymenoptera: Vespidae). Insect Soc 43(4):411–420
Kojima J, Jeanne RL (1986) Nests of *Ropalidia (Icarielia) nigrescens* and *R. (I.) extrema* from the Philippines, with reference to the evolutionary radiation in nest architecture within the subgenus *Icarielia* (Hymenoptera: Vespidae). Biotropica 18(4):324–336
Kojima J, Van Achterberg K (1997) Social wasps collected by Malaise trapping in Southeast Asia, with a note on relative abundance of swarm-founding species (Insecta: Hymenoptera: Vespidae). Nat History Bull Ibaraki Univ 1:1–13
Landolt PJ, Jeanne RL, Reed HC (1998) Chemical communication in social wasps. In: Vander Meer RK, Breed MD, Winston ML, Espelie KE (eds) Pheromone communication in social insects. Westview Press, Boulder, pp 216–235
Liebert AE, Hui J, Nonacs P, Starks PT (2008) Extreme polygyny: multi seasonal "hypergynous" nesting in the introduced paper wasp *Polistes dominulus*. J Insect Behav 21(2):72–81
Litte M (1981) Social biology of the polistine wasp *Mischocyttarus labiatus*: survival in a Colombian rain forest. Smithsonian Institution Press, Washington, D.C.
London KB, Jeanne RL (1998) Envelopes protect social wasps' nests from phorid infestation (Hymenoptera: Vespidae, Diptera: Phoridae). J Kansas Entomol Soc 71(2):175–182
London KB, Jeanne RL (2000) The interaction between mode of colony founding, nest architecture and ant defense in polistine wasps. Ethol Ecol Evol 12(1):13–25
Makino S (1985) List of parasitoids of polistine wasps. Sphecos 10:19–25
Makino S (2010) Gastral rubbing on nest pedicel by the founding queen of the hornet *Vespa analis* (Hymenoptera: Vespidae). Sociobiology 56(2):283–290
Mateus S, Noll FB (1997) *Nectarinella xavantinensis*, a new neotropical social wasp (Hymenoptera: Vespidae; Polistinae). J New York Entomol Soc 105(1–2):45–49
Matsuura M (1999) Size and composition of swarming colonies in *Provespa anomala* (Hymenoptera, Vespidae), a nocturnal social wasp. Insect Soc 46(3):219–223
Matsuura M, Yamane S (1990) Biology of the vespine wasps. Springer, New York
Matthews RW (1968) *Microstigmus comes*: sociality in a sphecid wasp. Science 160(3829):787–788

Mitra A, Gadagkar R (2011) Can Dufour's gland compounds honestly signal fertility in the primitively eusocial wasp *Ropalidia marginata*? Naturwissenschaften 98(2):157–161

Mitra A, Gadagkar R (2012) Queen signal should be honest to be involved in maintenance of eusociality: chemical correlates of fertility in *Ropalidia marginata*. Insect Soc 59(2):251–255

Mitra A, Saha P, Chaoulideer ME, Bhadra A, Gadagkar R (2011) Chemical communication in *Ropalidia marginata*: Dufour's gland contains queen signal that is perceived across colonies and does not contain colony signal. J Insect Physiol 57(2):280–284

Naumann MG (1968) A revision of the genus *Brachygastra* (Hymenoptera: Vespidae). Univ Kansas Sci Bull 47:929–1003

Oi CA, van Oystaeyen A, Oliveira RC, Millar JG, Verstrepen KJ, Van Zweden JS, Wenseleers T (2015) Dual effect of wasp queen pheromone in regulating insect sociality. Curr Biol 25:1–3

Oi CA, Millar JG, Van Zweden JS, Wenseleers T (2016) Conservation of queen pheromones across two species of vespine wasps. J Chem Ecol 42:1175

Owen J (1962) The behavior of a social wasp *Polistes fuscatus* (Vespidae) at the nest, with special reference to differences between individuals. Ph.D. thesis, University of Michigan

Page RE, Post DC, Metcalf RA (1989) Satellite nests, early males, and plasticity of reproductive-behavior in a paper wasp. Am Nat 134(5):731–748

Peso M, Elgar MA, Barron AB (2015) Pheromonal control: reconciling physiological mechanism with signalling theory. Biol Rev 90(2):542–559

Pickett KM (2003) A new species of social wasp in the genus *Apoica* Lepeletier (Hymenoptera: Vespidae: Polistinae: Epiponini). Proc Entomol Soc Wash 105(3):592–598

Pickett KM, Carpenter JM (2010) Simultaneous analysis and the origin of eusociality in the Vespidae (Insecta: Hymenoptera). Arthropod Syst Phylogeny 68(1):3–33

Pickett KM, Wenzel JW (2007) Revision and cladistic analysis of the nocturnal social wasp genus, *Apoica* Lepeletier (Hymenoptera: Vespidae; Polistinae, Epiponini). Am Museum Novitates 3562:1–30

Pickett KM, Carpenter JM, Dejean A (2009) "Basal" but not primitive: the nest of *Apoica arborea* de Saussure, 1854 (Insecta, Hymenoptera, Vespidae, Polistinae). Zoosystema 31(4):945–948

Platt TG, Queller DC, Strassmann JE (2004) Aggression and worker control of caste fate in a multiple-queen wasp, *Parachartergus colobopterus*. Anim Behav 67:1–10

Rau P (1941) The swarming of *Polistes* wasps in the temperate regions. Ann Entomol Soc Am 34:580–584

Raw A (1999) Two new species of social wasps (Hymenoptera, Vespidae) from Roraima, Northern Brazil. Revista Brasileira de Zoologia 16(4):987–990

Reeve HK (1991) *Polistes*. In: Ross KG, Matthews RW (eds) The social biology of wasps. Cornell University Press, Ithaca, pp 99–148

Reeve HK, Gamboa GJ (1983) Colony activity integration in primitively eusocial wasps: the role of the queen (*Polistes fuscatus*, Hymenoptera, Vespidae). Behav Ecol Sociobiol 13(1):63–74

Richards OW (1945) A revision of the genus *Mischocyttarus* de Saussure (Hymen., Vespidae). Trans R ent Soc London 95(7):295–462. + 294 plates

Richards OW (1978a) The Australian social wasps (Hymenoptera: Vespidae). Aust J Zool Suppl Series 61:1–132

Richards OW (1978b) The social wasps of the Americas excluding the Vespinae. British Museum (Natural History), London

Richards OW, Richards MJ (1951) Observations on the social wasps of South America (Hymenoptera Vespidae). Trans Royal Entomol Soc London 102:1–170. + 174 plates

Ross KG, Carpenter JM (1991) Population genetic structure, relatedness, and breeding systems. In: Ross KG, Matthews RW (eds) The social biology of wasps. Cornell University Press, Ithaca, pp 451–479

Saha P, Balasubramaniam KN, Kalyani JN, Supriya K, Padmanabhan A, Gadagkar R (2012) Clinging to royalty: *Ropalidia marginata* queens can employ both pheromone and aggression. Insect Soc 59(1):41–44

Santos EF, Noll FB (2013) Biological notes on the parasitism of *Apoica flavissima* Van der Vecht (Hymenoptera: Vespidae) by *Seminota marginata* (Westwood) (Hymenoptera: Trigonalidae): are social paper wasps primary or secundary hosts of Trigonalidae? Sociobiology 60(1):123–124

Schmid-Hempel P (1998) Parasites in social insects. Princeton University Press, Princeton

Schmolz E, Bruders N, Daum R, Lamprecht I (2000) Thermoanalytical investigations on paper covers of social wasps. Thermochim Acta 361(1–2):121–129

Schremmer F (1972) Beobachtungen zur Biologie von *Apoica pallida* (Olivier, 1791), einer neotropischen sozialen Faltenwespe (Hymenoptera, Vespidae). Insect Soc 19(4):343–357

Silveira OT, Carpenter JM (1995) A new species of *Agelaia* Lepeletier from Brazilian Amazonia (Hymenoptera: Vespidae; Polistinae). J New York Entomol Soc 103(1):69–72

Simões D, Noll FB, Zucchi R (1996) Duration of *Protopolybia exigua exigua* (de Saussure) nests and related aspects as influenced by phorid fly infestation (Vespidae, Polistinae, Epiponini). Sociobiology 28(1):121–129

Sonnentag PJ, Jeanne RL (2009) Initiation of absconding-swarm emigration in the social wasp *Polybia occidentalis*. J Insect Sci 9(11):1–11

Spradbery JP (1991) Evolution of queen number and queen control. In: Ross KG, Matthews RW (eds) The social biology of wasps. Cornell University Press, Ithaca, pp 336–388

Spradbery JP, Kojima J (1989) Nest descriptions and colony populations of eleven species of *Ropalidia* (Hymenoptera, Vespidae) in New Guinea. Jpn J Entomol 57(3):632–653

Strassmann J (1981) Evolutionary implications of early male and satellite nest production in *Polistes exclamans* colony cycles. Behav Ecol Sociobiol 8:55–64

Suryanarayanan S, Hermanson JC, Jeanne RL (2011) A mechanical signal biases caste development in a social wasp. Curr Biol 21:231–235

Szathmáry E, Maynard Smith J (1995) The major transitions in evolution. W. H. Freeman, New York

Toth AL, Sumner S, Jeanne RL (2016) Patterns of longevity across a sociality gradient in vespid wasps. Curr Opin Insect Sci 16:28–35

Trindade OSN, Azevedo GG, Smith DR, Silva JC (2012) Occurrence of the parasitoid *Seminota marginata* (Westwood, 1874) (Hymenoptera: Trigonalidae) in a nest of the social wasp, *Apoica (Apoica) flavissima* (Van der Vecht, 1973) (Hymenoptera: Vespidae). Braz J Biol 72(4):967–968

Van Oystaeyen A, Caliari Oliveira R, Holman L, Van Zweden JS, Romero C, Oi CA, d'Ettorre P, Khalesi M, Billen J, Wackers F, Millar JG, Wenseleers T (2014) Conserved class of queen pheromones stops social insect workers from reproducing. Science 343(6168):287–290

Warrant EJ (2008) Seeing in the dark: vision and visual behaviour in nocturnal bees and wasps. J Exp Biol 211(11):1737–1746

Wcislo WT, Arneson L, Roesch K, Gonzalez V, Smith A, Fernandez H (2004) The evolution of nocturnal behaviour in sweat bees, *Megalopta genalis* and *M. ecuadoria* (Hymenoptera: Halictidae): an escape from competitors and enemies? Biol J Linn Soc 83(3):377–387

Wenzel J (1991) Evolution of nest architecture. In: Ross KG, Matthews RW (eds) The social biology of wasps. Cornell University Press, Ithaca, pp 480–519

West-Eberhard MJ (1969) The social biology of polistine wasps, vol 140. Miscellaneous Publications of the Museum of Zoology, University of Michigan, Ann Arbor, pp 1–101

West-Eberhard MJ (1978a) Polygyny and the evolution of social behavior in wasps. J Kansas Entomol Soc 51(4):832–856

West-Eberhard MJ (1978b) Temporary queens in *Metapolybia* wasps: nonreproductive helpers without altruism? Science 200(4340):441–443

West-Eberhard MJ (1982) The nature and evolution of swarming in tropical social wasps (Vespidae, Polistinae, Polybiini). In: Jaisson P (ed) Social insects in the tropics, vol 1. Université Paris-Nord, Paris, pp 97–128

West-Eberhard MJ (1996) Wasp societies as microcosms for the study of development and evolution. In: Turillazzi S, West-Eberhard MM (eds) Natural history and evolution of paper wasps. Oxford University Press, New York, pp 290–317

West-Eberhard MJ, Carpenter JM, Gelin LF, Noll FB (2010) *Chartergellus golfitensis* West-Eberhard: a new species of Neotropical swarm-founding wasp (Hymenoptera: Vespidae, Polistinae) with notes on the taxonomy of *Chartergellus zonatus* Spinola. J Hymenopt Res 19(1):84–93

Willink A (1959) Las especies afines a *Parachartergus apicalis* (F.) (Hymenoptera, Vespidae, Polybiinae). Acta Zoologica Lilloana 17:263–292

Wilson EO (1990) Success and dominance in ecosystems: the case of the social insects. Oldendorf/Luhe, Ecology Institute

Yamane S (1985) Social relations among females in pre- and post-emergence colonies of a subtropical paper wasp, *Parapolybia varia* (Hymenoptera Vespidae). J Ethol 3:27–38

Yamane S (1996) Ecological factors influencing the colony cycle of *Polistes* wasps. In: Turillazzi S, West-Eberhard MJ (eds) Natural history and evolution of paper-wasps. Oxford University Press, Oxford, pp 75–97

Yamane S, Mateus S, Hozumi S, Kudo K, Zucchi R (2009) How does a colony of *Apoica flavissima* (Hymenoptera: Vespidae, Epiponini) maintain a constant temperature? Entomol Sci 12(3):341–345

Zavattari E (1906) Viaggio del Dr. Enrico Festa nel Darien, nell'Ecuador e regioni vicine, Imenotteri, I. Diploptera. Boll Musei Zool Anat Comp R Univ Torino 21(529):1–22

Zucchi R, Sakagami SF, Noll FB, Mechi MR, Mateus S, Baio MV, Shima SN (1995) *Agelaia vicina*, a swarm-founding polistine with the largest colony size among wasps and bees (Hymenoptera: Vespidae). J New York Entomol Soc 103(2):129–137

Chapter 3
The Foraging Behaviour of Neotropical Social Wasps

Mateus Detoni and Fábio Prezoto

Abstract The evolution of eusociality among the Hymenoptera is related to the development of colonies with big populations, which in turn presented these social insects with the need for a constant intake of nutrients necessary for their collective survival. Taking a closer look on how the social wasps (Hymenoptera: Vespidae) manage to provide their colony with resources is of great value to understand the evolution of social life, but also the impact insect societies have on the environments they live in. Here, we attempt to explore the foraging behaviour of vespids, summarizing its general aspects and defining historically established key concepts. We build on previous reviews while bringing updated references and focusing on the literature on Neotropical social wasps (Polistinae). In this chapter, we define and investigate a series of topics, including the foraging activity of colonies, the flight range of foragers, the intricacies of the trophallaxis behaviour, the search and obtainment of different types of resources, and their storage inside nests. Finally, we attempt to diagnose the issues and challenges faced by Neotropical wasp researchers and suggest a few ways investigations may take for the advancement of the field when studying the foraging behaviour.

Keywords Activity · Paper wasps · Polistinae · Predation

3.1 Introduction

Among the diversity of behaviours shown by the social wasps, foraging is probably the most popular topic in specialized literature. Through their foraging activity, wasps form a network of interactions that is ultimately responsible for their value in providing ecological services (such as pest predation, pollination or seed dispersal).

M. Detoni (✉)
Department of Zoology, University of Otago, Dunedin, New Zealand

F. Prezoto
Departamento de Zoologia, Universidade Federal de Juiz de Fora, Juiz de Fora, MG, Brazil

F. Prezoto et al. (eds.), *Neotropical Social Wasps*,
https://doi.org/10.1007/978-3-030-53510-0_3

There is a great deal of importance in investigating the foraging behaviour, for it leads to understanding the social wasps' evolutionary history, basic biology and the ecological context in which they are inserted.

"Foraging activity" can be defined for social wasps as the functional group of behaviours of collecting environmental resources essential for survival at the colony level. These resources can be used either in the nourishing of adult wasps and their offspring or in the construction and maintenance of nests. Since the Polistinae social wasps are opportunistic and generalist predators, collected resources vary in nature and origin. This allows wasps to explore a wide array of ecological interactions. The foraging behaviour of social vespids has been detailed in literature review papers (Raveret-Richter 2000) and chapters in books dedicated to the group (Spradbery 1973; Carpenter and Marques 2001; Jeanne and Taylor 2009). In this chapter, we attempt to summarize and describe the general aspects of social wasp foraging, building on those previous studies by updating the references to reflect the state of the art. Furthermore, while previous literature reviews have been mostly skewed towards temperate-zone organisms, we here focus our efforts on the Neotropical Polistinae, aiming to bring a fresh perspective to the field of social foraging research.

3.2 Activity Schedule and Rhythm

The Polistinae clade is almost exclusively made of diurnal species (Spradbery 1973). A flow of individuals committed to the foraging activity coming in and out of the nest are often visible during daytime. At night, nests are usually more cryptic, since most of their inhabitants inside the nest, showing low levels of activity. Even in colonies of the Epiponini tribe, which sometimes keep guards around nest openings during the night (e.g. Chavarría-Pizarro and West-Eberhard 2010), such individuals will remain mostly motionless unless disturbed. The remarkable exception to this rule within the Polistinae is the wasps from the genus *Apoica* Lepeletier, which is exclusively comprised of nocturnal species and therefore presents foraging activity that differs from the rest of the subfamily (Hunt et al. 1995; Warrant 2008).

Diurnality is the main trend among insects, especially in tropical zones. This is partially due to a stimulation of their activity by environmental factors that are typically higher during the day, such as luminosity and temperature (Huffaker and Gutierrez 1999). Indeed, foraging activity in Neotropical social wasps is normally positively correlated to environmental temperature and negatively correlated to the relative air humidity across different genera (e.g. Ribeiro-Júnior et al. 2006; Santos et al. 2009; Cavenazzi and Noll 2011; Castro et al. 2011; Detoni et al. 2015; Jacques et al. 2018). The influence of wind speed in the process is not as clear, although there is consensus that strong winds discourage foraging by hindering the wasps' ability to fly (Elisei et al. 2005, 2013). Luminosity regulates the foraging activity through physioethological thresholds (Spradbery 1973; Ribeiro-Júnior et al. 2006), which are required since wasps rely heavily on visual orientation during their foraging bouts. Furthermore, there is evidence that artificial light sources may attract large amounts of wasps around sunset, when foragers are usually returning to their nests (Barbosa et al. 2016). This argument reinforces the group's positive phototropism and its relation with foraging light thresholds.

The diurnal habit of the social wasps may, however, be explained by more than only their stimulation by abiotic factors. Firstly, the localization of their prey and other alimentary resources is oriented to a good extent by visual clues, thus requiring a source of light (as seen in Jeanne 1972). Secondly, most of the prey captured by Polistinae wasps (especially caterpillars) are also active during the day (Heinrich 1979), which makes prey-predator encounters more likely and is therefore essential for the wasps' success in protein foraging.

One of the main research interests in the social wasp foraging schedule is how does the foraging fluctuate in intensity through the day, a process that can be called "foraging activity rhythm". A colony's rhythm can be measured by counting how many foragers leave and return to the nest per time unit, generating a flow that varies according to the time of the day (Giannotti et al. 1995; Jacques et al. 2018). Spradbery (1973) has described a general pattern of foraging rhythm for diurnal social wasps: the activity begins around sunrise, being limited by the availability of solar light, and reaches its daily peak around noon and decreases through the afternoon until it ceases around sunset.

In theory, social wasp larvae will require a greater amount of energetic resources during the morning, after a period of deprivation during the night. That would explain why foraging tends to be more intense during the first half of the day, even if the amount of light available is similar to that in the afternoon (Gaul 1952). Alternatively, studies show that in many different species, the foraging activity peaks at the warmest hours of the day, which would also explain that positive correlation (Resende et al. 2001; Lima and Prezoto 2003; Elisei et al. 2005; Ribeiro-Júnior et al. 2006; Rocha and Giannotti 2007; Castro et al. 2011; Detoni et al. 2015).

There is a great difference between the foraging rhythm of the independent-founding social wasps (Mischocyttarini and Polistini) and the swarm-founding social wasps (Epiponini), the latter usually being more active. Colony size, as observed by Edwards (1980), is likely a variable that plays a role in that difference: while a *Mischocyttarus* or *Polistes* colony has a few dozen workers on average, nests of the Epiponini tribe commonly host hundreds, thousands or even millions of foragers (that last case being true for some nests of the extremely populous *Agelaia* colonies). Still, the difference in adult populations alone does not seem to explain rhythm differences across species. For example, *Polistes simillimus* Zikán (1951) (Polistini) and *Protopolybia sedula* (de Saussurre, 1854) have similar colony sizes, yet the rhythm of the swarm-founding *Pr. sedula* (average 105.4 forager departures and 108 returns per hour; Detoni et al. 2015) is clearly much more intense than that of *Po. simillimus* (average 18.4 forager departures and 20.9 returns per hour; Prezoto et al. 1994). A possible explanation for such phenomenon is that, due to the lower worker:brood ratio in Epiponini populations, the colony's metabolic needs are higher and, therefore, an intense foraging rhythm is required. In fact, the worker:brood ratio seems to be a central element in the social biology of the Polistinae (London and Jeanne 2003).

Unlike diurnal Neotropical social wasps, the *Apoica* genus is distinguishable from other swarm-founding wasps (and social wasps, in general) for their nocturnal habits. Hunt et al. (1995), when studying *Apoica pallens* (Fabricius, 1804) in Venezuela, noted the complete absence of foraging activity during daytime, a period which workers spent by ventilating the nest and guarding the colony against threats. After sunset, the workers start foraging; the activity begins at a slow pace, but then an "explosive departure" occurs, with hundreds of foragers leaving the nest in few minutes, in which "the air was filled with wasps" (Hunt et al. 1995). The intense flow is normalized in a few instants and remains stable through most of the night time, decreasing in the hours prior to dawn. Similar events were observed in the species *Apoica flavissima* (Vecht, 1972) by Nascimento and Tannure-Nascimento (2005), which also verified the typical influence of temperature and humidity on the species' foraging, as well as possible relationship to the phase of the moon. This exceptional pattern of activity has likely arisen as a strategy to avoid losses caused by direct competition (Smith et al. 2017).

3.3 Flight Range

Foraging behaviour shares dynamics across different groups of social insects. The need to not only locate and obtain food but also to bring it back to the nest is a challenge faced by all social foragers. The exploration of natural resources must, therefore, have the nest as its geographical reference. Studying the flight range of social wasps allows us to better understand the geographical reach of colonies when interacting with their ecosystems, especially regarding their impact as predators.

By looking at predator ants as an example, it is clear how social foraging is related to the nest site. Camponotini ants create trails that run for several dozen meters, on which they mark their progress through pheromones so that nestmates may easily find the way between nest and food. The species *Camponotus sericeiventris* Guérin, for example, usually builds one or more foraging trails radiating from the nest and reaching up to 86 meters in length, either vertically or horizontally (Elisei et al. 2012).

A major biological difference between ant foragers and wasps resides in the latter's ability to fly. Flight makes dispersion easier, if not more effective, especially when regarding the vertical exploration of ecosystems. On the other hand, the lack of attachment to the substrate while foraging means that a reliable pheromone-marked trail, as seen in ants, is not possible for wasps. That is undoubtedly one of the main drives for the wasps' strong dependence on visual orientation (Raveret-Richter and Jeanne 1985). As a result, there is a cost-benefit relationship regarding exploration: the further away from the nest a forager flies, the smaller the chance that it will be able to find its way back, which can lead to worker loss in the colony. A colony set in a poor environment, where its workers must increase their foraging flight range in order to obtain food, is therefore at severe disadvantage (Prezoto et al. 2016b).

"Flight range" is defined as the maximum distance a forager is able to fly with significant chances of returning to the nest and, therefore, being successful at foraging. By obtaining the flight range value, it is possible to deduct an effective radius in which foragers will be active around the colony (equal to ½ flight range, since that includes the flights from the nest and back) (Gobbi 1978; Santos et al. 1994).

Studies on the flight range of Brazilian social wasps are few and restricted to few taxa (Table 3.1). The method used to assess a species' flight range is consistent among publications and is made by following these steps: (1) intercepting social wasp workers from colonies in the pre-emergence stage; (2) marking the workers to allow individual distinction; (3) releasing the workers at multiple distance intervals

Table 3.1 Social wasp foraging flight range across studies

Taxon				Flight range (m)	Study
Vespidae					
Polistinae					
	Polistini				
		Polistes			
			Polistes canadensis	250	Santos et al. (1994)
			Polistes lanio	> 200	Giannotti et al. (1995)
			Polistes simillimus	100	Prezoto and Gobbi (2005)
			Polistes versicolor	300	Gobbi (1978)
	Epiponini				
		Angiopolybia			
			Angiopolybia pallens	48	Cruz et al. (2006)
		Polybia			
			Polybia emaciata	76	Galé et al. (2015)
			Polybia fastidiosuscula	81	Souza et al. (2011)
			Polybia occidentalis occidentalis	126	Santos et al. (2000)
			Polybia occidentalis venezuelana	50	Hernandéz et al., (2009)
			Polybia rejecta	<210	Dejean et al. (2017)
			Polybia scutellaris	150	Machado and Parra (1984)
			Polybia sericea	75	Bichara Filho et al. (2010)
		Protopolybia			
			Protopolybia exigua	75	Ribeiro Júnior et al. (2008)

The 70% return rate suggested by Santos et al. (2000) was used as criteria to set the values displayed here from the nest; and (4) assessing the proportion of workers that returned to the nest from each distance (Santos et al. 1994). Although this method allows the assessment of the maximum distance from which a worker can return to its nest, a return rate of at least 70% is used as a valid estimator of a species' effective flight range (Santos et al. 2000). Following that assumption, most species would have an effective flight range within 300 meters (Gobbi 1978).

3.4 Trophallaxis

Thus far we focused on how external factors influence on a colony's foraging activity, especially those linked to external environmental variables. In addition, social wasps are also capable of an internal and social behavioural control of foraging. Of these behaviours, the most fascinating – and possibly, the most relevant for the group's natural history – is trophallaxis.

The term trophallaxis was coined by author William Wheeler in 1918 to describe the reciprocal exchange of liquids between two insects through their respective mouths. The term's definition has been updated and broadened to include unidirectional exchanges as well (Hunt and Nalepa 1994). This behaviour has implications for various aspects of social life other than just foraging; trophallaxis has been studied in the context of establishing social cohesion and the hierarchy of social wasp colonies (Roubaud 1916; Jeanne 1991).

According to Spradbery (1973), social wasp trophallaxis is divided into two distinct categories according to the life stage of its participants: *adult-larva trophallaxis* and *adult-adult trophallaxis*. We will explain each of these individually, as they present distinct effects on a colony's foraging behaviour.

To initiate adult-larva trophallaxis, a reproductive wasp or worker will approach a larva, usually observable when the adult inserts its head inside a nest cell containing brood (see Fig. 3.1a). The adult wasp stimulates the larva by antennating its mouth parts (and also possibly by beating its gaster against the nest in *Polistes*). The wasp larva will then secrete saliva, which is transferred into the adult's gastric tract. Pioneer studies on adult-larva trophallaxis showed that larval saliva is highly attractive to adults (Du Buysson 1903; Janet 1903); later, chemical analysis allowed the characterization of saliva as mainly composed of sugars and amino acids. Indeed, Hunt and colleagues (1982) describe larval saliva as being nutritionally much superior to wasps than the richest floral nectar.

At first impression, this scenario seems to escape the logic of social foraging. Since wasp larvae are in constant need of nutritional intake in order to grow and develop, what would be the biological rationale behind giving away such a highly caloric resource to adult wasps, which are able to forage for self-sustenance? Researchers tried to solve this enigma by addressing the question from different perspectives. Ishay and Ikan (1968a, b) showed through radioactive marking that the sugary substances in larval saliva do not derive from the nectar collected from foragers but instead from proteolytic enzymes. This finding led to the conclusion

Fig. 3.1 (**a**) An adult *Mischocyttarus* sp. wasp inserts its heads inside a cell containing a larva; (**b**) a group of adult *Polistes* wasps engaged in trophallaxis; (**c**) an adult paper wasp (to the left) holds a mass of maxalated prey in her mandibles

that larvae provide rewards for adults feeding them, generating a positive feedback. However, other studies suggest that the process is not that simple. First, trophallaxis in *Polistes* seems to be not always reciprocal but rather a one-way exchange; this prevents larvae from rewarding foragers immediately after being fed (Suryanarayanan & Jeanne 2008). In addition, larva saliva provision seems to increase in frequency when colonies show minimal foraging activity (Ikan and Ishay 1966). In other words, larvae will dispense more resources despite being fed less often. Indeed, *Mischocyttarus* wasps seem to have higher sugar concentration in their saliva during the pre-emergence stage of the colony cycle, when worker populations are smaller and foraging is costlier when compared to post-emergence (Hunt 1988). In conjunction, these findings come together in the hypothesis that trophallaxis plays a major role in the social cohesion of a colony: in times when food is scarce or foraging is otherwise energetically costly, larvae will provide attractive resources to avoid being cannibalized by adults, a common strategy during periods of nutritional stress (Gadagkar 1991). Nevertheless, it is still possible that adult-larva trophallaxis has a role in foraging during periods of regular activity, and these nutritional interindividual dynamics constitute a cornerstone in the evolution of eusociality in Hymenopterans.

The second category, adult-adult trophallaxis (see Fig. 3.1b), is somewhat less studied than when immature wasps are involved in the process and its biological significance has been less explored by researchers. There is evidence for some groups that adult-adult trophallaxis is a mediator in determining the social hierarchy in social insects. In Polistinae, foraging load partitioning between adult nestmates seems to also play a role, or at least reflect, the hierarchical organization of a colony (O'Donnell 1995a; Tindo and Dejean 2000; Sinzato et al. 2003).

It has been suggested for some swarm-founding species that trophallaxis between adults could result in recruitment for foraging within the colony (Taylor et al. 2012). Conversely, a study with *Agelaia multipicta* (Haliday, 1836) does not support this hypothesis (Jeanne et al. 1995). Food recruitment in social wasps is a somehow elusive topic, and we recommend the reading of Jeanne and Taylor's (2009) discussion about this phenomenon. Since trophallaxis is often observed immediately after a forager returns to its nest with sugary substances, it has also been used as a non-invasive behavioural indicator to distinguish the forager's load (Giannotti et al. 1995).

3.5 Searching for Resources

A detailed narrative on the challenges faced by a social wasp worker when leaving the nest to forage is provided by Raveret-Richter and Jeanne (1991). The first step is locating the resource in the environment, whether it is sugary food, water, nest material or a prey. For the latter, the forager must still subjugate its prey, and, if it is too big to be carried back in one piece, it must be fragmented before the wasp can return to its nest. In this case, the wasp still needs to memorize where she left the remaining fragments, so it can return to claim them later – and potentially fight off opportunists to do so. The whole process involves a diverse series of behaviours, of which the most instigating include orientation during flights and the ability to locate resources.

Most behaviours linked to the foraging activity in social wasps are intrinsically connected to a learning process. Prior to being able to fly out in the open, young foragers-to-be make short trips around the nest in order to memorize the surrounding landscape. These visual cues will be used to find the colony upon returning from a foraging trip. Gaul (1951) suggests that temperate-zone vespids make a series of three flights, in which the forager executes progressively wider loops around the nest opening, culminating in a sudden departure for the actual foraging during the third flight. This behavioural display is named *orientation flight* (Spradbery 1973).

After their departure, foragers must be able to detect resources in the environment using sensorial stimuli. Authors attribute different weights to the importance of different cues for the localization and identification of food sources, but there is a consensus on the conjoined relevance of visual and olfactory cues (Raveret-Richter 2000). Pioneer studies on resource identification in social wasps stressed the role of olfactory cues (Iwata 1976), and evidence has been added to the case of chemical perception in wasps over the years. A good example is how wasps are able to detect leaves that were damaged by chewing herbivores (especially by caterpillars, which they usually prey upon) and choose to land on those instead of healthy leaves as a substrate in order to maximize prey encounters (Saraiva et al. 2017). Furthermore, olfactory cues seem to be especially important for the detection and obtainment of sugary substances. Some wasps may consider a site's scent more than its colour when choosing where to land (Mcpheron and Mills 2007) and are able to detect honeydew from long distances (Gaul 1952). Finally, carrion-feeding species (such as those in the tropical genus *Agelaia*) are likely to use mainly chemotactic cues to locate their food (F.P., pers. obs.).

Resources may also be located visually, which is especially true for prey. Wasps may even misdirect predation behaviours towards shapes that resemble prey (Jeanne 1972). Some researchers defend that visual orientation may be more important than olfactory, especially for the capture of highly active prey, since the wasp's sight may rely heavily on the perception of movement (Spradbery 1973; Jeanne and Taylor 2009).

Overall, a forager may use visual cues to search for prey in a larger scale, chemical cues as it approaches the prey in a smaller scale to choose a landing site. This integration of different stimuli was suggested and described in detail for *Polybia* wasps by Raveret-Richter and Jeanne in 1985.

After identifying and capturing their prey, foragers must bring the food back to their nests. Prey will often have a much larger body mass than foragers, and so wasps need to cut their prey in fragments and carry them in multiple trips. Sometimes, wasps will execute orientation flights around the fragmented prey to memorize its location (Takagi et al. 1980).

Inter- and intraspecific interactions between social wasps will often take place at a food source. Forager aggregates in *Agelaia* sp. may be evidence that local enhancement (use of social cues to facilitate the finding of good foraging sites) is being used to a colony's benefit (Forsyth 1978). Other species may use local enhancement to exploit a competitor's efforts. *Mischocyttarus immarginatus* (Richards, 1940) wasps have been observed to attract foragers of three other different species (*Polistes instabilis*, *Polybia diliguetana* and *Polybia occidentalis*) that quickly robbed prey fragments after the original predator left the site (Raveret-Richter 1990). Conversely, niche overlap between species may not affect their direct survival in some environments (Jacques et al. 2018). When wasps compete with ants for food, they might show exceptional behaviours to overcome direct competition; a notable example is the reported removing of ants from sites by dropping them from heights (Ricioli et al. 2017). These are only some of the examples on the ways the social wasp's search for resources might generate multi-level ecological interactions, constituting one of the group's most relevant traits.

3.6 Items Foraged

Social wasps are notably generalist and opportunistic in their foraging behaviour. This implies that wasp foragers are able to explore a wide range of resources while showing varied behavioural strategies to do so. These habits were probably developed in the group in the answer to the evolution of sociality, which requires a nutritional and energetic income much higher to that of individual life. Greater foraging behavioural plasticity, in contrast with specialization, could allow colonies to survive periods of food shortage and extend their lifespans (Jeanne and Taylor 2009).

These traits add to the complexity of social foraging and must be taken into account when studying social wasp diet and prey preferences.

The items foraged by wasps are usually divided into four groups (West-Eberhard 1969; Raveret-Richter 2000): (1) water, (2) nest-building materials (often called "vegetal fibre"), (3) sugary substances (also called "carbohydrates" or even "nectar") and (4) protein (also called "prey"). These groups have different functions for the maintenance of the colony, as discussed below. Raveret-Richter (2000) has reviewed the foraging of each of these items in detail for social wasps, and a good part of the literature on them has remained untouched for Neotropical paper wasps. We here provide a summarized description of these behaviours and discuss more recent findings for the Polistinae and advise the reader to refer to Raveret-Richter's work for a more extensive review.

3.6.1 Water

Social wasps imbibe water droplets during foraging activity, which is done from water bodies such as lakes and puddles or from accidental/temporary containers such as morning dew (Jeanne 1986). Wasp foragers may display unusual behavioural alternatives to avoid competition while foraging for water. For instance, wasps may choose to perform a short nimble landing on the water's surface to drink from it instead of joining other wasps on the water's edge (Prezoto et al. 2016a).

After its collection, water may be used in more than one way in the colony. In addition to being ingested by adults or larvae for metabolic purposes (Kasuya 1982), water is also added to vegetal fibre during nest construction and repair (Akre 1982). Workers will also place droplets on nest walls or inside cells. In conjunction with fanning (wing beating) behaviour, it makes up a behavioural strategy to thermoregulate the colony on hot days (Wilson 1971).

3.6.2 Nest-Building Materials

Vegetal fibre is the main construction material used by Polistinae, which has rendered them the popular name "paper wasps". Fibre is collected from natural vegetal structures such as branches or trunks, but may also be obtained from untreated, unpainted wood worked by man such as old furniture (Wenzel 1991). Foragers land on the substrate and cut a shred of the fibre using their mandibles, which are then maxalated and mixed with saliva into a rounded shape. At least for one species, foragers will regurgitate water on the original source before cutting of the fibre – probably making its removal easier and helping in keeping the material's cohesion (Jeanne 1986).

3.6.3 Sugary Substances

Sugary substances, or simply carbohydrates, are collected by foragers and used as a main source of energy by adult wasps, which feed almost exclusively on it (Jeanne and Taylor 2009). Carbohydrates are usually collected in the form of plant nectar or the honeydew exuded by homopterans (Evans and West-Ebehrard 1970; Moller et al. 1991). This resource is shared at the nest to both adult and immature wasps via trophallaxis.

Nectar can be collected from both extrafloral nectaries and flowers, although many of the latter are not accessible to wasps due to their morphology (Spradbery 1973; Köhler 2008). The flowers visited by wasps reflect the group's generalist/opportunistic habits: favoured species vary as different flowering seasons coincide with a need for high carbohydrate intake at the colonies (Köhler 2008). Additionally, it is during flower visitation that wasps may act as pollinators.

One of the most studied alternatives for plant nectar as a carbohydrate source is the collection of honeydew. This substance is a metabolic by product secreted by aphids and scale insects after the ingestion of sap. Homopteran families such as Aphidae, Psyllidae and Coccidae are often visited by wasp foragers for the collection of honeydew and may even benefit from wasp protection against predators (Letourneau and Chloe 1987; Moller et al. 1991).

Other alternatives for carbohydrates exploited by wasp foragers include tree sap (Jeanne and Taylor 2009), bee honey robbing (Edwards 1980) and ripe fruit. The latter seems to be especially attractive to foragers when at the over-ripe or rotting stage: wasps have been observed foraging at Brazilian cherries (Souza et al. 2013), Brazilian grapes (De Souza et al. 2010), cactus fruits (Santos et al. 2007), cashews (Santos and Presley 2010), grapes (Hickel and Schuck 1995), guavas (Brugger et al. 2011), mangoes (Barbosa et al. 2014) and prunes (Prezoto and Braga 2013).

3.6.4 Protein

Protein foraging in social wasps is the main interest in a myriad of studies on the group's ecology. In addition to its role in the social cohesion of a colony (see "Trophallaxis), protein foraging in Neotropical social wasps is the key behaviour behind the growing interest in exploiting the group as agents of pest control (Prezoto et al. 2019).

Even though adult wasps may eventually consume portions of prey, these are largely destined to larval nourishment. In addition to ingesting sugary substances, larvae require a constant protein intake in order to fully develop into adults. For that reason, it is generally expected that protein foraging intensifies during stages when the proportion of brood in a colony is higher (London and Jeanne 2003).

The main source of protein foraged by wasps comes from prey, namely, terrestrial arthropods (Prezoto et al. 2019). The standard method for assessing social wasp prey preference is the interception of foragers as they return to the nest and physically removing the load from their mandibles for analysis (Giannotti et al. 1995). This method is limited, however, because most wasps chew prey fragments prior to their transport (Fig. 3.1c). Visual taxon identification performed on prey matter is therefore restricted to higher taxonomical levels, such as families (Jeanne and Taylor 2009). *Polybia occidentalis* (Olivier, 1791) is a notable exception: since their prey is carried back almost intact, the species stands as a good model for cost-efficient diet assessment (Hunt et al. 1987).

Despite the challenges involved, various researchers have put effort into assessing social wasp diets. Jeanne and Taylor (2009) state that wasps, although generalists, tend to choose soft-bodied terrestrial arthropods as their prey. The biggest portion of these is represented by caterpillars (Lepidoptera) and adult flies (Diptera), with a smaller amount of Dermaptera, Orthoptera, Odonata, Hemiptera, Coleoptera, Hymenoptera and Arachnida being captured. The trend for targeting caterpillars during protein foraging has been verified in many Neotropical social wasps; Table 3.2 shows species and environments for which Lepidoptera larvae constituted at least 50% of their diets.

This preference for caterpillar as prey seems to be established across different genera of both individual- and swarm-founding social wasps, especially in environments altered by man. In environments dominated by fewer species, such as in monoculture plantations, the proportion of Lepidoptera in social wasp diets can reach values as high as 95–100% (Prezoto et al. 2006; Elisei et al. 2010). This supports the claim that social wasps, although generalist, may act as facultative specialists (Raveret-Richter 1990).

Table 3.2 Research showing species of social wasps that capture a majority (50% or more) of Lepidoptera as their prey in different Neotropical environments

Species	Environment	Study
Apoica pallens	Pasture areas	F. Prezoto, unpublished data
Mischocyttarus cassununga	Urban parks	M. Detoni, unpublished data
Polistes lanio	Farms	Giannotti et al. (1995)
Polistes simillimus	Farms	Prezoto et al. (1994)
Polistes versicolor	Farms; urban parks	Prezoto et al. (2006); Elisei et al. (2010)
Polybia dimidiata	Farms	Campos-Farinha and Pinto (1996)
Polybia ignobilis	Farms; urban parks	Silva et al. (1968), Gobbi and Machado (1986), Picanço et al. (1988), Marques (1996)
Polybia occidentalis	Farms; urban parks	Gravena (1983), Gobbi et al. (1984)
Polybia paulista	Farms; urban parks	Gobbi and Machado (1985), Campos-Farinha and Pinto (1996), Marques (1996)
Polybia sericea	Farms; urban parks	Silva et al. (1968); Marques (1996)

Nevertheless, a generalist diet still plays an important role in social wasp ecology. Diet composition may shift significantly away from lepidopterans to fit prey seasonality. As an example, research shows colonies preying on Diptera at high proportions for the species *Polybia platycephala* (Richards 1951) (Prezoto et al. 2005) and *Synoeca cyanea* (Fabricius, 1775) (Prezoto and Braga 2013).

The opportunistic aspect of social wasp foraging allows the exploitation of a range of alternatives for protein foraging. *Agelaia* wasps are notable for their carrion-feeding behaviour, a trait that sometimes is also seen in *Angiopolybia*, *Brachygastra*, *Parachartergus*, *Polybia*, *Protonectarina* and *Synoeca* (O'Donnell 1995b; Silveira et al. 2005). Wasp foragers may further act on corpse decomposition Gomes et al. (2007) observed *Polybia ignobilis* individuals capturing flies that were laying eggs on rotting carcasses. Another usual strategy is preying on alate ants and termites during nuptial flights, when they are most vulnerable (Holfling and Machado 1985; Raveret-Richter 2000). Finally, unusual behaviours have been recorded for social wasps such as robbing prey from spider webs and other arthropod predators (Chevalier 1924; Jeanne 1972), preying on vertebrate eggs (Warkentin 2000) and capturing prey from bait traps used in insect diversity studies (Maciel et al. 2014).

3.7 Resource Storage

To overcome the challenge of feeding a colony for extended periods of time, some vespids have developed strategies to store resources in their nests. This behaviour is comparable to the production of honey in social bees (Seeley 1989) or the storage of vegetal matter in subterranean ant nests (Taber 1999). In Polistinae these behaviours are somewhat simpler but still reflect the importance of resource storage in the success and evolution of eusociality (Hunt 1982, 1991).

In their "Observations on the social wasps of South America (Hymenoptera, Vespidae)", Richards and Richards (1951) performed an extensive analysis on the nests of different social wasp species and found two types of resources: liquid storages (honey-like sugary substances) and solid storages (maxalated prey fragments). Both types were found inside nest cells, although each one showed different distribution patterns through the nest. Other studies analysed liquid storages and found them to be of vegetal origin, usually nectar from flowers surrounding the colonies. These substances are sometimes called "wasp honey" or simply "honey" (Strassman 1979; Guimarães et al. 2008).

Resource storages are found in nests of both independent- and swarm-founding species, but seem to be more common in the latter (Richards and Richards 1951; Prezoto and Gobbi 2003). The nests of swarm-founding paper wasps have multiple layers of combs surrounded by an envelope, which probably provides food stocks with better protection against robbers when compared to the exposed, single-combed nests of the independent-founding *Polistes* and *Mischocyttarus*. Honey storage has been verified for the swarm-founding species *Metapolybia cingulata*

(Fabricius, 1804), *Polybia bicyttarella* (Richards, 1951), *Polybia bistriata* (Fabricius, 1804), *Polybia ignobilis* (Haliday, 1836), *Polybia occidentalis* (Olivier, 1791), *Polybia platycephala* (Richards, 1978), *Polybia scutellaris* (White, 1841), *Protonectarina* spp. and *Protopolybia emortualis* (de Saussurre, 1855) (Richards and Richards 1951; Hunt et al. 1987; Corujo et al. 2010; Rocha 2011) and for the independent-founding *Mischocyttarus cassununga* (von Ihering, 1903), *Mischocyttarus drewseni* (de Saussurre, 1854), *Mischocyttarus socialis* (de Saussurre, 1854) *Polistes simillimus* (Zikán, 1951) and *Polistes versicolor* (Olivier, 1791) (Richards and Richards 1951; Jeanne 1972; Prezoto and Gobbi 2003; Guimarães et al. 2008; Barbosa et al. 2017). The species *Brachygastra lecheguana* (Latreille, 1824) stands out for storing honey in remarkable amounts (Richards and Richards 1951).

Rossi and Hunt (1988) described two patterns for honey storage inside nests. Honey droplets may be placed inside empty cells (Strassman 1979; Machado 1984; Hunt et al. 1987) or cells shared by eggs or first instar larvae (and rarely by second instar larvae) (Rau 1928, 1939; Heldmann 1936; Jeanne 1972). Honey placed in cells also containing immature wasps was related to providing young larvae with additional nourishment by Jeanne (1972). This claim is supported by behavioural evidence, since adults may engage in trophallaxis with larvae immediately after visiting honey storages (Jeanne 1972; Guimarães et al. 2008). When stored in empty cells, honey is usually placed in older, unproductive combs (in swarm-founding species) or in the comb's peripheral cells (in independent-founding species). These storages are possibly long-term food reserves to grant the colony its sustenance at the final phases of the cycle, which often coincide with food shortage periods (Strassman 1979). This claim is supported by the fact that honey storages seem to be more conspicuous during dry seasons (Hunt et al. 1987). Finally, from behavioural observations, when individual wasps do not interact with nestmates after visiting honey storages, it can be inferred that the reserves are purposed for individual adult nourishment (Prezoto and Gobbi 2003).

Solid resource storages, on the other hand, seem to be a more widespread behaviour in Neotropical social wasps – especially in the genus *Polybia*. Studies with *Polybia platycephala* and *Polybia occidentalis* (Richards and Richards 1951; Rocha 2011) report a high number of nest cells containing fragments of flies (Chironomidae) and alate ants (Hymenoptera) and termites (Blattodea). The high proportion of alates found in storages is further evidence of the wasp's opportunistic foraging, since reproductive ants and termites are usually available as prey in short periods but in very high densities. Furthermore, the unusual abundance of prey could be the necessary stimulus for building solid resource storages.

Investigating resource storages in social wasp nests can provide insights outside of the wasp's life history. As an example, pollen analysis performed on honey storages allows researchers to identify flowers visited by wasp foragers without needing to deal with the challenges of tracking wasps during their bouts (Corujo et al. 2010). Overall, storage analysis can be a complementary key to understand the ecological interaction webs that wasps take part in.

3.8 Future Directions

Due to many factors, the research on the foraging behaviour of social wasps in the Neotropical environments – especially in Brazil, where most of its research groups are based at – unfortunately lags behind studies carried out in other climatic zones, not unlike social wasp research that sometimes lags behind ant and social bee studies. Underfunding, allied to the relatively small number of researchers devoted to study the group, is probably the main factor responsible for that issue. Curiously, the general public's negative perception of wasps as organisms (Sumner et al. 2018) could be one element involved in preventing wasp research to be funded while also pushing prospective researchers away from the field during their formation. Another considerable challenge faced by researchers is the impressive diversity of social wasps in the Neotropics. Many of these species have nothing but their taxonomical description available in literature, and even abundant, widely distributed ones lack published data on their life history. This often forces investigators to focus on general biology and behaviour prior to addressing the field's trending theories, which is usually looked down by editors and funding organizations.

Yet, as research on basic biology and behavioural ecology progresses, scientists must look ahead to explore the frontiers of sociobiology. Here, we discuss what we consider promising methods and theories for the future of social wasp foraging behaviour research in Neotropical areas.

3.8.1 Biological Control

The use of paper wasps as agents of agricultural pest control has been suggested as early as in the 1950s (Rabb and Lawson 1957). In the Neotropics, this possibility has been repeatedly brought up as the inspiration behind investigations on wasp foraging behaviour and even materialized itself in the form of experimental studies (Prezoto and Machado 1999). Hard experimental evidence obtained in controlled settings (such as pest predation rates and plant productivity in greenhouses) has only very recently been published (Southon et al. 2019), finally opening the way for exploiting social wasps in carefully designed pest control plans. Current research on the matter has been mostly limited to anecdote reports on wasps nesting or foraging within crops. The allocation of scientific resources to explore social wasp foraging in controlled experiments could be hugely beneficial to the field. Additional possibilities were recently opened as wasps have been suggested as controllers of dipterans of medical interest (Prezoto et al. 2016b). The current epidemiological relevance of *Aedes* mosquitoes, vectors of the dengue, Zika and Chikungunya fevers, could be used to motivate researchers and funding agents to invest in social wasps foraging as a control alternative.

3.8.2 Molecular Analysis

The overwhelming majority of studies in the foraging behaviour of Neotropical social wasps rely on observation data. Conversely, the so-called Omics Age brings a wide range of new ways to answer questions on sociobiology (Ghoul et al. 2017). In addition to applying well-established genetics methods to social wasps – such as searching for genes linked to foraging success (Ingram et al. 2011) or identified genetic-physiological thresholds for the perception of specific foraged items (Değirmenci et al. 2018) – we also suggest that researchers explore the innovative fields on ecogenetics in their investigations. For instance, metabarcoding the gut content of foragers and larvae (Krehenwinkel et al. 2016) is a way to overcome the challenges of tracking foragers outside their nests while obtaining reliable data on their interactions with both vegetal species and prey. If successfully done, this could reinforce arguments on the ecological services provided by social wasps.

3.8.3 Collective Personalities

Since the early 2000s, the study of animal personalities has been experiencing a sharp increase in interest among behavioural ecology researchers. Previously interpreted as statistical noise, consistent behavioural variations are now investigated for their theorized role in success and evolution of organisms (Sih et al. 2004). Moreover, the study of social behaviour and the personality of superorganisms – or collective personalities – has presented itself as an exciting new subarea in which sociobiology may be explored through research (Jandt et al. 2014; Wright et al. 2019). For foraging behaviour, relationships between personality traits and general ecology – such as aggressiveness and foraging success or parental care – could be verified for Neotropical social wasps, helping not only in the understanding of their life history but also of the evolution of social behaviour as a whole.

3.9 Concluding Remarks

The foraging behaviour of Neotropical social wasps is one of the most fascinating topics to be explored on the group's behavioural ecology. Here, we briefly presented some of the most unique aspects involved in the activity of wasp foragers and tried to show how its different particularities make up complex relationships with other traits such as nest building, pollination or social hierarchy. Studying the foraging behaviour of social wasps is one of the keys to understand the evolution of eusociality in Hymenoptera.

In our perspective, in the specific case of the Neotropical wasps, the group's diversity should transition from being a challenge to being an advantage to researchers. Exploring the foraging activity of different clades with both classic behavioural approaches and innovative methods will surely lead to the discovery of unheard behavioural strategies and shed a new light on the existing knowledge not only for the taxon but also for Neotropical systems as a whole.

Acknowledgements We thank Dr. Bruno C. Barbosa for his support during the writing of this chapter and for providing the photographs used to illustrate foraging behaviours. We also thank Dr. Jennifer M. Jandt for her valuable insights during the revision of our manuscript.

References

Akre RD (1982) Social wasps. In: Hermann H (ed) Social insects. Academic, New York, 385 pp

Barbosa BC, Paschoalini MF, Prezoto F (2014) Temporal activity patterns and foraging behavior by social wasp (Hymenoptera, Polistinae) on fruits of *Mangifera indica* L. (Anacardiaceae). Sociobiology 61(2):239–242

Barbosa BC, Detoni M, Maciel TT, Prezoto F (2016) Studies of social wasp diversity in Brazil: over 30 years of research, advancements and priorities. Sociobiology 63(3):858–880. https://doi.org/10.13102/sociobiology.v63i3.1031

Barbosa BC, Detoni M, Maciel TT, Prezoto F (2017) Resource storage in the Neotropical social wasp *Mischocyttarus socialis* (Saussure, 1854) (Vespidae: Polistini). Sociobiology 64(3):356–358. https://doi.org/10.13102/sociobiology.v64i3.1686

Bichara Filho CC, Santos GMM, Santos Filho AB, Santana-Reis VP, Cruz JD, Gobbi N (2010) Foraging behavior of the swarm-founding wasp *Polybia (Trichothorax) sericea* (Hymenoptera, Vespidae): daily resource collection activity and flight capacity. Sociobiology 55(3):899–907

Brugger BP, Souza LSA, Souza AR, Prezoto F (2011) Social wasps (*Synoeca cyanea*) damaging *Psidium sp.* (Myrtaceae) fruits in Minas Gerais state, Brazil. Sociobiology 57:533–535

Campos-Farinha AEC, Pinto NPO (1996) Natural enemies of *Chlosyne lacinia saundersii* Doubl. & hew. (Lepidoptera: Nymphalidae) in the state of São Paulo. Anais da Sociedade de Entomologia do Brasil 25:165–168

Carpenter JM, Marques OM (2001) *Contribuição ao estudo dos vespídeos do Brasil (Insecta: Hymenoptera, Vespoidea, Vespidae)*, vol 2. Cruz das Almas, Universidade Federal da Bahia. Publicações digitais. 147p

Castro MM, Guimarães DL, Prezoto F (2011) Influence of environmental factors on the foraging activity of *Mischocyttarus cassununga* (Hymenoptera, Vespidae). Sociobiology 58:133–141

Cavenazzi NCS, Noll FB (2011) Environmental factors influencing foraging activity in the social wasp *Polybia paulista* (Hymenoptera: Vespidae: Epiponini). Pysche 2011:542487. https://doi.org/10.1155/2011/542487

Chavarría-Pizarro L, West-Eberhard MJ (2010) The behavior and natural history of *Chartergellus*, a little-known genus of neotropical social wasps (Vespidae Polistinae Epiponini). Ethol Ecol Evol 22(4):317–343. https://doi.org/10.1080/03949370.2010.510035

Chevalier L (1924) Observations sur les guêpes. Bulletin de la Société des sciences naturelles de l'Ouest de la France 3:66–72

Corujo VLF, Basilio MA, Galati BJ (2010) Pollen content accumulated in nests of *Polybia scutellaris* (Hymenoptera, Vespidae). Grana 49:308–313. https://doi.org/10.1080/00173134.2010.525663

Cruz JD, Giannotti E, Santos G, Bichara-Filho C, Rocha A (2006) Nest site selection and flying capacity of neotropical wasp *Angiopolybia pallens* (Hymenoptera: Vespidae) in the Atlantic rain Forest, Bahia, Brazil. Sociobiology 47(3):739–750

De Souza AR, Venancio D, Prezoto F (2010) Social wasps (Hymenoptera: Vespidae: Polistinae) damaging fruits of *Myrciaria sp.* (Myrtaceae). Sociobiology 55:297–299

Değirmenci L, Thamm M, Scheiner R (2018) Responses to sugar and sugar receptor gene expression in different social roles of the honeybee (*Apis mellifera*). J Insect Physiol 106(1):65–70. https://doi.org/10.1016/j.jinsphys.2017.09.009

Dejean A, Rodríguez-Pérez H, Carpenter JM, Azémar F, Corbara B (2017) The predatory behavior of the Neotropical social wasp *Polybia rejecta*. Behav Process 140:161–168. https://doi.org/10.1016/j.beproc.2017.05.011

Detoni M, Mattos MC, Castro MM, Barbosa BC, Prezoto F (2015) Activity schedule and foraging in *Protopolybia sedula* (Hymenoptera, Vespidae). Revista Colombiana de Entomología 41(2):245–248

Du Buysson R (1903) Monographie des guêpes au *Vespa*. Annales de la Société Entomologique de France 72:260–288

Edwards R (1980) Social wasps: their biology and control. East Grinstead, Rentokil Ltd, 398p

Elisei T, Ribeiro-Júnior C, Guimarães DL, Prezoto F (2005) Foraging activity and nesting of swarm-founding wasp *Synoeca cyanea* (Fabricius, 1775) (Hymenoptera, Vespidae, Epiponini). Sociobiology 46(2):317–327

Elisei T, Nunes JV, Ribeiro C Jr, Fernandes AJ Jr, Prezoto F (2010) Uso da vespa social *Polistes versicolor* no controle de desfolhadores de eucalipto. Pesquisa Agropecuária Brasielira 45(9):958–964

Elisei R, Ribeiro Júnior C, Guimarães DL, Prezoto F (2012) Comportamento de forrageio de *Camponotus sericeiventris* Guérin (Hymenoptera, Formicidae) em ambiente urbano. EntomoBrasilis 5(2):170–172

Elisei T, Nunes JV, Ribeiro C Jr, Fernandes AJ Jr, Prezoto F (2013) What is the ideal weather for social wasp *Polistes versicolor* (Olivier) go to forage? Entomobrasilis 6(3):214–216. https://doi.org/10.12741/ebrasilis.v6i3.342

Evans HE, West-Ebehrard MJ (1970) The wasps. University of Michigan Press, Ann Arbor, 265 pp

Forsyth AB (1978) Studies on the behavioral ecology of polygynous social wasps. Doctorate thesis. Harvard University, Cambridge, 226 pp

Gadagkar R (1991) *Belonogaster, Mischocyttarus, Parapolybia,* and independent-founding *Ropalidia*. In: Ross KG, Matthews RW (eds) The social biology of wasps. Cornell University Press, Ithaca, pp 151–190

Galé YL, Morinson MV, Gracia PC (2015) Amplitud de vuelo y capacidad de carga de la avispa social *Polybia emaciata* Lucas, 1879 (Hymenoptera: Vespidae: Polistinae). Entomotropica 30(11):105–111

Gaul AT (1951) Addition to vespine biology VII. Orientation flight. Bull Brooklyn Entomol Soc 46:54–56

Gaul AT (1952) The awakening and diurnal flight activities of vespine wasps. Proc Royal Entomol Soc London 27:33–35

Ghoul M, Andersen SB, West SA (2017) Sociomics: using Omics approaches to understand social evolution. Trends Genet 33(6):4048–4419. https://doi.org/10.1016/j.tig.2017.03.009

Giannotti E, Prezoto F, Machado VLL (1995) Foraging activity of *Polistes lanio lanio* (Fabr.) (Hymenoptera, Vespidae). An. Soc. Entomol. Brasil 24(3):455–463

Gobbi N (1978) Determinação do raio de vôo de operárias de *Polistes versicolor* (Hymenoptera: Vespidae). Ciência & Cultura 30:364–365

Gobbi N, Machado VLL (1985) Material capturado e utilizado na alimentação de *Polybia* (*Myrapetra*) *paulista* Ihering, 1896 (Hymenoptera, Vespidae). Anais da Sociedade Entomológica do Brasil 14:189–195

Gobbi N, Machado VLL (1986) Material capturado e utilizado na alimentação de *Polybia* (*Trichothorax*) *ignobilis* (Haliday,1836) (Hymenoptera, Vespidae). Anais da Sociedade Entomológica do Brasil 15:117–124

Gobbi N, Machado VLL, Tavares-Filho JA (1984) Sazonalidade das presas utilizadas na alimentação de *Polybia occidentalis occidentalis* (Olivier, 1791) (Hym., Vespidae). Anais da Sociedade Entomológica do Brasil 13:63–69

Gomes L, Gomes G, Oliveira HG, Morlin JJ, Desuó IC, Silva IM, Shima SN, Von Zuben CJ (2007) Foraging by *Polybia (Trichothorax) ignobilis* (Hymenoptera, Vespidae) on flies at animal carcasses. Revista Brasileira de Entomologia 51:389–393

Gravena S (1983) Táticas de manejo integrado do bicho mineiro do cafeeiro *Perileucoptera coffeella* (Guérin-Méneville,1842): I – Dinâmica populacional e inimigos naturais. Anais da Sociedade de Entomologia do Brasil 12:61–71

Guimarães DL, Castro MM, Prezoto F (2008) Patterns of honey storage in colonies of the social wasp *Mischocyttarus cassununga* (Hymenoptera, Vespidae). Sociobiology 51(3):655–660

Heinrich B (1979) Foraging strategies of caterpillars. Oecologia 42(3):325–337

Heldmann G (1936) Über das Leben auf Wabe mi mehreren überwinterten Weibchen von *Polistes gallica* L. Biologisches Zentralblatt 56:389–400

Hernández J, Sarmiento CE, Fernández H (2009) Actividad de forrajeo de *Polybia occidentalis venezuelana* (Hymenoptera, Vespidae). Revista Colombiana de Entomología 35(2):230–234

Hickel ER, Schuck E (1995) Vespas e abelhas atacando a uva no Alto Vale do Rio do Peixe. Agropecuária Catarinense 8:38–40

Holfling JC, Machado VLL (1985) Análise populacional de colônias de *Polybia ignobilis* (Haliday, 1836) (Hymenoptera, Vespidae). Revista Brasileira de Entomologia 29(2):271–284

Huffaker CB, Gutierrez AP (1999) Ecological entomology. Wiley, New York, 756 p

Hunt JH (1982) Trophallaxis and the evolution of eusocial hymenoptera. In: Breed MD, Michener CD, Evans HE (eds) The biology of social insects. Westview Press, Boulder, pp 201–205

Hunt JH (1988) Lobe erection behavior and its possible social role in larvae of *Mischocyttarus* paper wasps. J Insect Behav 1:379–386

Hunt JH (1991) Nourishment and the evolution of the social Vespidae. In: Ross KG, Hunt JH, Baker I, Baker HG (eds) 1982. Similarity of amino acids in nectar and larval saliva: the nutritional basis for trophallaxis in social wasps, Evolution, vol 36, pp 1318–1322

Hunt JH, Nalepa CA (1994) Nourishment, evolution and insect sociality. In: Hunt JH, Nalepa CA (eds) Nourishment and evolution in insect societies. Westview, Boulder, pp 1–19

Hunt JH, Jeanne RL, Baker I, Grogan DE (1987) Nutrient dynamics of a swarm-founding social wasp species, *Polybia occidentalis* (Hymenoptera: Vespidae). Ethology 75:291–305

Hunt JH, Jeanne RL, Keeping MB (1995) Observations on *Apoica pallens*, a nocturnal Neotropical social wasp (Hymenoptera: Vespidae, Polistinae, Epiponini). Insect Soc 42:223–236

Ikan R, Ishay J (1966) Larval wasp secretions and honeydew of the Aphids, *Chaitophorus populi* feeding on *Populus euphratica* as sources of sugars in the diet of the oriental hornet, *Vespa orientalis* F. Isr J Zool 15:64–68

Ingram KK, Kleeman L, Peteru S (2011) Differential regulation of the foraging gene associated with task behaviors in harvester ants. BMC Ecol 11:19

Ishay J, Ikan R (1968a) Food exchange between adults and larvae in *Vespa orientalis* F. Anim Behav 16:298–303

Ishay J, Ikan R (1968b) Gluconeogenesis in the oriental hornet, *Vespa orientalis* F. Ecology 49:1

Iwata K (1976) Evolution of instinct: comparative ethology of hymenoptera. Amerind, Nova Deli, 535 pp

Jacques GC, Pikart TG, Santos VS, Vicente LO, Silveira LCP (2018) Niche overlap and daily activity pattern of social wasps (Vespidae: Polistinae) in kale crops. Sociobiology 65(2):312–319. https://doi.org/10.13102/sociobiology.v65i2.2670

Jandt J, Bengston S, Pinter-Wollman N, Pruitt JN, Raine NE, Dornhaus A, Sih A (2014) Behavioural syndromes and social insects: personality at multiple levels. Biol Rev 89:48–67. https://doi.org/10.1111/brv.1204

Janet C (1903) Observations sur les Guêpes. Naud, Paris, 82 pp

Jeanne RL (1972) Social biology of the Neotropical wasp *Mischocyttarus drewseni*. Bull Mus Comp Zool 144:63–150

Jeanne RL (1986) The organization of work in *Polybia occidentalis:* costs and benefits of specialization in a social wasp. Behav Ecol Sociobiol 19:333–341

Jeanne RL (1991) Polyethism. In: Ross KG, Matthews RW (eds) The social biology of wasps. Cornell University Press, Ithaca, 678 pp

Jeanne RL, Taylor BJ (2009) Individual and social foraging in social wasps. In: Harau S, Hrncir M (eds) Food exploitation by social insects: ecological, behavioral and theoretical approaches. CRC, Boca Raton, pp 53–79

Jeanne RL, Hunt JH, Keeping MG (1995) Foraging in social wasps: *Agelaia* lacks recruitment to food (Hymenoptera: Vespidae). J Kansas Entomol Soc 68:279–289

Kasuya E (1982) Central place water collection in the Japanese paper wasp, *Polistes chinensis antennalis*. Anim Behav 30:1010–1014

Köhler A (2008) Floral preferences of the polistine wasp *Polistes versicolor versicolor* Olivier, 1792 (Hymenoptera: Vespidae: Polistinae, Polistini) in Santa Cruz do Sul, Southern Brazil. Biociencias 16(2):162–165

Krehenwinkel H, Kennedy S, Pekár S, Gillespie RG (2016) A cost-efficient and simple protocol to enrich prey DNA from extractions of predatory arthropods for large-scale gut content analysis by Illumina sequencing. Methods Ecol Evol 8(1):126–134. https://doi.org/10.1111/2041-210X.12647

Letourneau DK, Chloe J (1987) Homoptera attendance by wasps and ants: the stochastic nature of interactions. Psyche 94:81–91

Lima MAP, Prezoto F (2003) Foraging activity rhythm in the neotropical swarm-founding wasp Polybia platycephala sylvestris (Hymenoptera: Vespidae) in different seasons of the year. Sociobiology 42(3):745–752

London KB, Jeanne RL (2003) Effects of colony size and stage of development on defense response by the swarm founding wasp *Polybia occidentalis*. Behav Ecol Sociobiol 54(539):546

Machado VLL (1984) Análise populacional de colônias de *polybia (Myrapetra) paulista* (Ihering 1896) (Hymenoptera, Vespidae). Revista Brasileira de Zoologia 2(4):187–201

Machado VLL, Parra RP (1984) Capacidade de retorno ao ninho de operarias de *Polybia (Myrapetra) scutellaris* (White,1841). Anais da Sociedade Entomológica do Brasil 13:13–18

Maciel TT, Ad'víncula HL, Barbosa BC, Prezoto F (2014) Forrageio Oportunista de Vespas Sociais em Armadilhas Atrativas. Anais do XXV Congresso Brasileiro de Entomologia

Marques OM (1996) Vespas sociais (Hymenoptera, Vespidae): características e importância em agrossistemas. Insecta 5:13–39

Mcpheron LJ, Mills NJ (2007) Learning colour and odour in a paper wasp, *Mischocyttarus flavitarsis*. Entomologia Generalis 30(2):105–118

Moller H, Tilley JAV, Thomas BW, Gaze PD (1991) Effect of introduced social wasps on the standing crop of honeydew in New Zealand beech forests. New Zealand Journal of Zoology 18:171–179

Nascimento FS, Tannure-Nascimento IC (2005) Foraging patterns in a nocturnal swarm-founding wasp, *Apoica flavissima* van der Vecht (Hymenoptera: Vespidae). Neotrop Entomol 34(2):177–181

O'Donnell S (1995a) Division of labor in post-emergence colonies of the primitively eusocial wasp *Polistes instabilis* de Saussure (Hymenoptera: Vespidae). Insect Soc 42:17–29

O'Donnell S (1995b) Necrophagy by Neotropical swarm-founding social wasps (Hymenoptera: Vespidae, Epiponini). Biotropica 27:133–136

Picanço M, Ribeiro LJ, Leite GLD, Gusmão. MR (1988) Seletividade de inseticidas a *Polybia ignobilis* (Halyday) (Hymenoptera, Vespidae) predador de *Ascia monuste orseis* (Godart) (Lepidoptera: Pieridae). Anais da Sociedade de Entomologia do Brasil 27:85–90

Prezoto F, Braga N (2013) Predation of *Zaprionus indianus* (Diptera: Drosophilidae) by the social wasp *Synoeca cyanea* (Hymenoptera: Vespidae). Fla Entomol 96:670–672. https://doi.org/10.1653/024.096.0243

Prezoto F, Gobbi N (2003) Patterns of honey storage in nests of the neotropical paper wasp *Polistes simillimus* Zikán, 1951 (Hymenoptera, Vespidae). Sociobiology 41(2):437–442

Prezoto F, Gobbi N (2005) Flight range extension in *Polistes simillimus* Zikán, 1951 (Hymenoptera, Vespidae). Braz Arch Biol Technol 48(6):947–950

Prezoto F, Machado VLL (1999) Ação de *Polistes (Aphanilopterus) simillimus* Zikán (Hymenoptera, Vespidae) no controle de *Spodoptera frugiperda* (Smith) (Lepidoptera, Noctuidae). Revista Brasileira de Zoociências 16(3):841–850

Prezoto F, Giannotti E, Machado VLL (1994) Atividade forrageadora e material coletado pela vespa social *Polistes simillimus* Zikán, 1951 (Hymenoptera, Vespidae). Insecta 3:11–19

Prezoto F, Lima MAP, Machado VLL (2005) Survey of preys captured and used by *Polybia platycephala* (Richards) (Hymenoptera: Vespidae, Epiponini). Neotrop Entomol 34(5):849–851

Prezoto F, Santos-Prezoto HH, Machado VLL, Zanúncio JC (2006) Prey captured and used in *Polistes versicolor* (Hymenoptera, Vespidae) nourishment. Neotrop Entomol 35(5):707–709

Prezoto F, Barbosa BC, Cappas JP, Santos ME (2016a) Water landing as a foraging strategy to water collection in a social wasp: *Polistes dominulus*. Curr Ethol 15(2)

Prezoto F, Barbosa BC, Maciel TT, Detoni M (2016b) Agroecossistemas e o serviço ecológico dos insetos na sustentabilidade. In: Resende LO, Prezoto F, Barbosa BC, Gonçalves EL (eds) Sustentabilidade: Tópicos da Zona da Mata Mineira. Real Consultoria em Negócios Ltd, Juiz de Fora, pp 19–30

Prezoto F, Maciel TT, Detoni M, Mayorquin AZ, Barbosa BC (2019) Pest control potential of social wasps in small farms and urban gardens. Insects 10(7):1–10. https://doi.org/10.3390/insects10070192

Rabb RL, Lawson FR (1957) Some factors influencing the predation of *Polistes* wasps on the tobacco hornworm. J Econ Entomol 50:778–784

Rau P (1928) The honey-gathering habitats of *Polistes wasps*. Biol Bull Marine Biol Lab, Woods Hole 54:503–519

Rau P (1939) Studies in the ecology and behavior of *Polistes* wasps. Bull Brooklyn Entomol Soc 34:36–44

Raveret-Richter MA (1990) Hunting wasp interactions: influence of prey size, arrival order, and wasp species. Ecology 71:1018–1030

Raveret-Richter MA (2000) Social wasp (Hymenoptera: Vespidae) foraging behavior. Annu Rev Entomol 45:121–150. https://doi.org/10.1146/annurev.ento.45.1.121

Raveret-Richter MA, Jeanne RL (1985) Predatory behavior of *Polybia sericea* (Olivier), a tropical social wasp (Hymenoptera: Vespidae). Behav Ecol Sociobiol 16:165–170

Raveret-Richter MA, Jeanne RL (1991) Hunting behavior, prey capture and ant avoidance in the tropical social wasp *Polybia sericea* (Hymenoptera: Vespidae). Insect Soc 38:139–147

Resende JJ, Santos GMM, Bichara Filho CC, Gimenes M (2001) Atividade diária de busca de recursos pela vespa social *Polybia occidentalis occidentalis* (Olivier, 1791) (Hymenoptera, Vespidae). Revista Brasileira de Zoociências 3(1):105–115

Ribeiro Júnior C, Elisei T, Guimarães DL, Prezoto F (2008) Flight range extension in the swarm – founding wasp *Protopolybia exigua* (Hymenoptera, Vespidae, Epiponini). Sociobiology 51(1):173–180

Ribeiro-Júnior C, Guimarães DL, Elisei T, Prezoto F (2006) Foraging activity rhythm of the Neotropical swarm-founding wasp *Protopolybia exigua* (Hymenoptera, Vespidae, Epiponini) in different seasons of the year. Sociobiology 47(1):115–123

Richards OW, Richards MJ (1951) Observations on the social wasps of South America (Hymenoptera, Vespidae). Trans Royal Entomol Soc London 102:1–170

Ricioli LS, Mendes G, Guillermo-Ferreira R (2017) Kick out the ants: a novel and striking behavior in ant-wasp interactions. Sociobiology 64(1):119–121. https://doi.org/10.13102/sociobiology.v64i1.1199

Rocha MP (2011) Biologia e ecologia comportamental da vespa social *Polybia platycephala* (Richards, 1978) *(Hymenoptera: Vespidae, Epiponini)*. Master's dissertation. Juiz de Fora, Universidade Federal de Juiz de Fora, 2011. 76 pp

Rocha AA, Giannotti E (2007) Foraging activity of *Protopolybia exigua* (Hymenoptera, Vespidae) in different phases of the colony cycle, at an area in the region of the Médio São Francisco River, Bahia, Brazil. Sociobiology 50:813–831

Rossi AM, Hunt JH (1988) Honey supplementation and its developmental consequences: evidence for food limitation in a paper wasp, *Polistes metricus*. Ecol Entomol 13:437–442

Roubaud E (1916) Recherches biologiques sur les Guepes solitaires et sociales d'Afrique. Annales des Sciences Naturelles, Zoologie 10, 1(1):–160

Santos GMM, Presley SJ (2010) Niche overlap and temporal activity patterns of social wasps (Hymenoptera: Vespidae) in a Brazilian cashew orchard. Sociobiology 56:121–131

Santos M, Marques OM, Carvalho A (1994) Raio de ação de *Polistes canadensis canadensis* (L., 1758) (Hymenoptera, Vespidae). Insecta 3(2):20–24

Santos GMM, Santana-Reis VPG, Resende JJ, Marco PD, Bichara-Filho CC (2000) Flying capacity of swarm - founding wasp *Polybia occidentalis occidentalis* Oliver, 1791 (Hymenoptera, Vespidae). Revista Brasileira de Zoociências 2(2):33–39

Santos GMM, Cruz JD, Bichara Filho CC, Marques OM, Aguiar CML (2007) Utilização de frutos de cactos (Cactaceae) como recurso alimentar por vespas sociais (Hymenoptera, Vespidae, Polistinae) em uma área de Caatinga (Ipirá, Bahia, Brasil). Revista Brasileira de Zoologia 24:1052–1056. https://doi.org/10.1590/S0101-81752007000400023

Santos GMM, Zanuncio JC, Pires EM, prezoto F, Pereira JMM, Serrão JE (2009) Foraging of *Parachartergus fraternus* (Hymenoptera: Vespidae: Epiponini) on cloudy and sunny days. Sociobiology 53(2B):431–441

Saraiva NB, Prezoto F, fonseca MG, blassioli-Moraes MC, Borges M, Laumann RA, Auad AM (2017) The social wasp *Polybia fastidiosuscula* Saussure (Hymenoptera, Vespidae) uses herbivore-induced maize plant volatiles to locate its prey. J Appl Entomol. https://doi.org/10.1111/jen.12378

Seeley TD (1989) The honey bee colony as a superorganism. Am Sci 77:546–553

Sih A, Bell A, Johnson JC (2004) Behavioral syndromes: an ecological and evolutionary overview. Trends Ecol Evol 19:372–378

Silva AGA, Gonçalves CR, Galvão DM, Gonçalves AJL, Gomes J, Silva MN, Simoni L (1968) Quarto catálogo dos insetos que vivem nas plantas do Brasil: Seus parasitos e predadores. Rio de Janeiro, Ministério da Agricultura, Depto. de Defesa e Inspeção Agropecuária, pt. 2, t.1, 621p

Silveira OT, Esposito MC, dos Santos JN, Gemarque FE (2005) Social wasps and bees captured in carrion traps in a rainforest in Brazil. Entomol Sci 8:33–39

Sinzato DMS, Prezoto F, Del-Claro K (2003) The role of males in a neotropical paper wasp, *Polistes ferreri* Saussure, 1853 (Hymenoptera, Vespidae, Polistinae). Revista Brasileira de Zoociências 5(1):89–100

Smith AR, Kitchen SM, Toney RM, Ziegler C (2017) Is nocturnal foraging in a tropical bee an escape from interference competition? J Insect Sci 17(2):1–7. https://doi.org/10.1093/jisesa/iex030

Southon RJ, Fernandes OA, Nascimento FS, Sumner S (2019) Social wasps are effective biocontrol agents of key lepidopteran crop pests. Proc R Soc B Biol Sci 286(1914). https://doi.org/10.1098/rspb.2019.1676

Souza AR, Rocha M, Netto P, Venâncio DFA, Prezoto F (2011) Preliminary homing ability study of *Polybia fasidiosuscula* (Hymenoptera: Vespidae) workers. Sociobiology 56(1):133–136

Souza GK, Pikart TG, Jacques GC, Castro AA, Souza MM, Serrão JE, Zanúncio JC (2013) Social wasps on *Eugenia uniflora* Linnaeus (Myrtaceae) plants in an urban area. Sociobiology 60:204–209. https://doi.org/10.13102/sociobiology.v60i2.204-209

Spradbery JP (1973) Wasps: an account of the biology and natural history of solitary and social wasps. University of Washington Press, Seattle, 408 p

Strassman JE (1979) Honey caches help female paper wasps (*Polistes annularis*) survive Texas winters. Science 204:207–209

Sumner S, Law G, Cini A (2018) Why we love bees and hate wasps? Ecol Entomol 43(6):836–845. https://doi.org/10.1111/een.12676

Suryanarayanan S, Jeanne RL (2008) Antennal drumming, Trophallaxis, and Colony development in the social wasp *Polistes fuscatus* (Hymenoptera: Vespidae). Ethology 114:1201–1209

Taber SW (1999) The world of the harvester ants. Texas A & M University Press, Austin, 213 pp

Takagi M, Hirose Y, Yamasaki M (1980) Prey-location learning in *Polistes jadwigae* Dalla Torre (Hymenoptera, Vespidae). Kontyû 48:53–58

Taylor BJ, Brus EJ, Jeanne R (2012) Introduction of a scented carbohydrate resource into the nest increases departure rate in *Polybia occidentalis*. Insect Soc 59(2):151–157

Tindo M, Dejean A (2000) Dominance hierarchy in colonies of *Belonogaster juncea juncea* (Vespidae, Polistinae). Insect Soc 47(2):158–163

Warkentin KM (2000) Wasp predation and wasp-induced hatching of red-eyed treefrog eggs. Animal Behav 60:503–510

Warrant EJ (2008) Seeing in the dark: vision and visual behaviour in nocturnal bees and wasps. J Exp Biol 211:1737–1746. https://doi.org/10.1242/jeb.015396

Wenzel JW (1991) Evolution of nest architecture. In: Ross KG, Matthews RW (eds) The social biology of wasps. Cornell University Press, Ithaca, 678 pp

West-Eberhard MJ (1969) The social biology of Polistine wasps. In: Miscellaneous publications n° 140. Museum of Zoology, University of Michigan, Ann Arbor, 101p

Wheeler WM (1918) A study of some ant larvae, with a consideration of the origin and meaning of the social habit among insects. Proc Am Philos Soc 57:193–343

Wilson EO (1971) The insect societies. Belknap Press, Harvard University Press, Cambridge, MA, 548 pp

Wright CM, Lichtenstein JLL, Doreing GN, Pretorius J, Meunier J, Pruitt J (2019) Collective personalities: present knowledge and new frontiers. Behav Ecol Sociobiol 73:31. https://doi.org/10.1007/s00265-019-2639-2

Chapter 4
The Choice of Sexual Partner in Social Wasps

André Rodrigues de Souza, Camila Folly Baptista, Fábio Santos Nascimento, and José Lino-Neto

Abstract In many animals, females (and sometimes males) are more or less choosy in relation to what type of partner to mate. This intersexual selection favors traits that make individuals sexually attractive. Traditionally, social insects have mostly been the targets of sociobiological studies, but the exciting recent literature on ants, bees, and wasps has provided a lot of insights about how sexual selection shapes the reproductive behavior and also the social organization in insect societies. In this chapter, we discuss the reasons why wasps are good models for studying sexual selection. In addition, we describe the different criteria that wasps use during the mate choice. Finally, we highlight future directions to extend knowledge about how sexual selection shapes the choice of the sexual partner in these insects.

Keywords Sexual selection · Mate choice · Visual signals

4.1 Introduction

The selection of a sexual partner is an important aspect of animal life. Technically speaking, it is referred to as intersexual selection. Throughout this process, traits that improve the probability of an individual to be chosen by the opposite sex as a sexual partner are favored. Intersexual selection is one of the two components of the theory of sexual selection (the other one is intrasexual selection) proposed by Darwin in 1871. Such a phenomenon, which has already been studied in various animals (reviewed by Andersson and Iwasa 1996), can result in the evolution of more or less choosy species in relation to what type of partner they mate.

A. R. de Souza (✉) · F. S. Nascimento
Departamento de Biologia, Faculdade de Filosofia Ciências e Letras de Ribeirão Preto, Universidade de São Paulo, Ribeirão Preto, SP, Brazil

C. F. Baptista · J. Lino-Neto
Departamento de Biologia Geral, Universidade Federal de Viçosa, Viçosa, Minas Gerais, Brazil

F. Prezoto et al. (eds.), *Neotropical Social Wasps*,
https://doi.org/10.1007/978-3-030-53510-0_4

Although well studied at a certain rate, sexual selection in social insects has received relatively little attention. This is probably due to the impact of the early contributions of Hamilton (1964a, b) as well as Trivers and Hare (1976), which motivated numerous sociobiological investigations focused on female cooperation. Their studies have marked the development of the theory of kin selection as the basis for explaining how helping behavior can be favored by natural selection and receiving since then widespread support (Queller 2016). As a result, investigations about sexual selection in social insects have fallen into the background. Nevertheless, more recent studies, including those with social wasps, have shown that sexual selection has shaped several traits of these insects, including different criteria for sexual partner choice (described below).

The perception about the importance of studying sexual selection in social insects has motivated a growing number of investigations, resulting in excellent reviews on the subject (Strassmann 2001; Baer 2003; Boomsma et al. 2005; Baer 2014; Beani et al. 2014; Baer 2015; Heinze 2016). In this chapter, we focused on *Polistes* paper wasps (but also some other wasps) to (i) discuss the reasons why wasps are good models for studying sexual selection, (ii) describe the different criteria that wasps use during the mate choice, and (iii) highlight some future directions to extend knowledge about how sexual selection shapes the choice of the sexual partner in these insects.

4.2 Social Wasps as Models for the Study of Sexual Selection

In studying sexual selection, model species must be able to be observed in the field since the main objective is to understand how the mechanism of selection works in the natural environment where species evolve and acquire their traits. Well-controlled experiments that can only be done in the laboratory may then be interpreted against a solid framework of ecology, selective pressures, and adaptations of the species to its natural evolutionary context (Andersson and Simmons 2006). *Polistes*' life cycle is relatively well known (Fig. 4.1), and the general pattern of sexual behavior in at least some species can be observed in the field (Beani et al. 1992). Typically, males are bred in late summer, they stay only for a few days in the colony, and then, they disperse to find sexual partners. Male aggregations are observed, repeatedly visiting conspicuous areas (= *landmarks*) and sunny places, comprising "meeting points" where females visit and mate. Interspecific differences in sexual behavior are observed regarding the occurrence of defense of territories and the degree of aggression among males. Thus, the *Polistes* mating system ranges from swarms of little-aggressive males which collectively patrol common routes and do not defend territories to *leks* formation, in which highly aggressive males defend individual territories against cospecifics. Male sexual behavior in some *Polistes* has been reviewed by Beani (1996). Females, in turn, generally mate only with one male (monandric; Strassmann 2001) for a short time after their emergence – they never copulate again! Female multiple mating (polyandry) is observed

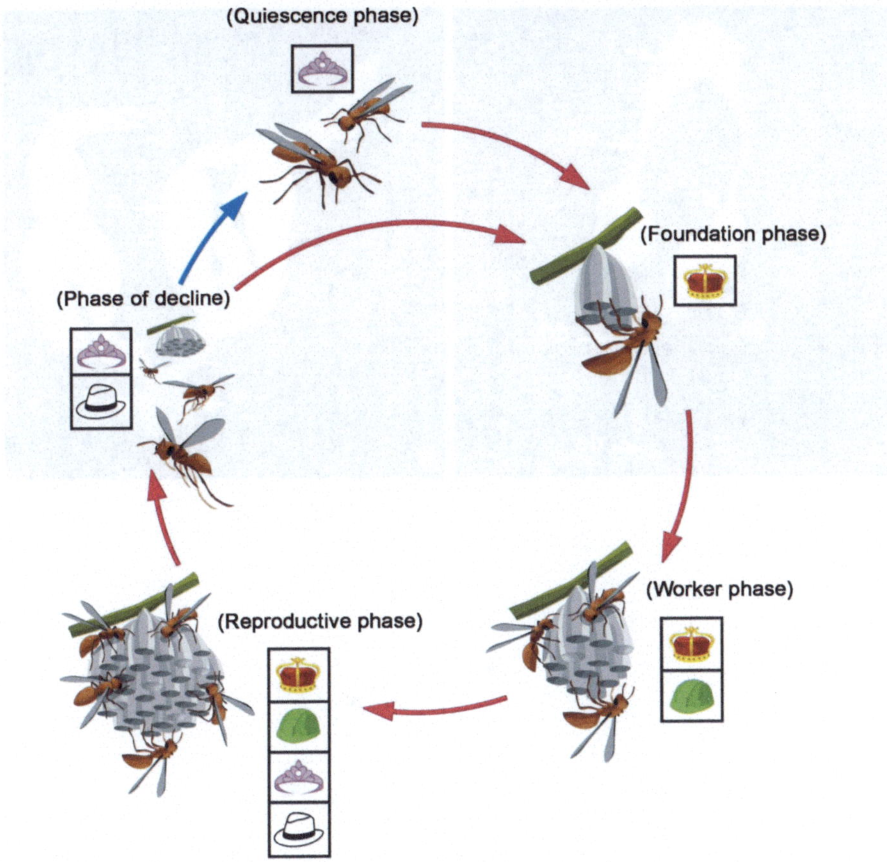

Fig. 4.1 Typical colonial cycle of *Polistes versicolor*. *Foundation phase:* The cycle begins when environmental conditions are favorable (red arrows). At this stage, a female initiates the foundation of a nest (alone or cooperatively). Oviposition occurs immediately after the construction of each nest cell by the founder (crown icon), which performs, in addition to oviposition, all maternal behavior (foraging, feeding the larvae, building, and nest defense). *Worker phase:* The first female offspring are female workers (helmet icon). These individuals remain in the colony and assume all maternal behavior, while the female founder, now called queen, is devoted mainly to oviposition. *Reproductive phase:* After producing several generations of workers, the colony also produces males (hat icon) and potential future queens, also called potential founders or *gynes* (icon tiara). These reproductive forms leave the colony to mate; males leave the colony permanently while females return. *Phase of decline:* Reproductive forms still emerge (to a lesser extent), but the queen and workers die, and little by little the nest runs out of immature forms. Even at this stage, if environmental conditions remain favorable, potential future queens leave the colony where they were raised to start their own colonies, becoming founders and then queens. If environmental conditions are no longer favorable (blue arrow), potential future queens enter the *quiescence phase:* Potential future queens remain in a kind of reproductive pause, sheltering in their own nest or in weather-protected cavities until favorable environmental conditions return when the foundation phase begins. The colonial cycle lasts from 3 to 10 months. The reproductive roles of the castes are flexible. For example, subordinate females in cooperative foundations (foundress association) may behave as workers. Also, a worker can replace the queen in case of her death or disappearance (nest inheritance). Data on the natural history of *P. versicolor* was obtained from Gobbi et al. (2006) and references cited by them. The cycle design was adapted from Hunt et al. (2011)

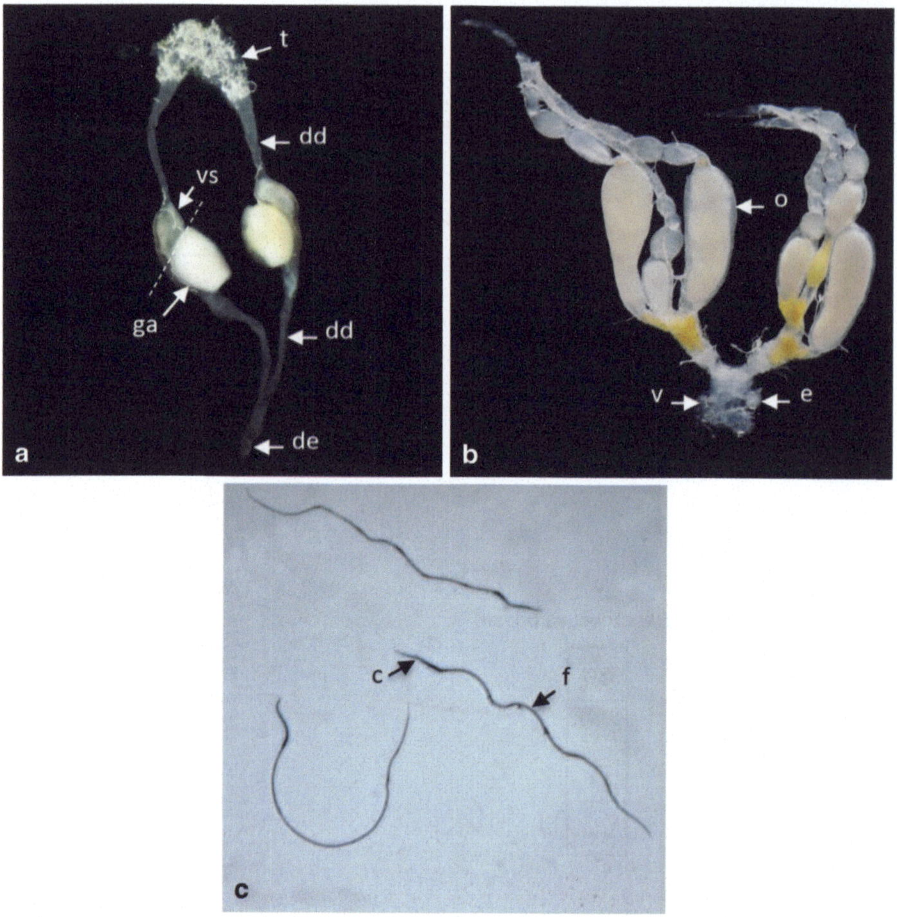

Fig. 4.2 Reproductive systems and sperm of *P. versicolor*. (**a**) Reproductive system of the sexually mature male: Sperm are produced in the testes which degenerate during sexual maturity (t). Sperm migrate through the anterior portion of the vas deferens (dd) to the seminal vesicles (vs), where they are stored until ejaculation. Afterwards, the accessory glands (ga) are inserted. Their content along with the sperm compose the ejaculate which passes through the posterior region of the vas deferens to the ejaculatory duct (de). (**b**) Reproductive system of the sexually mature female: Sperm obtained during copulation are stored in a specialized region, connected to the vagina (v), called spermatheca (e). Oocytes (o) develop as they approach the vagina, and as they pass through it, fertilization may occur. (**c**) Sperm obtained from seminal vesicles of a sexually mature male. After hematoxylin staining it is possible to discriminate the head (c) and flagella (f) of each sperm

in some *Polistes* species (Seppä et al. 2011; Southon et al. 2019). Inseminated females store sperm in a specialized organ, the spermatheca (Fig. 4.2). These sperm are used to produce several generations of workers and, at the end of the colonial cycle, reproductive females (potential future queens, also called gynes). Males are produced from unfertilized eggs (parthenogenesis).

In addition, the sexual behavior of *Polistes* can be studied in large cages (Beani and Turillazzi 1988; Beani and Zaccaroni 2015) and also in small glass arenas (Liebert et al. 2010; Izzo and Tibbetts 2012; de Souza et al. 2014), which allows well-controlled observations and experiments.

Model species should also have conspicuous sexual dimorphism, as this is an evidence of strong sexual selection (Andersson and Simmons 2006). *Polistes* have several sexually dimorphic characteristics. For example, males and females have body color patches with different sizes and shapes (Izzo and Tibbetts 2012; De Souza et al. 2014; Cappa et al. 2016; De Souza et al. 2016), and some of these patches are larger and more variable in males than in females. Below it is described how such body patches comprise sexually selected signals mediating the choice of the sexual partner.

In addition to the factors mentioned above, particularities related to the sex determination system, the effect of kinship relations on the social dynamics of the colonies, and the differences in the reproductive potential of females make *Polistes* wasps excellent models to investigate how selection has shaped the choice of the sexual partner.

4.3 Criteria for Selecting a Sexual Partner

The criteria for choosing a sexual partner in *Polistes* social wasps are slowly being unveiled, as we know.

4.3.1 Kinship

In Hymenoptera in general, males are haploid and females diploid. In these insects, normal female development requires heterozygosity at one or more specific *loci*, constituting the complementary sex determination mechanism (Whiting 1943; Crozier 1971; Whiting 1924; Van Wilgenburg et al. 2006). Therefore, mating between highly related individuals (e.g., between siblings) can result in the production of genetic "errors" known as diploid males. Diploid males are produced when diploid individuals are homozygous for one or more sex-related *loci* (Fig. 4.3) (Whiting 1933). These individuals would normally be females, but due to homozygosity, they develop in males. Diploid males impose fitness costs on colony productivity because they consume resources but are typically sterile (Liebert et al. 2004). Because of the costs of incestuous copulation, natural selection should favor behaviors that prevent its occurrence. In this sense, copulation between closely related individuals can be prevented by (i) intersexual nestmate recognition and subsequent rejection of related partners and/or (ii) a mechanism that minimizes the likelihood of finding a sexual partner closely related.

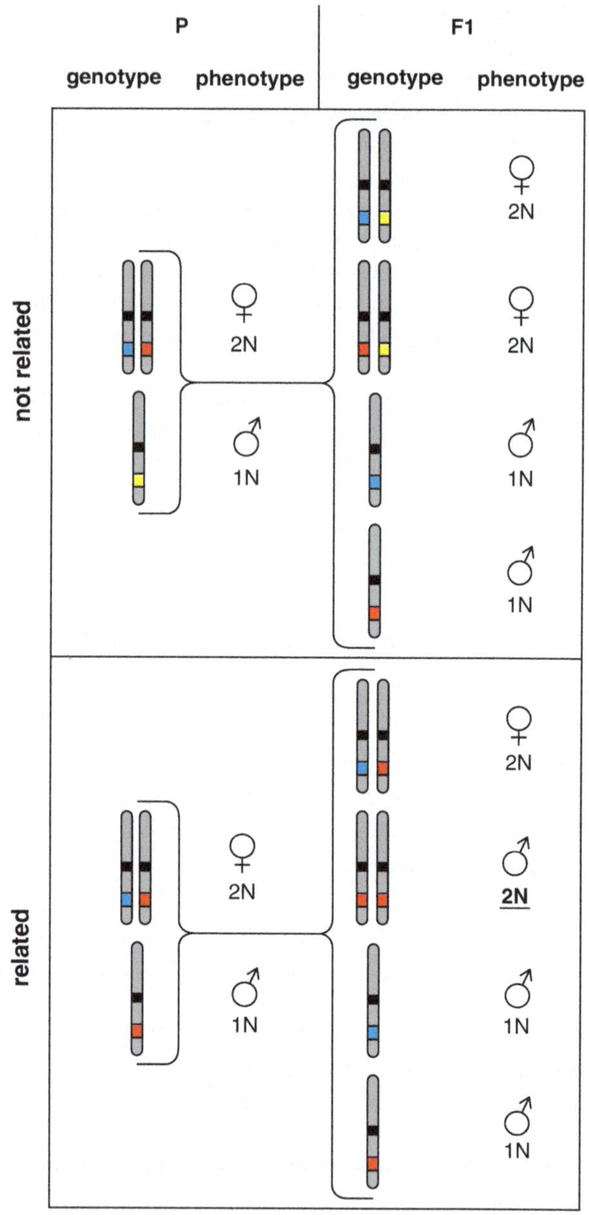

Fig. 4.3 Chromosomal representation of the complementary sex determination system in Hymenoptera, where males are typically haploid (1 N) and females are diploid (2 N). Note that in this scheme, sex is determined by a single *locus* with multiple alleles (colored regions on chromosomes) in a population. Heterozygosity results in a female phenotype, and the hemi- or homozygosity produces a male phenotype. Crossbreeding between siblings can result in the production of diploid (**2 N**) males which do not contribute to colonial productivity and are functionally sterile

Same-sex nestmate recognition has been well studied in *Polistes*. Both males and females are known to possess such ability (Shellman-Reeve and Gamboa 1985; Ryan and Gamboa 1986; Starks et al. 1998). This recognition is based on colony-specific chemical signatures determined by a unique combination of hydrocarbons present in the cuticle of individuals and also on the surface of nests (reviewed by Gamboa 2004; see Chap. 11 on cuticular hydrocarbons). The ontogeny of nestmate recognition may involve a number of mechanisms (Cappa et al. 2020). Thus, by recognizing such combinations of hydrocarbons, wasps would be able to avoid copulation with related sexual partners. However, evidence of intersexual nestmate recognition and subsequent nestmate avoidance has been demonstrated only for *Polistes fuscatus* (Ryan and Gamboa 1986), in which controlled behavioral trials have shown that copulation between relatives is less frequent compared with those between unrelated individuals. The other species of wasps studied do not seem to avoid copulation with related sexual partners (*Polistes*, Liebert et al. 2010; Beani and Turillazzi 1988; De Souza et al. 2017a; *Ropalidia*, Sen et al. 2010; Shilpa et al. 2010), even when they are able to discriminate related and unrelated sexual partners (Liebert et al. 2010). This suggests that the absence of sexual preference does not result from a limitation in the discriminatory ability. So how do these species deal with the possibility of incestuous copulation? As stated above, incest can also be prevented through a mechanism that minimizes the likelihood of finding a related partner. Copula in *Polistes* occurs predominantly outside the nest, constituting a mechanism that can help to prevent incest. In addition, emerging males are still sexually immature, as it takes a few days for the sperm to migrate from the testes to the seminal vesicles (Fig. 4.2a). Thus, eventual copulations with newly emerged males, sometimes reported, are unlikely to result in insemination. Also, males, leaving the colony, take on a nomadic life when they seek females to mate. This removal from the colony may also decrease the occurrence of incest. Sometimes the colony may specialize in producing only reproductive males or only reproductive females, thus avoiding incest. Finally, in the population, virtually all reproductive males and females are produced at a relatively short interval in summer, resulting in many potential sexual partners. This type of reproductive assembly results in low likelihood of mating between related sexual partners. This set of biological, ecological, and behavioral traits is believed to relax the selection of sexual partner in relation to kinship (De Souza et al. 2017a).

But after all, do wasps actually avoid incestuous copulation? Diploid males, a consequence of mating with related partners, have not been found in native social wasp populations (Liebert et al. 2006; Nagamati Junior et al. 2010). This suggests that the above mechanisms are effective in preventing incest among these insects. However, diploid males have been reported in invasive populations (Liebert et al. 2006), suggesting the occurrence of copulation between relatives. Incest may be common in early invasive populations, especially when small and highly related, thus, reflecting low allelic diversity. Nonetheless, it is also seen as an adaptive form so that under those conditions, some female offspring are produced until allelic diversity increases (Liebert et al. 2010).

4.3.2 *Number of Partners*

In female social Hymenoptera, the number of sexual partners varies according to a number of costs and benefits (Strassmann 2001). In *Polistes* and *Ropalidia* females are typically monandric (Strassmann 2001), resulting in a high degree of kinship among colony members (compared to polyandric females). Of note, low and high polyandry have evolved in a few *Polistes* (Seppä et al. 2011; Southon et al. 2019). Monandry enhances social harmony, because the greater the relationship between cooperating females, the greater the indirect fitness obtained (inclusive fitness theory; Hamilton 1964a, 1964b). As a result, helper females emerged from monandric colonies have more to gain (mean sister relationship = 75%) compared to females emerged from polyandric colonies (mean sister relationship <75%). High kinship is especially important for fostering harmony in primitively eusocial groups. This is because reproductive and non-reproductive females are totipotent and can behave as a helper, start their own colony, or even reproduce directly in the colony where they were born, if there is an opportunity.

In the highly eusocial wasps Vespinae, queens can be monandric or polyandric, as the morphological difference between breeding and non-breeding females is so pronounced that workers lose the ability to start their own colony. Wasp polyandry results in increased genetic polyethism (allowing a more efficient division of labor) and colonial immunity (Saga et al. 2020), which is especially important due to the large number of genetically similar individuals in highly eusocial insect colonies. It also promotes policing among workers, preventing them from laying eggs that would give rise to males, thus favoring queens (Strassmann 2001; Crozier and Fjerdingstad 2001). These benefits of polyandry have been described for highly eusocial wasps, but whether they also hold for *Polistes* is unknown.

Primitively eusocial wasp males, such as *Ropalidia*, follow the male stereotype and can fertilize several females (polygyny) (Shilpa et al. 2012). This ability is especially important for these primitively eusocial wasps, since not all inseminated females will in fact be queens. When a single female starts a colony, she will be the queen of this colony. However, when the colony is started by more than one female, a dominance hierarchy is established. Then, one or a few individuals hold all or most of the direct reproduction (functional queens), while the others have little or no participation in oviposition (functional workers) (*Polistes*: De Souza et al. 2008; Grazinoli et al. 2010). Thus, mating with a future foundress does not guarantee reproductive success. Nevertheless, polygyny may increase the chances that at least some females will use sperm. Male multiple mating (polygyny) likely holds for *Polistes* as well, but it remains to be investigated.

4.3.3 Caste

In primitively eusocial wasps, the reproductive output can differ dramatically among females in the same colony, despite they are all morphologically simillar and reproductivelly totipotent (De Souza et al. 2008; Grazinoli et al. 2010). Queens tend to assume most direct reproduction (principal egg layers), while workers tend to assume non-reproductive roles (e.g., foraging or nest defense). Since males produce a limited amount of sperm (the testes degenerate after sexual maturity; Fig. 4.2), they are expected to be able to discriminate and avoid copulation with female workers. Inseminated *Polistes dominula* females begin the foundation of colonies in spring, producing several generations of workers, and by the end of summer, they also produce reproductive forms (males and potential future queens). A few days after emergence, males leave the nest and congregate in *lek* aggregations at strategic points located in areas of high wasp density (Beani and Turillazzi 1988). During this period, future founders, still virgin, visit the *leks* to mate, while workers forage near the *leks* increasing the chances of their encounter with males (Beani and Turillazzi 1988; Beani 1996). Therefore, males can interact with both future founders and workers. In studying the sexual behavior of males in the laboratory, Cappa et al. (2013) showed that they discriminate castes, preferring to court future foundresses (future queens). Such behavior is presumably adaptive, especially since after the mating season (late summer), females hibernate for a few months before starting the new colonies. As foundresses usually survive the weather conditions in winter and workers do not, the sexual preference of *P. dominula* males seems to be a mechanism to ensure copulation with more appropriate sexual partners.

4.3.4 Ornaments

The use of ornamentation as a criterion for choosing a sexual partner in social wasps is one of the most recently studied topics. Neotropical wasp males *Polistes simillimus* have a black spot on the vertex around the ocellae (Fig. 4.4). This spot is sexually dimorphic, being always small and slightly variable in females, but highly variable in males (Fig. 4.4). Experimentally speaking, when wasps interact in a small arena, males with a higher proportion of black pigment are more likely to be chosen as sex partners. Likewise, males experimentally manipulated to have higher black pigmentation are preferably chosen as a sexual partner over males manipulated to have lower proportion of black pigmentation. This means that *P. simillimus* females discriminate these spots during the choice of sexual partner (De Souza et al. 2014). Such ornaments are considered signs of quality as they are associated with the male's immunocompetence and longevity (De Souza et al. 2018).

Similar behavior has been previously reported for the social wasp of the temperate region, the *P. dominula*. Males have a pair of yellow abdominal spots, located on the lateral surface of the second abdominal tergite (Izzo and Tibbetts 2012; De Souza et al. 2017b; De Souza et al. 2020). Such patches are always small and round in females, but in males they range from round to extremely irregular ones. When wasps interact in a small arena, males with rounded spots are more likely to be selected by the females. This same result is also observed when spots on males are experimentally manipulated, characterizing them as sexual ornaments. Such ornaments are considered signs of quality as they are associated with the male's fighting ability (Izzo and Tibbetts 2012), territory holding (Beani et al. 2019) and nutrition during the larval period (Izzo and Tibbetts 2015).

Fig. 4.4 Variation in the proportion of black pigment in male (above) and female (below) heads of *P. simillimus*. Note that males have larger and more variable spots than females

4.4 Future Directions

Despite the growing number of publications related to sex partner choice in *Polistes* social wasps, there is still much to understand. Here are some promising topics:

1. In several species from temperate places, females have visual signals of quality used to mediate aggressive interactions related to the intracolonial dominance hierarchy (Cervo et al. 2015). These signs are known to be good predictors of survival, reproductive success, and rank (Tibbetts et al. 2015). Although these signs have presumably evolved in a female-female context, it is possible that males will preferentially discriminate and mate with females whose visual signals indicate high quality and have greater reproductive success.
2. It has been seen that the preference of males for future foundresses rather than workers in a temperate species (Cappa et al. 2013) seems to be an adaptation to acquire mating with more appropriate females (since workers do not survive to found new colonies after winter). But, what about such preference in neotropical species? In these regions winter is not harsh, so workers can survive to found new colonies.
3. How to explain the preference of females for sexual partners with certain ornaments (Izzo and Tibbetts 2012; De Souza et al. 2014) when apparently males offer nothing but the ejaculate? Possibly, females could choose mates with better reproductive potential if the ornaments indicate quality ejaculate. This hipothesis was recently addressed in *P. dominula* (De Souza et al. 2020) providing negative results.

4. Quality signals, which reflect physiological and ontogenic aspects of individuals (Tibbetts 2010; Izzo and Tibbetts 2015), are known to be affected by the presence of parasites (Tibbetts et al. 2011). So, is male sexual ornamentation also altered by parasitism? This hipothesis was recently addressed in *P. dominula* (De Souza et al. 2020) also providing negative results.

References

Andersson M, Iwasa Y (1996) Sexual selection. Trends Ecol Evol 11(2):53–58

Andersson M, Simmons LW (2006) Sexual selection and mate choice. Trends Ecol Evol 21(6):296–302

Baer B (2003) Bumblebees as model organisms to study male sexual selection in social insects. Behav Ecol Sociobiol 54:521–533

Baer B (2014) Sexual selection in social insects. In: Shuker DM, Simmons LW (eds) The evolution of insect mating systems. Oxford University Press, New York, pp 261–274

Baer B (2015) Female choice in social insects. In: Peretti AV, Aisemberg A (eds) Cryptic female choice in arthropods. Springer International Publishing, Cham, pp 461–477

Beani L (1996) Lek-like courtship in paper-wasps: 'a prolonged, delicate, and troublesome affair'. In: Turillazzi S, West-Eberhard MJ (eds) Natural history and evolution of paper-wasps. Oxford University Press, New York, pp 113–125

Beani L, Turillazzi S (1988) Alternative mating tactics in males of *Polistes dominulus* (Hymenoptera: Vespidae). Behav Ecol Sociobiol 22(4):257–264

Beani L, Zaccaroni M (2015) Experimental male size manipulation in *Polistes dominula* paper wasps. Ethol Ecol Evol 27(2):185–199

Beani L, Cervo R, Lorenzi CM, Turillazzi S (1992) Landmark-based mating systems in four *Polistes* species (Hymenoptera: Vespidae). J Kansas Entomol Soc 65(3):211–217

Beani L, Dessì-Fulgheri F, Cappa F, Toth A (2014) The trap of sex in social insects: from the female to the male perspective. Neurosci Biobehav Rev 46(4):519–533

Beani L, Bagnères AG, Elia M, Petrocelli I, Cappa F, Lorenzi MC (2019) Cuticular hydrocarbons as cues of sex and health condition in Polistes dominula wasps. Insec Soc 66(4):543–553

Boomsma JJ, Baer B, Heinze J (2005) The evolution of male traits in social insects. Annu Rev Entomol 50:395–420

Cappa F, Bruschini C, Cervo R, Turillazzi S, Beani L (2013) Males do not like the working class: male sexual preference and recognition of functional castes in a primitively eusocial wasp. Anim Behav 86(4):801–810

Cappa F, Beani L, Cervo R (2016) The importance of being yellow: visual over chemical cues in gender recognition in a social wasp. Behav Ecol 47:1182

Cappa F, Cini A, Signorotti L, Cervo R (2020) Rethinking recognition: social context in adult life rather than early experience shapes recognition in a social wasp. Philos Trans R Soc B 375(1802): 20190468

Cervo R, Cini A, Turillazzi S (2015) Visual recognition in social wasps. In: Aquiloni L, Tricarico E (eds) Social recognition in invertebrates. Springer International Publishing, Cham, pp 125–145

Crozier RH (1971) Heterozygosity and sex determination in haplodiploidy. Am Nat 105(945):399–412

Crozier RH, Fjerdingstad EJ (2001) Polyandry in social Hymenoptera – disunity in diversity? Ann Zool Fenn 38(3):267–285

Darwin C (1871) The descent of man and selection in relation to sex. John Murray, London, 423 p

De Souza AR, Rodrigues IL, Rocha IVA, Reis WAA, Lopes JF, Prezoto F (2008) Foraging behavior and dominance hierarchy in colonies of the neotropical social wasp *Polistes ferreri* (Hymenoptera, Vespidae) in different stages of development. Sociobiology 52(2):293–303

De Souza AR, Júnior CAM, Nascimento FS, Lino-Neto J (2014) Sexy faces in a male paper wasp. PLoS One 9:e98172
De Souza AR, Turillazzi S, Lino-Neto J, Santini G (2016) Colder environments may select for darker paper wasps. Biol J Linn Soc 120(3):700–704
De Souza AR, Barbosa BC, da Silva RC, Prezoto F, Lino-Neto J, Nascimento FS (2017a) No evidence of intersexual kin recognition by males of the neotropical paper wasp *Polistes versicolor*. J Insect Behav 30:180
De Souza AR, Lino-Neto J, Tibbetts EA, Turillazzi S, Beani L (2017b) The leks of *Polistes dominula* paper wasps: tiny abdominal spots play a critical role in male attacks toward potential rivals. Ethol Ecol Evol 29(4):410–419
De Souza AR, Simões TG, Rantala MJ, Santos EF, Lino-Neto J, Nascimento FS (2018) Sexual ornaments reveal the strength of melanization immune response and longevity of male paper wasps. J Insect Physiol 109:163–168
De Souza AR, Santos EF, Nascimento FSD, Stanyon R, Lino-Neto J, Beani L (2020) Sexually dimorphic traits and male fertility in a paper wasp. Biol J Linn Soc 130(3):555–562
Gamboa GJ (2004) Kin recognition in eusocial wasps. Ann Zool Fenn 41(6):789–808
Gobbi N, Noll FB, Penna MAH (2006) "Winter" – aggregations, colony cycle, and seasonal phenotypic change in the paper wasp *Polistes versicolor* in subtropical Brazil. Naturwissenschaften 93(10):487–494
Grazinoli DJ, De Souza AR, Prezoto F (2010) Dominance hierarchy and division of work in colonies of *Polistes simillimus* (Hymenoptera, Vespidae). Sociobiology 56(2):507–514
Hamilton WD (1964a) The genetical evolution of social behavior I. J Theor Biol 7:1–16
Hamilton WD (1964b) The genetical evolution of social behavior II. J Theor Biol 7:17–52
Heinze J (2016) The male has done his work – the male may go. Curr Opin Insect Sci 16:22–27
Hunt JH, Mutti NS, Havukainen H, Henshaw MT, Amdam GV (2011) Development of an RNA interference tool, characterization of its target, and an ecological test of caste differentiation in the eusocial wasp *Polistes*. PLoS One 6(11):e26641
Izzo AS, Tibbetts EA (2012) Spotting the top male: sexually selected signals in male *Polistes dominulus* wasps. Anim Behav 83:839–845
Izzo AS, Tibbetts EA (2015) Heightened condition dependence of a sexually selected signal in male *Polistes dominulus* paper wasps. Ethology 121(6):586–592
Liebert AE, Johnson RN, Switz GT, Starks PT (2004) Triploid females and diploid males: underreported phenomena in *Polistes* wasps? Insect Soc 51(3):205–211
Liebert AE, Gamboa GJ, Stamp NE, Curtis TR, Monnet KM, Turillazzi S, Starks PT (2006) Genetics, behavior and ecology of a paper wasp invasion: *Polistes dominulus* in North America. Ann Zool Fenn 43:595–624
Liebert AE, Wilson-Rich N, Johnson CE, Starks PT (2010) Sexual interactions and nestmate recognition in invasive populations of *Polistes dominulus* wasps. Insect Soc 57(4):457–463
Nagamati Junior K, Simokomaki K, Gruber CV, Del Lama MA (2010) Sociogenetic structure of *Polistes* (*Aphanilopterus*) *versicolor* Olivier, 1791 colonies (Hymenoptera, Vespidae, Polistini). Genet Mol Biol 33(4):669–675
Queller DC (2016) Kin selection and its discontents. Philos Sci 83(5):861–872
Ryan RE, Gamboa GJ (1986) Nestmate recognition between males and gynes of the social wasp *Polistes fuscatus* (Hymenoptera, Vespidae). Ann Entomol Soc Am 79(4):572–575
Saga T, Okuno M, Loope KJ, Tsuchida K, Ohbayashi K, Shimada M, Okada Y (2020). Polyandry and paternity affect disease resistance in eusocial wasps. Behav Ecol. https://doi.org/10.1093/beheco/araa062
Sen R, Samudre S, Shilpa MC, Tarak RC, Gadagkar R (2010) Middle aged wasps mate through most of the year, without regard to body size, ovarian development and nestmateship: a laboratory study of the primitively eusocial wasp *Ropalidia marginata*. Insect Soc 57(1):95–103
Seppä P, Fogelqvist J, Gyllenstrand N, Lorenzi MC (2011). Colony kin structure and breeding patterns in the social wasp, Polistes biglumis. Insec Soc 58(3):345–355
Shellman-Reeve JS, Gamboa GJ (1985) Male social wasps (*Polistes fuscatus*, Hymenoptera: Vespidae) recognize their male nestmates. Anim Behav 33(1):331–333

Shilpa MC, Ruchira S, Gadagkar R (2010) Nestmateship and body size do not influence mate choice in males and females: a laboratory study of a primitively eusocial wasp *Ropalidia marginata*. Behav Process 85(1):42–46

Shilpa MC, Sen R, Samudre S, Gadagkar R (2012) Males, but not females, mate with multiple partners: a laboratory study of a primitively eusocial wasp *Ropalidia marginata*. Insect Soc 59(1):61–65

Southon RJ, Bell EF, Graystock P, Wyatt CD, Radford AN, Sumner S (2019) High indirect fitness benefits for helpers across the nesting cycle in the tropical paper wasp Polistes canadensis. Mol Ecol 28(13):3271–3284

Starks PT, Fischer DJ, Watson RE, Melikian GL, Nath SD (1998) Context- dependent nestmate discrimination in the paper wasp, *Polistes dominulus*: a critical test of the optimal acceptance threshold model. Anim Behav 56(2):449–458

Strassmann JE (2001) The rarity of multiple mating by females in the social Hymenoptera. Insect Soc 48(1):1–13

Tibbetts EA (2010) The condition dependence and heritability of signaling and nonsignaling color traits in paper wasps. Am Nat 175(5):495–503

Tibbetts EA, Skaldina O, Zhao V, Toth AL, Skaldin M, Beani L, Dale J (2011) Geographic variation in the status signals of *Polistes dominulus* paper wasps. PLoS One 6(12):e28173

Tibbetts EA, Forrest T, Vernier C, Jinn J, Madagame A (2015) Socially selected ornaments and fitness: signals of fighting ability in paper wasps are positively associated with survival, reproductive success, and rank. Evolution 69(11):2917–2926

Trivers RL, Hare H (1976) Haplodiploidy and the evolution of the social insects. Science 191(4224):249–263

Van Wilgenburg E, Driessen G, Beukeboom LW (2006) Single locus complementary sex determination in Hymenoptera: an "unintelligent" design? Front Zool 3(1):1–15

Whiting AR (1924) The inheritance of sterility and of other defects induced by abnormal fertilization in the parasitic wasp, *Habrobracon juglandis* (Ashmead). Genetics 10(1):33–58

Whiting PW (1933) Selective fertilization and sex-determination in Hymenoptera. Science 78(2032):537–538

Whiting PW (1943) Multiple alleles in complementary sex determination of *Habrobracon*. Genetics 28(5):365–382

Chapter 5
Nesting Habits of Neotropical Social Wasps

Bruno Corrêa Barbosa, Tatiane Tagliatti Maciel, and Fábio Prezoto

Abstract With the main functions of providing a microenvironment for the offspring and providing protection for the colony against vertebrate and invertebrate predators, the complex architecture of social wasp nests varies from group to group. The nests can have from one to multiple combs covered or not by a protective envelope, presenting an important taxonomic value for the classification of the groups to which they belong. Here, we propose a brief overview of the nesting habits, foundation strategies, and the distinction and identification of the genera of social wasps (Polistinae) occurring in Brazil.

Keywords Nesting habits · Nesting substrate · Paper wasps

When talking about social insects, we can say that social wasps are superorganisms because they function as if they were a single organism when they are together, with a diversity of behaviors which is directly reflected in varying nesting strategies exhibited by their different genera, such as nest architecture and material used in nest construction (Wenzel 1998).

In addition, the main function of social wasp nests is colony protection against natural enemies (Jeanne 1975; Wenzel 1998), as well as accumulating auxiliary functions such as weather protection structure, offspring care, and, in some species cases, food storage (cf. Jeanne 1991; Prezoto and Gobbi 2003; Guimaraes et al. 2008; Rocha 2012).

Like bees, social wasp nests are very diverse and are made up of combs, cells, and protective envelopes and may contain a single discovered comb with a few dozen brood cells fixed by a pedicel (Fig. 5.1a) or several layers of combs with brood cells overlapped and encased in the protective envelopes (Fig. 5.1b).

Some characteristics regarding studies of nesting habits of social wasps may hinder a researcher's observation process, such as species aggressiveness and

B. C. Barbosa (✉) · T. T. Maciel · F. Prezoto
Laboratório de Ecologia Comportamental e Bioacústica (LABEC), Programa de Pós-graduação em Biodiversidade e Conservação da Natureza, Universidade Federal de Juiz de Fora, Juiz de Fora, Minas Gerais, Brazil
e-mail: barbosa.bc@outlook.com; tatitagliatti@hotmail.com; fabio.prezoto@ufjf.edu.br

F. Prezoto et al. (eds.), *Neotropical Social Wasps*,
https://doi.org/10.1007/978-3-030-53510-0_5

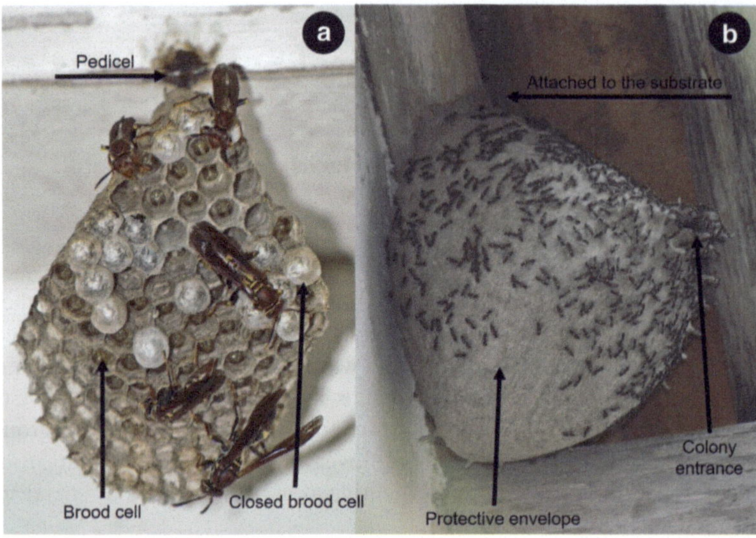

Fig. 5.1 Examples of social wasp nest types: (**a**) open nest fixed by a pedicel (*Polistes versicolor*); (**b**) closed nest attached to a substrate (*Polybia scutellaris*)

especially because nests are usually covered by envelopes and are difficult to access. For these reasons, except very specific cases involving direct observations of Epiponini behavior, they can still be considered a poorly studied group (e.g., Mateus 2005; Gelin et al. 2008; Kudô et al. 2016).

These difficulties are not found for Mischocyttarini and Polistini due to the fact that their nests are open, facilitating behavioral observations in loco; therefore, they are the best-studied groups (e.g., Oliveira et al. 2006; De Souza et al. 2012; Castro et al. 2014; Barbosa et al. 2016).

In this chapter, we will present several subjects and curiosities that are part of the nesting habits of neotropical social wasps, starting with nesting strategies and ending with an illustrated key for the genres of Brazilian social wasps.

5.1 Foundation Strategies

Social wasps are traditionally divided into two large groups considering their foundation strategy: an independent foundation and a swarming foundation species (Jeanne 1972; Jeanne 1980) (Fig. 5.2a, d).

The independent foundation is only practiced by species of the Polistini and Mischocyttarini tribes in the neotropical region, usually with relatively small colonies when compared to swarming species (Carpenter and Marques 2001). The independent foundation can occur in two ways: with nest building initiated by a single inseminated female (haplometrose) and a group of them (pleometrose) (Wenzel 1991; West-Eberhard 1969) (Fig. 5.2a, b).

Fig. 5.2 (**a**) *Polistes simillimus* colony with haplometrose foundation; (**b**) *Polistes versicolor* colony with pleometrose foundation; (**c**) swarming behavior (*Synoeca virginea*); (**d**) a swarming social wasp colony (*Polybia paulista*) under construction

Other females join the process after the foundation begins in some species where haplometrosis occurs; however, the original founder maintains oviposition exclusivity through aggressive behavior such as persecution and subjugation (Carpenter and Marques 2001). The founder participates in building the nest and foraging until the first workers emerge and is then restricted to reproduction.

One behavior associated with the independent foundation strategy is usurpation. This behavior can come from a founder abandoning her own nest, which may be by choice or after being expelled by co-founders; the usurpation itself consists of a foreign founder invading an existing nest and becoming dominant (breeding) through aggressive interactions with resident females (Iwahashi 1989; Turillazzi 1992). Field (1992) elaborated hypotheses about the conditions which stimulate usurpation behavior to the detriment of founding a new nest, such as ecological constraints, temporal limitations, behavioral stimuli, and phenotypic tendencies.

Swarming foundation only occurs in Epiponini in the neotropical region (Fig. 5.2c, d). Swarming behavior can occur for two reasons; one is for the reproductive division of the colony which is in the production phase of new sexed individuals, also called the "reproductive swarm" (cf. Richards and Richards 1951). In this case, the workers (sterile females) of an established colony go out to select a new nesting site and indicate it to the colony by pheromone trails, called trail making (cf. Naumann 1975), which causes the swarm to move and form a new colony (West-Eberhard 1982; Carpenter and Marques 2001). The queens in this foundation strategy are already defined from the beginning, being responsible for oviposition; the swarms are polygenic, i.e., they have multiple functional queens (Hölldobler and Wilson 1977; Mateus 2005).

The second reason for swarming is forced migration or an evacuating swarm (cf. Richards and Richards 1951), in which the swarm is induced by adverse mechanisms such as predation where predators damage structural parts of the nest or parts of the substrate or by biotic factors such as rain and wind which can damage the nest, making it impossible to rebuild the site, in addition to other accidental facts.

Different foundation strategies in swarming have advantages over independent foundation, especially in relation to the size that swarming colonies reach, enabling greater defense against predators and adverse conditions (Jeanne 1991). However, the independent foundation also has positive points, as its small size increases the chance of being overlooked by predators, as well as requiring a much smaller nesting area, making it possible to colonize a larger group of substrates.

There are records of various sizes that a colony can reach in terms of both nest size and number of colony individuals. The largest numbers among all social wasps belong to the *Agelaia* genus, which has the largest nest size and the largest concentration of individuals which can reach more than 1 million adults in a colony and can contain 956 thousand cells weighing over 13 kg (cf. Zucchi et al. 1995; Oliveira 2008).

5.2 Materials Used in Nest Construction

The materials used in building social wasp nests are indicators of the wasps' interaction with the environment and have a strong influence on the architecture and physical appearance of nests (Wenzel 1991). A significant part of the adult resource-gathering effort is directed toward collecting nest building materials (Giannotti et al. 1995).

Neotropical social wasps build nests from plant fibers collected by scraping their mandible on the plant substrate (Fig. 5.3a). The foragers collect the material, crush, and mix it with water and salivary secretion resulting in a mass which is then applied in the construction and repair of new nest and envelope cells (Fig. 5.3b), as well as reinforcing the pedicel. As a result, this obtained mass is a paperlike fiber, being responsible for the popular name paper wasps. The thickness can vary considerably, being thick in most Epiponini and thinner in Polistini and Mischocyttarini (Gallo et al. 1988; Wenzel 1991; Wenzel 1998).

There are exceptions such as the *Polybia emaciata* Lucas, 1879 species which uses nesting clay, and *Mischocyttarus iheringi* Zikán, 1935 specie which uses unprocessed plant materials such as algae, bryophytes and trichomes (Fig. 5.3c) (cf. Wenzel 1991; Barbosa et al. 2016).

Fig. 5.3 (**a**) Collection of nest building or repair material (*Parachartergus fraternus*); (**b**) nest building behavior (*Parachartergus fraternus*); (**c**) *Mischocyttarus iheringi* nest; this species uses an unprocessed plant material for camouflage

5.3 Choice of Nesting Substrate

The choice of nesting site is the focus of studies on social wasps, given the importance of the theme and its ecology (e.g., Corbara et al. 2009; Alvarenga et al. 2010; Castro et al. 2014; Barbosa 2015; Virgínio et al. 2016). Factors taken into consideration when choosing a location include concealability, weather protection, and the risk of predation (Richards and Richards 1951; Downing and Jeanne 1986; Dejean et al. 2010).

Species commonly build nests under or on leaves, branches, and logs and within natural cavities in a natural environment (Jeanne 1991). Social wasp nesting in vegetation seems to be generally influenced by broadleaf and evergreen plants (Fig. 5.4a) concentrated in understory species (e.g., Dejean et al. 1998; Souza et al. 2014; Barbosa 2015), which keeps the nests more concealed and protected against abiotic factors due to the density, such that the vegetation morphology and height are more relevant for choosing the social nesting site than any specific plant species.

In parallel, social wasp nests are commonly found in urban areas, especially in man-made structures. This relationship indicates the synanthropism of the group which is most observed in *Mischocyttarus* and *Polistes* species, sometimes preferring human to natural substrates (e.g., Yeison-López et al. 2012; Barbosa 2015). Although there are several materials in man-made substrates such as concrete, metals, plastics, and glass, wood stands out due to its plant origin, suggesting that social wasps are still undergoing a process of synanthropic adaptation, as they tend to nest in substrates of plant origin.

Fig. 5.4 (**a**) *Polybia fastidiosuscula* colony nested on *Dracaena fragrans*, a broadleaf and evergreen plant species; (**b**) difference in concealability of nests on human substrates for species of independent and swarming foundations, highlighting a *Polistes versicolor* nest next to a *Polybia fastidiosuscula* nest

It can also be noted that the use of man-made substrates is limited for some species due to the need for the nest to be fixed onto the substrate and its concealability, and which the size of the nest influences, i.e., independently found wasp nests are generally small and remain concealed, while on the other hand swarming species present larger, much more visible colonies and lose their concealability, as observed in Fig. 5.4b.

5.4 Colonial Cycle and Seasonality

Less-defined annual seasons in tropical regions also lead to less-defined phases of the colonial cycle. West-Eberhard (1969) showed the existence of asynchrony in the social wasp cycle in tropical regions, since nests can be started at any time of the year; for this very reason, the cycle is strictly related to the environment in which they are inserted, and may go through distinct phases, as characterized by Reeve (1991):

- *Foundation*: it begins with the construction of the nest and goes until the emergence of the first female offspring wasp (Fig. 5.5a).
- *Development*: it begins with the appearance of the first female offspring and extends until the appearance of reproductive forms (new queens). This phase is also known as the colony growth period which can last several months, and the colony increases in size several times (both in physical and population structures) (Fig. 5.5b).
- *Decline*: this period is characterized by ceasing nest expansion activities; irreversible reduction of the population, mainly immature forms (eggs, larvae, and pupae); and the appearance of reproductive forms (males) (Fig. 5.5c).
- *Abandonment*: the colony will only be made up of adult individuals at some point during the decline phase, who eventually abandon it and may disperse throughout the region to found new colonies or form an aggregate.

Fig. 5.5 (**a**) Foundation phase, nest building and first egg laying; (**b**) development phase, emergence of the first lineage and colony expansion; (**c**) decline phase, decrease in new comb construction, population, and male production; (**d**) aggregation, optional behavior performed in cold seasons

- *Aggregated*: it has a very variable duration (from a few days to several months), being an optional event which is mainly dependent on climatic conditions. This phenomenon can be frequently observed during the dry season. Aggregation is similar to mammalian hibernation and is performed by a group of adults, usually relatives, gathering over the nest to spend times of drought protected (Fig. 5.5d).

Cycles in tropical regions often do not follow the sequence suggested by Reeve (1991) and can repeat, skip, or recycle phases as needed by colonies, increasing or decreasing colony life, such as in records of a *Polybia scutellaris* colony which is over 4 years old. However, it is still possible to see some seasonality in the cycles even in asynchronous nests, even though the weather seasons are not so defined; Jeanne (1972) argues that this is due to intrinsic control.

According to the author, aggressive queen-worker interactions are essential to maintain the queen's reproductive exclusivity; as the colony expands, the dominant female cannot control all workers, leading to a large proportion of non-working females and males in the nest, which causes marked declines in productivity, leading to hunger and ultimately nest abandonment. In addition to the influence of environmental seasonality and intrinsic factors, extrinsic factors such as direct predation interference can occur, in which nests can be destroyed and adults disperse and start a new foundation (Maciel et al. 2016).

5.5 Nest Architecture

Several classifications of neotropical social wasp nests have been proposed over the years due to the diversity of their shapes; the classical and most widely accepted system by Polistinae researchers was created by Saussure (1853–1858) and was later revised by Richards and Richards (1951). This classification is separated into three major divisions which will be presented next.

Stelocyttarus nests They attach to the nesting substrate via pedicels or pillars; they can be formed by one or more combs, and subsequent combs can be linked to the previous ones by means of peduncles. Although the peduncles are of equal origin to the envelope, they are made of a stronger resinous material. Stelocyte nests are further divided into two subtypes: Gymnodornous, which usually have only one comb and do not have a protective envelope, and Calyptodomous, which have a protective envelope involving their combs and great structural diversity. The Gymnodornous nest type is represented in *Mischocyttarus*, *Polistes*, and some *Agelaia* species, while Calyptodomous nests can be seen in *Angiopolybia*, *Charteginus*, *Leipomeles*, *Parachartergus*, *Pseudopolybia*, and some *Agelaia* and *Protopolybia* species.

Astelocyttarus nests Only one comb is attached directly to the substrate, protected by a dome-shaped envelope; nest enlargement occurs through expansion of the initial comb under the substrate as well as the protective envelope. The genera which build astelocyte nests are *Asteloeca*, *Clypearia*, *Marimbonda*, *Metapolybia*, *Nectarinella*, and *Synoeca*.

Phragmocyttarus nests They are initially composed of a comb connected directly to the substrate from which a protective envelope is constructed; the subsequent comb is added to the envelope surface which is enlarged to cover the new comb and so on. The genera which present this type of nest are *Brachygastra*, *Epipona*, *Chartergus*, *Polybia*, and *Protonectarina*.

5.6 Generic Key for Nests of Genera Found in Brazil

As there is a known difficulty in identifying the morphological characteristics of social wasp nests which occur in Brazil, this chapter will conclude by presenting an illustrated key to the neotropical social wasp genera based on the structural characteristics of the nests:

1. Nests are usually built in natural cavities such as tree trunks and termites (Fig. 5.6a) or in abandoned houses; they are exposed nests, unenveloped and with brood cells of size up to 4 mm. .... *Agelaia*
   - Nests fully exposed and fixed to the substrate; generally exposed nests with or without envelope and brood cell size of 5 mm (Fig. 5.6b) **2**

Fig. 5.6 (**a**) Cavity (termite) housing *Agelaia vicina* colony; (**b**) *Polybia bifasciata* nest; (**c**) *Mischocyttarus rotundicollis* nest; (**d**) *Apoica pallens* nest; (**e**) *Polistes similimus* nest

2. Uncoated nests with fully exposed brood cells (Fig. 5.6c). **3**
   - Enveloped nests with hidden/concealed brood cells (Fig. 5.6b) **5**
3. Pedicel absent (Fig. 5.6d); single layer of brood cells facing downward in a horizontal position, with the largest portion in contact with the substrate and completely enveloped therein; usually the nests are light in color and have a hat or shower shape. .... *Apoica*
   - Pedicel present (Fig. 5.6c), which may be single or multiple; composed of only one layer of brood cells; variable cell orientation (vertical or horizontal) found fixed on leaves or human buildings; colored nests which can vary..... **4**
4. Brood cells smaller than 4 mm in diameter (Fig. 5.6c); nests are composed of short vegetable fibers. .... *Mischocyttarus*
   - Brood cells larger than 4 mm in diameter (Fig. 5.6e); nests are composed of long and woody fibers. .... *Polistes*
5. Nest with a single sessile comb, built directly on the surface of the trunk....... **6**
   - All combs hung by pedicels or at least with sessile initiation.................. **10**

Fig. 5.7 (**a**) *Metapolybia* sp nest, the darkest and brightest areas are transparent windows elaborated with a secretion; (**b**) *Synoeca cyanea* nest; (**c**) *Parachartergus fraternus* nest; (**d**) *Pseudopolybia* nest; (**e**) *Angiopolybia pallens* nest; (**f**) *Protopolybia exigua* nest. Red arrow points to the nest entrance

6. Nest entrance in the lower half of envelope (Fig. 5.7c, e)................................ **7**
   - Nest entrance in the upper half or central part of the envelope (Fig. 5.7b)..**8**
7. Substrate around the nest entrance with small droplets of sticky substance; cells are 3.5–3.7 mm wide.. *Nectarinella*
   - Substrate around nest entrance and no droplets of sticky substance; 1.8-mm-wide cells.. *Leipomeles*

8. Envelope forms eaves (Fig. 5.7a) extending laterally; envelope with transparent windows of some secretion. .... *Metapolybia*
   - Envelope does not form eaves.. **9**
9. Some areas of the envelope show added bark particles and other plant materials; envelope with transparent windows is made up of some secretion; the entrance is positioned below the middle of the nest.............................. *Clypearia* or *Asteloeca*
   - Nests with brood cells parallel to the substrate and larger than 4 mm in diameter; envelope with transverse undulations except for *Synoeca virginea* (Fig. 5.7b) .... *Synoeca*
10. Envelope against or covering substrate. **11**
    - Envelope does not touch the substrate.. **14**
11. Combs are supported by one or more pedicels.. **12**
    - First sessile comb or suspended by a broad leaf-shaped pedicel; when smaller combs are present, they are built over the upper comb envelope (Fig. 5.8a) **15**
12. Nests with brood cells horizontal to substrate ....
    .... *Parachartergus* (Fig. 5.7a) or *Chartergellus*
    - Combs supported by an approximately central pedicel........................... **13**
13. First and following combs supported by a single central pedicel; envelope has more than one layer (Fig. 5.7d).. *Pseudopolybia*
    - First comb is supported by more than one pedicel; the nest entrance is located in the lower portion of the tubular envelope projection in some species (Fig. 5.7e) .... *Angiopolybia*
14. Nests with one or more layers of brood cells are 1 mm in diameter, usually they are attached to leaves by multiple pedicels, which are very small; they are light-colored, very fragile, and thin or have no leaflike wrapper playing a protective role (Fig. 5.7f). .... *Protopolybia*
    - Nest entrance located on the comb floor through an opening between the enclosure and the substrate; often star-shaped nests.................. *Charterginus*
15. Nests involving and/or attached to tree trunks, branches, or leaves, with a single opening in the lower region; the envelope is mostly fragile (Fig. 5.8a)..............
    .... *Polybia*
    - Nest built differently from the above **16**
16. Rigid nest with a hard and fibrous envelope. **17**
    - Nest with a soft and friable envelope... **18**

Fig. 5.8 (**a**) *Polybia platycephala* nest; (**b**) *Chartergus* nest; (**c**) *Epipona tatua* nest; (**d**) *Brachygastra lecheguana* nest; (**e**) *Protonectarina sylveirae* nest

17. Nest with a flat or conical bottom; entrance is located in the center of the nest and between the combs (Fig. 5.8b).. *Chartergus*

    • Nest with bottom projecting ever further sideways; entrance is located in the bottom projected portion of the nest (Fig. 5.8c).............................. *Epipona*

18. Nests involving tree trunks or branches with more than one layer of brood cells, small or medium in size; some species have thin and fragile envelope; other species may use clay in the envelope, making it very sturdy and well expressed, and may have irregular, ever-present auburns of varying color, usually lighter, resembling a gas canister (Fig. 5.8d). .... *Brachygastra*

    • Nests involving tree trunks or branches with more than one layer of brood cells; these are small, ever-present envelopes of varying coloration, usually darker and more resistant (Fig. 5.8e)............................. *Protonectarina*

References

Alvarenga RB, Castro MM, Santos-Prezoto HH, Prezoto F (2010) Nesting of social wasps (Hymenoptera, Vespidae) in urban gardens in southeastern Brazil. Sociobiology 55:445–452

Barbosa BC (2015) Vespas sociais (Vespidae: Polistinae) em um fragmento urbano: riqueza, estratificação e redes de interação. 60p. Dissertação de Mestrado (Mestrado em Ciências Biológicas), Universidade Federal de Juiz de Fora, Juiz de Fora, Minas Gerais, Brasil.

Barbosa BC, Dias ML, Vieira KM, Prezoto F (2016) Cryptic nest of *Mischocyttarus iheringi* (Hymenoptera: Vespidae: Polistinae) with description of camouflage. Fla Entomol 99(1):135–138

Carpenter JM, Marques OM (2001) Contribuição ao estudo de vespídeos do Brasil (Insecta, Hymenoptera, Vespoidea, Vespidae). Cruz das Almas, Universidade Federal da Bahia, Brasil. Série: Publicações digitais 2

Castro MM, Avelar DLG, De Souza AR, Prezoto F (2014) Nesting substrata, colony success and productivity of the wasp *Mischocyttarus cassununga*. Revista Brasileira de Entomologia 58:168–172

Corbara B, Carpenter JM, Céréghino R, Leponce M, Gibernau M, Dejean A (2009) Diversity and nest site selection of social wasps along Guianese forest edges: assessing the influence of arboreal ants. C R Biol 332(5):470–479

De Souza AR, Ribeiro B, Silva NJJ, Prezoto F (2012) Paint marking social wasps: an evaluation of behavioral effects and toxicity. Entomologia Experimentalis et Applicata (Print) 144:244–247

Dejean A, Cordaba B, Carpenter JM (1998) Nesting site selection by wasp in the Guianese rain forest. Insectes sociaux 45:33–41

Dejean A, Carpenter JM, Gibernau M, Leponce M, Corbara B (2010) Nest relocation and high mortality rate in a Neotropical social wasp: impact of an exceptionally rainy La Niña year. C R Biol 333:35–40

Downing HA, Jeanne RL (1986) Intra- and interspecific variation in nest architecture in the paper wasp Polistes (Hymenoptera, Vespidae). Insect Soc 33:422–443

Field J (1992) Patterns of nest provisioning and parental investment in the solitary digger wasp *Ammophila sabulosa*. Ecol Entomol 17(1):43–51

Gallo D, Nakano O, Silveira Neto S, Carvalho RPL, Batista GC, Berti Filho E, Parra JRP, Zucchi RA, Alves SB, Vendramim JD (1988) Manual de entomologia agrícola. Agronômica Ceres, São Paulo, 649p

Gelin LF, Cruz JDD, Noll FB, Giannotti E, Santos GMDM, Bichara-Filho CC (2008) Morphological caste studies in the neotropical swarm-founding Polistinae wasp *Angiopolybia pallens* (Lepeletier) (Hymenoptera: Vespidae). Neotrop Entomol 37(6):691–701

Giannotti E, Prezoto F, Machado VLL (1995) Foraging activity of *Polistes lanio lanio* (Fabricius) (Hymenoptera: Vespidae). Anais da sociedade entomológica do Brasil 24(3):455–463

Guimaraes DL, Castro MMD, Prezoto F (2008) Patterns of honey storage in colonies of the social wasp *Mischocyttarus cassununga* (Hymenoptera, Vespidae). Sociobiology 51(3):655–660

Hölldobler B, Wilson EO (1977) The number of queens: an important trait in ant evolution. Naturwissenschaften 64(1):8–15

Iwahashi O (1989) Society of *Ropalidia fasciata*. In: Iwahashi O, Yamane S (eds) Society of *Ropalidia* wasps. Tôkai University Press, Tokyo, pp 3–209. (in Japanese)

Jeanne RL (1972) Social biology of the Neotropical wasp *Mischocyttarus drewseni*. Bull Museum Compar Zool 144:63–150

Jeanne RL (1975) The adaptiveness of social wasps nest architecture. Q Rev Biol 50:267–287

Jeanne RL (1980) Evolution of social behavior in the Vespidae. Annu Rev Entomol 25:371–395

Jeanne RL (1991) The swarm-founding Polistinae. p 191–231. In: Ross KG, Matthews RW (eds) The social biology of wasps. Cornell University Press, Ithaca, p 678

Kudô K, Oliveira LA, Mateus S, Zucchi R, Nascimento FS (2016) Nestmate larval discrimination by workers in the swarm-founding wasp *Polybia paulista*. Ethol Ecol Evol:1–11

Maciel TT, Barbosa BC, Prezoto F (2016) Opportunistic predation of a colony of *Polybia platycephala* (Richards) (Hymenoptera, Vespidae) by *Labidus praedator* (smith) (Hymenoptera, Formicidae). Sociobiology 63(1):724–727

Mateus S (2005) Análises dos comportamentos envolvidos na organização social e no processo de enxameio de *Parachartergus fraternus* (Hymenoptera, Polistinae, Epiponini). 143p, 2005. Tese (Doutorado em Entomologia), Tese (Doutorado em Entomologia) FFCLRP – USP. Ribeirão Preto, 143p

Naumann MG (1975) Swarming behavior: evidence for communication in social wasps. Science 189(4203):642–644

Oliveira, OAL de (2008) Biologia comportamental de *Agelaia vicina*: forrageio, comportamento das operárias e desenvolvimento dos ninhos (Hymenoptera: Vespidae; Epiponini). 57p. Dissertação de Mestrado (Mestrado em Biologia Animal), Universidade Estadual Paulista, São José do Rio Preto, São Paulo, Brasil.

Oliveira SA, Santos JFL, Prezoto F (2006) Dominance hierarchy in different stages of development in the colonies of the neotropical eusocial paper wasp *Polistes versicolor* (Hymenoptera, Vespidae). Sociobiology, EUA 48(2):515–526

Prezoto F, Gobbi N (2003) Patterns of honey storage in nests of the Neotropical paper wasp *Polistes simillimus* Zikán, 1951 (Hymenoptera, Vespidae). Sociobiology 41(2):437–442

Reeve HK (1991) In: Ross K, Matthews R (eds) The social biology of wasps. Cornell University Press, Ithaca, pp 99–148

Richards OW, Richards MJ (1951) Observations on the social wasps of South America (Hymenoptera, Vespidae). Trans Entomol Soc London 102:1–170

Rocha MP (2012) Ecologia Comportamental de *Polybia platycephala* (Richards, 1978) (Hymenoptera, Vespidae, Epiponini). Mestrado em Ciências Biológicas (Zoologia), Universidade Federal de Juiz de Fora

Saussure HF (1853–1858) Études sur la familie des Vespides. 2. Monographie des guêpes sociales, ou de la tribu des Vespiens. Paris: Masson, and Geneva: J. Cherbuliez

Souza MM, Pires P, Elpino-Campos A, Louzada J (2014) Nesting of social wasps (Hymenoptera: Vespidae) in a riparian forest of Rio das Mortes in southeastern Brazil. Acta Scientiarum. Biol Sci 36:189–196

Turillazzi S (1992) Biology and evolution of social insects. In: Billen J (ed) . Leuven University Press, Leuven, pp 263–272

Virgínio F, Maciel TT, Barbosa BC (2016) Hábitos de Nidificação de *Polistes canadensis* (Linnaeus) (Hymenoptera: Vespidae) em Área Urbana. EntomoBrasilis 9(2):81–83

Wenzel JW (1991) Evolution of nest architecture, p. 480–519. In: Ross KG, Matthews RW (eds) The social biology of wasps. Cornell University Press, Ithaca, p 678

Wenzel JW (1998) A generic key to the nests of hornets, yellow jackets, and paper wasps worldwide (Vespidae: Vespinae, Polistinae). Am Mus Novit 3224:1–39

West-Eberhard MJ (1969) The social biology of Polistine wasps. Miscellaneous. Publications of the Museum of Zoology University, Ann Arbor

West-Eberhard MJ (1982) The nature and evolution of swarming in tropical social wasps (Vespidae, Polistinae, Polybiini). In: Jaisson P (ed) Social insects in the tropics, vol 2. Univ. Paris-Nord, Paris, pp 97–128

Yeison-López G, Canchila S, Durán A, Álvarez G (2012) Hábitos de nidificación de avispas sociales (Hymenoptera: Vespidae: Polistinae) en un área urbana del Caribe colombiano. Revista Colombiana de entomologia 38(2):347–350

Zucchi R, Sakagami SF, Noll FB, Mechi MR, Mateus S, Baio MV, Shima SN (1995) *Agelaia vicina*, a swarm-founding polistine with the largest colony size among wasps and bees (Hymenoptera: Vespidae). J New York Entomol Soc 103:129–137

Chapter 6
Castes and Polymorphisms in Neotropical Social Wasps

Marjorie da Silva, Sidnei Mateus, and Fernando Barbosa Noll

Abstract Neotropical social wasps have a set of chimeric characteristics that make them unique and difficult to fit into current theories on the evolution of social behavior (Noll and Wenzel, Biol J Linn Soc 93: 509–22, 2008). Among them, the presence of more than one functional queen (polygyny), absence of strong morphological differentiation, and flexibility between casts result in relaxation of ovarian control and breeding opportunity for workers (West-Eberhard, Science 200: 441–3, 1978; Noll, Sociobiology 60: 347–54, 2013). Adding to these particularities, the Polistinae wasps, especially those belonging to the Epiponini, reached a great radiation and evolutionary success in the Neotropical region due to the unique mode of social organization: colony foundation by a swarm (Jeanne, The swarm-founding Polistinae. In: Ross KG, Matthews RW (eds) The social biology of wasps. Cornell University Press, Ithaca, pp 191–231, 1991), where several or many reproductive females (queens) tolerate each other laying eggs. This chapter deals in general with aspects of the complex caste delimitation, the division of labor in the colony, the recruitment and foraging, and the lack of colony in social wasps, with emphasis on swarm founders.

Keywords Defense · Foraging · Polistinae · Polygyny · Syndromes

6.1 Castes

The origin of the morphological variation that establishes the social role (caste) is one of the definitive elements of the most sophisticated insect societies (Noll and Wenzel 2008). The study of castes differences in Epiponini species in Brazil began with the works of Richards (1978). Among Brazilian authors, studies of

M. da Silva (✉) · F. B. Noll
Universidade Estadual Paulista "Júlio de Mesquita Filho", São José do Rio Preto, SP, Brazil
e-mail: fernando.noll@unesp.br

S. Mateus
Faculdade de Filosofia, Ciências e Letras de Ribeirão Preto – USP, Ribeirão Preto, SP, Brazil
e-mail: sidneim@ffclrp.usp.br

F. Prezoto et al. (eds.), *Neotropical Social Wasps*,
https://doi.org/10.1007/978-3-030-53510-0_6

Shima et al. (1994, 1996a, b, 1998, 2000) were the pioneers. Understanding the mechanisms that lead to this differentiation is therefore of great importance for understanding the evolution of eusociality (Jeanne and Suryanarayanan 2011). The degree of distinction between breeding and worker caste varieties is often used as part of the definition of the degree of sociality (Noll et al. 2004), with highly social species presenting a complex division of labor, caste morphological differentiation, and a large number of individuals (Bourke 1999).

In Vespidae, this degree of differentiation ranges from no difference in the subfamily Stenogastrinae to large differences in size and morphology in Vespinae (Jeanne and Suryanarayanan 2011). In species presenting distinct morphological castes, the developmental difference between the future queens and workers should begin in the larval stage (Jeanne and Suryanarayanan 2011). In the independent-founding species of Polistinae, although the castes are morphologically identical, they present physiological differences that also should have origin in the larval stage (O'Donnell 1998a). This fact is supported by evidence that differences in mRNA expression and levels of hexameric storage proteins are already evident in the fifth and last larval instar in *Polistes metricus* Say (Jeanne and Suryanarayanan 2011; Hunt et al. 2007; Toth et al. 2007). In the case of Epiponini species, although described as highly social (Noll et al. 2004), the distinction between caste is not very evident and often difficult to identify (Richards and Richards 1951; Richards 1978, cited as Polybiini, junior synonym of Epiponini, see Carpenter 1993, 1997). According to Jeanne (2003), the complexity in this tribe is more related to the number of behaviors exhibited by the workers than to the presence of morphological differences.

There is no evidence of a genetic basis for caste differentiation in wasps (Jeanne and Suryanarayanan 2011). Differences in size among castes appeared independently in several taxa of this tribe, and caste evolved in different ways in the various lineages of Epiponini (Noll et al. 2004). A phylogenetic interpretation showed several distinct syndromes, representing a more complex scenario than previously thought (Noll et al. 2004; Noll and Wenzel 2008), challenging the defined standards. The four syndromes are (1) absence of caste differentiation (no differences in size or shape associated with reproduction); (2) physiological castes (absence of morphometric differences between castes, but females can be separated by degree of ovarian development); (3) larger queens, but with the same shape; and (4) queens who are different in size and shape from the workers (Noll et al. 2004; Chavarría 2013; Fig. 6.1 and Table 6.1).

As expected for highly social insects, pre-imaginal determination, given by nutritional differences during larval development (Hunt et al. 1996; Sakagami et al. 1996) or changes in the longitudinal axis of the body due to different programs in developmental parameters in the pre-adult stage (Jeanne et al. 1995), has been reported for the genera *Agelaia* (Fig. 6.2), *Apoica, Chartergus, Polybia, Protopolybia,* and *Pseudopolybia* (Noll et al. 2004). On the other hand, breeding females in *Chartergellus, Metapolybia, Parachartergus,* and *Synoeca* are morphologically similar to non-breeding and present no physiological discontinuities. In these cases, castes are flexible, and the determination is imaginal, given by the disputes between

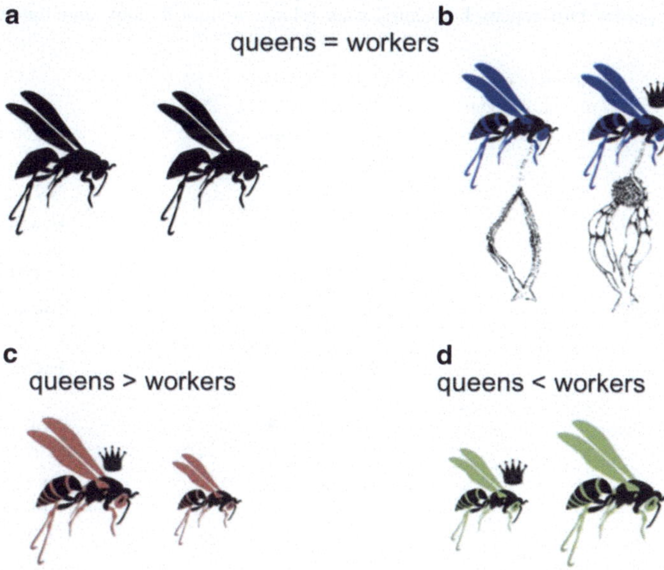

Fig. 6.1 The four syndromes related to the determination of castes in Polistinae. In (**a**, **b**) castes do not present morphological differences in size or shape. However, in (**b**) females can be separated by the degree of development of the ovaries (physiological caste). In (**c**) queens are larger in size, but there is no difference in the shape, and in (**d**) queens differ in size and shape from workers. The crowns identify the reproductive caste (queens)

adult females instead by larval manipulation (West-Eberhard 1981; Mateus et al. 2004; Noll and Wenzel 2008; Chavarría 2013). Theoretically, in species with imaginal determination, all females when emerging could develop the ovaries and reach the status of queen, a condition that could lead to conflicts among the individuals of the colony, since the workers could also lay eggs (Hart and Ratnieks 2005).

In fact, in many colonies, workers with functional ovaries (named intermediates by Richards and Richards 1951) have been identified. The role of intermediate is debatable, being considered as producers of trophic eggs, males (Richards 1971), or even young queens (Forsyth 1978; West-Eberhard 1978; Gastreich et al. 1993). The level of ovarian development is inversely proportional to the number of queens, that is, when a few females are present, the intermediates present more developed ovaries (Fig. 6.3) and vice versa (Richards 1971; West-Eberhard 1978). Nevertheless, intermediates are absent in many other taxa (revised in Noll et al. 2004). Anyway, the presence of this type of female is part of a complex scenario related to the evolution of castes (Noll 2013).

The origin of queens that are highly tolerant of each other brings some consequences. The first is that many females could aspire for a chance of reproduction, which would lead to the totipotency of the caste (Strassmann et al. 2002). As a consequence, it would be expected to find a generalized ovarian development in members of polygynic societies, and, in fact, in several Epiponini species, workers who lay eggs are found (Noll et al. 2004). Several basal genera of Epiponini fit into

Table 6.1 Epiponini Differences between castes related to morphology and the presence of intermediates in

Species	Absence of morphological differentiation	Queens are larger in size	Castes differ in shape	Presence of intermediates	Reference
Agelaia areata (Say)		X	X		Jeanne and Fagen (1974)
Agelaia flavipennis (Ducke)					Evans and West-Eberhard (1970)
Ag. fulvofasciata (DeGeer)		X			Richards (1978)
Ag. lobipleura (Richards)		X			Richards (1978)
Ag. multipicta (Haliday)		X	X		Noll et al. (1997a)
Ag. pallipes		X	X		Richards (1978), Noll et al. (1997a)
Ag. vicina (de Saussure)		X	X		Sakagami et al. (1996), Baio et al. (1998)
Ag. yepocapa (Richards)		X			Hunt et al. (2001)
Angiopolybia pallens (Lepeletier)	X			X	Richards (1978), Noll et al. (2004)
Apoica flavissima van der Vecht			X		Shima et al. (1994)
Ap. gelida van der Vecht			X		Richards (1978)
Ap. pallens (Fabricius)			X		Richards (1978), Jeanne et al. (1995)
Asteloeca traili (Cameron)	X				Noll et al. (2004)
As. ujhelyii (Ducke)	X				
Brachygastra augusti (de Saussure)		X		X	Baio et al. (2003a)[b]
B. bilineolata Spinola		X		X	Richards (1978)
B. lecheguana (Latreille)	X			X	Shima et al. (2000)
B. moebiana (de Saussure)	X			X	Richards (1978)

(continued)

Table 6.1 (continued)

Species	Absence of morphological differentiation	Queens are larger in size	Castes differ in shape	Presence of intermediates	Reference
B. scutellaris (Fabricius)			X	X	Richards and Richards (1951), Carpenter and Ross (1984)
Chartergellus communis[a] Richards	X			X	Richards (1978), Mateus et al. (1999)
Charterginus fulvus Fox		X		X	Noll et al. (2004)
Chartergus artifex (Christ)		X			Richards (1978)
C. globiventris de Saussure		X			Noll and Zucchi (2002)
C. metanotalis Richards		X			Noll et al. (2004)
Clypearia sulcata (de Saussure)	X				Noll et al. (2004)
Epipona guerini (de Saussure)		X			Hunt et al. (1996)
E. tatua (Cuvier)		X			Richards (1978), Noll et al. (2004)
Leipomeles dorsata (Fabricius)		X			Noll et al. (2004)
Metapolybia aztecoides Richards	X				Noll et al. (2004)
M. cingulata (Fabricius)	X				Richards and Richards (1951), Carpenter and Ross (1984)
M. docilis Richards		X			Baio et al. (2003b)
Nectarinella championi (Dover)		X		X	Noll et al. (2004)
Parachartergus colobopterus (Lichtenstein)	X				Strassmann et al. (1991)
Pa. fraternus (Gribodo)	X			X	Richards (1978)
Pa. smithii (de Saussure)	X			X	Mateus et al. (1997)
Polybia belemensis Richards		X			Richards (1978)

(continued)

Table 6.1 (continued)

Species	Absence of morphological differentiation	Queens are larger in size	Castes differ in shape	Presence of intermediates	Reference
Po. bicytarella Richards	X				Richards and Richards (1951)
Po. bistriata (Fabricius)		X			Richards and Richards (1951)[c]
Po. catillifex Möbius		X			Richards and Richards (1951)
Po. chrysothorax (Lichtenstein)	X				Richards (1978)
Po. dimidiata (Olivier)			X		Richards (1978), Shima et al. (1996a)
Po. emaciata Lucas		X			Richards (1978)
Po. erythrothorax Richards	X				Richards (1978)
Po. jurinei de Saussure		X			Richards (1978)
Po. liliacea (Fabricius)			X		Noll et al. (2004)
Po. micans Ducke	X				Richards (1978)
Po. occidentalis (Olivier)		X			Richards (1978), Noll et al. (2000)
Po. paulista (von Ihering)		X			Noll and Zucchi (2000)
Po. platycephala sylvestris Richards		X			Richards and Richards (1951)
Po. quadricincta de Saussure	X				Richards (1978)
Po. rejecta (Fabricius)		X			Noll et al. (2004)
Po. ruficeps (Schrottky)		X			Richards (1978), Noll et al. (2004)
Po. scutellaris (White)		X			Richards (1978), Noll et al. (1997b), Noll and Zucchi (2000)
Po. singularis Ducke		X			Richards (1978)
Po. spinifex Richards		X			Noll et al. (2004)
Po. striata (Fabricius)		X			Richards (1978)

(continued)

Table 6.1 (continued)

Species	Absence of morphological differentiation	Queens are larger in size	Castes differ in shape	Presence of intermediates	Reference
Protonectarina sylveirae (de Saussure)		X	X?	X	Shima et al. (1996b), Tanaka Jr. et al. (2010)
Protopolybia exigua (de Saussure)		X		X	Noll et al. (1996), Noll and Zucchi (2002)
Pr. minutíssima (Spinola)		X		X	Richards and Richards (1951)
Pr. sedula (de Saussure)		X		X	Richards and Richards (1951), Richards (1978)
Pseudopolybia compressa (de Saussure)		X			Richards (1978)
Ps. difficillis (Ducke)			X		Jeanne (1996)
Ps. vespiceps (de Saussure)	X				Shima et al. (1998)
Synoeca chalibea de Saussure	X				Richards (1978)
Sy. cyanea (Fabricius)		X			Noda et al. (2003)
Sy. surinama (Linnaeus)	X				Richards (1978), Noll et al. (2004)

Data from literature
[a]The yellow coloration of the genus has a paler tone in the queens
[b]Richards (1978) found no differences
[c]Carpenter and Ross (1984) found queen slightly smaller than workers, but not significantly

the proposed scenario for polygynic groups (West-Eberhard 1978, 1981), that is, caste flexibility due to the absence of morphological differences between them and ovarian development in all females. This condition is found mainly in *Angiopolybia*, *Pseudopolybia*, *Parachartergus*, *Chartergellus,* and *Leipomelles* (Noll et al. 2004). Thus, the distribution of this characteristic (presence of workers laying eggs) actually constitutes a plesiomorphic rather than an apomorphic state (Noll 2013).

Due to the absence of morphologically distinct castes, the social way of life in Epiponini has already been considered as little complex. However, Jeanne (2003) suggested the presumption of self-organization rather than the presence of distinct morphologies as a characteristic that would define complex societies. In this case, the presence of self-organization would remove the need for evolution in two distinct forms in Epiponini (Noll 2013). Indeed, as suggested by West-Eberhard (2003), the origin of morphologically distinct castes is much more a condition for colony stability and defense than indicative of social complexity. From this perspective, the complex Epiponini societies may have undergone a phase of totipotent and

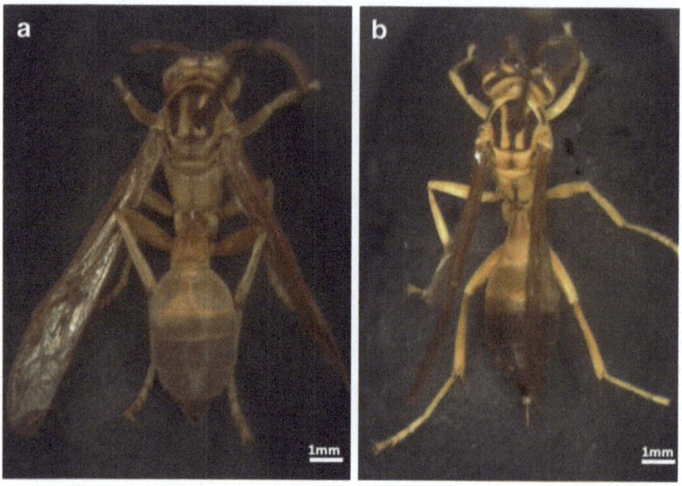

Fig. 6.2 Queen (**a**) and worker (**b**) of *Agelaia pallipes* (Olivier), species that presents a clear morphological difference between the castes

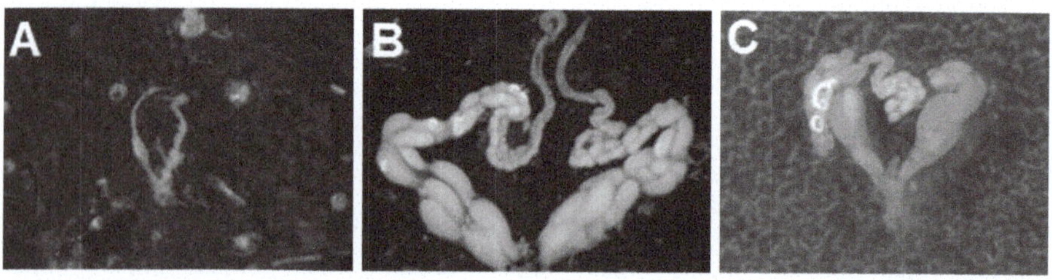

Fig. 6.3 Different degrees of ovarian development in Epiponini. Pictures of *Protonectarina sylveirae* (de Saussure). (**a**) Worker, (**b**) queen, and (**c**) intermediate

monomorphic females for several morphologically distinct queen systems in different lineages. Such a theory proved to be perfectly congruent when optimized in a cladogram of Epiponini genera (Noll et al. 2004).

Reproductive control in social wasps can be done in two ways: through physical aggression (primitively eusocial wasps) or through chemical control. Within the colonies of primitively eusocial wasps (*Polistes*, *Mischocyttarus*, *Belonogaster*, *Parapolybia,* and *Ropalidia*), there is a hierarchy of dominance, where the female with greater ovarian development is solely in charge of egg laying (Fletcher and Ross 1985). Because any female can lay eggs, the queen status is maintained within the colony through physical aggression and oophagy. In *Polistes*, dominance interactions are more frequent than in *Mischocyttarus* (Spradbery 1991), and, in *Ropalidia* there is little or no direct interaction, with this genus presenting the least dominance repertoire among Polistinae wasps (Spradbery 1991; Kardile and Gadagkar 2002; Bhadra et al. 2010). After establishment of dominance, a reduction in the levels of aggression is observed (West-Eberhard 1986; Hughes et al. 1987), suggesting that the reproductive dominance can also be maintained by chemical signals (West-Eberhard 1986; Dapporto et al. 2007; Bhadra et al. 2010), possibly by mixtures of hydrocarbons secreted by the Van der Vecht gland (Dapporto et al. 2007). The factors of endocrine control and oogenesis are not yet clear, however. Dominance behavior alone seems to be insufficient to maintain reproductive control (Röseler 1991). In the absence of the original queen, one of the female co-founders of the nest or the female presenting the highest ovarian development may assume a role in reproduction (Spradbery 1991). The queen monopolizes the production of eggs but can also forage occasionally and starts the construction of new cells (Chavarría 2013).

For Epiponini, physical aggression was observed only during the phase of establishment and selection of queens (Mateus 2005); however, physical aggression on the part of the queen against the workers, to maintain their status within the colony, is not common. For this reason, it is believed that reproductive control is also made through chemical signaling or hormonal control (Naumann 1970; Forsyth 1978; Fletcher and Ross 1985; West-Eberhard 1989a; Spradbery 1991), although no study has proven this phenomenon. In addition to chemical signals, queens use ritualized postures to communicate their status within the colony (West-Eberhard 1978; Spradbery 1991). The queens, in Epiponini, do little more than lay eggs and perform no other task in the colony (Herman et al. 2000).

If the queens in Epiponini do not exercise a strong reproductive control, the workers do. The workers test and remove queens from the colony (West-Eberhard 1978, 1981; Herman et al. 2000; Platt et al. 2004) and favor the production of queens in colonies where there is a low number of queens and the production of males in colonies with high number of queens (Queller et al. 1993). Moreover, workers favor production of males by queens because they are more genetically related to the "sons" of the queens than to the "sons" of other workers (Hastings et al. 1998), which seems good to maintain the "workers' interest" (Strassmann et al. 1997, 1998). This type of control is probably done in species with imaginal determination, although other control mechanisms may be related (Chavarría 2013). Studies in colonies of *Protopolybia sedula* (de Saussure) (Naumann 1970, still as *P. pumila* in his works), *Metapolybia aztecoides* Richards (West-Eberhard 1973, 1978), *Parachartergus colobopterus* (Lichtenstein) (Herman et al. 2000), *Asteloeca ujhelyii* (Ducke) (Nascimento et al. 2004), *Parachartergus fraternus* (Gribodo) (Mateus 2005), and *Metapolybia docilis* Richards (Chavarría 2009) demonstrated that worker behavior may be an important component in the mechanism of delimitation of functions in the colony.

A peculiar feature that occurs in Polistinae is the presence of polygyny (presence of several queens coexisting in the same nest). This characteristic appeared three times within the subfamily – in the genera *Polybioides* and *Ropalidia,* and in the tribe Epiponini (Carpenter 1991) – and is associated with the colony's foundation mode by swarming (the colony is constructed by a large number of workers accompanied by more than one queen) and complex social organization (Carpenter 1991; Jeanne 1991; Fig. 6.4). Epiponini species may exhibit permanent polygyny or variation in queen numbers (decrease) during the colony cycle, resulting in monogyny

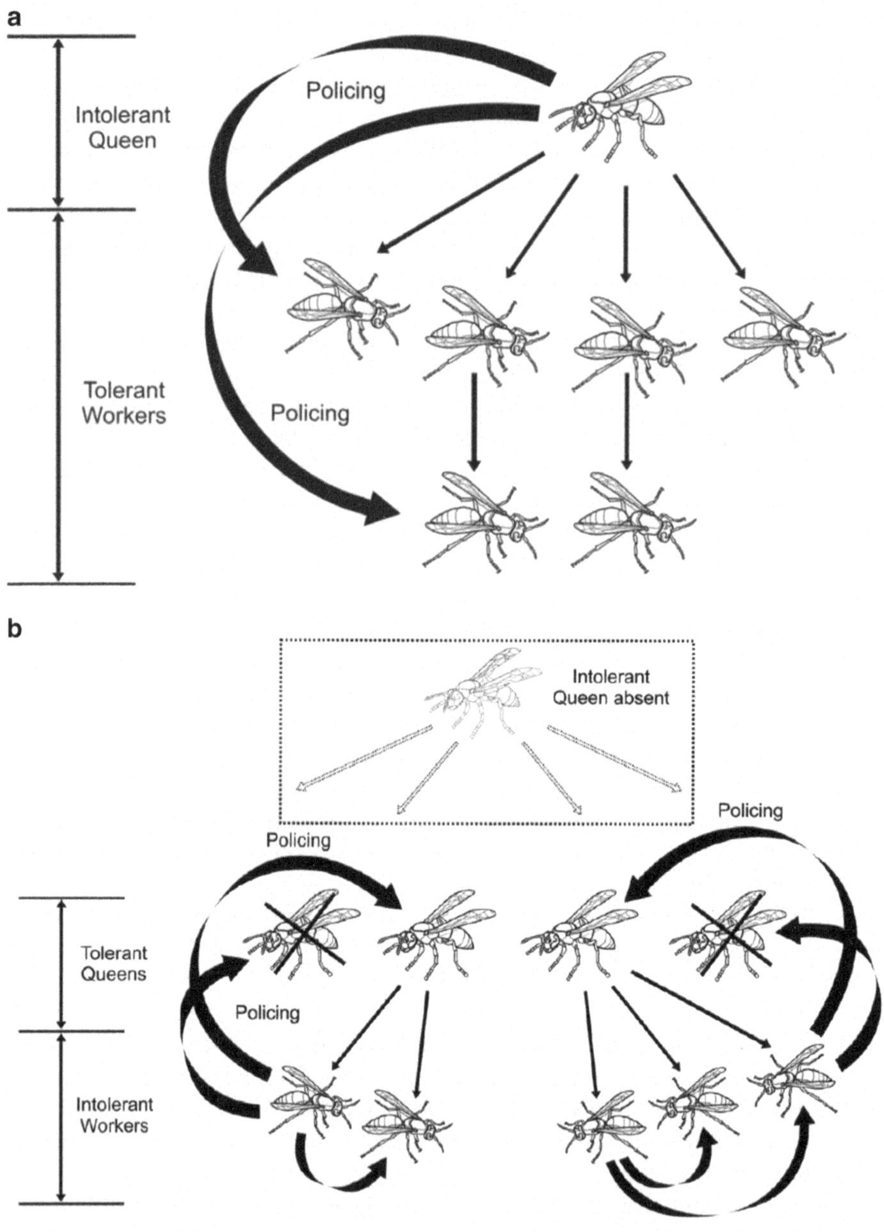

Fig. 6.4 Evolution of polygyny in Epiponini. (**a**) In a plesiomorphic society with short-term monogyny, the dominant female is intolerant of the reproduction of other females. The queen performs policing, while the workers are tolerant to the reproduction of other females. Occasionally, the workers reproduce. (**b**) In Epiponini no intolerant queens are present; instead, reproduction is performed by several tolerant females. The role of policing is adopted by workers, who suppress each other, as well as select among breeders. (Modified from Noll 2013)

(West-Eberhard 1978; Richards 1978; Jeanne 1991). The presence of polygyny in the most basal clades indicates that the presence of multiple functional queens was already present in the common ancestor of the tribe (Carpenter 1991; Noll 2013).

It is not difficult to imagine that the origin of polygyny was molded by the same factors as swarm foundation and nest architecture and that ant predation was an important factor (Jeanne 1979). There is a consensus that eusociality of wasps originated in the tropics (Evans and West-Eberhard 1970), where ants are the most important predators of the colonies (Richards and Richards 1951). From this point of view, it is appropriate to state that the most peculiar characteristics of the Epiponini lifestyle are the result of the improvement of defense against ants (Richards 1971, 1978; Jeanne 1975, 1991; Starr 1990; Simões et al. 1996). Thus, the rapid abandonment of the nest through the migratory behavior (swarming) would be an important strategy, assured by polygyny, that led to the selection of queens morphologically undifferentiated (less body mass, easier to fly) and numerically more abundant (greater opportunity for reestablishment of the colony in a new place) (Mateus 2005). Considering that the loss of a single queen could be fatal, the presence of several queens could be advantageous for the colony in terms of survival (Noll 2013).

The presence of several queens laying eggs could represent a problem regarding "conflict of interest" occurring in the colonies of social insects (Noll 2013). Although degrees of relatedness may be low in colonies of the independent-founding wasps *Polistes* (Pickett et al. 2006), an additional problem arises in Epiponini: the division of reproduction among several queens would result in an even lower degree of kinship among the females of one colony (Hastings et al. 1998), which would not reinforce the sterility of the workers. However, kinship analyses for Epiponini colonies have indicated that the degree of relationship between females is relatively high (Queller et al. 1988, 1993; Strassmann et al. 1992; Gastreich et al. 1993; Hughes et al. 1993; Hastings et al. 1998). These results are in line with what West-Eberhard (1978, 1981) observed in colonies of *Metapolybia aztecoides*. After the swarm, at the beginning of the cycle, several queens and workers are present. Throughout the development of the colony, some of the queens disappear or begin to display worker functions, and, later, the number of queens is reduced until there remains a few or only one. Queen production and swarms occur predominantly after the establishment of monogyny or oligogyny, restricting the production of future queens to only a few individuals. This pattern is observed in several other species, such as *Polybia occidentalis* (Olivier) (West-Eberhard 1978; Queller et al. 1993), *Parachartergus colobopterus* (Strassmann et al. 1991), *Polybia emaciata* Lucas (Strassmann et al. 1992), *Protopolybia exigua* (de Saussure) (Gastreich et al. 1993), *Brachygastra mellifica* (Say) (Hastings et al. 1998), and *Agelaia multipicta* (Haliday) (West-Eberhard 1990). The term cyclic oligogyny was coined to define the decrease in the number of queens and the production of new ones during the period when there are few of them in the colony. That way, the relationship between the queen and daughters is increased, as predicted by the kinship selection theory (Hamilton 1964a, b,

1972). The high kinship is also maintained by the fact that queens in Epiponini are inseminated only once (Goodnight et al. 1996). The possibility that the wasps of this tribe are inflexible in terms of loss of these elements of sociality should not be ruled out either (Noll 2013).

6.2 Communication

6.2.1 Alarm Pheromone

Swarm-founding wasps and their populous colonies are engaged in sophisticated communication systems to coordinate the activities of all members of the colony (Jeanne 1991). The defense behavior in *Polybia occidentalis* involves two steps. First, in response to a sudden movement or mechanical disturbance to the nest, a large number of individuals are quickly recruited to the outer surface of the envelop. Second, a number of these individuals immediately fly and attack the intruder or predator. Alarm recruitment has been demonstrated experimentally in *Polybia occidentalis* (Jeanne 1981, 1991). The workers' venom contains a pheromone that immediately recruits a large number of adults, which move to the surface of the envelope. Outside the nest, the attack occurs immediately against dark-colored objects that move close to the nest. The chemical alarm communication certainly occurs in most of the swarming wasps, but it is still necessary to investigate whether it occurs in colonies with low numbers of individuals (Jeanne 1981, 1991). The alarm and defense behavior of the nest in *Synoeca surinama* (Linnaeus) is coordinated by visual factors and by odors, which are recognized by the individuals of the same colony. In the presence of intruders, these factors together incite and produce the wasp attack (Castellón 1981).

6.2.2 Chemical Communication During Migration

Communication during migration depends on the chemical signals left by the scout workers on different substrates indicating a direction to follow (Naumann 1975; Jeanne 1975; Forsyth 1978; West-Eberhard 1982; Mateus 2005; Mateus 2011). For migration to occur, three major challenges need to be overcome. First, the scouts select a location for the construction of the new nest. Second, these workers create a "trail" of chemical signals, left on prominent objects (such as leaves) along the way to the new site. Finally, the rest of the colony follows the chemical trail to reach the site of the new nest (Sonnentag and Jeanne 2009; Mateus 2011).

The chemical communication during swarming in Epiponini seems to be related to a gland present in the fifth gastral sternite. Reports of the presence of this gland were made by Richards (1971) and later studied by Jeanne and Post (1982) and Jeanne et al. (1983). The "dragging behavior", which consists of rubbing the gaster on a substrate to make the chemical pathway that is followed by the rest of the colony during migration, was first described by Naumann (1975) and observed in species of the genera *Agelaia*, *Angiopolybia*, *Leipomeles,* and *Polybia*. Jeanne (1975, 1981) experimentally confirmed the findings of Naumann (1975) observing the wasps followed an artificial trail made with substances removed from the gland of the fifth gastral sternite. However, this seems not to be the only strategy used. Studies with *Apoica pallens* (Fabricius) (Howard et al. 2002) showed a different strategy of chemical communication during the swarm. This species does not make a chemical trail on a substrate and does not previously select the location of the new nest. After leaving the old nest, the population forms clusters in leaves of plants, and the recruitment is done when the wasps elevated the abdomen and exposes the base of the fifth and sixth sternites. This behavior suggests that the wasps are emitting a pheromone in the air ("calling display") to group the population and lead it to a site to build the new nest (Howard et al. 2002).

In *Parachartergus fraternus*, although the presence of scouts doing "dragging behavior" during migration has been observed by some authors (Jeanne et al. 1983; Smith et al. 2002), no evidence of external glands was found in the gastral sternites (Jeanne et al. 1983). In this species, the communication of the end point of the chemical pathway was established by dozens of scouts, creating an area of easy recognition due to the strong odor of the substance applied to the substrate. In the moments that preceded the migration, it was observed that the number of scouts sprinkling venom on the substrate of the new nest site is high. Simultaneously, in the old nest, the scouts who returned there make lateral movements with the abdomen ("dragging behavior") in the substrate and oral contacts to signal and stimulate the population to migrate. The strategy of marking the place of the new nest by sprinkling venom possibly prevents the action of possible predators until the colony is established (Mateus 2005, 2011). Venom spraying is well known in the *Parachartergus* (Jeanne et al. 1995; West-Eberhard 1982), which has specialized muscles in the venom reservoir (Maschwitz and Kloft 1971). Chemical communication during migration in the Epiponini has been described as very efficient, since the loss of individuals during the process has been reported as low. The main causes of loss are related to predation or to the fact that individuals are newly emerged and not fully capable (Bouwma et al. 2003; Mateus 2011).

There are two distinct types of migration in Epiponini, "diffuse swarms" and "clumped swarms". For both types, use of chemical communication as a signal of the path to the new nest was observed for all species studied (West-Eberhard 1982). In diffuse swarms, observed in *Agelaia areata* (Say) (Jeanne 1975), *Synoeca surinama* (West-Eberhard 1982), and *Parachartergus fraternus* (Mateus 2011), the wasps spread over an area of more than 100 m in length by 20–50 m in width, in search of the chemical track. In the clumped swarm, several clusters are formed on a migration route, and, as the population advances along the migration path using the chemical trail, the clusters disintegrate behind the advance (West-Eberhard 1982). This type of migration was observed in *Parachartegus apicalis* (Fabricius), *Polybia ignobilis* (Haliday), *Polybia raui* Bequaert, *Polybia occidentalis* (Forsyth 1981), and *Polybia sericea* (Olivier) (Jeanne 1981).

6.2.3 Recruitment for Food Forage

Wasps can be considered as the great flying hunters and gatherers of the world of social insects. As far as we know, all social wasps are omnivorous, feeding on both animal proteins and sugar-rich carbohydrate sources. The sources of foraged proteins are predominantly living arthropods, which provide nutrients for immature growth. Carbohydrate sources serve primarily as an energy source for adults (Jeanne and Taylor 2009). Unlike other social Hymenoptera, bees and ants, in wasps there is no sophisticated communication system to inform nest mates of location of the source or when and where to forage for food. Thus, the foragers of social wasps seem to act individually (Jeanne and Taylor 2009).

Some hypotheses have been suggested by Jeanne et al. (1995). One is that wasps do not have sufficient genetic variability to give rise to structures or behavior to inform nest mates about the location of resources, which seems unlikely. Swarm-founding wasps exhibited sufficient genetic variability for the evolution of trail pheromones, which guide swarms to new nests (Jeanne 1981, 1991), a mechanism that should probably be able to function in the recruitment of nest mates to food sources. The lack of a signal-based recruitment system may be due also to a social constraint. A critical colony size may be necessary for the benefits of recruitment to outweigh the costs (Beckers et al. 1989), and it can be argued that the number of foragers in most social wasp species is too low to effectively defend food sources against competitors such as some species of ants and stingless bees. A third possibility is that ecological factors may make recruitment an unproductive strategy in social wasps. For recruitment to be worth it, the resources need to be clustered in space and persist for considerable periods of time. However, according to Jeanne and Taylor (2009), it may be premature to generalize that all social wasps do not recruit for food sources, since only a few species have been studied carefully and there are many other species to be studied.

6.2.4 Division of Labor

The division of labor is a central theme in the study of social insects (Garcia and Noll 2013). The ecological success of these groups can be credited to their ability to coordinate their colonies for feeding, breeding, cleaning, and nesting (Beshers and Fewell 2001; Wilson 1979; Wilson and Hölldobler 2005). In ants and termites, the division of labor among castes is regulated by ontogenetic and physiological mechanisms (Miura 2005). In Epiponini, as in bees, the division of labor is associated with age polyethism[1] and specialization of tasks (Chavarría and Noll 2013; Garcia and Noll 2013).

Workers of independent foundation species begin to forage at the age of several days and combine foraging activities with activities within the nest throughout their lives. Foragers bring food (nectar and prey) and material (vegetable matter and water) to build the nest, and division of labor among them may occur by preference for a particular type of activity (polyethism) and in the way of processing the load (division of tasks). The forager can perform all the necessary handling alone or divide its load of material totally or partially with other members of the colony. The frequency of load partitioning increases with colony growth (Post et al. 1988; Rusina 2006), and the food division is often more frequent than the construction material division (Rusina et al. 2011). In *Polistes fuscatus* (Fabricius), for example, three functional groups are present: (1) workers who prefer to hunt than to build and tend to pass protein to other individuals, (2) workers who deliver construction material and do not pass prey to other workers, and (3) non-forage workers who engage exclusively in activities within the nest (Post et al. 1988). A similar pattern was found for *P. dominula* (Christ), *P. nimpha* (Christ), and *P. gallicus* (Linnaeus). In the latter two species, however, the constructors were involved in establishing and maintaining the dominance structure of the workers (Rusina 1999, 2006). Dominant workers in *Mischocyttarus mastigophorus* Richards collect building material and obtain protein food from prey forage more frequently than subordinate individuals (O'Donnell 1998b).

In Epiponini, the high level of social organization resulted in the presence of age-related polyethism and in the specialization of foragers. Handling of the delivered material (building material, water, prey, and nectar) is divided into two sub-tasks: collection and use. Foragers of the two functional groups bring food (prey and nectar) and construction material (wood pulp and water) to the nest and pass to other workers. A forager can deliver a specific type or both types of material throughout its active period. If a forager changes to a different type of load, the change usually occurs within the same group (food or nest material) (Rusina et al. 2011). Construction activity in these wasps is not only subdivided into delivery of pulp and construction but also involves the transportation of water by specialized workers. In addition, pulp delivered by foragers is generally too large to be used directly in the construction, so they have to pass at least part of the material to the workers in the nest who then distribute the material among themselves (Jeanne 1986; O'Donnell and Jeanne 1990; Karsai and Wenzel 1998). According to Jeanne (2003), the main difference between the independent and swarm-founding species is in the way of dealing with the material collected and in the fact that the foragers transfer the material to another worker (and continue to forage), allowing the development of the age

[1] The phenomenon whereby an animal shows different forms of behavior at different ages. In social insects, certain members of the colony may perform different tasks as they get older. In Epiponini genera *Polybia*, *Protopolybia*, and *Agelaia*, for example, young and middle-aged workers perform nest tasks (building, brood care, nest maintenance, defense), while old workers forage.

polyethism and a more complex organization of the colony. In general, young workers perform tasks within the nest, such as cleaning and caring for the offspring. Meanwhile, the middle-aged workers are concerned with the construction and repair of the nest. Older workers, in turn, act as foragers, carrying material for construction and water and food for the colony, in addition to defending the nest (O'Donnell 2001). Studies with *Polybia*, *Protopolybia,* and *Agelaia* found that young and middle-aged workers perform tasks in the nest such as construction, care of the offspring, nest maintenance, and defense, while older workers forage (Simões 1977; Forsyth 1978; Jeanne et al. 1988, 1992). Chavarría and Noll (2013) observed a similar number of young and old female workers performing tasks such as cell inspection, construction, and foraging in *Metapolybia miltoni* Andena and Carpenter colonies. However, younger workers tend to forage less than the older ones (Chavarría and Noll 2013).

Specialization, that is, the presence of workers who perform a single type of task, is rare in most social wasps (Robinson 1992; Sendova-Franks and Franks 1999; O'Donnell 1998c; Karsai and Wenzel 2000; Beshers and Fewell 2001; Johnson 2003). Karsai and Wenzel (2000) did not find specialization in colonies of *Metapolybia aztecoides* and *M. mesoamerica* Smethurst and Carpenter. The same was observed by Chavarría and Noll (2013) for *Metapolybia miltoni.* Mateus (2005) also found flexibility in performing tasks in *Parachartergus fraternus*, where a worker who acted as a scout during the pre-swarm was subsequently observed laying eggs and, when dissected, the presence of ovaries developed with mature oocytes was found. Other scouts of the same nest performed construction tasks after nest establishment (Mateus 2005). In *Polybia occidentalis* and *Metapolybia* sp., there was greater stability in the preference for collection of a type of material when colonies were large (Jeanne 1986; Karsai and Wenzel 1998, 2000). Data obtained for *Polybia aequatorialis* (Zavattari) indicate that workers with specialization for the same activity have a close genetic relationship (O'Donnell 1996, 1998c). When there is little variation of age in the population, the workers should be allocated to tasks according to the needs of the colony, not being very relevant to the age of the individuals (Chavarría and Noll 2013). On the other hand, in colonies with individuals of different ages, polyethism seems to be more important in the delimitation of tasks, as demonstrated in the works of Jeanne et al. (1988, 1992). In short, the workers' choice of task depends on the circumstances of the colony, and the workers can be allocated to this or that task according to need (Karsai and Wenzel 2000).

Workers in any insect society retain some degree of behavioral flexibility, which is decisive for the survival of the colony in changing situations (Robinson 1992). Likewise, caste flexibility is an important factor for colony survival in the swarm-founding wasps, allowing these insects to respond efficiently to changes that may occur (Chavarría and Noll 2013). As evidenced by Noll and Wenzel (2008), cast dimorphism evolved at least eight times in Epiponini, and social organization derives directly from an ancestor with incipient caste dimorphism in most taxa. Given this fact, general patterns are insufficient to explain the different strategies adopted by the species of this tribe (Beshers and Fewell 2001; Chavarría and Noll 2013).

6.2.5 *Colony Defense*

"I suspect that social wasps pay much more attention to ants in their environment than we think". The phrase by Jeanne (1991) reinforces the idea that in the tropics, the ants represent the most important predators of the colonies of social wasps (Richards and Richards 1951). Predation by these insects may have strongly influenced the evolution of these wasps (Jeanne 1979). The rapid abandonment of the nest and swarm, the complex architecture of nests, and the presence of breeds without clear morphological differences suggest evolutionary adaptations against ant predation as well as optimization of work (Jeanne 1975; Wenzel 1998). While independent-founding species (*Polistes* and *Mischocyttarus*) rely heavily on secreted chemical repellents to keep ants away from their offspring, the swarming wasps depend primarily on the envelope surrounding the combs, coupled with the presence of workers guarding the nest entrance (Jeanne 1975). In addition, in the face of an ant attack, abandoning the nest by the rapid migration of the adult population is certainly an important and unique strategy in Epiponini (Mateus 2005).

In addition to envelope protection, workers actively guard the nest (Jeanne 1975, 1991). When the ants get too close, the wasps can resort to a series of behaviors to drive away the invaders. *Protopolybia exigua* and *Polybia occidentalis*, for example, exhibit "wing buzzing" and "burst" behaviors, where female workers swing their bodies up and backward by buzzing their wings rapidly and producing a snap (Chadab 1979). In *P. exigua*, this behavior is shown when visual or olfactory stimuli, as well as the odor of formic acid, are detected by the wasps and act as a warning signal to the companions who can quickly evacuate the nest if the ants reach this one (Chadab 1979). When an ant reaches a *P. occidentalis* nest, it can be stung by a worker who immediately uses the jaws to trap and carry the ant away from the nest (Jeanne 1991). As described by Chadab (1979), buzzing and burst behaviors cause small bursts of air directed at the ants whose antennae curve at each flight wrist. These pulses paralyze the ants momentarily, and the display of these behaviors persistently by the wasps results in the retreat of the ants (Jeanne 1991). Workers of *Clypearia sulcata* (de Saussure), *Metapolybia aztecoides,* and *Polybia occidentalis* use the jaws to scrape the substrate of the nest where the ants passed in order to remove the pheromone trail left by them (Chadab 1979; West-Eberhard 1989b; Jeanne 1991). Thus, the chances of ants reaching the nest in the future are reduced (Jeanne 1991).

Although most species bet on the active defense of the nest against invasive ants, a different strategy has evolved in some species. *Nectarinella xavantinensis* Mateus and Noll, *N. championi* (Dover), and *Leipomelles dorsata* (Fabricius) protect access to the nest with the placement of "sticky pillars" in areas near the entrance of the nest, apparently produced from an oral secretion (Jeanne 1991; Wenzel 1991; Mateus and Noll 1997; Carpenter and Marques 2001). Wasps of these species are quite small, and defending the nest actively against many ant species would be impossible (Jeanne 1991). Envelope protection, active defense, and building sticky traps can be very effective against small ants, but they are not enough against a group of ants, in particular the army ants. The majority of social wasps seem to have no effective defense against army ants, and the tactic against their arrival is the immediate abandonment of the nest by the entire adult population so that they do not end up as prey of the invaders, which will occur with the abandoned larvae and pupae (Jeanne 1991). In some species of Polistinae, however, ways of escaping of predation by these ants have evolved. One of them is to build very high nests in the canopy, as do *Agelaia areata*, *Epipona tatua* (Cuvier), and *Chartergus metanotalis* Richards (Jeanne 1991; Carpenter and Marques 2001). Other species with more resistant envelopes like *Synoeca septentrionalis* Richards, *S. chalibea* de Saussure, *Polybia emaciata*, and *P. jurinei* de Saussure prevent the entry of army ants using the body itself to block the entrance of the nest.

Curiously, despite the antagonistic relations between the ants and the social wasps, some species developed interspecific associations. *Polybia rejecta* (Fabricius), a very aggressive social wasp, often builds its nest in association with colonies of the ant *Azteca* (Richards 1978; Somavilla et al. 2012). This association may be one of the few defenses that tropical wasps have against invasions by ants of the army ants (Richards and Richards 1951). Similarly, wasps exhibit aggressive behavior when the ant colony is disturbed, protecting the ants primarily against predatory mammals such as anteaters (Virgínio et al. 2015). *Synoeca chalybea* can also construct nests in association with ants of the genus *Azteca* (Souza et al. 2013).

In addition to barring the ants, the presence of the envelope also seems to be an effective defense against parasites. When these are found, they are far less numerous when compared to the large infestations that may occur in non-enveloped nests of independent-founding species. Forsyth (1978) analyzed 141 colonies of *Metapolybia azteca* and *Polybia occidentalis* finding parasites in only 7% of them and less than 1% of offspring infestation. On the other hand, Nelson (1968) observed that about 60% of the colonies of *Polistes metricus* analyzed by him were parasitized by the moth *Chalcoela iphitalis*, with some nests reaching 90–95% infestation of the offspring.

The most famous defense mechanism of social wasps is undoubtedly the painful sting. In spite of the bad reputation, the sting apparatus in its origin has as primary function the capture of prey (Macalintal and Starr 1996). However, they assumed an important defense role against bigger enemies, especially in social lineages (Macalintal and Starr 1996). The behavior of stinging would have evolved due to the predation pressure exerted by vertebrates attracted by the increase in colony size, particularly in the tropics, in open and seasonally dry forests (Starr 1985; Carpenter and Marques 2001). This form of defense is undoubtedly quite effective in discouraging potential predators, although, it is not uncommon for wasp colonies to be attacked by vertebrates such as birds, bats, and primates (Jeanne 1975, 1991). Nevertheless, different mechanisms were adopted by other species in an attempt to minimize the risk of being attacked by these animals. Species with small colonies construct nests camouflaged for the purpose of hindering the visualization by vertebrates and parasites. *Leipomeles dorsata*, for example, constructs its nest at the bottom of a broad leaf and covers the envelope with green details in order to mimic plant venation (Williams 1928; Richards 1978; Jeanne 1991). Other species build large and sturdy nests, inaccessible to most vertebrates.

In *Apis mellifera* Linnaeus, defense and foraging activities are carried out by different groups of workers (Breed et al. 1990), that is, the workers who act in the defense of the colony are specialized in this task. In Epiponini, Jeanne et al. (1992) analyzing colonies of *Polybia occidentalis* did not find a subgroup of specialized workers in this task, concluding that the wasps that defended the nest also performed other activities inside the nest and of foraging, for example (Fig. 6.5a). In fact, there is a positive correlation since forager workers are more likely to act in defense than non-forager ones (Jeanne et al. 1992). Still for *P. occidentalis*, the likelihood of a wasp displaying defense response increases significantly among workers aged between 8 and 12 days, stabilizing after the 13th day (Jeanne et al. 1992; Fig. 6.5b). During a mechanical stimulus, young wasps also leave the nest in response to the attack signal, but only the older workers effectively attack the intruder. This fact makes it clear that the behaviors of "leaving the nest" and "attacking" represent different responses (Jeanne et al. 1992) and that the participation of a large number of individuals in the exit response suggests that this behavior functions as an aposematic sign to the predator (Jeanne et al. 1992).

Thanks to its defensive potential, the sting apparatus is considered a key factor that favored the origin of social behavior several times within Aculeata. Added to this is the fact that only females act as workers in the societies of wasps, bees, and ants (Silveira and Silveira 1994). An evolutionary analysis of the sting apparatus in Vespidae showed that some variations in the characteristics of this structure are related to the highly social way of life (Silva et al. 2014). In addition to the defense function, the use of the sting apparatus in a phylogenetic reconstruction supports the unique origin of eusociality, which would have occurred in the ancestor of Vespinae + Polistinae (Silva et al. 2014).

6.3 General Conclusion

The peculiar features presented by Neotropical social wasps, specially the swarm-founding Epiponini, resulted in a great diversity of species. Among the representatives of this tribe, all stages of caste differentiation are present, from castes morphologically very distinct to forms in which there are no external differences. Physiological differentiation also follows this rule, giving rise, in some species, to

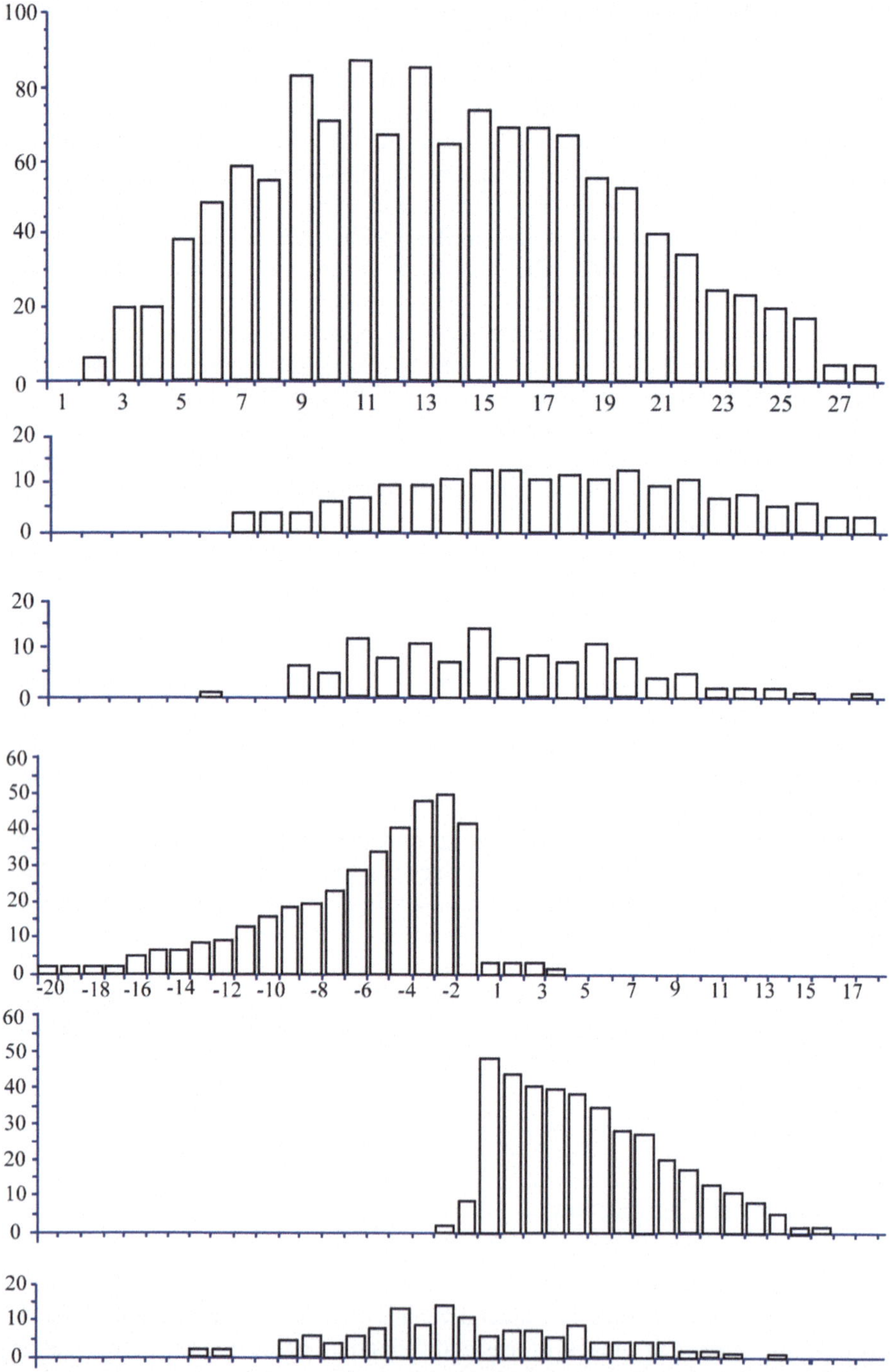

Fig. 6.5 Defense behavior according to age. (**a**) Frequency of task performance according to the age of the worker. Number of tasks related to nest maintenance, foraging trips, and defensive responses. (**b**) Frequency of task performance as a function of relative age (age is relative to the transition of each individual from the work in the nest to the foraging). (Modified from Jeanne et al. 1992)

females with ovary developed but not fertilized, so-called intermediates. Because of polygyny, many females are able to lay eggs, tolerating each other, and the egg layers are tested and occasionally removed by sterile workers. Chemical communication guides the swarms and defense of the nest, and the tasks performed inside and outside the nest are divided according to the age of the wasps.

References

Baio MV, Noll FB, Zucchi R, Simões D (1998) Non-allometric caste differences in *Agelaia vicina* (Hymenoptera, Vespidae, Epiponini). Sociobiology 32:465–476

Baio MV, Noll FB, Zucchi R (2003a) Morphological caste differences, variation according to colony cycle, and non-sterility of workers in *Brachygastra augusti* (Hymenoptera, Vespidae, Epiponini), a Neotropical swarm-founding wasp. J N Y Entomol Soc 111:243–253

Baio MV, Noll FB, Zucchi R (2003b) Shape differences rather than size differences between castes in the Neotropical swarm-founding wasp *Metapolybia docilis* (Hymenoptera: Vespidae, Epiponini). BMC Evol Biol 3:1–9

Beckers R, Goss S, Deneubourg JL, Pasteels JM (1989) Colony size, communication and ant foraging strategy. Psyche 96:239–256

Beshers SN, Fewell JH (2001) Models of division of labor in social insects. Annu Rev Entomol 46:413–440

Bhadra A, Mitra A, Deshpande S, Chandrasekhar K, Naik D, Hefetz A, Gadagkar R (2010) Regulation of reproduction in the primitively eusocial wasp *Ropalidia marginata*: on the trail of the queen pheromone. J Chem Ecol 36:424–431

Bourke A (1999) Colony size, social complexity and reproductive conflict in social insects. J Evol Biol 12:245–257

Bouwma AM, Bouwma PE, Nordheim EV, Jeanne RL (2003) Adult mortality rates in young colonies of a swarm-founding social wasp (*Polybia occidentalis*). J Zool 260:11–16

Breed MD, Robinson GE, Page RE Jr (1990) Division of labor during honey bee colony defense. Behav Ecol Sociobiol 27:395–401

Carpenter JM (1991) Phylogenetic relationships and the origin of social behavior. In: Ross KG, Matthews RW (eds) The social biology of wasps. Cornell University Press, Ithaca, pp 7–32

Carpenter JM (1993) Biogeographic patterns in the Vespidae (Hymenoptera): two views of Africa and South America. In: Goldblatt P (ed) Biological relationships between Africa and South America. Yale University Press, New Haven, pp 139–155

Carpenter JM (1997) A note on the names of paper wasp tribes (Insecta: Hymenoptera: Vespidae). Ibaraki Univ Nat Hist Bull 1:15–16

Carpenter JM, Marques OM (2001) Contribuição ao Estudo de Vespídeos do Brasil (Insecta, Hymenoptera, Vespoidea, Vespidae). [CD - ROM]. Universidade Federal da Bahia, série Publicações Digitais, 2, Cruz das Almas

Carpenter JM, Ross KG (1984) Colony composition in four species of Polistinae from Suriname, with a description of the larva of *Brachygastra scutellaris* (Hymenoptera, Vespidae). Psyche 91:237–250

Castellón EG (1981) Alarma e defesa no ninho de *Synoeca surinama* (L) (Hymenoptera: Vespidae). Acta Amazônica 11:377–382

Chadab R (1979) Army-ant predation on social wasp. PhD Dissertation, University of Connecticut, Storrs

Chavarría L (2009) Comportamento, atividade e interações sociais entre raonhas e operárias de *Metapolybia docilis* (Vespidae: Polistinae Epiponini). Dissertação de mestrado em Ciências (Entomologia) - Universidade de São Paulo, Ribeirão Preto, São Paulo, 132 f

Chavarría L (2013) Sobre a produção e seleção de rainhas em diferentes fases do ciclo colonial em Epiponini (Vespidae: Polistinae). Tese de doutorado em Ciências (Entomologia) - Universidade de São Paulo, Ribeirão Preto, São Paulo, 139 f

Chavarría L, Noll FB (2013) Age polyethism in the swarm-founding wasp *Metapolybia miltoni* (Andena and Carpenter) (Hymenoptera: Vespidae; Polistinae, Epiponini). Sociobiology 60:214–216

Dapporto L, Santini A, Dani FR, Turillazzi S (2007) Workers of a *Polistes* paper wasp detect the presence of their queen by chemical cues. Chem Senses 32:795–802

Evans HE, West-Eberhard MJ (1970) The wasps. University of Michigan Press, Michigan, 265p

Fletcher DJC, Ross KG (1985) Regulation of reproduction in eusocial Hymenoptera. Annu Rev Entomol 30:319–343

Forsyth AB (1978) Studies on the behavioral ecology of polygynous social wasps. Dissertation, Harvard University

Forsyth AB (1981) Swarming activity of polybiine social wasps (Hymenoptera: Vespidae: Polybiini). Biotropica 13:93–99

Garcia ZJ, Noll FB (2013) Age and morphological changes in the Epiponini wasp *Polybia paulista* Von Ihering (Hymenoptera: Vespidae). Neotrop Entomol 42:293–299

Gastreich KR, Strassmann JE, Queller DC (1993) Determinants of high genetic relatedness in the swarm-founding wasp, *Protopolybia exigua*. Ethol Ecol Evol 5:529–539

Goodnight KF, Strassmann JE, Klinger CJ, Queller DC (1996) Single mating and its implications for kinship structure in a multiple-queen wasp, *Parachartergus colobopterus*. Ethol Ecol Evol 8:91–198

Hamilton WD (1964a) The genetical evolution of social behaviour I. J Theor Biol 7:1–16

Hamilton WD (1964b) The genetical evolution of social behaviour I. J Theor Biol 7:17–52

Hamilton WD (1972) Altruism and related phenomena, mainly in the social insects. Annu Rev Ecol Syst 3:193–232

Hart AG, Ratnieks FLW (2005) Crossing the taxonomic divide: conflict and its resolution in societies of reproductively totipotent individuals. J Evol Biol 18:383–395

Hastings MD, Queller DC, Eischen F, Strassmann JE (1998) Kin selection, relatedness and worker control of reproduction in a large-colony epiponine wasp, *Brachygastra mellifica*. Behav Ecol 9:573–581

Herman RA, Queller DC, Strassmann JE (2000) The role of queens in colonies of the swarm-founding wasp *Parachartergus colobopterus*. Annu Behav 59:841–848

Howard KJ, Smith AR, O'Donnell S, Jeanne RL (2002) Novel method of swarm emigration by the epiponine wasp, *Apoica pallens* (Hymenoptera Vespidae). Ethol Ecol Evol 14:365–371

Hughes CR, Beck MO, Strassmann JE (1987) Queen succession in the social wasp *Polistes annularis*. Ethology 76:124–132

Hughes CR, Queller DC, Negrón-Stomayor JA, Strassmann JE, Solis C, Gastreich KR (1993) The maintenance of high genetic relatedness in multi-queen colonies of social wasps. In: Keller L (ed) Queen number and sociality in insects. Oxford University Press, New York, pp 153–170

Hunt JH, Schmidt DK, Mulkey SS, Williams MA (1996) Caste dimorphism in *Epipona guerini* (Hymenoptera: Vespidae): further evidence for larval determination. J Kansas Entomol Soc 69:362–369

Hunt JH, O'Donnell S, Chernoff N, Brownie C (2001) Observations on two neotropical swarm-founding wasps, *Agelaia yepocapa* and *A. panamaensis* (Hymenoptera: Vespidae). Ann Entomol Soc Am 94:555–562

Hunt JH, Kensinger BJ, Kossuth JA, Henshaw MT, Norberg K, Wolschin F et al (2007) A diapause pathway underlies the gyne phenotype in *Polistes* wasps, revealing an evolutionary route to caste-containing insect societies. Proc Natl Acad Sci U S A 104:14020–14025

Jeanne RL (1975) The adaptiveness of social wasp nest architecture. Q Rev Biol 50:267–287

Jeanne RL (1979) A latitudinal gradient in rates of ant predation. Ecology 60:1211–1224

Jeanne RL (1981) Chemical communication during swarm emigration in the social wasp *Polybia sericea* (Olivier). Anim Behav 29:102–113

Jeanne RL (1986) The organization of work in *Polybia occidentalis*: costs and benefits of specialization in a social wasp. Behav Ecol Sociobiol 19:333–341

Jeanne RL (1991) The swarm-founding Polistinae. In: Ross KG, Matthews RW (eds) The social biology of wasps. Cornell University Press, Ithaca, pp 191–231

Jeanne RL (1996) Non-allometric queen-worker dimorphism in *Pseudopolybia difficilis*. J Kansas Entomol Soc 69:370–374

Jeanne RL (2003) Social complexity in the Hymenoptera, with special attention to the wasps. In: Kikuchi T, Azuma N, Higashi S (eds) Genes, behaviors and evolution of social insects, proceedings of the 14th congress of the IUSSI. Hokkaido University Press, Sapporo, pp 81–130

Jeanne RL, Fagen R (1974) Polymorphism in *Stelopolybia areata* (Hymenoptera, Vespidae). Psyche 81:155–166

Jeanne RL, Post DC (1982) Richards´ gland and associated cuticular modifications in social wasps of the genus *Polybia* (Hymenoptera: Vespidae: Polistinae: Polybiini). Insects Soc 29:280–294

Jeanne RL, Suryanarayanan S (2011) A new model for caste development in social wasps. Commun Integr Biol 4:373–377

Jeanne RL, Taylor BJ (2009) Individual and social foraging in social wasps. In: Jarau S, Hrncir M (eds) Food exploitation by social insects: ecological, behavioral, and theoretical approaches. CRC Press, Boca Raton, pp 53–73

Jeanne RL, Downing HA, Post DC (1983) Morphology and function of sternal glands in polistine wasps (Hymenoptera: Vespidae). Zoomorphology 103:149–164

Jeanne RL, Downing HA, Post DC (1988) Age polyethism and individual variation in *Polybia ocidentalis*, and advance eusocial wasp. In: Jeanne RL (ed) Interindividual behavioral variability in social insects. Westview Press, Boulder, pp 323–357

Jeanne RL, William NM, Yandell BS (1992) Age polyethism and defense in tropical social wasps (Hymenoptera: Vespidae). J Insect Behav 5:211–227

Jeanne RL, Graf CA, Yandell BS (1995) Non-size-based morphological castes in a social insect. Naturwissenschaften 82:296–298

Johnson BR (2003) Organization of work in the honeybee: a compromise between division of labor and behavioral flexibility. Proc R Soc Lond 270:147–152

Kardile SP, Gadagkar R (2002) Docile sitters and active fighters in paper wasps: a tale of two queens. Naturwissenschaften 89:176–179

Karsai I, Wenzel JW (1998) Productivity, individual-level and colony-level flexibility, and organization of work as consequences of colony size. Proc Natl Acad Sci U S A 95:8665–8669

Karsai I, Wenzel JW (2000) Organization and regulation of nest construction behavior in *Metapolybia* wasp. J Insect Behav 13:111–140

Macalintal EA, Starr CK (1996) Comparative morphology of the stinger in the social wasp genus *Ropalidia* (Hymenoptera: Vespidae). Mem Entomol Soc Wash 17:108–115

Maschwitz UWJ, Kloft W (1971) Morphology and function of the venom apparatus of insects-bees, wasps, ants, and caterpillars. In: Bücherl EE (ed) Venomous animals and their venoms. Academic Press, New York and London, pp 1–60

Mateus S (2005) Análise dos comportamentos envolvidos na organização social e no processo de enxameio de *Parachartergus fraternus* (Hymenoptera, Polistinae, Epiponini). Tese de doutorado em Ciências (Entomologia) – Universidade de São Paulo, Ribeirão Preto, São Paulo, Brasil. 160 f

Mateus S (2011) Observations on forced colony emigration in *Parachartergus fraternus* (Hymenoptera: Vespidae: Epiponini): New Nest Site Marked with Sprayed Venom. Psyche, Volume 2011, Article ID 157149, 8 pages

Mateus S, Noll FB (1997) *Nectarinella xavantinensis,* a new Neotropical social wasp (Hymenoptera: Vespidae; Polistinae). J N Y Entomol Soc 105:45–49

Mateus S, Noll FB, Zucchi R (1997) Morphological caste differences in Neotropical swarm-founding Polistinae wasps: *Parachartergus smithii* (Hymenoptera, Vespidae). J N Y Entomol Soc 105:129–139

Mateus S, Noll FB, Zucchi R (1999) Caste differences and related bionomic aspects of *Chartegellus communis*, a Neotropical swarm-founding Polistinae wasps (Hymenoptera: Vespidae: Polistinae: Epiponini). J N Y Entomol Soc 107:390–405
Mateus S, Noll FB, Zucchi R (2004) Caste flexibility and variation according to the Colony cycle in the swarm-founding wasps, *Parachartergus fraternus* (Gribodo) (Hymenoptera; Vespidae: Epipoini). J Kansas Entomol Soc 77:470–483
Miura T (2005) Developmental regulation of caste-specific characters in social-insect polyphenism. Evol Dev 7:122–129
Nascimento FS, Tannure-Nascimento IC, Zucchi R (2004) Behavioral mediators of cyclical oligogyny in the Amazonian swarm-fouding wasp *Asteloeca ujhelyii* (Vespidae, Polistinae, Epiponini). Insect Soc 51:17–23
Naumann MG (1970) The nesting behavior of P*rotopolybia pumila* in Panama (Hymenoptera: Vespidae). PhD Dissertation, University of Kansas, Lawrence
Naumann MG (1975) Swarming behavior: evidence for communication in social wasps. Science 189:642–644
Nelson JM (1968) Parasites and symbionts of nests of *Polistes* wasps. Annu Entomol Soc Am 61:1528–1539
Noda SCM, Shima SN, Noll FB (2003) Morphological and physiological caste differences in *Synoeca cyanea* (Hymenoptera, Vespidae, Epiponini) according to the ontogenetic development of the colonies. Sociobiology 41:547–570
Noll FB (2013) “Marimbondos”: a review on the neotropical swarm-founding polistines. Sociobiology 60:347–354
Noll FB, Wenzel JW (2008) Caste in the swarming wasps: “Queenless” societies in highly social insects. Biol J Linn Soc 93:509–522
Noll FB, Zucchi R (2000) Increasing caste differences related to life cycle progression in some neotropical swarm-founding polygynic polistine wasps (Hymenoptera, Vespidae, Epiponini). Ethol Ecol Evol 12:43–65
Noll FB, Zucchi R (2002) Castes and the influence of the colony cycle in swarm-founding polistine wasp (Hymenoptera, Vespidae, Epiponini). Insect Soc 48:01–13
Noll FB, Mateus S, Zucchi R (1996) Morphological caste differences in neotropical swarm-founding Polistinae wasps V: *Protopolybia exigua exigua* (Hymenoptera, Vespidae). J N Y Entomol Soc 104:62–69
Noll FB, Simões D, Zucchi R (1997a) Morphological caste differences in the Neotropical swarm-founding Polistinae wasps: *Agelaia n. multipicta* and *A. p. pallipes* (Hymenoptera, Vespidae). Ethol Ecol Evol 9:361–372
Noll FB, Zucchi R, Mateus S (1997b) Morphological caste differences in the Neotropical swarm-founding and polygynous Polistinae wasp *Polybia scutellaris*. Stud Neotrop Fauna Environ 32:76–80
Noll FB, Yamane S, Zucchi R (2000) Morphological caste differences in the Neotropical swarm-founding polistine wasps. IX. *Polybia (Myrapetra) occidentalis* (Hymenoptera, Vespidae). Entomol Sci 3:491–497
Noll FB, Wenzel JW, Zucchi R (2004) Evolution of caste in Neotropical swarm-founding wasps (Hymenoptera: Vespidae; Epiponini). Am Mus Novit 3467:1–24
O’Donnell S (1996) RAPD markers suggest genotypic effects on forager specialization in a eusocial wasp. Behav Ecol Sociobiol 38:83–88
O’Donnell S (1998a) Reproductive caste determination in eusocial wasps (Hymenoptera: Vespidae). Annu Rev Entomol 43:323–346
O’Donnell S (1998b) Dominance and polyethism in the eusocial wasp *Mischocyttarus mastigophorus* (Hymenoptera: Vespidae). Behav Ecol Sociobiol 43:327–331
O’Donnell S (1998c) Genetic effects on task performance, but not on age polyethism, in a swarm-founding eusocial wasp. Anim Behav 55:417–426
O’Donnell S (2001) Worker biting interactions and task performance in a swarm-founding eusocial wasp (*Polybia occidentalis*, Hymenoptera: Vespidae). Behav Ecol 12:353–359
O’Donnell S, Jeanne RL (1990) Forager specialization and the control of nest repair in *Polybia occidentalis* Olivier. Behav Ecol Sociobiol 27:359–364

Pickett M, Carpenter JM, Wheeler WC (2006) Systematics of *Polistes* (Hymenoptera: Vespidae), with a phylogenetic consideration of Hamilton's haplodiploidy hypothesis. Ann Zool Fenn 43:390–406

Platt TG, Queller DC, Strassmann JE (2004) Aggression and worker control of caste fate in a multiple-queen wasp *Parachartergus colobopterus*. Anim Behav 67:1–10

Post DC, Jeanne RL, Erickson HE (1988) Variation in behavior among workers of the primitively social wasp *Polistes fuscatus variatus*. In: Jeanne RL (ed) Interindividual behavior variability in social insects. Boulder, Westview, pp 283–321

Queller DC, Hughes CR, Strassmann JE (1988) Genetic relatedness in colonies of tropical wasps with multiple queen. Science 242:1155–1157

Queller DC, Negrón-Stomayor HCR, Strassmann JE (1993) Queen number and genetic relatedness in a Neotropical wasp, *Polybia occidentalis*. Behav Ecol 4:7–13

Richards OW (1971) The biology of the social wasps (Hymenoptera: Vespidae). Biol Rev 46:483–528

Richards OW (1978) The social wasps of the Americas excluding the Vespinae. British Museum (Natural History), London, 580p

Richards OW, Richards MJ (1951) Observations on the social wasps of South America (Hymenoptera, Vespidae). Trans R Entomol Soc Lond 102:1–170

Robinson GE (1992) Regulation of division of labor in insect societies. Annu Rev Entomol 37:637–665

Röseler PF (1991) Reproductive competition during colony establishment. In: Ross KG, Matthews RW (eds) The social biology of wasps. Cornell University, Ithaca, pp 309–335

Rusina LY (1999) Task partitioning in Polistine wasps (Hymenoptera, Vespidae). Vestnik Zool 33:105–108

Rusina LY (2006) Polistine wasps in natural and anthropogenic landscapes of the lower Dnieper area. Kherson Gos. Univ. [in Russian]

Rusina LY, Firman LA, Yu RY, Starr CK (2011) Pulp partitioning and worker specialization in Polistine wasps (Hymenoptera, Vespidae, Polistinae). Entomol Rev 91:820–827

Sakagami SF, Zucchi R, Yamane S, Noll FB, Camargo JMF (1996) Morphological caste differences in *Agelaia vicina*, the neotropical swarm-founding polistine wasp with the largest colony size among social wasps (Hymenoptera: Vespidae). Sociobiology 28:207–223

Sendova-Franks AB, Franks NR (1999) Self- assembly, self- organization and division of labour. Phil Trans R Soc Lond 1388:1395–1405

Shima SN, Yamane S, Zucchi R (1994) Morphological caste differences in some Neotropical swarm-founding Polistinae wasps I. *Apoica flavissima* (Hymenoptera, Vespidae). Jpn J Entomol 62:811–822

Shima SN, Yamane S, Zucchi R (1996a) Morphological caste differences in some Neotropical swarm-founding Polistinae wasps. II. *Polybia dimidiata* (Hymenoptera; Vespidae). Jpn J Entomol 64:131–144

Shima SN, Yamane S, Zucchi R (1996b) Morphological caste differences in some Neotropical swarm-founding Polistinae wasps. III. *Protonectarina sylveirae* (Hymenoptera; Vespidae). Bull Fac Educ Ibaraki Univ 45:57–67

Shima SN, Noll FB, Zucchi R, Yamane S (1998) Morphological caste differences in the Neotropical swarm-founding Polistinae wasps IV: *Pseudopolybia vespiceps*, with preliminary considerations on the role of intermediate females in the social organization of the Epiponini (Hymenoptera; Vespidae). J Hymenopt Res 7:280–295

Shima SN, Noll FB, Zucchi R (2000) Morphological caste differences in the Neotropical swarm-founding Polistinae wasp, *Brachygastra lecheguana* (Hymenoptera: Vespidae, Polistinae, Epiponini). Sociobiology 36:41–52

Silva M, Noll FB, Carpenter JM (2014) The usefulness of the sting apparatus in phylogenetic reconstructions in vespids, with emphasis on the Epiponini: more support for the single origin of eusociality in the Vespidae. Neotrop Entomol 43:134–142

Silveira O, Silveira A (1994) Comparative morphology of skeletal parts of the sting apparatus in Neotropical Polistinae social wasps (Hymenoptera, Vespidae, Polistinae). Sociobiology 25:295–327

Simões D (1977) Etologia e diferenciação de casta em algumas vespas sociais (Hymenoptera, Vespidae). Tese de Doutorado em Ciências – Universidade de São Paulo, Ribeirão Preto, São Paulo, 169 f
Simões D, Noll FB, Zucchi R (1996) Duration of *protopolybia exigua* nest and related aspects as influence by phorid fly infestation (Vespidae, Polistinae, Epiponini). Sociobiology 28:212–229
Smith AR, O'Donnell S, Jeanne RL (2002) Evolution of swarm communication in eusocial wasps (Hymenoptera: Vespidae). J Insect Behav 15:751–764
Somavilla A, Oliveira ML, Silveira OT (2012) Guia de identificação dos ninhos de vespas sociais (Hymenoptera, Vespidae, Polistinae) na Reserva Ducke, Manaus, Amazonas, Brasil. Rev Bras Entomol 56:405–414
Sonnentag PJ, Jeanne RL (2009) Initiation of absconding swarm emigration in the social wasp *Polybia occidentalis*. J Insect Sci 9, article 11
Souza MM, Pires EP, Prezoto F (2013) Nidification of *Polybia rejecta* (Hymenoptera: Vespidae) associated to *Azteca chartifex* (Hymenoptera: Formicidae) in a fragment of Atlantic Forest, in the state of Minas Gerais, southeastern Brazil. Biota Neotrop 13:390–392
Spradbery JP (1991) Evolution of queen number and queen control. In: Ross KG, Matthews RW (eds) The social biology of wasps. Cornell University Press, Ithaca, pp 336–388
Starr CK (1985) Enabling mechanisms in the origin of sociality in the Hymenoptera - the sting's the thing. Ann Entomol Soc Am 78:836–840
Starr CK (1990) Holding the fort: colony defense in some primitively social wasps. In: Evans DL, Schmidt JO (eds) Insect defenses, adaptive mechanisms and strategies of prey and predators. State University of New York Press, New York, pp 421–463
Strassmann JE, Queller DC, Solis CR, Hughes CR (1991) Relatedness and queen number in the neotropical wasp, *Parachartergus colobopterus*. Ann Behav 42:461–470
Strassmann JE, Gastreich KR, Queller DC, Hughes CR (1992) Demographic and genetic evidence for cyclical changes in queen number in a neotropical wasp, *Polybia emaciata*. Am Nat 140:363–372
Strassmann JE, Solis CR, Hughes CR, Goodnight KF, Queller DC (1997) Colony life history and demography of a swarm founding social wasp. Behav Ecol Sociobiol 40:71–77
Strassmann JE, Goodnight KF, Klinger CJ, Queller DC (1998) The genetic structure of swarms and the timing of their production in the queen cycles of neotropical wasps. Mol Ecol 7:709–718
Strassmann JE, Sullender BW, Queller DC (2002) Caste totipotency and conflict in a large-colony social insect. Proc R Soc L Ser B 269:263–270
Tanaka GM Jr, Soleman RA, Noll FB (2010) Morphological and physiological variation between queens and workers of *Protonectarina sylveirae* (de Saussure) (Hymenoptera, Vespidae, Epiponini). Rev Brasil Entomol 54:104–109
Toth AL, Varala K, Newman TC, Miguez FE, Hutchison SK, Willoughby DA et al (2007) Wasp gene expression supports an evolutionary link between maternal behavior and eusociality. Science 318:441–444
Virgínio NF, Maciel TT, Barbosa BC (2015) Nidificação de *Polybia rejecta* (Fabricius) (Hymenoptera: Vespidae) associada à *Azteca chartifex* Forel (Hymenoptera: Formicidae) em ecótono de Bioma Caatinga/Mata Atlântica, no Estado do Rio Grande do Norte. EntomoBrasilis 8:242–245
Wenzel JW (1991) Evolution of nest architecture. In: Ross KG, Matthews RW (eds) The social biology of wasps. Cornell University, Ithaca, pp 7–29
Wenzel JW (1998) A generic key to the nests of hornets, yellowjackets, and paper wasps worldwide (Vespidae: Vespinae, Polistinae). Am Mus Novit 3224:1–39
West-Eberhard MJ (1973) Monogyny in "polygynous" social wasps. In: Proceedings of the 7th international congress of International Union Study for the Social Insects, London, pp 396–403
West-Eberhard MJ (1978) Temporary queens in *Metapolybia* wasps: non-reproductive helpers without altruism? Science 200:441–443
West-Eberhard MJ (1981) Intragroup selection and the evolution of insect societies. In: Alexander RD, Tinkle DW (eds) Natural selection and social behavior. Chiron Press, New York, pp 3–17

West-Eberhard MJ (1982) The nature and evolution of swarming in tropical social wasps (Vespidae, Polistinae, Polybiini). In: Jaisson P (ed) Social insects in the tropics, vol 1. Universit'e de Paris-Nord, Paris, pp 97–128

West-Eberhard MJ (1986) Dominance behavior in *Polistes canadensis* (L.), a tropical social wasp. Monit Zool Ital 20:263–281

West-Eberhard MJ (1989a) The establishment of reproductive dominance in social wasp colonies. In: Proceedings of the 8th International Congress of the International Union for the Study of Social Insects, Wageningen, pp 223–227

West-Eberhard MJ (1989b) Scent-trail diversion, a novel defense against ants by tropical social wasps. Biotropica 21:280–281

West-Eberhard MJ (1990) The genetic and social structure of polygynous social wasp colonies (vespidae: Polistinae). In: Veeresh GK, Mallik D, Viraktamath CS (eds) Social insects and the environment. Oxford and IBH, New Delhi, pp 254–255

West-Eberhard MJ (2003) Developmental plasticity and evolution. Oxford University Press, New York, 816p

Williams FX (1928) Studies in tropical wasps – their hosts and associates (with descriptions of new species). Bull Exp Station Hawaii Sugar Planters' Assoc (Entomol) 19:1–179

Wilson EO (1979) The evolution of caste systems in social insects. Proc Am Philos Soc 123:204–210

Wilson EO, Hölldobler E (2005) Eusociality: origin and consequences. PNAS 102:13367–13371

Chapter 7
The Biology of Swarm-Founding Epiponine Wasp, *Polybia paulista*

Kazuyuki Kudô

Abstract The tribe Epiponini belongs to the swarm-founding polistine wasps and has achieved a remarkable ecological success in Neotropical regions. 20 out of 26 genera of the subfamily Polistinae are Epiponini, suggesting that these wasps have undergone an evolutionary radiation. Social wasps are quite common in Brazil, wherein 304 species in 22 genera are found. In particular, the species *Polybia paulista* is a common wasp, being found in the states of São Paulo, Goiás, Mato Grosso, Paraná, and Minas Gerais. Although several earlier studies in 1980s investigated mortality and development of colonies in this species, studies in the last two decades intensively have addressed various aspects of biology in this species, especially by morphological, genetical, population, and chemical approaches. In this chapter, I will review the biology of *P. paulista* inclusively by gathering studies for the last four decades.

Keywords Neotropical regions · polygyny · Epiponini · swarm-founding · *Polybia*

The subfamily Polistinae (26 genera and 958 species; Pickett and Carpenter 2010) comprises species with diverse social organization and can be divided into two groups according to colony foundation modalities, i.e., the independent founders and swarm founders. Colonies of independent-founding species are initiated by one or several inseminated queens, independently of any workers. Dominance behavior and reproductive skew among founding queens have been reported for species belonging to all five genera (*Polistes*, *Mischocyttarus*, *Belonogaster*, *Parapolybia*, and several species of *Ropalidia*) (Gadagkar 1991, 1996). Generally, reproductive dominance is based on direct physical attacks by the queens (Pardi 1946), while egg-laying queens of *R. marginata* rarely dominate or interact aggressively with other co-founding queens (Gadagkar 2001).

In the swarm-founding species, a colony is initiated by a swarm consisting of a large number of workers accompanied by a smaller number of queens (Jeanne

K. Kudô (✉)
Laboratory of Insect Ecology, Faculty of Education, Niigata University, Niigata, Japan
e-mail: kudok@ed.niigata-u.ac.jp

F. Prezoto et al. (eds.), *Neotropical Social Wasps*,
https://doi.org/10.1007/978-3-030-53510-0_7

1991). Swarm-founding Polistinae (the other species of *Ropalidia*, *Polybioides*, and the tribe Epiponini) widely occur in tropical regions, although the ranges of several species exceed the limits of the tropics (Jeanne 1991). Particularly, all 20 genera of the tribe Epiponini are very common in Neotropical regions (Jeanne 1991, 2003). Aggressive interactions between queens are rare, and the clear reproductive specialization in queens exists. Moreover, there is no evidence that queens attack, harass, or otherwise physically intimidate workers into action (Herman et al. 2000, but see Kelstrup et al. 2014a).

In the swarm-founding epiponines, several lines of studies have been done during the last few decades (Jeanne 2003). The first one is morphological caste difference. A striking feature is that caste dimorphism among epiponines is not as great, even in species with a great number of adults. In *Agelaia vicina*, whose colonies can exceed 10^6 adults (Sakagami 1996), queens are only 5% larger than workers (Sakagami 1996; Baio et al. 1998). In *Parachartergus smithii*, *Protopolybia exigua*, *Pseudopolybia vespiceps*, and recently *Synoeca septentrionalis*, there is no evidence of morphological differences among castes (Mateus et al. 1997; Shima et al. 1998; Santos et al. 2018). In contrast, queens are significantly smaller than workers in *Polybia dimidiata* (Shima et al. 1996). From these evidences, Jeanne (2003) concluded that the degree of morphological differences correlates with colony size very weakly in the swarm-founding wasps, particularly Epiponini.

The second is colony genetic structure related to the maintenance of polygyny. A colony alternates between polygyny and oligogyny, eventually monogyny, in the course of its life cycle (West-Eberhard 1978). New queens are only produced when the number of old queens is reduced to one or very few, and consequently new queens are highly related, often as full sisters. The process, called cyclical oligogyny, has the effect of elevating genetic relatedness between daughter offspring of the new queens. All analyzed species showed lower effective number of queens contributed to queens than to worker production, suggesting that the Epiponini exhibits cyclical oligogyny (see Table 7.1). On the contrary, division of labor among workers includes a genetic component in *Polybia aequatorialis*, suggesting that colony-level selection on variation in division of labor is a possible factor favoring

Table 7.1 Comparison of the effective number of queens that produced queens and workers in epiponine wasps

	Effective no. of queens that produced		
Species	Queens	Workers	References
Parachartergus colobopterus	1.2	1.2–4.3	Strassmann et al. (1991, 1998)
Polybia emaciata	1.5	3.6	Strassmann et al. (1992)
Polybia occidentalis	1.4	1.4–4.8	Queller et al. (1993a), Strassmann et al. (1998)
Protopolybia exigua	0.9	0.9–1.9	Gastreich et al. (1993), Strassmann et al. (1998)
Brachygastra mellifica	1.2	20	Hastings et al. (1998)
Polybia paulista	1.2	21	Kudô et al. (2005a, b)

the evolutionary maintenance of high genotypic variability (though low relatedness) (O'Donnell 1996).

The third is organization of tasks by workers, which has been summarized in reviews by Jeanne (1991, 2003). Jeanne (2003) pointed out that "full task partitioning of all material-handling tasks in the swarm founders is what enabled clear-cut age polymorphism to evolve in the epiponines." *Polybia occidentalis* has been best studied in this respect. In this species, the handling of all four materials (pulp, water, prey, and nectar) is fully partitioned into two tasks, collection and utilization. In addition, three team tasks can be identified in colonies, i.e., nest construction, nest cooling, and swarm emigration.

The last one is chemical communication among nestmates. According to a review by Jeanne (2003), there is evidence for existence for three chemical signals in the Epiponini, i.e., alarm pheromones contained in the venom, emigration trail pheromones, and queen pheromones (Landolt et al. 1998). Recently, in addition, a few studies have analyzed the correlation between cuticular hydrocarbons (CHCs) and fertility in epiponine wasps. Kelstrup et al. (2014a) showed in *Polybia micans* that queens and workers exhibit remarkable differences in their CHC profiles, with two compounds (3-methyl-pentacosane and *n*-pentacosane) being far more abundant in queens. These compounds increase in female during ovarian development and with high levels of juvenile hormone. CHC profiles of queens and workers are also distinct in *Synoeca surinama* (Kelstrup et al. 2014b) and *S. septentrionalis* (Santos et al. 2018). Furthermore, a recent study also showed that the CHC compositions of fifth instar larvae in *Polybia paulista* are significantly different among colonies, which may function as a nestmate larval discrimination (Kudô et al. 2017; see this chapter).

This chapter reviews the biology of an epiponine wasp, *Polybia paulista* (Fig. 7.1). *P. paulista* is one of the common swarm-founding wasps in Brazil, being found in the states of São Paulo, Goiás, Mato Grosso, Paraná, and Minas Gerais, also present in Paraguay and Argentina (Richards 1978). In the early 1980s, some Brazilian researchers have published ecological studies on mortality and development of colonies in *P. paulista* (Machado 1980, 1984; Simões and Mechi 1983). Since then, however, any studies had not been made in *P. paulista* until Itô et al. (1997) described the initial development of colonies. After the Itô's study, biologi-

Fig. 7.1 A mature nest of (left) and workers around the entrance of the nest (right)

cal information in *P. paulista* has been accumulated in various aspects in the last two decades, i.e., morphological caste differences, nest materials, colony genetic structure, parasitism, nestmate recognition, and individual and colony mortality. Thus, it must be important to introduce the biology of *P. paulista* by gathering those studies for the last four decades. I hope this review will help many people who are interested in social wasps understand the evolution and maintenance of highly eusociality in wasps.

7.1 Colony Cycle

It is perhaps difficult to describe the colony cycle of *P. paulista* correctly, because swarming can occur in several contexts and, when a swarm is collected in the field, it is impossible to determine the context that gave rise to it (see also Jeanne 1991). Colony initiation occurs at any months during the year, but is observed intensively in rainy, nest-growing months (November to April, hereafter "summer") in Ribeirão Preto (Machado 1984; Simões and Mechi 1983; Kudô et al. 2003). Swarms build several stacked combs in quick succession. Particularly, the first three combs and the envelope covering those combs are quickly completed within only few days (commonly, 3–4 days) after colony initiation. After colony initiation, while many individuals are actively engaged in nest construction, a large number of individuals remain tightly clustered adjacent to the incipient nest (Fig. 7.2a). Wasps that sit around the nests are very aggressive, especially just after colony initiation, and the cluster size gradually decreases with colony development. Individuals that are outside of the clusters orient the head toward the opposite direction from their nests (Fig. 7.2a). The cluster disappears within 10 days after colony initiation, as the wasps move inside the nest or to the envelope. Days spent until the disappearance of the cluster tended to be shorter in summer colonies than in winter colonies (May to October) (Kudô et al. 2003).

Successive nest construction is stopped within a month from colony initiation. Days for the initial nest development in summer colonies (12.6 days) tended to be shorter than those in winter colonies (24.5 days) (Kudô et al. 2003). The authors examined colony composition at pause of nest development in a summer colony; the nest consisted of five combs and had ca. 3400 cells, 577 eggs, and 320 first and/or second instar larvae. Presence of second instar larvae suggests that nest development was paused after few days from larval emergence.

After the nest is built by the founding swarm, there is no addition of cells to the nest for weeks or even months. Jeanne (1991) mentioned that when nest expansion does occur in swarm-founding wasps, it is typically a discrete event, lasting only several days. In a winter colony of *P. paulista*, I observed construction of new cells and the envelope covering the comb that was just 3 weeks after the pause of initial nest development (Kudô et al. 2003).

Noll and Zucchi (2000) defined three consecutive phases in the colony cycle of *P. paulista*: (1) pre-emergence of workers, (2) emergence of workers, and (3) male production. According to the definition, no adult offspring had been produced in the

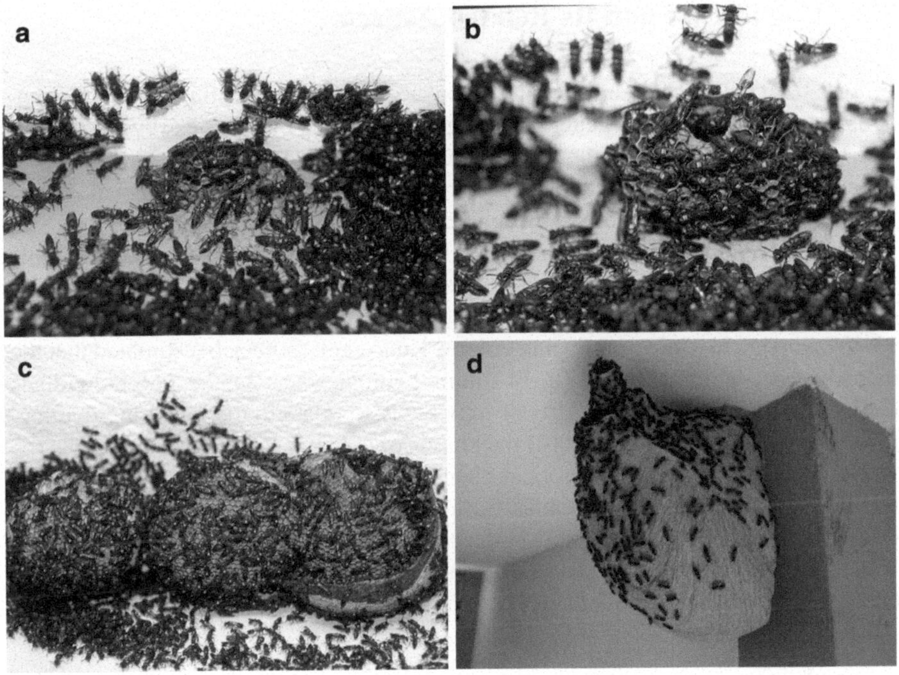

Fig. 7.2 Initial comb construction and workers outside of the clusters orienting the head toward the opposite direction from the nest (**a**), construction of new cells (second comb) on the envelope (**b**); precocious completion of first several combs (**c**); and reinforcement of primary comb(s) (upper part of the nest) (**d**)

pre-emergence phase. After producing several generations of workers, the colony has adequate number of workers and finally starts to produce males. In *P. paulista*, I observed several "reproductive swarming" (division of the colony into two or more groups) that occurred after the presence of many males on the nest envelope. This may support the definition by Noll and Zucchi (2000). However, Forsyth (1978) claims that male production of *P. occidentalis*, a consubgeneric species of *P. paulista*, is not directly correlated with colony size or stage of development of the colony but occurs in colonies that have old queens. On the other hand, Jeanne (1991) suggested that colonies of *P. occidentalis* in Costa Rica appear to occupy two nests successively during a single colony cycle, i.e., the colony produces one cohort of workers in the first nest, emigrates to found a new nest, produces several cycles of worker brood, and then reproduces.

7.2 Colony Size and Its Related Aspects

Colony size (number of adult females) in swarm-founding polistines varies tremendously among species and genera (see Fig. 6.4 depicted by Jeanne (1991)). Colony size was less than 100 individuals in *Marimbonda* spp. (Jeanne 1991), while it was estimated over million individuals in a colony of *Agelaia vicina* (Zucchi et al. 1995). In this regard, notable ranges are recognized in the genus *Polybia*, whose colony size spans at least 2.5 order of magnitude: less than 100 individuals in some species while several thousands of individuals in other species. Noll and Zucchi (2000) reported colony size in 2 *P. paulista* nests: 4747 workers in a worker emergence colony and 13,229 workers in a male-producing colony. Kudô et al. (2005a, b) estimated that the mean number of adult females in 14 summer colonies was 3457 ± 666 individuals (±SE). Colony population in *P. paulista* is likely to relate to the following two factors, i.e., (1) nesting season and (2) the phases of colony cycle. In order to know the worker demography and its seasonal change, I carried out mark-recapture experiments for four *P. paulista* colonies from the middle of summer (January) to the middle of winter (August) in 2001 at the campus of the Universidade de São Paulo, Ribeirão Preto (Kudô et al. 2011a, b). The study showed that the colonies experienced substantial reductions in worker number from summer to winter (Fig. 7.3).

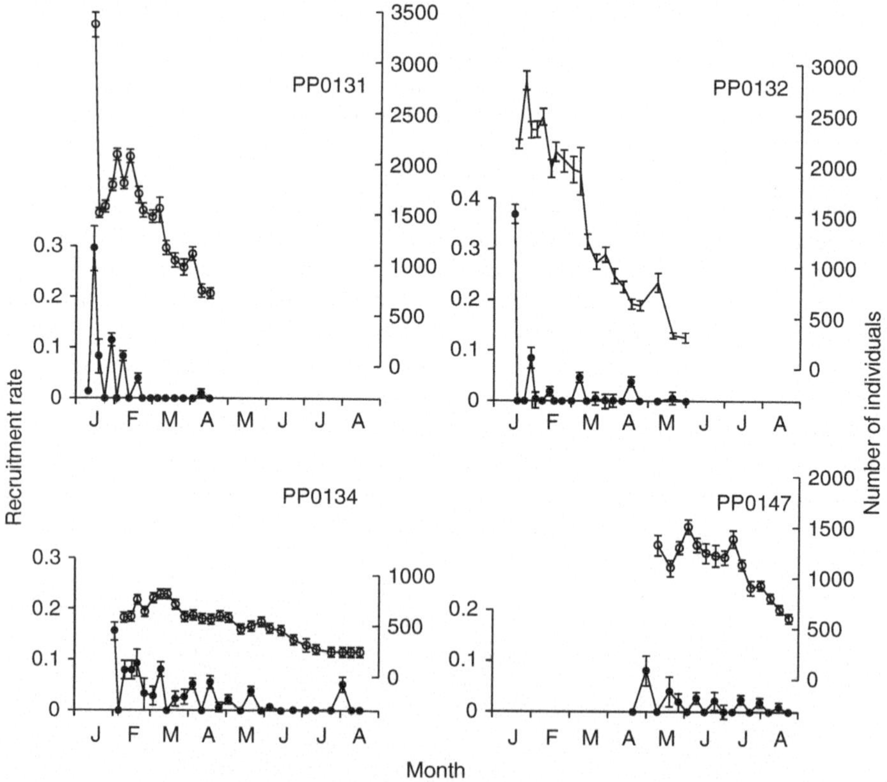

Fig. 7.3 Seasonal change in the number of workers (open circles) and recruit rate (closed circle) in four colonies (Kudô et al. 2011a)

This reduction may be related to lower foraging rates (Kudô et al. 2003; Canevazzi and Noll 2011) and a few or absence of brood in winter colonies of *P. paulista*. Actually, 60–70% colonies do not rear broods in the city during winter, and few nests contain nectar deposits (Kudô unpublished data). On the contrary, there is no study to compare the difference in colony population among the phases of colony cycle. However, as colony size just prior to "reproductive swarm" is large adequately, it is likely that there is a relation between colony size and the phase of colony cycle.

7.3 Nests

7.3.1 Sites

P. paulista nests are naturally built under broad leaves such as palm trees. The nests are sometimes built on the small twigs (a few centimeters) of trees, but such nests are easily destroyed by strong wind and heavy rain. Rather than natural substances for nest building, *P. paulista* prefer to build their nests under man-made structures. These circumstances together with broad leaves provide rain-sheltered conditions, which are beneficial not only in protecting nests from direct rain but also reducing the amount of oral secretion for nest maintenance (Simões and Mechi 1983; Kudô et al. 1998).

In addition to rain-sheltered places, swarms select a particular orientation as nest sites. Kudô and Zucchi (2009) showed that swarms selected various orientations as nest sites, but there was a particular trend that swarms in the winter period preferred to build northward-facing nests. Northward-facing nests are warmer through the gain of direct solar heat during the winter period; consequently, choosing northward-facing sites may be advantageous for swarms in terms of a shortened brood development and shortened time need to increase metabolic rates during warm-up for flight. Because architecture of *P. paulista* nests has insulation effect from ambient temperature, thermal conditions inside nests are stable even in the summer period (Hozumi et al. 2005; see later).

7.3.2 Materials

Wenzel (1991) mentioned that *Polybia* species exhibit great diversity in nest building materials: long woody fibers, plant hairs, short vegetable chips, and mud. The diversity is remarkable when compared with those of *Polistes* (long woody fibers and plant hairs), the most primitive genus in Polistinae (Wenzel 1991). SEM observations supported his description, and *P. paulista* wasps used vegetable chips, which are a dominant material, plant hairs, and mud and/or inorganic particles as nest building materials (Fig. 7.4a, b; Kudô et al. 2001). I have seen that *P. paulista* workers collected nest materials from decayed woods, which support that the workers collected short vegetable chips.

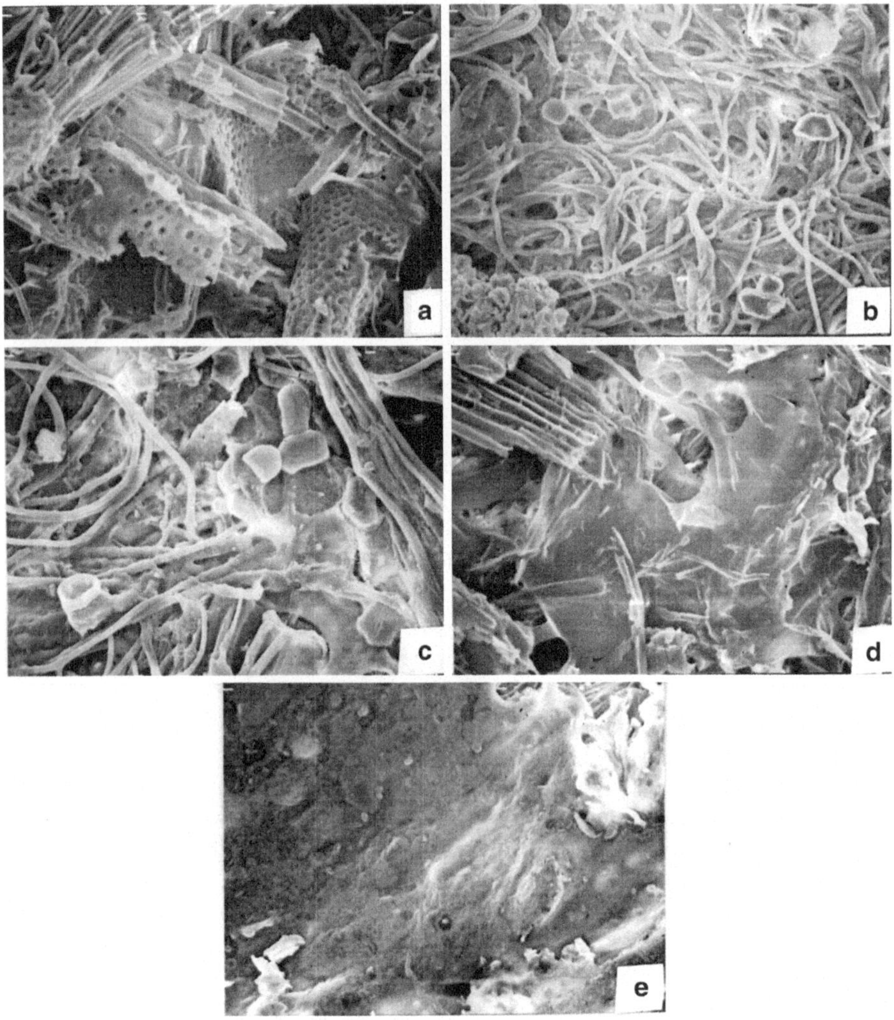

Fig. 7.4 Scanning electron micrographs of the outer surface of the upper part of the envelope (**a**), the cell walls (**b**, **c**), the outer surface of the upper part of the envelope cysts in the nest (**d**), the nest entrance on the envelope (**e**). See minute vegetable chips in **a**, plant hairs in **b**, mud and/or inorganic particles in **b**, oral secretion binding nest materials in **c** and **d**, and oral secretion covering the nest surface in **e** (Kudô et al. 2001)

Polistine wasps use oral secretion, of which a major proportion consists of proteinaceous elements, to glue nest materials (Fig. 7.4c, d). This salivary secretion is also used to physically maintain their nests (Fig. 7.4e). Kudô et al. (2001) showed that *P. paulista* used only a small amount of oral section for the construction and maintenance of their nests, compared with nests of independent-founding wasps. It has been first mentioned by Schwarz (1931) and later measured by Jeanne (1986) that the reduced role of oral secretion in *Polybia* species may be due in part to the division of labor. That is, increasing division of labor may shift the responsibility of the secretion production to progressively fewer wasps.

7.3.3 Architecture

Nest architecture of *P. paulista* is characterized by a figure (14.48) of Wenzel (1991): (1) sessile initiation (cell walls are built directly from the substrate with no paper foundation) (Fig. 7.2a), (2) construction of the envelope covering the primary comb(s) (Fig. 7.2b), (3) construction of new cells (second comb) on the envelope (Fig. 7.2b), (4) precocious completion of first several combs (Fig. 7.2c), and (5) reinforcement of primary comb(s) (upper part of the nest) (Fig. 7.2d) (Itô 1997; Kudô et al. 2003). There is no variation in nest architecture within the species except for a case that some swarms build multiple initial combs, of which combs were fused into a single one within a few days (Fig. 7.2c) (Kudô et al. 2003). This architecture is called "a phragmocyttarous nest" (Richards and Richards 1951; Jeanne 1975).

7.3.4 Thermoregulation

Hozumi et al. (2005) measured diel changes of a field nest of *P. paulista*. They showed that nest temperature followed the ambient temperature, but during the day the nest temperatures were lower than the ambient temperature, and this relationship was reversed during the night. The authors also showed a virtual lack of thermal effects from the presence of adult wasps. From these results, insulation effect of phragmocyttarous nests (narrow compartments between combs and the presence of envelope) makes stable thermal conditions of *P. paulista* nests.

7.4 Morphological Differences of Female Wasps Within a Colony

Because swarm-founding polistine wasps are highly eusocial, morphological caste differences exist. Commonly, queens are larger than workers in these wasps, but the degree varies among species. According to Jeanne (1991), colony size and caste dimorphism in swarm-founding polistines seem to be directly related. Richards (1978) noted that caste differences in *P. paulista* are indistinct. Noll and Zucchi (2000) made morphological analyses of female wasps between the castes in this species and showed that caste differences progressively increase because larger queens appeared in more abundance in later stages of the colony cycles.

Parasitism affects worker size in *P. paulista* (Kudô et al. 2004). Two kinds of parasites have been recognized in the gaster of female wasps, i.e., a strepsipteran possibly *Xenos myrapetrus* and an undescribed gregarine (Kudô et al. 2004, 2018, 2019) (Fig. 7.5). Kudô et al. (2004) compared differences in worker size between

Fig. 7.5 External view of a gaster of stylopyzed workers (**upper left**), a worker parasitized by a single strepsipteran (possibly *Xenos myrapetrus*) (**upper right**), and a worker parasitized by multiple strepsipterans (**bottom left**), and the gametocysts of an undescribed gregarine (**bottom right**). Thanks to W. Goi for the photos

uninfected workers and workers infected by strepsipterans or gregarines. Uninfected workers were larger than workers infected by strepsipterans, but smaller than workers infected by gregarines. Regarding the latter result, the authors suggested that larvae infected by gregarines, for which developmental time is extended (shown in *Polybia occidentalis*; Haward and Jeanne 2004), solicit more food from adults. The effect of parasitism on worker size was also shown in consubgeneric species, *P. occidentalis* (Kudô et al. 2011b).

7.5 Colony Genetic Structure

The evolution of eusociality in insects is understood in terms of kin selection, where workers gain inclusive fitness from rearing sexuals related to them (Hamilton 1964a, b; Hamilton 1972). In polygynous species, however, several or many queens reproduce in a colony, thus decreasing the relatedness of workers to brood. In this respect, epiponine wasps have attracted special attention, because all of the species have tens or even hundreds of queens. It has been empirically or genetically confirmed in polygynous polistines that relatedness within nestmates can be elevated by a mechanism known as cyclical monogyny, under which new queens are produced only after the number of old queens is reduced to one (West-Eberhard 1978; Hughes et al. 1993; Queller et al. 1993a; Strassmann et al. 1997, 1998; Hastings et al. 1998; Tsuchida et al. 2000) (Fig. 7.6). By using microsatellite markers, Kudô et al. (2005a, b) estimated effective queen number (the number of queens actively producing female offspring in a nest) that produced queens and workers. Queens were produced mostly by a single mother (effective number of queens = 1.2), while workers were produced by many queens (21) (Fig. 7.7). Such a lower effective number of queens contributed to queens than to worker production, suggesting that *P. paulista* also exhibits cyclical oligogyny. Regarding abundance of larger queens in later stages of the colony cycles, Noll and Zucchi (2000) suggested that queen elimination by a mechanism of cyclical oligogyny occurred in *P. paulista*. Kudô et al.

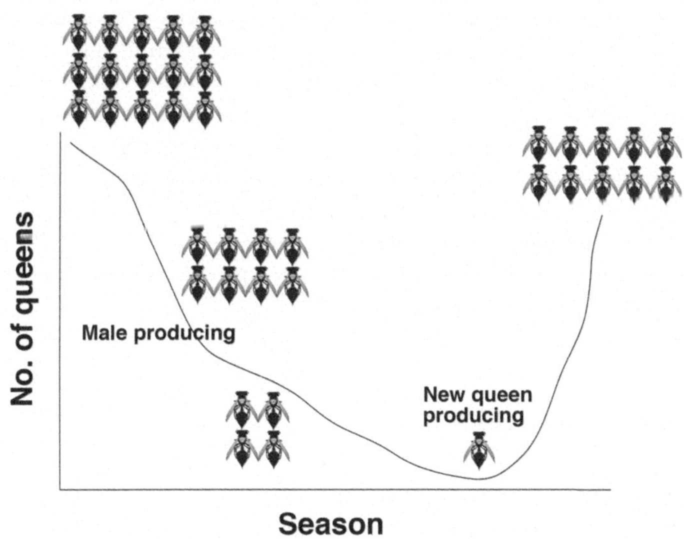

Fig. 7.6 Schematic drawing of cyclical monogyny where new queens are only produced when the number of old queens is reduced to one, while males are produced in colonies with a higher queen number. Redrawn from Tsuchida (2011)

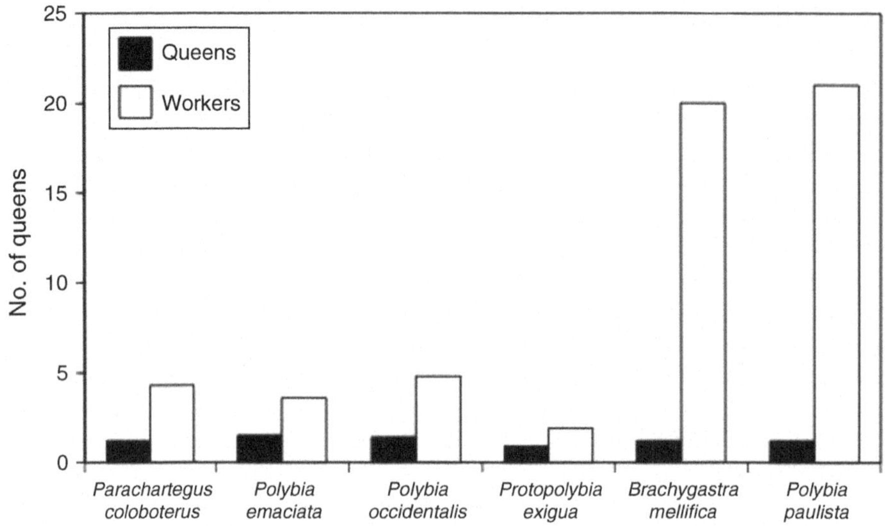

Fig. 7.7 Comparison of the effective number of queens that produced queens and workers in Epiponine wasps. (See Table 7.1)

(2005a, b) also tested another likely factor that can increase relatedness within nestmates under polygyny, i.e., comb partitioning by queens in *P. paulista*. If queens concentrate their egg laying on one or a subset of the available combs, then workers may be able to rear closer relatives by focusing their work on the comb where they emerged. In *P. occidentalis*, pupae within combs are significantly more closely related to each other than they are to pupae in other combs (Queller et al. 1993b). However, comb partitioning by queens was not supported in *P. paulista*.

In addition to female wasps, genetic relatedness among males was estimated in *P. paulista*. Kudô et al. (2013) reported that relatedness within males averaged 0.431 ± 0.050, which did not differ significantly from 0.5 (a possibility that a single mother contributes to male production). However, there were several alleles at least in one locus, suggesting that more than one queen produced males in each colony, as the cyclical monogyny predicts.

7.6 Nestmate Recognition

Nestmate recognition is well developed in many social insect species. However, social insect recognition systems can be dynamic and modulated in response to context-specific cues. In ants, for example, nestmate recognition can vary with colony-specific factors, such as colony size (Stuart 1991), stage (Balas and Adams 1996), and queen number (Starks et al. 1998; Vander Meer and Alonso 2002), in addition to environmental factors, such as proximity to other colonies (Heinze et al. 1996), experience with the intruding colony (Sanada-Morimura et al. 2003), and the time of year (D'Ettorre et al. 2004; Brandt et al. 2005). In social wasps, the effect of queen number on nestmate recognition has been investigated in an epiponine wasp, *Parachartergus colobopterus*, for the first time (Gastreich et al. 1990). Gastreich et al. (1990) conducted both laboratory and field recognition bioassays of this species and found no evidence that females had the ability to discriminate nestmates from non-nestmates. The authors proposed that if recognition cues are genetic, nestmate discrimination would be harder to achieve when relatedness is low. In *P. paulista*, Kudô et al. (2007) conducted a field study to examine whether workers discriminate between nestmate and alien workers. Most workers were accepted from their own colony, while all of alien workers were aggressively chased, bitten, stung, and ultimately rejected by recipients. Therefore, it was concluded that *P. paulista* workers are able to discriminate nestmates from non-nestmates.

Kudô and Zucchi (2008) examined whether *P. paulista* workers exhibit seasonality on tolerance of alien workers. There are growing evidences that social insect colonies change the acceptance rate of alien-conspecific individuals over the seasons, possibly due to the reduced costs of defense from intraspecific and social parasitism (honeybee, Bell et al. 1974, Downs and Ratnieks 2000; ants, Brunner et al. 2002, D'Ettorre et al. 2004, Brandt et al. 2005; wasp, Gamboa et al. 1991). However, the results by Kudô and Zucchi (2008) did not support it; *P. paulista* colonies did not accept any workers from alien colonies during winter as well as summer (Fig. 7.8).

In *P. paulista*, it has been examined whether young wasps acquire the colony-specific odor shortly after their emergence. Nestmates are characterized by a distinct chemical label (colony odor), which is determined by the interaction of genetic and environmental cues (Downs and Ratnieks 1999). Early in adult life, each colony member must learn these chemical labels and act to determine the colonial membership of other individuals encountered (Gamboa et al. 1986). Kudô and Zucchi

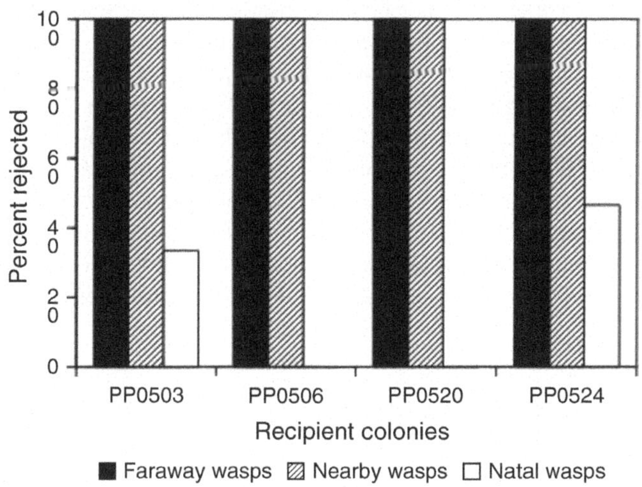

Fig. 7.8 Percentages of faraway wasps, nearby wasps, and natal wasps (control) rejected from four recipient colonies (Kudô and Zucchi 2008)

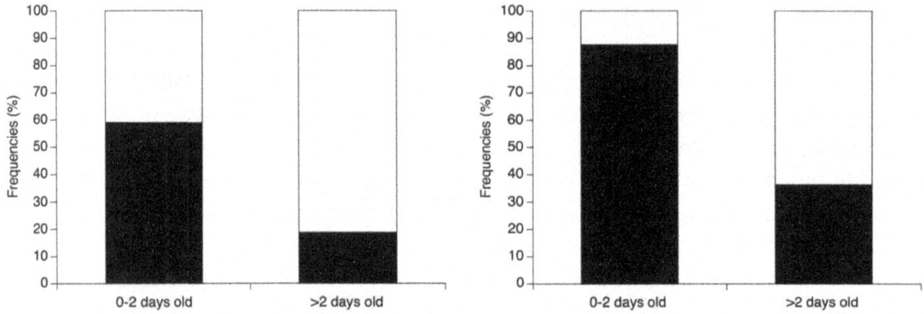

Fig. 7.9 Comparisons in frequencies of acceptance into alien recipient colonies for two age cohorts of donor males (left) and females (right) (Kudô and Zucchi 2006; Kudô and Zucchi 2008)

(2006) and Kudô et al. (2007) tested the possibility whether newly emerged individuals are accepted from alien colonies in *P. paulista*. As expected, the percent acceptance of young male and female wasps into alien colonies decreased as a function of their age (Fig. 7.9). These studies suggest that newly emerged individuals of *P. paulista* express colony-specific chemical odors shortly after emergence, as demonstrated by chemical analyses in independent-founding wasps (Paneck et al. 2001; Lorenzi et al. 2004). However, there is a result that the acceptance rate of newly emerged female wasps by alien colonies was pretty lower by male-producing colonies than worker-producing colonies (Kudô et al. 2010). Kudô et al. (2010) mentioned that the male-producing colonies abandoned soon after the field experiments, suggesting that those recipient workers rejected the introduced female wasps due to the absence of colony expansion.

Recently, whether the workers of *P. paulista* recognized their own larvae when presented outside the nest comb was investigated (Kudô et al. 2017). Behavioral tests showed that workers did not bite non-nestmate larvae more frequently than their nestmate larvae. However, they spent significantly more time licking with their nestmate larvae than with non-nestmate larvae. Kudô et al. (2017) also compared the cuticular chemical profiles of fifth instar larvae from different colonies. Analyses of the cuticular extracts of fifth instar larvae from the experimental colonies significantly allocated (100%) in their predicted groups, suggesting that discrimination of own larvae by the workers may be mediated by colony-specific, larval-borne cues. This was the first study that nestmate larval recognition could be evolved not only in independent-founding wasps (Paneck and Gamboa 2000; Cotoneschi et al. 2007) but also in colonies with many matrilines, as *P. paulista* colonies have.

7.7 Conclusions and Perspectives

P. paulista exhibits seasonal changes in colony population, which is tightly linked with foraging activities of workers. In summer between November and April, colonies produce lots of adults quickly, which is ensured by high foraging rates by workers. However, such activities may be sometimes limited because high rates of parasitism by gregarines during summer may make a part of workers idleness, as shown in *P. occidentalis*. In contrast, great proportion of colonies does not rear any brood during winter. Although it has been supported that colonies of *P. paulista* exhibit cyclical monogyny, it is unknown that changes in queen number are associated with the seasonality. Reducing to a single queen will impose some potential costs to the colony, such as the sudden loss of the remaining queens and/or loss of colony productivity. If this is the case in *P. paulista*, the reduction of queen number may not occur in the colony's growing season, i.e., summer. However, my unpublished work shows that there is no association between relatedness among workers (queen number) and seasonality.

Nestmate recognition by workers of *P. paulista* is stable throughout the year. Reeve (1989) developed the optimal acceptance threshold model to describe nestmate recognition plasticity, whereby the acceptance threshold varies according to context, to balance the fitness costs of accepting non-nestmates and rejecting nestmates. For example, in the independent-founding wasp *Polistes fuscatus*, female wasps were the most aggressive toward both nestmates and non-nestmates late in the colony cycle, possibly so as to defend nectar in the nest from conspecific intruders (Gamboa et al. 1991). In swarm-founding wasps, it has been not known that intruders enter to conspecific alien colonies to steal nectar and/or brood in the nest. It should be investigated whether CHC compositions of workers in *P. paulista* change among different seasons or colony development. Moreover, the correlation between CHCs and fertility should be needed in future works.

In the Neotropical regions, 70% of the species of social wasps (mostly Polistinae) are endemic (Noll et al. 2018). The most important threats to social wasps are the loss of their biodiversity in response to anthropogenic activities. Habitat loss due to landscape fragmentation, pesticides, and invasive species is related to reasons implicating in the species decline. Until just only 10 years ago, there had been so many diverse social wasp colonies not only in downtown area in Ribeirão Preto city but also in the campus of the Universidade de São Paulo, which is relatively conserved area. Since then, however, diversification and abundance of social wasps including *Polybia paulista* are seriously reduced. In fact, a journal *Neotropical Entomology* had a special section in 2018 ("Social insects in the Neotropics"), which informs us the importance of diversity of social insects in the Neotropics. As well as other social wasps, *Polybia paulista* should be conserved and understood in their unknown biology.

Acknowledgments I thank Research Fellowships of the Japan Society for the Promotion of Science for Young Scientists (No. 02415) and for abroad (2005–2006), whose funding has supported my studies of *Polybia paulista*. I thank Fabio S. Nascimento, Koji Tsuchida, Sôichi Yamane, Sidnei Mateus, and Ronaldo Zucchi for their thoughtful discussion of some of ideas included here.

References

Baio MV, Noll FB, Zucchi R, Simões D (1998) Non-allometric caste differences in *Agelaia vicina* (Hymenoptera, Vespidae, Epiponini). Soc Forces 32:465–476

Balas MT, Adams ES (1996) Nestmate discrimination and competition in incipient colonies of fire ants. Anim Behav 51:49–59

Bell WJ, Breed MD, Richards KD, Michener CD (1974) Social stimulatory and motivational factors involved in intraspecific nest defense of a primitively eusocial halictine bee. J Comp Physiol 93:173–181

Brandt H, Heinze J, Schmitt T, Foitzik S (2005) A chemical level in the coevolutionary arm race between an ant social parasite and its hosts. J Evol Biol 18:576–586

Brunner E, D'Ettorre P, Heinze J (2002) Phenotypic plasticity or co-evolution? Testing the acceptance threshold model in a host of a social parasite. Actes Coll Ins Soc 50:155

Canevazzi NCDS, Noll FB (2011) Environmental factors influencing foraging activity in the social wasp *Polybia paulista* (Hymenoptera: Vespidae: Epiponini). Psyche J Entomol

Cotoneschi C, Dani FR, Cervo R, Sledge MF, Turillazzi S (2007) *Polistes dominulus* (Hymenoptera: Vespidae) larvae possess their own chemical signatures. J Insect Physiol 53:954–963

D'Ettorre P, Brunner E, Wenseleers T, Heinze J (2004) Knowing your enemies: seasonal dynamics of host-social parasite recognition. Naturwissenschaften 91:594–597

Dani FR, Turillazzi S (2017) Chemical communication and reproduction partitioning in social wasps. J Chem Ecol

Downs SG, Ratnieks FLW (1999) Recognition of conspecifics by honeybee guards uses nonheritable cues acquired in the adult stage. Anim Behav 58:643–648

Downs SG, Ratnieks FLW (2000) Adaptive shifts in honey bee (*Apis mellifera* L.) guarding behavior support predictions of the acceptance threshold model. Behav Ecol 11:326–333

Forsyth AB (1978) Studies on the behavioral ecology of polygynous social wasps. PhD Dissertation, Harvard University, Cambridge

Gadagkar R (1991) *Belonogaster*, *Mischocyttarus*, *Parapolybia*, and independent-founding *Ropalidia*. In: Ross KG, Matthews RW (eds) The social biology of wasps. Cornell University Press Comstock Cornell, Ithaca, pp 149–190

Gadagkar R (1996) The evolution of eusociality, including a review of the social status of *Ropalidia marginata*. In: Turillazzi S, West-Eberhard MJ (eds) . Oxford University Press, Oxford, pp 248–271

Gadagkar R (2001) The social biology of *Ropalidia marginata*. Harvard University Press, Cambridge. 368 pp.

Gamboa GJ, Reeve HK, Pfennig DW (1986) The evolution and ontogeny of nestmate recognition in social wasps. Annu Rev Entomol 31:431–454

Gamboa GJ, Foster RL, Scope JA, Bitterman AM (1991) Effects of stage of colony cycles, context, and intercolony distance on conspecific tolerance by paper wasps (*Polistes fuscatus*). Behav Ecol Sociobiol 29:87–94

Gastreich KR, Queller DC, Hughes CR, Strassmann JE (1990) Kin discrimination in the tropical swarm-founding wasp, *Parachartergus colobopterus*. Anim Behav 40:598–601

Gastreich KR, Strassmann JE, Queller DC (1993) Determinants of high genetic relatedness in the swarm-founding wasp, *Protopolybia exigua*. Ethol Ecol Evol 5:529–539

Hamilton WD (1964a) The genetical evolution of social behaviour. I. J Theor Biol 7:1–16

Hamilton WD (1964b) The genetical evolution of social behaviour. II. J Theor Biol 7:17–52

Hamilton WD (1972) Altruism and related phenomena, mainly in the social insects. Annu Rev Ecol Syst 3:193–232

Hastings MD, Queller DC, Eischen F, Strassmann JE (1998) Kin selection, relatedness, and worker control of reproduction in a large-colony epiponine wasp, *Brachygastra mellifica*. Behav Ecol 9:573–581

Haward KJ, Jeanne RL (2004) Rates of brood development in a social wasp: effects of colony size and parasite infection. Insect Soc 51:179–185

Heinze J, Foitzik S, Hippert A, Hölldobler B (1996) Apparent dear-enemy phenomenon and environmental-based recognition cues in the ant *Leptothorax nylanderi*. Ethology 102:510–522

Hughes CR, Queller DC, Strassmann JE, Solís CR, Negón-Sotomayor JA, Gastreich KR (1993) The maintenance of high genetic relatedness in multiple-queen colonies of social wasps. In Keller (ed) Queen number ans sociality in insects. Oxford University Press Oxford pp153–170

Herman RA, Queller DC, Strassmann JE (2000) The role of queens in colonies of the swarm-founding wasp *Parachartergus colobopterus*. Anim Behav 59:841–848

Hozumi S, Sô Y, Miyano S, Mateus S, Zucchi R (2005) Diel changes of temperature in the nests of two *Polybia* species, *P. paulista* and *P. occidentalis* (Hymenoptera, Vespidae) in the subtropical climate. J Ethol 23:153–159

Itô Y (1997) Initial stage of nest construction in a Neotropical swarm-founding wasp, Polybia paulista (Hymenoptera: Vespidae). Sociobiology 29:227–235

Itô Y, Noll FB, Zucchi R (1997) Initial stage of nest construction in a Neotropical swarm-founding wasp, *Polybia paulista* (Hymenoptera: Vespidae). Soc Forces 29:227–235

Jeanne RL (1975) The adaptiveness of soicial wasp nest architecture. The quarterly review of biology 50: 267–287

Jeanne RL (1986) The evolution of the organization of work in social insects. Monit Zool Ital 20:119–133

Jeanne RL (1991) The swarm-founding Polistinae. In: Ross KG, Matthews RW (eds) The social biology of wasps. Cornell University Press Comstock Cornell, Ithaca, pp 191–231

Jeanne RL (2003) Social complexity in the Hymenoptera, with special attention to the wasps. In: Kikuchi T, Azuma N, Higashi S (eds) Genes, behaviors and evolution of social insects. Hokkaido University Press, Sapporo, pp 81–130

Kelstrup HC, Hartfelder K, Nascimento FS, Riddiford LM (2014a) Reproductive status, endocrine physiology and chemical signaling in the Neotropical, swarm-founding eusocial wasp *Polybia micans*. J Exp Biol 217:2399–2410

Kelstrup HC, Hartfelder K, Nascimento FS, Riddiford LM (2014b) The role of juvenile pheromone signalling in the caste-flexible wasp, *Synoeca surinama*. Front Zool 11:e78

Kudô K, Zucchi R (2006) Effect of a foreign male's age on acceptance into a conspecific colony in a Neotropical wasp. Entomol Sci 9:355–357

Kudô K, Zucchi R (2008) Nestmate recognition in a Neotropical swarm-founding wasp: no effect of seasonality on tolerance of alien conspecifics. Ethol Ecol Evol 20:43–50

Kudô K, Zucchi R (2009) Nest site selection by a Neotropical swarm-founding wasp: seasonal alternation of nest orientation. J Ethol 27:275–278

Kudô K, Yamne S, Yamamoto H (1998) Physiological ecology of nest construction and protein flow in pre-emergence colonies of *Polistes chinensis* (Hymenoptera Vespidae): effects of rainfall and microclimates. Ethol Ecol Evol 10:171–183

Kudô K, Yamane S, Mateus S, Tsuchida K, Itô Y, Miyano S, Yamamoto H, Zucchi R (2001) Nest materials and some chemical characteristics of nests of a new world swarm-founding polistine wasp, *Polybia paulista* (Hymenoptera Vespidae). Ethol Ecol Evol 13:351–360

Kudô K, Zucchi R, Tsuchida K (2003) Initial nest development in the swarm-founding paper wasp, *Polybia paulista* (Hymenoptera: Vespidae, Epiponini): cases of building multiple initial combs. J N Y Entomol Soc 111:151–158

Kudô K, Yamane S, Mateus S, Tsuchida K, Itô Y, Miyano S, Zucchi R (2004) Parasitism affects worker size in the Neotropical swarm-founding social wasp, *Polybia paulista* (Hymenoptera, Vespidae). Insect Soc 51:221–225

Kudô K, Ishiguro N, Tsuchida K, Tsujita S, Sô Y, Zucchi R (2005a) Polymorphic microsatellite loci for the swarm-founding wasp *Polybia paulista* (Hymenoptera: Vespidae). Entomol Sci 8:5–7

Kudô K, Tsujita S, Tsuchida K, Goi W, Yamane S, Mateus S, Itô Y, Miyano S, Zucchi R (2005b) Stable relatedness structure of the large-colony swarm-founding wasp *Polybia paulista*. Behav Ecol Sociobiol 58:27–35

Kudô K, Tsuchida K, Mateus S, Zucchi R (2007) Nestmate recognition in a Neotropical polygynous wasp. Insect Soc 54:29–33

Kudô K, Hozumi S, Mateus S, Zucchi R (2010) The acceptance rate of young wasps by alien colonies depends on colony development stages in the swarm-founding wasp, *Polybia paulista* von Ihering (Hymenoptera: Vespidae). Neotrop Entomol 39:918–920

Kudô K, Koji S, Mateus S, Zucchi R, Tsuchida K (2011a) Worker demography in a large-colony, swarm-founding wasp. Popul Ecol 53:297–306

Kudô K, Ohka H, Zucchi R (2011b) Modification of morphological characteristics by endoparasites in workers of the swarm-founding wasp *Polybia occidentalis*. In: Stewart EM (ed) Social insects: structure, function, and behavior, Nova Science Publishers, New York, pp 83–96

Kudô K, Sugawara S, Shinohara K, Mateus S, Zucchi R, Hozumi S, Tsuchida K (2013) Estimates of genetic relatedness among males in a polygynous wasp. Neotrop Entomol 42:137–130

Kudô K, Oliveira LA, Mateus S, Zucchi R, Nascimento FS (2017) Nestmate larval discrimination by workers in the swarm-founding wasp *Polybia paulista*. Ethol Ecol Evol 29:170–180

Kudô K, Hasegawa M, Mateus S, Zucchi R, Nascimento FS (2018) Levels of parasitism of *Xenos myrapetrus* (Strepsiptera, Stylopidae) and its seasonal changes in the swarm-founding wasp, *Polybia paulista* (Hymenoptera, Vespidae). Neotrop Biodivers 4:74–76

Kudô K, Hasegawa M, Mateus S, Zucchi R, Nascimento FS (2019) Effect of seasonality on rates of gregarine infection in workers of a social wasp *Polybia paulista* (Hymenoptera, Vespidae). Neotrop Entomol 48:368–372

Landolt PJ, Jeanne RL, Reed HC (1998) Chemical communication in social wasps. In: Vender Meer RK, Breed M, Winston M, Espelie K (eds) Pheromone communication in social insects. Westview Press, Boulder, pp 216–235

Lorenzi MC, Sledge MF, Laiolo P, Sturlini E, Turillazzi S (2004) Cuticular hydrocarbon dynamics in young adult *Polistes dominulus* (Hymenoptera: Vespidae) and the role of linear hydrocarbons in nestmate recognition systems. J Insect Physiol 50:935–941

Machado VLL (1984) Análise populacional e a sazonalidade das colônias de *Polybia* (*Myrapetra*) *paulista* (Ihering, 1896). Rev Bras Zool 2:187–201

Mateus S, Noll FB, Zucchi R (1997) Morphological caste differences in the Neotropical swarm-founding polistine wasps: *Parachartergus smithii* (Hymenoptera, Vespidae). J N Y Entomol Soc 105:129–139

Noll FB, Zucchi R (2000) Increasing caste differences related to life cycle progression in some Neotropical swarm-founding polygynic polistine wasps (Hymenoptera Vespidae Epiponini). Ethol Ecol Evol 12:43–65

Noll FB, Nascimento FS, Vasconcelos H (2018) Special section: social insects in the Neotropics. Neotrop Entomol 47:729–730

O'Donnell S (1996) RAPD markers suggest genotypic effects on forager specialization in an eusocial wasp. Behav Ecol Sociobiol 38:83–88

Paneck LM, Gamboa GJ (2000) Queens of the paper wasp *Polistes fuscatus* (Hymenoptera: Vespidae) discriminate among larvae on the basis of relatedness. Ethology 106:159–170

Paneck LM, Gamboa GJ, Espelie KE (2001) The effect of a wasp's age on its cuticular hydrocarbon profile and its tolerance by nestmate and non-nestmate conspecifics (*Polistes fuscatus*, Hymenoptera: Vespidae). Ethology 107:55–63

Pardi L (1946) Ricerche sui Polistini VII. La "dominazione" ed il ciclo ovarico annuale di *Polistes gallicus* (L.). Bollettino dell'Istitudo di Entomologia dell'Universitá di Bologna 15:25–84

Pickett KM, Carpenter JM (2010) Simultaneous analysis and the origin of eusociality in the Vespidae (Insecta: Hymenoptera). Arthropod Syst Phylo 68:3–33

Queller DC, Strassmann JE, Solís CR, Hughes CR, DeLoach DM (1993a) A selfish strategy of social insect workers that promotes social cohesion. Nature 365:639–641

Queller DC, Negrón-Sotomayor J, Strassmann JE, Hughes CR (1993b) Queen number and genetic relatedness in a Neotropical wasp *Polybia occidentalis*. Behav Ecol 4:7–13

Reeve HK (1989) The evolution of conspecific acceptance thresholds. Am Nat 133:407–435

Richards OW (1978) The social wasps of the Americans excluding the Vespidae. British Museum (Natural History), London

Richards OW, Richards MJ (1951) Observations on the social wasps of South America (Hymenoptera Vespidae). Trans R Entomol Soc Lond 10:1–170

Sanada-Morimura S, Minami S, Yokokawa M, Hirota T, Satoh T, Obara Y (2003) Encounter-induced hospitality to neighbors in the ant *Pristomyrmex pungens*. Behav Ecol 14:713–718

Santos EB, Shemilt S, de Carvalho CAL, Martin SJ (2018) Life history and chemical ecology of the swarm-founding wasp *Synoeca septentrionalis* (Hymenoptera: Vespidae, Epiponini). PLoS One 13:e0194689

Schwarz HF (1931) The nest habits of the diplopterous wasp *Polybia occidentalis* variety *scutellaris* (white) as observed at Barro Colorado, canal zone. Am Mus Novit 471:1–27

Shima SN, Yamane S, Zucchi R (1996) Morphological caste differences in some Neotropical swarm-founding polistine wasps II: *Polybia dimidiata* (Hymenoptera, Vespidae). Jpn J Entomol 64:131–144

Shima SN, Noll FB, Zucchi R, Yamane S (1998) Morphological caste differences in the Neotropical swarm-founding polistine wasps IV: *Pseudopolybia vespiceps*, with preliminary considerations on the role of intermediate females in the social organization of the Epiponini (Hymenoptera, Vespidae). J Hymenopt Res 7:280–295

Simões D, Mechi MR (1983) Estudo sobre a fenologia de *Polybia* (*Myrapetra*) *paulista* Ihering, 1986 (Hymenoptera, Vespidae). Naturalia 8:185–191

Sakagami SF, Zucchi R, Yamane Sô, Noll FB, Camargo JMP (1996) Morphological caste differences in Agelaia vicina, the neotropical swarm-founding polistine wasp with the largest colony size among social wasps (Hymenoptera: Vespidae). Sociobiology 28:207–223

Starks PT, Watson RE, Dipaola MJ, Dipaola CP (1998) The effect of queen number on nestmate discrimination in the facultatively polygynous ant *Pseudomyrmex pallidus* (Hymenoptera: Vespidae). Ethology 104:573–584

Strassmann JE, Queller DC, Solís CR, Hughes CR (1991) Relatedness and queen number in the Neotropical wasp, *Parachartergus colobopterus*. Anim Behav 42:461–470

Strassmann JE, Gastreigh KR, Queller DC, Hughes CR (1992) Demographic and genetic evidence for cyclical changes in queen number in a Neotropical wasp, *Polybia emaciata*. Am Nat 140:363–372

Strassmann JE, Goodnight KF, Klingler CJ, Queller DC (1998) The genetic structure of swarms and the timing of their production in the queen cycles of Neotropical wasps. Mol Ecol 7:709–718

Stuart RJ (1991) Nestmate recognition in leptothorax ants: testing for effects of queen number, colony size and species of intruder. Anim Behav 42:277–284

Tsuchida (2011) Social structures and conflict resolution in primitively eusocial wasps. In: From genes to animal behavior - social structures, personalities, communication by color (Inoue-Murayama M, Kawamura S eds). Springer, Tokyo pp:83–110

Tsuchida K, Itô Y, Katada S, Kojima J (2000) Genetical and morphological colony structure of the Australian swarm-founding polistine wasp, Ropalidia romandi (Hymenoptera: Vespidae). Insectes Soc 47:113–116.

Vander Meer RK, Alonso LE (2002) Queen primer pheromone affects conspecific fire ant (*Solenopsis invicta*) aggression. Behav Ecol Sociobiol 51:122–130

Wenzel JW (1991) Evolution of nest architecture. In: Ross KG, Matthews RW (eds) The social biology of wasps. Cornell University Press Comstock Cornell, Ithaca, pp 480–519

West-Eberhard MJ (1978) Temporary queens in *Metapolybia* wasps: non-reproductive helpers without altruism? Science 200:441–443

Zucchi R, Sakagami SF, Noll FB, Mechi MR, Mateus S, Baio MV, Shima SN (1995) *Agelaia vicina*, a swarm-founding polistines with the largest colony size among wasps and bees (Hymenoptera: Vespidae). J N Y Entomol Soc 103:129–137

Chapter 8
Causes and Consequences of Reproductive Conflicts in Wasp Societies

Cintia Akemi Oi, Tom Wenseleers, and Ricardo Caliari Oliveira

Abstract Wasps are a true model in studies on the origin and evolution of cooperative behavior and the mechanisms that help to stabilize sociality and resolve internal conflicts. Indeed, the wide variety of social organizations found in the group – ranging from solitary to highly social – provides unique opportunities to test how cooperation evolved and how conflicts are suppressed in societies with different degrees of social complexity. As the individuals in wasp colonies are not genetic clones of each other, inclusive fitness theory predicts that there should often be significant scope for conflicts between the queen and worker castes arising from specific genetic relatedness asymmetries within the colony. In many species, for example, the workers retain the ability to develop their ovaries and lay unfertilized male-destined eggs, thereby allowing them to challenge the reproductive monopoly of the queen. The amount of worker reproduction that is tolerated within the colony is a function of the genetic relatedness patterns within the colony and the costs and benefits of cheating, which under some circumstances can drive social enforcement mechanisms, whereby eggs laid by workers are selectively cannibalized or "policed" by the queen herself or by other workers. In some wasp species, such policing is so effective that workers are better off not trying to reproduce in the first place because nearly all the workers' eggs would be policed anyway. The fact that policing can align the evolutionary interests of the queens and workers facilitates the evolution of queen signaling systems (i.e. queen pheromones, that act as honest signals for the presence of a healthy and fertile queen, resulting in workers refraining from reproducing in many social species). In many other species though, queen-worker conflict can be severe, with workers sometimes even engaging in matricide – killing their own mother queen to be able to reproduce without interference. This chapter provides an overview of both current and past research on reproductive conflicts within the Vespidae wasps and how their study has been instrumental in testing some key predictions of inclusive fitness theory.

Keywords Reproductive conflict · Social insects · Wasps · Inclusive fitness

C. A. Oi (✉) · T. Wenseleers · R. C. Oliveira
Laboratory of Socioecology and Social Evolution, KU Leuven, Leuven, Belgium
e-mail: tom.wenseleers@kuleuven.be; ricardo.oliveira@evobio.eu

F. Prezoto et al. (eds.), *Neotropical Social Wasps*,
https://doi.org/10.1007/978-3-030-53510-0_8

8.1 Inclusive Fitness Theory as a Basis to Understand Reproductive Conflicts

Wasps are a key model in the study of insect sociality. This is due to the fact that this group of insects presents a remarkable diversity of lifestyles – from solitary to truly social, "eusocial" species. Although the majority of species are solitary, the evolution of sociality in some lineages is thought to have greatly facilitated their ecological success (Wilson 1971; Grimaldi and Engel 2005). Wilson (1971) considered organisms eusocial when they possess three major characteristics: (1) a reproductive division of labor between a reproductive and nonreproductive caste, (2) overlap of generations, and (3) cooperative care of the young. Whether this definition should or should not include species with mostly behaviorally defined castes, such as most Polistinae and Stenogastrinae wasps, where workers are usually totipotent and retain the ability to mate, is a point of debate (Boomsma and Gawne 2018; Sumner et al. 2018). We will here follow the commonly adopted convention to refer to those as *primitively eusocial* and to the species with morphologically defined castes and an obligate worker caste, i.e., with workers that have lost the ability the mate, as *highly eusocial* (cf. Michener 1974; O'Donnell 1998; Wenseleers and van Zweden 2017). By this token, all of the Vespinae wasps are highly eusocial, whereas most of the Polistinae are primitively eusocial, with the possible exception of a few swarm-founding Polistinae, such as *Agelaia*, which can achieve colony sizes of over half a million individuals and which have quite pronounced caste dimorphism and have been suggested to possess an obligate worker caste (O'Donnell 1998; Wenseleers and van Zweden 2017; Piekarski et al. 2018). The Stenogastrinae hover wasps and some members of the Crabronid wasp *Microstigmus* are considered *facultatively eusocial*, given that many nests also just contain a single breeding female (Field 2008; Turillazzi 2012; Matthews 1991). Recent phylogenomic evidence indicates that the presence of preimaginally determined castes, which do not need to be reflected in overall morphology, was likely the ancestral condition for all of the Polistinae and Vespinae (Piekarski et al. 2018), implying that a subset of the daughters expressed a distinct, likely subfertile, phenotype at the very outset of eusociality. This has been explained on the basis that the queen-worker castes were likely built on preexisting ground plans and pathways in solitary ancestors, e.g., related to ancestral breeding cycles (Piekarski et al. 2018; Hunt and Amdam 2005; Hunt et al. 2007b). Recent phylogenetic data also show that social behavior in the Stenogastrinae evolved independently from that in the Polistinae and Vespinae (Piekarski et al. 2018).

The origin of a sterile worker caste that refrains from reproducing and instead helps to rear the queen's offspring presents an important evolutionary paradox and an apparent contradiction with the expectation that organisms should be selected to maximize their personal reproduction (Ratnieks et al. 2011). Indeed, Darwin himself referred to this question as a "special difficulty" that seemed to jeopardize his theory of evolution via natural selection (Ratnieks et al. 2011). A formal resolution of this paradox only came with the development of Hamilton's (1964) inclusive fitness theory, which showed that individuals can be selected to altruistically sacrifice their own reproduction if their altruism would be directed toward relatives, as these can indirectly help to propagate copies of the altruist's own genes, thereby allowing the altruistic trait to persist in the population (Wenseleers et al. 2010; Gardner et al. 2011). In formulating this theory, Hamilton was heavily influenced by the social insects, and he noticed that insect sociality appeared to be unusually common among the haplodiploid Hymenoptera, where females (workers and queens) develop from fertilized diploid eggs and males are usually derived from unfertilized, haploid eggs. He pointed out that as a result of this genetic system, any female would genetically be more highly related to full sisters (relatedness coefficient $r = 0.75$) than to its own daughters ($r = 0.5$, Fig. 8.1) and suggested this could predispose them to evolve helping behavior and give up the option to mate and reproduce on their own (Hamilton 1964). However, later authors noted that his argument was not quite correct, since haplodiploidy also causes females to be less related to their full brothers ($r = 0.25$) than to their own sons ($r = 0.5$), resulting in an average relatedness to siblings (brothers and sisters) that is the same as that to own offspring, 0.5, if they produce both sexes in an equal ratio (Trivers and Hare 1976). Later analyses have therefore concluded that haplodiploidy can only slightly increase the likelihood for eusociality to evolve and only does so under very specific conditions, e.g., if the workers would be able to bias the sex ratio toward highly related sisters (Grafen 1986; Seger 1983; Quiñones and Pen 2017), if the workers would be able to combine the rearing of sisters with the production of their own male offspring (Craig 1979, Fig. 8.1), or if there is male-biased dispersal (Johnstone et al. 2011). Indeed, several recent studies have concluded that haplodiploidy as such was most likely not a strong driving force in the evolution of eusociality but that high relatedness, linked with lifetime monogamy, and a high benefit to cost ratio of helping surely were (Hughes et al. 2008; Davies and Gardner 2018; Boomsma 2009; Boomsma and Gawne 2018; Liao et al. 2015; Boomsma et al. 2011; Quiñones and Pen 2017).

Another important insight that came from Hamilton's work was the realization that inclusive fitness theory would not just be able to explain under what conditions sociality would be able to evolve but that it also predicted the exact conditions under which social conflicts could be expected. Insect societies, including many social wasps, have proven an excellent test bed to test these theories, given that the genetic asymmetries within their colonies produce much scope for conflict, both among the queen and workers as well as among the workers themselves (reviewed in Ratnieks et al. 2006).

8.2 The Scope for Conflict in Insect Societies

Empirical study of conflict within insect societies began with Trivers and Hare (1976), who used inclusive fitness theory to show that in social Hymenoptera there is strong potential conflict between the mother queen and her daughter workers over

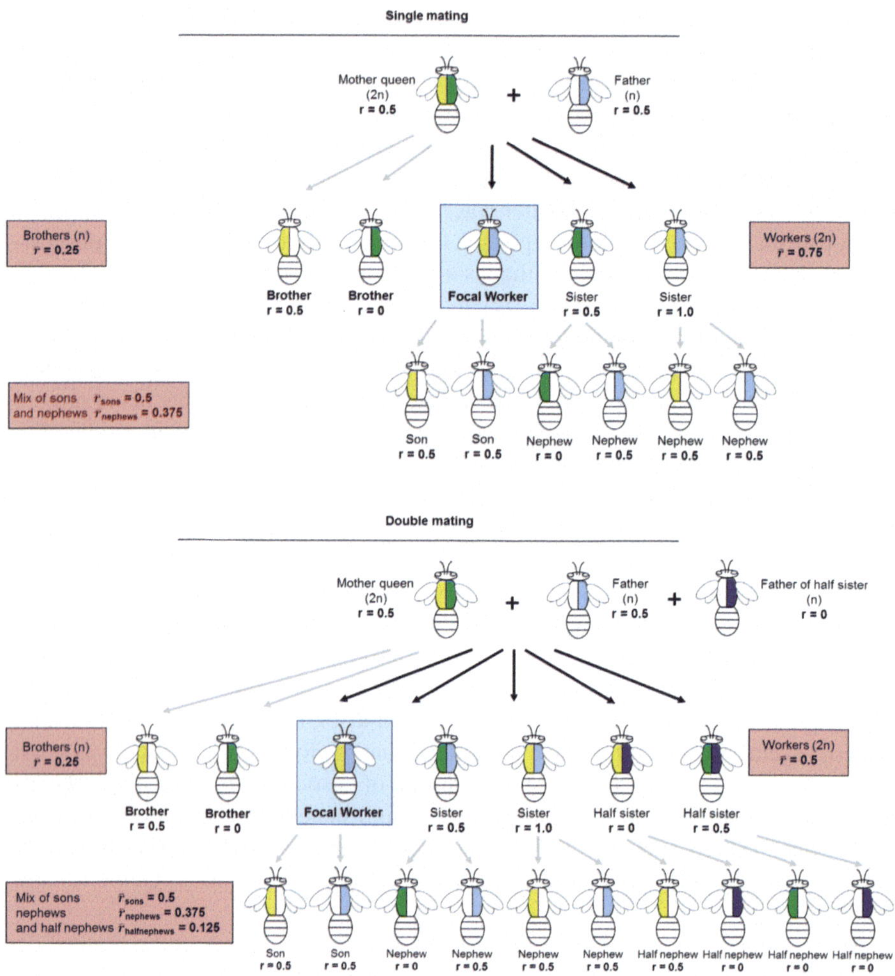

Fig. 8.1 Genetic relatedness coefficients in social Hymenoptera if the colony is headed by a single-mated (top) or double-mated queen (bottom). The relatedness coefficients *r* are the genetic relatedness values between a focal worker and individual sisters, brothers, sons, or nephews, whereas $\bar{r}$ is the average genetic relatedness of a focal work to each of these classes of individuals. The different thorax colors represent distinct genetic backgrounds. In this diagram, the production of diploid (2n) females, derived from fertilized eggs, are indicated with black arrows, whereas the production of haploid (n) males, derived from unfertilized eggs, are indicated with gray arrows

the rearing of young males versus queens. Specifically, they noted that in colonies headed by a single once-mated queen, workers would on average be three times more related to sisters (r = 0.75) than to brothers (r = 0.25), whereas the queen would always be related equally to sons and daughters (r = 0.5). Hence, it was predicted that the queen and workers would have conflicting evolutionary optima, with the workers being selected to invest three times more in females than in males but the queen being favored to invest equally in both (Trivers and Hare 1976). Since then, researchers have documented many other areas where conflict can be expressed, including in the context of conflict over male parentage (Ratnieks 1988; Bourke and Franks 1995; Hammond and Keller 2004; Wenseleers et al. 2004b; Wenseleers and Ratnieks 2006a; Ratnieks and Wenseleers 2008; Bourke 1988), queen rearing (Keller 1997; Tarpy et al. 2004; Ratnieks et al. 2006), and caste fate (Bourke and Ratnieks 1999; Wenseleers et al. 2003; Wenseleers and Ratnieks 2004) or, in societies lacking morphologically distinct queens and workers, over breeding role and the choice between helping and nesting independently (Hart and Ratnieks 2005; Hart and Monnin 2006; Reeve 1991; Reeve and Keller 2001; Tibbetts 2007; Strassmann et al. 2002). Below we will review how some of these conflicts have been documented in wasps and how this has led to some of the best tests of inclusive fitness theory to date. We start with two conflicts that theoretically are predicted to occur but which in wasps have only received relatively limited study, namely, conflict over sex allocation and queen rearing, and then move on to review types of conflicts that have been investigated very extensively in wasps, namely, conflicts over breeding roles in primitively eusocial wasps and conflicts over male parentage. We then discuss how in some species these conflicts are resolved via policing mechanisms and chemical signaling systems, which likely evolved hand in hand, while in others queen-worker conflict remains unresolved, resulting, for example, in worker matricide, where the workers kill their own mother queen to eliminate all reproductive competition and be able to reproduce on their own.

8.3 Conflict over Sex Allocation

Sex ratio theory predicts that workers in social Hymenoptera should be selected to invest in queens versus males in proportion to the relatedness asymmetry to both sexes (Bourke and Franks 1995; Crozier and Pamilo 1996; Ratnieks et al. 2006). This relatedness asymmetry becomes smaller in species where queens mate with multiple males, where colonies are headed by multiple queens, or when the majority of the males in the colony are produced by the workers (Bourke and Franks 1995; Crozier and Pamilo 1996; Ratnieks et al. 2006). Hence, in these situations, the sex ratio favored by the workers becomes closer to the 1 to 1 investment sex ratio favored by the mother queen(s), resulting in reduced potential conflict. Theory also predicts that intraspecific variation in the kin structure of colonies is expected to favor facultative sex ratio biasing or "split sex ratios," whereby workers should produce more queens in colonies with single-mated queens or low queen numbers, where relatedness asymmetry is high, and more males in colonies with multiple-mated queens or high queen numbers, where relatedness asymmetry is low (Boomsma and Grafen 1991; Ratnieks 1991).

These sex ratio predictions have been tested most extensively in ants, and data from many species support the hypothesis that workers selfishly bias sex allocation in their own favor or facultatively bias sex allocation in response to the colony kin structure (reviewed in Bourke and Franks 1995; Crozier and Pamilo 1996; West 2009). For example, a meta-analysis comprising data from 22 species of ants, bees, and wasps provided strong evidence for facultative sex allocation biasing in response to variation in relatedness asymmetry (or mate or queen number, which directly correlates with relatedness asymmetry) (Meunier et al. 2008). Key supporting evidence in this analysis came from primitively eusocial wasps, as in both *Polistes fuscatus* (Noonan 1978) and in five species of swarm-founding epiponine wasps (Queller et al. 1993; Hastings et al. 1998), sex ratios were found to be split and follow variation in queen number, thereby suggesting worker control of sex allocation. In *Polistes metricus*, Metcalf (1980) also reported that sex ratios were split due to some colonies being orphaned, causing the workers in those colonies to rear their own males and the remaining colonies to be selected to compensate for this male bias by producing more female-biased sex ratios. A different outcome though was recorded in *Polistes chinensis*, where Tsuchida (2003) also recorded male-biased allocation in orphaned nests but found that workers in queenright nests did not compensate by increasing female investment but rather by engaging in high levels of worker reproduction in the presence of the mated foundress, causing over half of all the adult males to be the workers' sons, whilst producing an investment sex ratio that was close to the queen's optimum (Tsuchida et al. 2003; Wenseleers and Ratnieks 2006b). This shows how different types of social conflict can interact with each other and how the workers can sometimes win one type of conflict at the cost of losing another. Such outcomes as well as multiple stable queen-worker equilibria or ever continuing queen-worker arms races have been predicted by models that considered how multiple traits in queens and different types of social conflicts coevolve with each other (Helms 1999; Wenseleers et al. 2013). Many other factors though are suspected to influence optimal sex allocation in social insects (West 2009). Suzuki (1986), for example, reported investment sex ratios for 11 Polistini and 2 Ropalidiini paper wasps and found an approximately equal investment sex ratio in species that produce males and reproductive females simultaneously, implying queen-controlled sex allocation in those species but a female-biased investment sex ratio and worker control in both protandrous and protogynous species that produce either males or females first in the season. These data illustrate the importance of considering the full dynamics of the production schedule of both sexes, which in most models of social Hymenopteran sex allocation have so far been neglected (but see Avila et al. 2019 for a promising approach to do so).

Whether investment sex ratios across different Vespine wasps correlate with queen mating frequency and relatedness asymmetry remains largely unknown, as sex ratios are quite hard to measure due to the long period over which sexuals are produced and the fact that in Vespine wasps, two different sizes of males are produced, derived from sexual and worker comb (Bonckaert et al. 2011b). Johnson et al. (2009), however, provided tentative evidence that queen polyandry reduces queen-worker conflict over sex allocation in the North American yellowjacket *Vespula maculifrons*, as the predicted optimal investment sex ratio of the mother queens and the workers, given the observed queen mating frequency, was very close to each other, 0.5 and 0.56, and the observed investment sex ratio intermediate and not significantly different from either. Likewise, the numerical sex ratios reported for the polyandrous yellowjackets *V. germanica* and *V. vulgaris* in Edwards (1980) and Spradbery (1971) translate into investment sex ratio values of 0.63 and 0.56, which are again intermediate between the queen's (0.5) and the workers' optima (0.65 and 0.65, given observed queen mating frequencies, Bonckaert et al. 2007; Foster and Ratnieks 2001b). On the other hand, Archer (1998) observed that workers selectively eliminate some of the male larvae in *V. vulgaris*, thereby suggesting that even in these polyandrous species, workers do bias sex allocation in their own favor to a limited extent. Hence, observed levels of polyandry are not fully effective at eliminating all potential conflict in these yellowjackets. Detailed comparative studies of investment sex ratios in other Vespine wasps, such as in *Dolichovespula* and *Vespa*, as far as we are aware are not available but a priori would be expected to show stronger potential female bias under worker control due to the effective mating frequency in those genera being much lower than in the large-colony *Vespula*, often being close to one (Foster et al. 2001; Foster and Ratnieks 2001a; Hughes et al. 2008; van Zweden et al. 2014). This prediction remains to be tested. In *D. arenaria*, however, split sex ratios have been suggested to occur, even though two studies reported opposite effects, with Foster and Ratnieks (2001a) citing decreased female investment in function of paternity but Loope (2016) finding the opposite pattern. Hence, these studies require further replication.

8.4 Conflict over Queen Rearing

Another potential conflict that can occur in insect societies is when there are different degrees of relatedness between individual workers and different queens raised in the colony. In wasps, this can occur when queens are multiple mated, as is common in many Vespine wasps (Foster and Ratnieks 2001a; Hughes et al. 2008), or when the colony contains multiple mother queens or foundresses, which is common in many Polistinae and Stenogastrinae paper wasps (Turillazzi and West-Eberhard 1996; Ross and Matthews 1991; Hughes et al. 2008). In these cases, one would expect that workers could be selected to nepotistically aid only in the rearing of new queens to which they are most related, i.e., full-sister queens. In a swathe of species, from the honeybee to stingless bees, ants and Polistine, and Vespine wasps, such nepotism has been looked for, but the consensus is that it is quite rare in social insects overall (Keller 1997; Tarpy et al. 2004; Goodisman et al. 2007; Wenseleers 2007). In wasps, nepotism and within-colony kin discrimination was, for example, found to be absent in the multiqueen colonies of the swarm-founding wasp *Parachartergus colobopterus* (Strassmann et al. 1997) and in the polyandrous colonies of the yellowjacket *Vespula maculifrons* (Goodisman et al. 2007). It is thought that even though potential conflict over queen rearing exists, recognitional errors often make it too costly in practice to be evolutionarily favored (Keller 1997; Wenseleers 2007; van Zweden and d'Ettorre 2010). Indeed, cuticular hydrocarbons from Polistinae and Vespinae wasps failed to distinguish reliably among patrilines in the hornet *Vespa crabro*, even though they could distinguish among nestmate workers of different matrilines in the paper wasp *Polistes dominulus* (Dani et al. 2004). In line with this last finding, Leadbeater et al. (2014) reported a limited degree of kin-biased behavior in *P. dominulus*, thereby providing the first evidence for within-nest kin discrimination in a primitively eusocial wasp.

8.5 Conflict over Breeding Role in Species with Totipotent Females

In most Polistinae and Stenogastrinae wasps and in the Crabronid genus *Microstigmus*, queen-worker dimorphism is either weak or absent (Noll and Wenzel 2008; Turillazzi 2012; Matthews 1991), and all or most females are thought to be "totipotent," i.e. being able to mate and produce both male and female offspring (Reeve et al. 1998; Tibbetts 2007; Strassmann et al. 2002). Totipotency provides significant scope for conflict over breeding role, as it allows any individual in the colony to replace the existing queen (Reeve 1991; Jandt et al. 2014; Hart and Monnin 2006; Hart and Ratnieks 2005), to become an additional queen (Strassmann et al. 2002), or, except in the swarm-founding Polistinae, to found a nest independently (Reeve and Keller 2001). The relatedness benefits for females to mate and breed are similar to those of becoming a morphologically specialized queen (Bourke and Ratnieks 1999; Wenseleers et al. 2003). That is, in the face of reproductive competition with other females, a female benefits if she can become the dominant breeder and produce own offspring (sons and daughters, $r = 0.5$) rather than a worker rearing the less-related offspring of other females ($r << 0.5$). The cost of helping can be significantly reduced though if there are potential direct fitness benefits associated with delayed nest inheritance (Leadbeater et al. 2011) or if there are insurance-based advantages of helping (Reeve and Nonacs 1997; Field et al. 2000; Shreeves et al. 2003), e.g., so-called assured fitness returns resulting from the fact that when a helper dies, the offspring she contributed may be brought to maturity by surviving nestmates (Gadagkar 1990), or "bet hedging"-type benefits of helping (Kennedy et al. 2018). Indeed, in both Polistinae paper wasps and Stenogastrinae hover wasps, insurance-based advantages (Reeve and Nonacs 1997; Field et al. 2000; Shreeves et al. 2003; Field 2008) as well as advantages associated with delayed nest inheritance (Field et al. 1999; Field et al. 2006; Shreeves and Field 2002; Field 2008; Monnin et al. 2009; Leadbeater et al. 2011) have both been shown to be important in driving helping behavior. In the paper wasp *Polistes dominula*, for example, females that inherit the dominant position were shown to produce more offspring than solitary females (Leadbeater et al. 2011).

There is extensive evidence for conflict over breeding role in wasps with totipotent females. In Polistinae with independent nest founding, dominance interactions

Fig. 8.2 Aggressive contests regulate reproduction in many primitively eusocial paper wasps. Here, aggressive "boxing" behavior decides which of these two *Polistes satan* females will be able to become the main egg-layer in the colony. Photo: Ricardo Caliari Oliveira

frequently determine who will take the dominant breeding spot (Fig. 8.2, Reeve 1991; Jandt et al. 2014), and subordinates sometimes challenge the dominant breeder to attempt to overthrow it (Hart and Monnin 2006; Hart and Ratnieks 2005; Jandt et al. 2014). Conflict over the decision to either help or reproduce is also apparent from the fact that some first-generation females, which normally act as workers, have been observed to leave the colony early to become a queen the next year (in *Polistes annularis*, Strassmann 1989; *P. fuscatus*, Reeve et al. 1998; and *P. dominula*, Tibbetts 2007). Similarly, first-generation females of *P. exclamans* may leave the nest to find satellite nests elsewhere the same season (in *P. exclamans*, Strassmann 1981). In swarm-founding Epiponini, conflict over breeding role ("caste fate conflict") has been suggested to result in an excess of females mating to be readopted in the natal nest when colonies have few mother queens left (Henshaw et al. 2000; Queller et al. 1993; Strassmann et al. 2002). This excess is then reduced as a result of queen-queen competition or active elimination by workers (Strassmann et al. 2002; Platt et al. 2004), thereby driving the characteristic cycles in queen number that occur in this group (Noll and Wenzel 2008). Such queen overproduction only occurs in colonies with few mother queens, presumably because the workers as a collective then favor their production (Strassmann et al. 2002) and because the potential benefits for females to mate and reproduce are largest, as there would be few competing queens already present. It is possible that the high levels of queen production observed in Epiponine wasp colonies with few queens is therefore not just a consequence of split sex ratios as Queller et al. (1993) proposed but that caste fate conflict is an important additional driver of this pattern.

Several mechanisms have been suggested by which overt conflict over breeding role could be avoided in primitively eusocial wasps. In North American populations of *Polistes dominula* and *P. exclamans* paper wasps, signals of dominance and fighting ability, so-called badges of status, have been suggested to be encoded by specific patterns on the foundresses' clypeus, which correlate with body size (Tibbetts and Dale 2004; Tibbetts and Sheehan 2011). In this way, conventional settlements can resolve conflict and prevent overt aggression. Furthermore, individuals that were painted to emulate the wasps' having a higher rank than their true rank were found to be punished by nestmates, thereby helping the signaling system to remain honest (Tibbetts and Dale 2004). Clypeal status badges, however, were not found to be used in European populations of *P. dominula* (Green and Field 2011; Green et al. 2013), while clypeal patterning was shown to be correlative with fertility in the Brazilian paper wasp *P. satan* (Tannure-Nascimento et al. 2008). Another potential mechanism that the dominant breeder could use to reduce the chances that it's offspring would reproduce would be to provide them with limited food to keep them smaller and less fecund. In neither paper wasps nor hover wasps, however, this subfertility hypothesis has so far received any empirical support (Field and Foster 1999; Couchoux and Field 2019). Some models have also suggested that social contracts could help to reduce conflict among cofounding females in primitively eusocial wasps (Reeve and Keller 2001; Nonacs and Hager 2011). In so-called transactional skew models, dominants were argued to offer a sufficiently large reproductive share to the subordinate, lest the subordinate would leave to found her own nest (Reeve and Keller 2001; Nonacs and Hager 2011). Although initial data from some *Polistes* paper wasps appeared to support social contract-type transactional skew models (e.g., Reeve et al. 2000; Tibbetts and Reeve 2000, reviewed in Nonacs and Hager 2011), many later studies have supported other models, such as ones where reproductive shares arise from selfish competition among individuals (so-called "tug of war" models) (Nonacs and Hager 2011) or follow biological market theory, in which dominants will compete for the helpers and pay a price for it, sharing part of reproduction with the subordinates, and subordinates choose who they will help depending on the options they are offered (Grinsted and Field 2017a, b). Over the past decade, the exact support for different classes of reproductive skew models has become quite contentious, however, and many of the formulated models have become quite hard to test due to the difficulty of measuring the underlying model parameters (Nonacs and Hager 2011). Finally, in some species of independently founding and swarm-founding wasps, a transition from aggressive regulation of reproduction via dominance interactions to one based on chemical signaling has been suggested (e.g., Mitra et al. 2011; Mitra and Gadagkar 2011; Dapporto et al. 2007b, reviewed in Jeanne 2020 this book). However, this would only be expected to be evolutionarily stable in those species where the queen and workers' interests are strongly aligned (Oi et al. 2015b).

8.6 Conflict over Male Parentage

A final area in which significant scope for conflict is expected in wasp societies is in the context of male parentage (Trivers and Hare 1976; Bourke 1988; Wenseleers and Ratnieks 2006a). Conflict over male parentage has been intensively studied across many ants, bees, and wasps and has yielded some of the best support for inclusive fitness theory to date (Wenseleers et al. 2004b; Wenseleers and Ratnieks

2006a, b; Ratnieks and Wenseleers 2008). The scope for conflict over male parentage stems from the fact that workers, though unable to mate, usually still possess functional ovaries and are therefore capable of laying unfertilized, male eggs (Wenseleers and Ratnieks 2006a). However, since both the queen and each of the workers are genetically most related to their own sons ($r = 0.5$), they are in conflict over who should produce the males in the colony. This can favor the queen to try to stop the workers from reproducing via aggression or egg eating. This process is known as "queen policing" and appears particularly common and effective in species with small colonies, including small-colony Polistine and Vespine wasps (e.g., *Polistes chinensis* and *P. dominulus*, Saigo and Tsuchida 2004; Liebig et al. 2005; *Dolichovespula sylvestris*, Wenseleers et al. 2005b; *D. norwegica*, Bonckaert et al. 2011a; *Vespula rufa*, Wenseleers et al. 2005a), as well as in Halictid bees and bumble bees (Ratnieks et al. 2006; Wenseleers and Ratnieks 2006a; Zanette et al. 2012).

Conflict over male parentage not only occurs between the queen and workers but also among the different workers themselves. Individual workers, for example, may benefit from producing highly related sons, but the workers as a collective may rather prefer that the queen would produce all the male offspring. This conflict is most pronounced in species where colonies are headed by multiple-mated queens or by multiple related queens, as in both cases the workers can end up being more related to the sons of the queen than to the sons of other workers (Starr 1984; Ratnieks 1988; Pamilo 1991) (Fig. 8.1). In these situations, nonreproductive workers have been shown to benefit from trying to suppress the reproduction of other workers via targeted aggression or the selective cannibalism of worker-laid eggs, a process that was dubbed "worker policing" (Ratnieks 1988; Ratnieks and Visscher 1989). In line with prediction, worker policing was first documented to occur in the multiple-mated honeybees (Ratnieks and Visscher 1989; Ratnieks et al. 2006) but has since been discovered to occur also in several social wasps, including in the multiple-mated large-colony yellowjackets *Vespula vulgaris* and *Vespula germanica* (Foster and Ratnieks 2001b; Bonckaert et al. 2007, Fig. 8.3) as well as in the predominantly single-mated European hornet *Vespa crabro* (Foster et al. 2002) and both monandrous and polyandrous ant species (Wenseleers and Ratnieks 2006a; Ratnieks et al. 2006).

The fact that worker policing is not purely restricted to species with multiple-mated queens suggests that the relatedness difference between queen's sons versus workers' sons is not the only evolutionary driver of this behavior. Indeed, several studies have pointed out that worker policing likely also has other benefits, e.g., linked to the fact that workers may kill worker-laid eggs as part of a strategy to cause a female-biased sex allocation ratio (Foster and Ratnieks 2001c), that workers may aggress egg-laying workers to increase their work rates and improve colony productivity (Ratnieks 1988; Wenseleers and Ratnieks 2006a), or that the policing may in fact be carried out by reproductive workers, who may gain direct fitness benefits from eliminating the eggs laid by other workers ("selfish policing", (Wenseleers et al. 2005b; Bonckaert et al. 2011a; Zanette et al. 2012). In a large meta-analysis of the male parentage in over 100 ants, bees, and wasps, however,

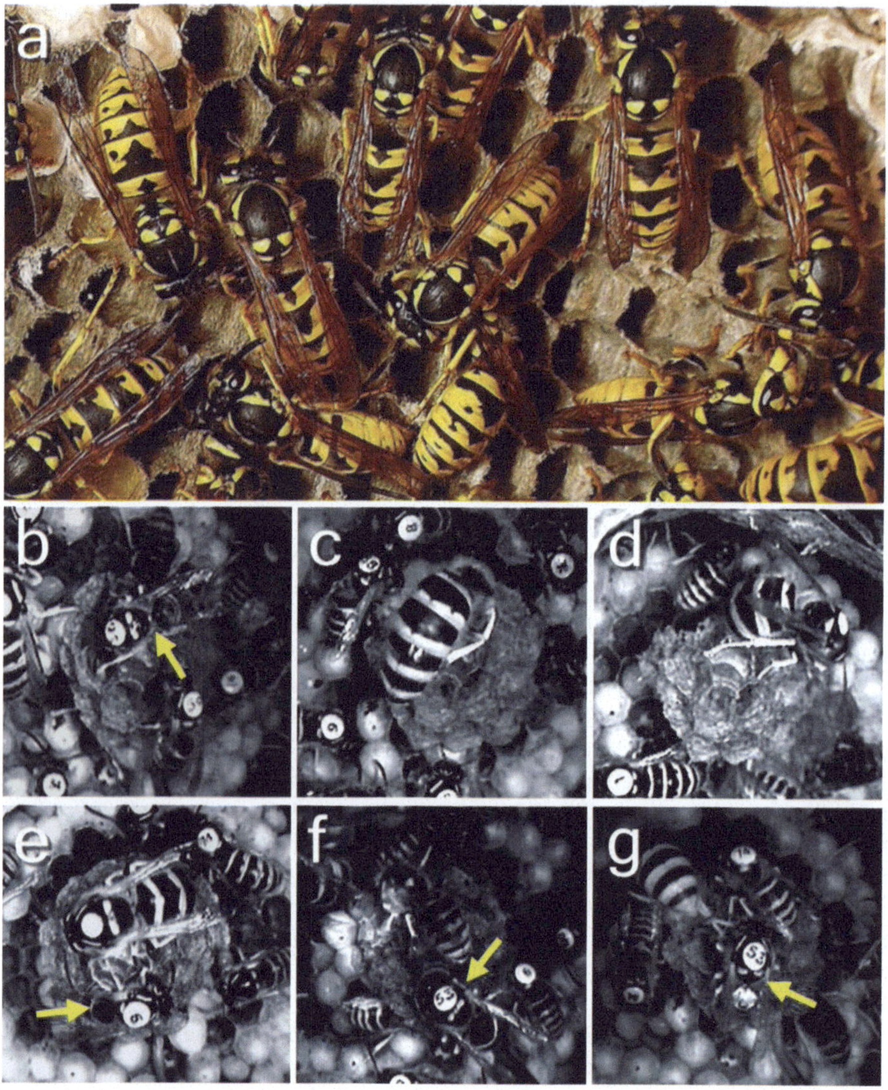

Fig. 8.3 Three types of policing regulate conflict over male parentage in social wasps. **a** In species where colonies are headed by a multiple-mated queen, as in the German wasp *Vespula germanica* (pictured) and the common wasp *Vespula vulgaris*, regular "worker policing" occurs (Foster and Ratnieks 2001b; Bonckaert et al. 2007). As in the honeybee (Ratnieks and Visscher 1989), such policing involves nonreproductive worker selectively detecting and removing eggs laid by "cheater" workers that try to reproduce inside the colony. By contrast, in species where colonies are normally headed by a single-mated queen, such as in the tree wasp *Dolichovespula sylvestris* (pictured in the video stills in panels b to g) (Wenseleers et al. 2005b), the Norwegian wasp *Dolichovespula norwegica* (Bonckaert et al. 2011a), and the small-colony yellowjacket *Vespula rufa* (Wenseleers et al. 2005a), it is usually either the queen herself that polices worker-laid eggs ("queen policing," panels b to d show a worker laying an egg, the queen then removing it, after which she deposits her own egg in the same cell) or reproductive workers that selectively destroy the eggs laid by other competing workers ("selfish policing," panels e to g show a worker laying an egg, after which another worker destroys the egg to then lay her own egg). Photos by Tom Wenseleers

colony kin structure did significantly correlate with the percentage of adult males that were workers' sons in the direction predicted by inclusive fitness theory (Wenseleers and Ratnieks 2006a, Fig. 8.4a). In wasps, for example, it was the case that in species where colonies were headed by a multiple-mated queen, as in large-colony yellowjackets (*Vespula* spp.) (Bonckaert et al., 2007; Foster and Ratnieks 2001b; Ross and Matthews 1991), or by multiple related queens, as in the swarm-founding Polistinae (Epiponinae) (Hastings et al. 1998; Henshaw et al. 2002; Henshaw et al. 2000), only few or none of the adult males were workers' sons, suggesting the presence of highly effective worker policing (Fig. 8.4a). By contrast, in species where colonies were usually headed by a single, once-mated queen, and where workers were collectively most related to the sons of other workers, a significant fraction of the adult males could be produced by the workers (Wenseleers and Ratnieks 2006a, Fig. 8.4a). In the Asian paper wasp *Polistes chinensis*, for example, genotyping data showed that over half of all adult males were worker produced (Tsuchida et al. 2003). In addition, in a British population of the Saxon wasp *Dolichovespula saxonica*, patterns of male parentage suggested that workers show a facultative expression of worker policing, with workers only policing the eggs laid by other workers in colonies with a multiple-mated queen, but not in those with a single-mated one (Foster and Ratnieks 2000). Such facultative worker policing, however, did not appear universal and was not found in a Danish population of the Saxon wasp (Bonckaert et al. 2011b).

In a number of Vespine wasps, several of the aforementioned forms of policing have been shown to co-occur within the same species. For example, in the tree wasp *Dolichovespula sylvestris*, the Norwegian wasp *D. norwegica*, and the small-colony yellowjacket *Vespula rufa*, eggs laid by rare cheater workers were found to be policed both by the queen (queen policing) and by competing reproductive workers (selfish policing) (Wenseleers et al. 2005b; Bonckaert et al. 2011a; Wenseleers et al. 2005a, Fig. 8.3). Interestingly, selfish worker policing, though driven purely by direct reproductive competition, was exclusively targeted toward the eggs laid by other workers, as the workers spared the eggs laid by the queen (Wenseleers et al. 2005b; Bonckaert et al. 2011a). Presumably, this is linked to the fact that if the reproductive workers would remove the eggs laid by the queen, it would become too costly from a colony-level perspective; as such, reproductive competition would then severely compromise the successful rearing of new brood in the colony, which would entail large indirect fitness costs.

Lastly, potential conflict over male parentage could also be caused by intraspecific worker parasitism, whereby workers could try to deposit male eggs in other colonies to have them reared by nonrelatives. Such "egg dumping" or intraspecific worker parasitism has been well-documented in bees (Beekman and Oldroyd 2008), and several studies have also tried to document it in primitively and highly eusocial wasps (Sumner et al. 2007; Oliveira et al. 2016). In the common wasp *Vespula vulgaris*, it was found that workers from queenright colonies left the colony at a higher rate than those in queenless colonies, presumably to evade worker policing in their natal colony, and that drifted workers activated their ovaries at ca. five times the rate than natal workers (Oliveira et al. 2016). RFID tracking technology also demonstrated very high rates of drifting in the paper wasp *Polistes canadensis*, even though

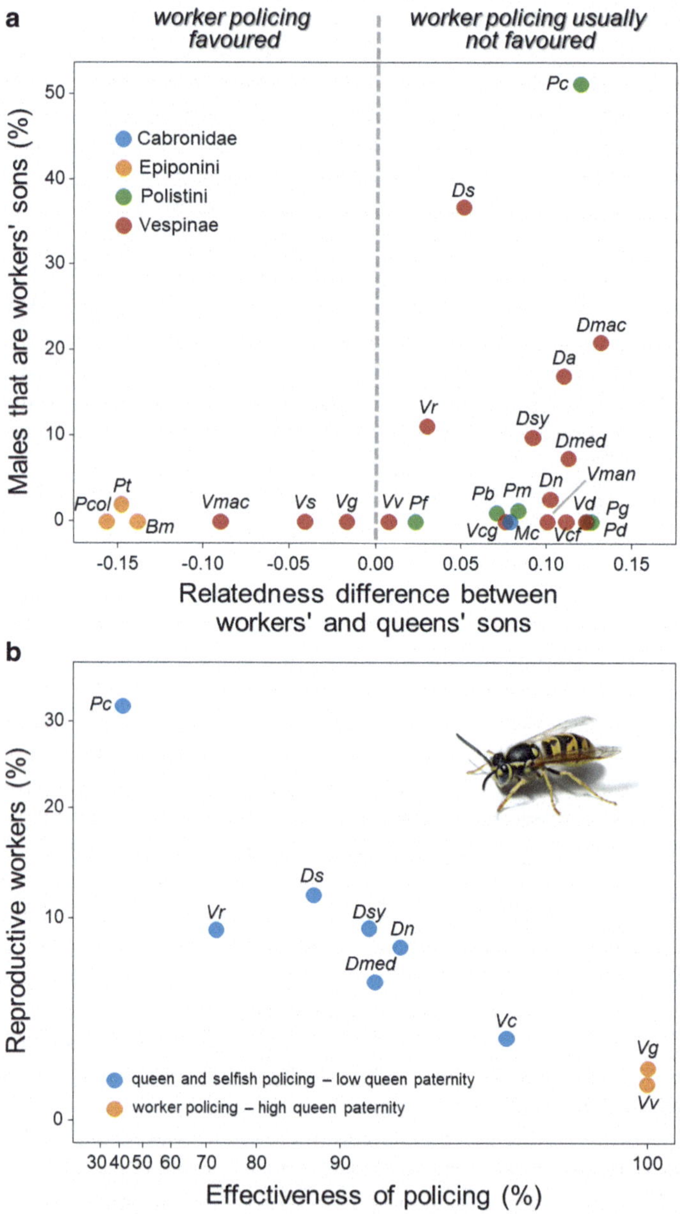

Fig. 8.4 Worker reproduction in wasp societies in function of relatedness (**a**) and policing (**b**). (**a**) Patterns of male parentage support the relatedness theory for the evolution of worker policing. In wasps, the percentage of adult males that are workers' sons, derived, e.g., from genotyping data, correlates with the relatedness difference of the workers to other workers' vs. the queen's sons. This is in line with the prediction that in species where colonies are headed by a multiple-mated queen or by multiple related queens, workers should be selected to suppress the reproduction of other workers due to the fact that they would then be more related to the sons of the queen than to

rates of worker ovary activation were not higher among the drifters than among the natal workers (Sumner et al. 2007).

8.7 Resolution of Conflict over Male Parentage via Policing

In general terms, all the forms of social policing described above can be regarded as mechanisms that should reduce the incentive for individuals to act selfishly and in the long term should benefit the functional cohesion of the group (Ratnieks 1988; Leigh 1977; Frank 2003; Agren et al. 2019). Hence, over evolutionary time, the mere presence of these forms of policing would be expected to increase the level of cooperation and reduce the percentage of workers in the colony that try to cheat by laying their own eggs (Wenseleers et al. 2004a, b; Ratnieks and Wenseleers 2008; Agren et al. 2019). A comparative study that looked at the relationship between policing and the percentage of workers that attempted to reproduce inside the colony supported this prediction (Wenseleers and Ratnieks 2006b). In particular, across nine species of wasps, the percentage of workers with active ovaries was found to be a strongly decreasing function of how effectively worker-laid eggs were policed by nestmates (Wenseleers and Ratnieks 2006b), as this disincentivizes them to even try to reproduce (Wenseleers et al. 2004a, Fig. 8.4b, b). In species where queens are predominantly single-mated, such as in *Dolichovespula* spp. and in the small-colony yellowjacket *Vespula rufa*, worker-laid eggs are policed mainly by the queen or by other egg-laying workers (selfish policing). Yet, these forms of policing are not 100% effective (Wenseleers et al. 2005a, b; Bonckaert et al. 2011a), implying that workers in these species can still have a significant incentive to reproduce inside the colony (Wenseleers et al. 2004b), often causing a significant fraction of the workers to activate their ovaries (Fig. 8.4b). By contrast, in two polyandrous large-colony

Fig. 8.4 (continued) the sons of other workers (Ratnieks 1988; Pamilo 1991) (Data from Wenseleers and Ratnieks 2006a). (**a**) More effective worker policing favors greater cooperation. Across nine species of wasps, the percentage of workers that attempt to reproduce inside the colony is a strongly decreasing function of how effectively worker-laid eggs are policed by nestmates, as this disincentivizes them to even try to reproduce (Wenseleers et al. 2004a, b). Species with queen polyandry (orange), where nonreproductive workers police eggs laid by other workers, have more effective worker policing and have fewer workers attempting to reproduce inside the colony than species with low queen paternity (blue), where it is mainly the queen and reproductive workers that remove eggs laid by other workers ("queen policing" and "selfish policing") (Data from Wenseleers and Ratnieks 2006b). Species abbreviations: Panel **a**: Crabronidae – Mc, *Microstigmus comes*; Epiponini – Pt, *Polybioides tabidus*; Bm, *Brachygastra mellifica*; Pt, *Parachartergus colobopterus*; Polistini – Pb, *Polistes bellicosus*; Pc, *P. chinensis*; Pd, *P. dorsalis*; Pf, *P. fuscatus variatus*; Pg, *P. gallicus*; Pm, *P. metricus*; Vespinae – Da, *Dolichovespula arenaria*; Dm, *D. maculata*; Dmed, *D. media*; Dn, *D. norwegica*; Ds, *D. saxonica*; Dsy, *D. sylvestris*; Vcf, *Vespa crabro flavofasciata*; Vcg, *V. crabro gribodi*; Vd, *V. ducalis*; Vman, *V. mandarinia*; Vg, *Vespula germanica*; Vmac, *V. maculifrons*; Vr, *V. rufa*; Vs, *V. squamosa*; Vv, *V. vulgaris*. Panel **b**: Pc, *Polistes chinensis*; Vr, *Vespula rufa*; Dmed, *Dolichovespula media*; Dn, *D. norwegica*; Ds, *D. saxonica*; Dsy, *D. sylvestris*; Vc, *Vespa crabro*; Vg, *Vespula germanica*; Vv, *V. vulgaris* yellowjackets, *Vespula germanica* and *V. vulgaris*, nearly all worker-laid eggs are removed by worker policing (Foster and Ratnieks 2001b; Bonckaert et al. 2007). Hence, in those species workers have almost no incentive to even try to reproduce in presence of the queen, causing only few to activate their ovaries (Foster and Ratnieks 2001b; Bonckaert et al. 2007, Fig. 8.4b). These data are interesting in that they document a form of "enforced altruism," whereby workers refrain from reproducing because it is made unprofitable by the presence of policing (Wenseleers and Ratnieks 2006b; Ratnieks and Wenseleers 2008). Paradoxically, this also causes levels of cheating to be the lowest and the level of cooperation to be the highest in the species with the lowest genetic relatedness, i.e., in colonies of the polyandrous yellowjackets *V. germanica* and *V. vulgaris*, because low relatedness selects more strongly for worker policing (Ratnieks 1988). This pattern, however, flips when the same nine wasp species were studied in the absence of the queen, as worker reproduction is then limited merely by genetic relatedness and the indirect fitness costs of having an excess of workers reproduce inside the colony (Wenseleers et al. 2004b). Hence, in that situation, a negative relationship was observed between the percentage of reproductive workers in the colony and genetic relatedness (Wenseleers and Ratnieks 2006b; Ratnieks and Wenseleers 2008). In yellowjackets, the breakdown of cooperation seen in queenless colonies is particularly apparent. In the common wasp *Vespula vulgaris*, for example, only around 1% of the workers try to reproduce in the presence of the queen, but this increases to 30% following the loss of the mother queen (Wenseleers and Ratnieks 2006b). This is far more than what is required for the successful reproduction of the colony, as shown by the fact that in queenless colonies, the workers deposit multiple eggs in the cells of their combs (Fig. 8.5). Such conflicts, where individual selfishness causes a cost to all, are known as "tragedies of the commons" and have also been documented in the context of conflict over caste fate (Wenseleers and Ratnieks 2004).

8.8 Queen Pheromones Can Regulate Conflict over Male Parentage

Given the widely divergent reproductive interests of workers in colonies with or without a queen (Ratnieks and Wenseleers 2008), it is imperative that the workers can unequivocally assess the queen's presence in order to optimally modulate their reproduction and behavior. In species with small colonies, individuals are in close contact with one another, and the queen alone is able to control the workers' reproduction, usually by being aggressive toward them or by policing worker-laid eggs (Ratnieks and Wenseleers 2005). This is the case, for example, in *Polistes* wasps and permanently queenless ant species where the dominant reproductive individual limits mating opportunities and reproduction of subordinates via targeted aggression (Reeve 1991; Monnin and Peeters 1998; Peeters 1993; Liebig et al. 2005; Saigo and Tsuchida 2004). Nevertheless, when the number of individuals reaches a certain threshold, the queen is no longer able to effectively keep workers from reproducing merely via the use of aggression (Keller and Nonacs 1993; Ratnieks and Reeve 1992). In such cases,

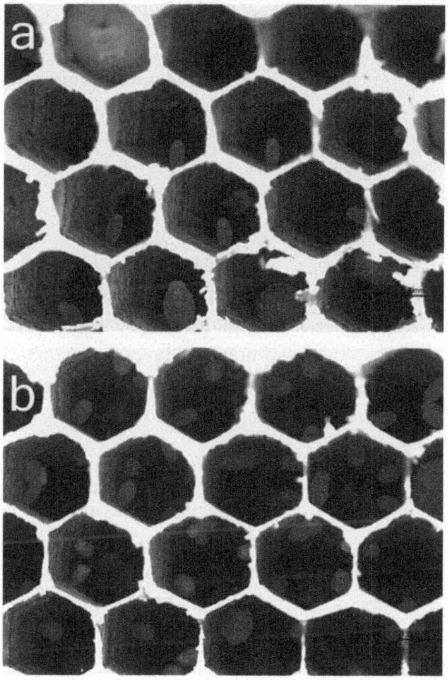

Fig. 8.5 Cooperation breaks down in queenless wasp colonies. (**a**) In queenright colonies of most social wasp species, such as in the common wasp *Vespula vulgaris* (pictured), the queen typically deposits just a single egg per cell, in line with the needs of the colony. (**b**) However, following the death of the mother queen, cooperation breaks down, and far more workers start reproducing than is necessary for the functioning of the colony, which is apparent from the fact that the workers then start depositing multiple eggs per cell. (Photos by Cintia Akemi Oi)

the workers assess whether or not the queen is present based on specific chemicals that she emits, which signal her presence and fertility status. When workers respond to such signals by inhibiting their ovary activation, such chemicals are referred to as *queen signals* or *queen pheromones* (Howard and Blomquist 2005; Holman et al. 2010, 2013; Van Oystaeyen et al. 2014; Monnin 2006; Liebig 2010; Peeters and Liebig 2009; Holman 2010; Liebig et al. 2000; D'Ettorre et al. 2004; Oliveira et al. 2015; Fig. 8.5). The first queen pheromone that was identified in social insects was in the well-studied honeybee *Apis mellifera* (Butler 1959; Wossler and Crewe 1999; Hoover et al. 2003). In this species, a honeybee-specific blend of pheromones derived from the queen's mandibular gland (QMP), containing the keto acid 9-oxo-decenoic acid as major compound, was shown to inhibit worker reproduction (Butler 1959; Hoover et al. 2003). Recent work, however, has shown that specific cuticular wax compounds act as more conserved queen pheromones across several groups of Hymenoptera, including wasps, ants, bumblebees, and the honeybee (Van Oystaeyen et al. 2014, reviewed in Oi et al. 2015b; Princen et al. 2019a). The fact that structurally related and sometimes even identical hydrocarbon compounds were found to suppress worker reproduction across these different independently evolved lineages of social insects implies very strong evolutionary conservation and suggested that these signals likely evolved from preexisting fertility-linked compounds that were already present in the common solitary ancestors of all extant social Hymenoptera (Van Oystaeyen et al. 2014; Oi et al. 2015b).

In wasps, nearly all fertility or queen-characteristic compounds have been shown to consist of linear or branched saturated alkanes or alkenes (Table 8.1). The only possible exception is provided by the Stenogastrinae wasp *Liostenogaster*

Table 8.1 Evidence for the identity of compounds being used as queen or fertility signals in different groups of wasps

Species	Identified queen pheromones or compounds overrepresented in queens or dominant egg-layers (source)	Type of evidence	Reference
Stenogastrinae			
Eustenogaster fraterna	*n*-C_{23} (cuticle of egg-layers vs. non-egg-layers)	C	Turillazzi et al. (2004)
Liostenogaster flavolineata	$C_{20}OH$, *n*-C_{23}, *n*-C_{24}, *n*-C_{25} (cuticle of egg-layers vs. non-egg-layers)	C	Turillazzi et al. (2004)
Liostenogaster vechti	$C_{31:1}$ (cuticle of egg-layers vs. non-egg-layers)	C	Turillazzi et al. (2004)
Parischnogaster striatula	*n*-C_{26}, *n*-C_{27}, *n*-C_{28}, *n*-C_{29} (cuticle of egg-layers vs. non-egg-layers)	C	Turillazzi et al. (2004)
Vespinae			
Vespa crabro	*n*-C_{24}, *n*-C_{26}, *n*-C_{27}, 3-MeC_{27}, *n*-C_{29} (cuticle of queens vs. workers)	C	Butts et al. (1991), Butts et al. (1995)
Vespula maculifrons	*n*-C_{29}, 3-MeC_{29}, *n*-C_{31}, 3-MeC_{31} (cuticle of queens vs. workers)	C	Butts et al. (1991)
Vespula squamosa	13 + 15-MeC_{29} (cuticle of queens vs. workers)	C	Butts et al. (1991)
Vespula vulgaris	***n*-C_{27}**, *n*-C_{28}, ***n*-C_{29}**, 3-MeC_{27}, **3-MeC_{29}**, 11,17-, 13,17-, 15,19-diMeC_{31} (cuticle of queens vs. workers)	C, BS	Bonckaert et al. (2012), Van Oystaeyen et al. (2014)
Dolichovespula maculata	$C_{27:1}$, $C_{29:1}$ (cuticle of queens vs. workers)	C	Butts et al. (1991)
Dolichovespula saxonica	***n*-C_{29}, *n*-C_{31}, 3-MeC_{29}, 3-MeC_{31}** (cuticle of queens vs. workers)	C, BS	Bonckaert et al. (2011b), van Zweden et al. (2014;), Oi et al. (2016)
Ropalidiini			
Ropalidia marginata	3 + 5-MeC_{29}, 3 + 7 + 9-MeC_{31}, 13 + 15 + 17-MeC_{33} (Dufour's gland of queens vs. workers)	C, BE	Bhadra et al. (2010), Mitra et al. (2011), Mitra and Gadagkar (2011)
Belonogaster longitarsus	*n*-C_{30}, $C_{31:1}$, *n*-C_{32}, *n*-C_{33}, 11,15,19-trimeC_{33}, 14-, 15-meC_{34}, 11,15,19-trimeC_{35}, 12-, 13-, 16-meC_{36}, 12,16-dimeC_{36}, 12,16,20-trimeC_{36}, 13,17-dimeC_{37}, 14,18-dimeC_{38}, 13-, 15-, 17-meC_{39} (cuticle of gynes vs. workers)	C	Kelstrup et al. (2017), Kelstrup et al. (2014a)
Polistini			

(continued)

Table 8.1 (continued)

Species	Identified queen pheromones or compounds overrepresented in queens or dominant egg-layers (source)	Type of evidence	Reference
Polistes dominula	*n*-C_{29}, *n*-C_{31}, 11 + 13-MeC_{27}, 5-MeC_{29} (cuticle and eggs of dominant foundresses vs. subordinate foundresses), 9-$C_{29:1}$, 9-$C_{31:1}$, $C_{33:2}$, $C_{35:2}$ (cuticle of dominant vs. subordinate foundresses), *n*-C_{31}, 2-MeC_{32}, *n*-C_{33}, 7 + 13 + 15 + 17-MeC_{33} (cuticle of foundresses vs. workers)	C	Bonavita-Cougourdan et al. (1991), Sledge et al. (2001), Dapporto et al. (2007a), Liebig et al. (2005)
Polistes gallicus	*n*-C_{30}, *x*-MeC_{31}, 11,17 + 9,19-diMeC_{31}, 3,13-diMeC_{31}, 13,15+ 9,11 + 11,21-diMeC_{33} (queen vs. worker Van der Vecht organ secretion)	C	Dapporto et al. (2007b)
Polistes metricus	*n*-C_{29}, 9 + 11 + 13 + 15-MeC_{29} (cuticle of queens vs. workers)	C	Layton et al. (1994)
Polistes satan	*n*-C_{23}, 3-MeC_{23}, *n*-C_{24}, *n*-C_{25}, 3-MeC_{25}, *n*-C_{26}, 4-MeC_{26}, 3-MeC_{26}, *n*-C_{27}, 5-MeC_{27}, 3-MeC_{27}, 5-MeC_{28}, 3-MeC_{28}, *n*-C_{29}, 3-MeC_{29}, *n*-C_{31}, 13,17-diMeC_{33}, 13-, 12-MeC_{34}, 13-MeC_{35}, 11,15-diMeC_{37} (cuticle of dominant egg-layers vs. subordinates)	C, BS	Tannure-Nascimento et al. (2008), Oi et al. (2019)
Epiponini			
Polybia micans	*n*-C_{25}, 3-MeC_{25}, *n*-C_{26}, $C_{27:1}$ (cuticle of queens vs. workers)	C	Kelstrup et al. (2014a)
Synoeca surinama	$C_{25:1}$, $C_{29:1}$, *n*-C_{29}, $C_{31:1}$, *n*-C_{31} (cuticle of queens vs. workers)	C	Kelstrup et al. (2014b)

Type of evidence: C, correlational; BS, bioassays with synthetic compounds; BE, bioassays of glandular or cuticular extracts. Species for which specific fertility-suppressing queen pheromones have been identified based on bioassays with particular compounds are underlined and are shown in bold

flavolineata, where egg-laying females also upregulate the production of icosanol (Table 8.1). An actual queen signaling function of these compounds, however, has been demonstrated via bioassays with pure compounds in only two Vespine wasps – the common wasp *Vespula vulgaris* (Van Oystaeyen et al. 2014, Fig. 8.6) and the Saxon wasp *Dolichovespula saxonica* (Oi et al. 2016) – where a partly overlapping set of hydrocarbons was found to suppress worker reproduction (Van Oystaeyen et al. 2014; Oi et al. 2016). In the common wasp, it was further shown that one of the worker reproduction-inhibiting queen pheromones, 3-methylnonacosane, also acted as a queen egg marking pheromone that helps workers to recognize queen-laid eggs from worker-laid eggs and selectively police worker-laid eggs (Fig. 8.6, Oi et al. 2015a). Such queen egg marking signals had long been proposed to exist (Endler et al. 2004; Ratnieks 1995; Oldroyd et al. 2002; Martin et al. 2004), but previous work had not succeeded in identifying the active compounds in any species (Martin et al. 2005).

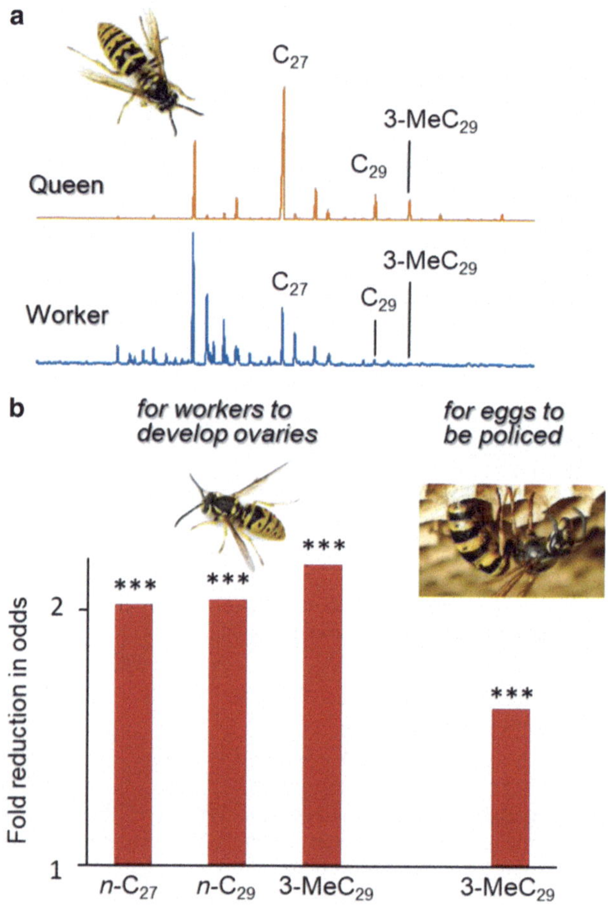

Fig. 8.6 The dual role of queen pheromones in regulating sociality in the common wasp. In the common wasp *Vespula vulgaris*, specific hydrocarbons that are abundant on the queen's cuticle and eggs, as established by gas chromatography-mass spectrometry analysis (**a**), appear to have a dual role in regulating the social order in the colony (**b**), with heptacosane (n-C_{27}), nonacosane (n-C_{29}), and 3-methyl nonacosane (3-MeC_{29}) acting as a worker reproduction inhibiting queen pheromone (**b**, left) (Van Oystaeyen et al. 2014) and also acting as a queen egg marking signal that helps workers to recognize queen-laid eggs from worker-laid eggs and selectively destroy or "police" worker-laid eggs (**b**, right) (Oi et al. 2015a). This was shown by the fact that administering these pheromones to queenless worker groups and applying them onto the surface of worker-laid eggs reduce the odds of the workers to activate their ovaries or reduce the odds of worker-laid eggs to be policed compared to in a solvent-only control treatment (*** = $p < 0.001$). (Modified based on data in Van Oystaeyen et al. (2014) and Oi et al. (2015a))

At an ultimate level, there is an ongoing debate over whether queen pheromones could ever "chemically castrate" the workers (the so-called "queen control" hypothesis) or if workers merely respond to the queen pheromones in their own best interests (the "honest signaling" hypothesis) (Keller and Nonacs 1993; Grüter and Keller

2016; Oi et al. 2015b). Evidence to date seems to give more support to the latter hypothesis, with workers merely modulating their reproductive physiology and behavior in accordance with their own evolutionary interests (Oi et al. 2015b). First, queen pheromones appear highly conserved across species, and this goes against the rapid evolution of the bioactive queen substances expected under the queen control hypothesis, where workers would be expected to rapidly evolve resistance to the chemicals that the queen emits (Heinze and d'Ettorre 2009; Van Oystaeyen et al. 2014). Further in line with honest signaling hypothesis is that queen pheromones in several species of ants, bees, and wasps have been shown to be positively correlated with fecundity (Liebig 2010; Peeters and Liebig 2009; Monnin 2006). This implies they are indeed honest signals of fertility. In some species, such as in the common wasp *Vespula vulgaris*, signal honesty has been shown to be maintained partly by the fact that queen fecundity and the production of specific hydrocarbon queen signals are under identical endocrine control ("hormonal pleiotropy," Oliveira et al. 2017). Finally, the fact that worker reproduction was found to be modulated by the presence or absence of the queen in exact accordance with inclusive fitness theory and was found to be affected by the colony kin structure in the predicted direction shows that workers respect the queen's signals only to the extent that it serves the workers' own inclusive fitness interests (e.g. Wenseleers and Ratnieks 2006b; van Zweden et al. 2014; Wenseleers et al. 2013; Ratnieks and Wenseleers 2008; Oi et al. 2015b).

Exactly how the hydrocarbon queen signaling system evolved from preexisting pathways in solitary ancestors is still not known with certainty, but it has been suggested that the queen signals could be derived from one of several preexisting pathways or unrelated signaling systems, e.g., they could (1) initially have been mere by-products of ovary activation, (2) be derived from ancestral oviposition-deterrent pheromones, or (3) be derived from ancestral sex pheromones (Oi et al. 2015b; Van Oystaeyen et al. 2014; Oliveira et al. 2015). In support of the by-product hypothesis, it has been shown that in both solitary and social species, the activation of the ovaries causes distinct changes in the cuticular hydrocarbon profiles of individuals (Blomquist and Bagnères 2010; Liebig 2010; Peeters and Liebig 2009; Holman 2012). In addition, in the primitively eusocial paper wasp *Polistes satan*, it was shown that dominant breeders overproduce particular hydrocarbons but that in bioassays these compounds did not inhibit worker reproduction (Oi et al. 2019). This supports the hypothesis that queen signals in highly eusocial species were derived from fertility cues that initially lacked any signaling function. Queen pheromones could also have been co-opted from oviposition-deterrent pheromones. Such pheromones are known in several solitary wasp species, where they mark the oviposition site with chemicals that signal to other females that the particular site is already in use (e.g., Stelinski et al. 2007; Anderson et al. 2002). Although these pheromones are directed to inhibiting other females from egg-laying, the lack of evidence for such pheromones from other groups of insects with a different social organization weakens this hypothesis (Oi et al. 2015b). Finally, queen pheromones could have been derived from sex pheromones and be initially directed at the male sex. In some bees, for example, the exact same compounds were shown to act both as a sex pheromone and as a queen pheromone or fertility-linked cue, with honeybee QMP being the best-known example (Oliveira et al. 2015). In honeybees, there is also some evidence that the queen-worker signaling system may have evolved by exploiting preexisting receiver physiology, since QMP has been found to also inhibit reproduction in the distantly related fruit fly and in bumblebees (Princen et al. 2019b), but in wasps there is so far no evidence to also support such an evolutionary scenario.

8.9 Overt Conflict over Male Parentage: Worker Matricide

While the sections above illustrate clear cases where conflict over male parentage is effectively resolved via the presence of policing mechanisms and queen signaling systems that coevolved with them, there are also clear examples in wasps where conflict over male parentage remains unresolved. A spectacular example is provided by worker matricide, where workers kill their own mother queen to be able to reproduce without interference, which has been documented not only in wasps but also in some ants and bumblebees (Ratnieks 1988; Bourke 1994; Foster and Ratnieks 2001a). Various direct observations of worker matricide exist from several Vespine wasps, including the European hornet *Vespa crabro* and several *Dolichovespula* species (Bourke 1994; Foster and Ratnieks 2001a; Loope 2015). The exact timing and conditions under which such matricide should be favored can be calculated from inclusive fitness theory (Bourke 1994; Almond et al. 2019). In particular, theory predicts that worker matricide should be favored most strongly in species with low queen paternity, where queen-worker conflict over male parentage is strongest and the workers are collectively mostly related to the sons of other workers (Ratnieks 1988; Bourke 1994). In line with this prediction, workers in the Vespine wasp *Dolichovespula arenaria* were found to kill their mother queen preferentially in colonies headed by a singly mated queen, that is, in colonies in which workers stood to gain the greatest fitness (Loope 2015). It was also shown that this was costly, as matricide resulted in the production of fewer queens (Loope 2016). Furthermore, in a comparative analysis of nine Vespine wasps, Foster et al. (2001) reported colonies to be queenless more frequently in species with low queen paternity, in accordance with worker matricide there being more common. Together, these studies again provide strong support for inclusive fitness theory.

8.10 Outstanding Questions

As we have shown, studies of reproductive conflicts in social wasps have yielded some remarkable insight into the basic structure of social conflicts and how they can be resolved. At the same time, several questions remain. In the areas of the study of social conflicts in wasp societies, the determinants of variation in investment sex ratio and the amount of overt conflict over male parentage in independently founding paper wasps and some monandrous Vespine wasps remain elusive (Foster et al. 2000; Foster and Ratnieks 2001a; Hammond and Keller 2004; Wenseleers and Ratnieks 2006a; Suzuki 1986; Tsuchida et al. 2003). To obtain a good match between empirically observed patterns and empirical data, it is likely that we will need new theoretical models that better take into account the dynamics of colony growth and sex allocation as well as the way in which different types of social conflict and traits in queens and workers coevolve with each other. At present, such theory still needs further development and testing as well as tailoring to the specific biology of particular species (e.g. Foster and Ratnieks 2001c; Wenseleers et al. 2013; Avila et al. 2019). The fact that model predictions have been shown to partly depend on the underlying genetic architecture of the traits may complicate such an endeavor (Olejarz et al. 2015, 2016; Davies and Gardner 2018). Another area that will be very rewarding to study further is the actual origin of sociality, which over the last decade has again become center stage. Indeed, fully understanding the factors that drove the major evolutionary transition to a eusocial lifestyle is surely one of the most important outstanding questions in evolutionary biology today. Social wasps are an excellent model system to study this question (Taylor et al. 2018). In fact, at an empirical level, research on primitively and facultatively eusocial wasps has given us some fantastic insights already, e.g., documenting direct fitness benefits of helping (Field et al. 2000, 2006, 2008), which has been matched by similar findings in other taxa (Korb and Heinze 2016; Downing et al. 2016), and showing how the evolution of helping behavior is subject to biological market theory (Grinsted and Field 2017b). At a theoretical level, social evolution theory and bet hedging theory have recently been combined in a unified framework (Kennedy et al. 2018), and the various preadaptations that can favor the evolutionary transition to eusociality have been subjected to systematic study, both from a proximate angle (Hunt 2012) and an ultimate perspective (Quiñones and Pen 2017). The detailed testing of some of these new models using comparative studies will remain an important task for the future. Lastly, much is yet to be unraveled about the genomic basis of social traits in wasps (Taylor et al. 2018). With many full genomes of species of key phylogenetic placement and contrasting social structure expected to become available in the near future (Taylor et al. 2018), the study of social wasps are expected to provide key insight in the genomic basis of a diverse set of traits, ranging from group living and altruistic behavior to reproductive plasticity, preimaginal caste differentiation, swarm founding, and queen-worker conflict over male parentage (Jandt and Toth 2015; Patalano et al. 2015; Taylor et al. 2018). Key questions that such data will be able to address is the extent to which sociality was built on novel genetic innovations (Ferreira et al. 2013) as opposed to the mere repurposing of existing pathways (Hunt and Amdam 2005; Hunt et al. 2007a) and how primitive and advanced eusocial systems, including queen-worker signaling systems and various conflict-resolving mechanisms, may have been built from preexisting traits in solitary ancestors (Johnson and Linksvayer 2010; Rehan and Toth 2015; Toth and Rehan 2017).

8.11 Concluding Remarks

Our aim in this chapter was to provide an overview on the current understanding of the causes and consequences of reproductive conflicts and conflict resolution mechanisms in social wasps and to present some new insights and perspectives for future research directions. Despite being among the most interesting organisms to study questions with respect to the origin and maintenance of sociality, we feel that wasps are still understudied compared to other social insect groups. We hope this chapter will incentivize students and young researchers to be as excited as we are to study and understand more about this fascinating group.

References

Agren JA, Davies NG, Foster KR (2019) Enforcement is central to the evolution of cooperation. Nat Ecol Evol 3:1018–1029. https://doi.org/10.1038/s41559-019-0907-1

Almond EJ, Huggins TJ, Crowther LP, Parker JD, Bourke AFG (2019) Queen longevity and fecundity affect conflict with workers over resource inheritance in a social insect. Am Nat 193:256–266. https://doi.org/10.1086/701299

Anderson P, Hilker M, Meiners T (2002) Oviposition pheromones in herbivorous and carnivorous insects. In: Chemoecology of insect eggs and egg deposition. Blackwell, Oxford, Berlin, pp 235–263

Archer M (1998) Worker versus sexual, and sex ratio investments in the social wasp *Vespula vulgaris* (L.) (Hymenoptera: Vespinae) in England. J Hymenopt Res 7:257–267

Avila P, Fromhage L, Lehmann L (2019) Sex-allocation conflict and sexual selection throughout the lifespan of eusocial colonies. Evolution 73:1116–1132

Beekman M, Oldroyd BP (2008) When workers disunite: Intraspecific parasitism by eusocial bees. Annu Rev Entomol 53:19–37. https://doi.org/10.1146/annurev.ento.53.103106.093515

Bhadra A, Mitra A, Deshpande SA, Chandrasekhar K, Naik DG, Hefetz A, Gadagkar R (2010) Regulation of reproduction in the primitively eusocial wasp *Ropalidia marginata*: on the trail of the queen pheromone. J Chem Ecol 36:424–431

Blomquist GJ, Bagnères A-G (2010) Insect Hydrocarbons Biology, Biochemistry, and Chemical Ecology. Cambridge University Press, Cambridge

Bonavita-Cougourdan A, Theraulaz G, Bagnères AG, Roux M, Pratte M, Provost E, Clément JL (1991) Cuticular hydrocarbons, social organization and ovarian development in a polistine wasp: *Polistes dominulus* Christ. Comp Biochem Physiol B 100:667–680

Bonckaert W, Vuerinckx K, Billen J, Hammond RL, Keller L, Wenseleers T (2007) Worker policing in the German wasp *Vespula germanica*. Behav Ecol 19:272–278

Bonckaert W, Tofilski A, Nascimento FS, Billen J, Ratnieks FL, Wenseleers T (2011a) Co-occurrence of three types of egg policing in the Norwegian wasp *Dolichovespula norwegica*. Behav Ecol Sociobiol 65:633–640

Bonckaert W, van Zweden JS, d'Ettorre P, Billen J, Wenseleers T (2011b) Colony stage and not facultative policing explains pattern of worker reproduction in the Saxon wasp. Mol Ecol 20:3455–3468. https://doi.org/10.1111/j.1365-294X.2011.05200.x

Bonckaert W, Drijfhout FP, d'Ettorre P, Billen J, Wenseleers T (2012) Hydrocarbon signatures of egg maternity, caste membership and reproductive status in the common wasp. J Chem Ecol 38:42–51

Boomsma JJ (2009) Lifetime monogamy and the evolution of eusociality. Philos Trans R Soc B 364:3191–3207

Boomsma JJ, Gawne R (2018) Superorganismality and caste differentiation as points of no return: how the major evolutionary transitions were lost in translation. Biol Rev 93:28–54

Boomsma J, Grafen A (1991) Colony-level sex ratio selection in the eusocial Hymenoptera. J Evol Biol 4:383–407

Boomsma JJ, Beekman M, Cornwallis CK, Griffin AS, Holman L, Hughes WO, Keller L, Oldroyd BP, Ratnieks FL (2011) Only full-sibling families evolved eusociality. Nature 471E4–E5; author reply E9–10. https://doi.org/10.1038/nature09832

Bourke AF (1988) Worker reproduction in the higher eusocial Hymenoptera. Q Rev Biol 63:291–311

Bourke AFG (1994) Worker Matricide in Social Bees and Wasps. J Theor Biol 167:283–292. https://doi.org/10.1006/jtbi.1994.1070

Bourke AF, Franks NR (1995) Social evolution in ants. Princeton University Press, Princeton

Bourke AF, Ratnieks FL (1999) Kin conflict over caste determination in social Hymenoptera. Behav Ecol Sociobiol 46:287–297

Butler CG (1959) The source of the substance produced by a queen honeybee (*Apis Mellifera* L.) which inhibits development of the ovaries of the worker of her colony. Proc R Entomol Soc Lond Ser A 34:137–138. https://doi.org/10.1111/j.1365-3032.1959.tb00249.x

Butts DP, Espelie KE, Hermann HR (1991) Cuticular hydrocarbons of 4 species of social wasps in the subfamily Vespinae - *Vespa crabro* L, *Dolichovespula maculata* (L), *Vespula squamosa* (Drury), and *Vespula maculifrons* (Buysson). Comp Biochem Physiol B 99:87–91

Butts DP, Camann MA, Espelie KE (1995) Workers and queens of the European hornet *Vespa crabro* L. have colony-specific cuticular hydrocarbon profiles (Hymenoptera, Vespidae). Insect Soc 42:45–55

Couchoux C, Field J (2019) Parental manipulation of offspring size in social groups: a test using paper wasps. Behav Ecol Sociobiol 73:36

Craig R (1979) Parental manipulation, kin selection, and the evolution of altruism. Evolution 33:319–334

Crozier RH, Pamilo P (1996) Evolution of social insect colonies: sex allocation and kin selection. Ecology and evolution. Oxford University Press, Oxford

D'Ettorre P, Heinze J, Schulz C, Francke W, Ayasse M (2004) Does she smell like a queen? Chemoreception of a cuticular hydrocarbon signal in the ant *Pachycondyla inversa*. J Exp Biol 207:1085–1091. https://doi.org/10.1242/jeb.00865

Dani FR, Foster KR, Zacchi F, Seppä P, Massolo A, Carelli A, Arévalo E, Queller DC, Strassmann JE, Turillazzi S (2004) Can cuticular lipids provide sufficient information for within–colony nepotism in wasps? Proc R Soc Lond Ser B 271:745–753

Dapporto L, Romana Dani F, Turillazzi S (2007a) Social dominance molds cuticular and egg chemical blends in a paper wasp. Curr Biol 17:R504–R505

Dapporto L, Santini A, Dani FR, Turillazzi S (2007b) Workers of a *Polistes* paper wasp detect the presence of their queen by chemical cues. Chem Senses 32:795

Davies NG, Gardner A (2018) Monogamy promotes altruistic sterility in insect societies. R Soc Open Sci 5:172190

Downing P, Cornwallis C, Griffin A (2016) How to make a sterile helper. BioEssays 39:1600136

Edwards R (1980) Social wasps. Their biology and control. Rentokil Ltd., East Grinstead

Endler A, Liebig J, Schmitt T, Parker JE, Jones GR, Schreier P, Hölldobler B (2004) Surface hydrocarbons of queen eggs regulate worker reproduction in a social insect. Proc Natl Acad Sci U S A 101:2945–2950

Ferreira PG, Patalano S, Chauhan R, Ffrench-Constant R, Gabaldon T, Guigo R, Sumner S (2013) Transcriptome analyses of primitively eusocial wasps reveal novel insights into the evolution of sociality and the origin of alternative phenotypes. Genome Biol 14:R20. https://doi.org/10.1186/gb-2013-14-2-r20

Field J (2008) The ecology and evolution of helping in hover wasps (Hymenoptera: Stenogastrinae). In: Ecology of social evolution. Springer, Heidelberg, pp 85–107

Field J, Foster W (1999) Helping behaviour in facultatively eusocial hover wasps: an experimental test of the subfertility hypothesis. Anim Behav 57:633–636
Field J, Shreeves G, Sumner S (1999) Group size, queuing and helping decisions in facultatively eusocial hover wasps. Behav Ecol Sociobiol 45:378–385
Field J, Shreeves G, Sumner S, Casiraghi M (2000) Insurance-based advantage to helpers in a tropical hover wasp. Nature 404:869–871. https://doi.org/10.1038/35009097
Field J, Cronin A, Bridge C (2006) Future fitness and helping in social queues. Nature 441:214–217. https://doi.org/10.1038/nature04560
Foster K, Ratnieks F (2000) Facultative worker policing in a wasp. Nature 407:692–693. https://doi.org/10.1038/35037665
Foster KR, Ratnieks FL (2001a) Paternity, reproduction and conflict in vespine wasps: a model system for testing kin selection predictions. Behav Ecol Sociobiol 50:1–8
Foster KR, Ratnieks FLW (2001b) Convergent evolution of worker policing by egg eating in the honeybee and common wasp. Proc R Soc Lond Ser B 268:169–174
Foster KR, Ratnieks FLW (2001c) The effect of sex-allocation biasing on the evolution of worker policing in hymenopteran societies. Am Nat 158:615–623. https://doi.org/10.1086/323588
Foster KR, Ratnieks FL, Raybould AF (2000) Do hornets have zombie workers? Mol Ecol 9:735–742
Foster KR, Ratnieks FL, Gyllenstrand N, Thorén PA (2001) Colony kin structure and male production in Dolichovespula wasps. Mol Ecol 10:1003–1010
Foster KR, Gulliver J, Ratnieks FL (2002) Worker policing in the European hornet Vespa crabro. Insect Soc 49:41–44
Frank SA (2003) Repression of competition and the evolution of cooperation. Evolution 57:693–705
Gadagkar R (1990) Evolution of eusociality: the advantage of assured fitness returns. Philos Trans R Soc Lond Ser B 329:17–25. https://doi.org/10.1098/rstb.1990.0146
Gardner A, West SA, Wild G (2011) The genetical theory of kin selection. J Evol Biol 24:1020–1043
Goodisman MA, Kovacs JL, Hoffman EA (2007) Lack of conflict during queen production in the social wasp *Vespula maculifrons*. Mol Ecol 16:2589–2595
Grafen A (1986) Split sex ratios and the evolutionary origins of eusociality. J Theor Biol 122:95–121
Green JP, Field J (2011) Interpopulation variation in status signalling in the paper wasp Polistes dominulus. Anim Behav 81:205–209
Green JP, Leadbeater E, Carruthers JM, Rosser NS, Lucas ER, Field J (2013) Clypeal patterning in the paper wasp Polistes dominulus: no evidence of adaptive value in the wild. Behav Ecol 24:623–633
Grimaldi D, Engel M (2005) Evolution of the Insects. Cambridge University Press, Cambridge
Grinsted L, Field J (2017a) Biological markets in cooperative breeders: quantifying outside options. Proc R Soc B 284:20170904
Grinsted L, Field J (2017b) Market forces influence helping behaviour in cooperatively breeding paper wasps. Nat Commun 8:13750. https://doi.org/10.1038/ncomms13750
Grüter C, Keller L (2016) Inter-caste communication in social insects. Curr Opin Neurobiol 38:6–11
Hamilton WD (1964) The genetical evolution of social behaviour. I & II. J Theor Biol 7:1–52
Hammond RL, Keller L (2004) Conflict over male parentage in social insects. PLoS Biol 2:e248
Hart A, Monnin T (2006) Conflict over the timing of breeder replacement in vertebrate and invertebrate societies. Insect Soc 53:375–389
Hart AG, Ratnieks F (2005) Crossing the taxonomic divide: conflict and its resolution in societies of reproductively totipotent individuals. J Evol Biol 18:383–395
Hastings MD, Queller DC, Eischen F, Strassmann JE (1998) Kin selection, relatedness, and worker control of reproduction in a large-colony epiponine wasp, Brachygastra mellifica. Behav Ecol 9:573–581
Heinze J, d'Ettorre P (2009) Honest and dishonest communication in social Hymenoptera. J Exp Biol 212:1775–1779
Helms KR (1999) Colony sex ratios, conflict between queens and workers, and apparent queen control in the ant *Pheidole desertorum*. Evolution 53:1470–1478

Henshaw MT, Strassmann JE, Queller DC (2000) The independent origin of a queen number bottleneck that promotes cooperation in the African swarm-founding wasp, *Polybioides tabidus*. Behav Ecol Sociobiol 48:478–483

Henshaw MT, Queller DC, Strassman JE (2002). Control of male production in the swarm-founding wasp, Polybioides tabidus. J Evol Biol 15:262–268 https://doi.org/10.1046/j.1420-9101.2002.00392.x

Holman L (2010) Queen pheromones: The chemical crown governing insect social life. Commun Integr Biol 3:558–560. https://doi.org/10.4161/cib.3.6.12976

Holman L (2012) Costs and constraints conspire to produce honest signaling: insights from an ant queen pheromone. Evolution 66:2094–2105

Holman L, Jorgensen CG, Nielsen J, d'Ettorre P (2010) Identification of an ant queen pheromone regulating worker sterility. Proc R Soc Lond Ser B 277:3793–3800. https://doi.org/10.1098/rspb.2010.0984

Holman L, Lanfear R, d'Ettorre P (2013) The evolution of queen pheromones in the ant genus *Lasius*. J Evol Biol 26:1549–1558. https://doi.org/10.1111/jeb.12162

Hoover SR, Keeling C, Winston M, Slessor K (2003) The effect of queen pheromones on worker honey bee ovary development. Naturwissenschaften 90:477–480. https://doi.org/10.1007/s00114-003-0462-z

Howard RW, Blomquist GJ (2005) Ecological, behavioral, and biochemical aspects of insect hydrocarbons. Annu Rev Entomol 50:371–393. https://doi.org/10.1146/annurev.ento.50.071803.130359

Hughes WO, Oldroyd BP, Beekman M, Ratnieks FL (2008) Ancestral monogamy shows kin selection is key to the evolution of eusociality. Science 320:1213–1216

Hunt JH (2012) A conceptual model for the origin of worker behaviour and adaptation of eusociality. J Evol Biol 25:1–19

Hunt JH, Amdam GV (2005) Bivoltinism as an antecedent to eusociality in the paper wasp genus *Polistes*. Science 308:264–267

Hunt JH, Kensinger BJ, Kossuth JA, Henshaw MT, Norberg K, Wolschin F, Amdam GV (2007a) A diapause pathway underlies the gyne phenotype in *Polistes* wasps, revealing an evolutionary route to caste-containing insect societies. Proc Natl Acad Sci 104:14020–14025

Jandt JM, Toth AL (2015) Physiological and genomic mechanisms of social organization in wasps (Family: Vespidae). In: Advances in insect physiology, vol 48. London, Academic Press, pp 95–130

Jandt J, Tibbetts E, Toth A (2014) Polistes paper wasps: a model genus for the study of social dominance hierarchies. Insect Soc 61:11–27

Johnson BR, Linksvayer TA (2010) Deconstructing the superorganism: social physiology, groundplans, and sociogenomics. Q Rev Biol 85:57–79

Johnson EL, Cunningham TW, Marriner SM, Kovacs JL, Hunt BG, Bhakta DB, Goodisman MA (2009) Resource allocation in a social wasp: effects of breeding system and life cycle on reproductive decisions. Mol Ecol 18:2908–2920

Johnstone RA, Cant MA, Field J (2011) Sex-biased dispersal, haplodiploidy and the evolution of helping in social insects. Proc R Soc B 279:787–793

Keller L (1997) Indiscriminate altruism: unduly nice parents and siblings. Trends Ecol Evol 12:99–103

Keller L, Nonacs P (1993) The role of queen pheromones in social insects: queen control or queen signal? Anim Behav 45:787–794. https://doi.org/10.1006/anbe.1993.1092

Kelstrup HC, Hartfelder K, Nascimento FS, Riddiford LM (2014a) Reproductive status, endocrine physiology and chemical signaling in the Neotropical, swarm-founding eusocial wasp *Polybia micans*. J Exp Biol 217:2399–2410

Kelstrup HC, Hartfelder K, Nascimento FS, Riddiford LM (2014b) The role of juvenile hormone in dominance behavior, reproduction and cuticular pheromone signaling in the caste-flexible epiponine wasp, *Synoeca surinama*. Front Zool 11:78

Kelstrup HC, Hartfelder K, Esterhuizen N, Wossler TC (2017) Juvenile hormone titers, ovarian status and epicuticular hydrocarbons in gynes and workers of the paper wasp *Belonogaster longitarsus*. J Insect Physiol 98:83–92

Kennedy P, Higginson AD, Radford AN, Sumner S (2018) Altruism in a volatile world. Nature 555:359

Korb J, Heinze J (2016) Major Hurdles for the Evolution of Sociality. Annu Rev Entomol 61:297–316. https://doi.org/10.1146/annurev-ento-010715-023711

Layton JM, Camann MA, Espelie KE (1994) Cuticular lipid profiles of queens, workers, and males of social wasp *Polistes metricus* Say are colony specific. J Chem Ecol 20:2307–2321

Leadbeater E, Carruthers JM, Green JP, Rosser NS, Field J (2011) Nest inheritance is the missing source of direct fitness in a primitively eusocial insect. Science 333:874–876. https://doi.org/10.1126/science.1205140

Leadbeater E, Dapporto L, Turillazzi S, Field J (2014) Available kin recognition cues may explain why wasp behavior reflects relatedness to nest mates. Behav Ecol 25:344–351. https://doi.org/10.1093/beheco/art113

Leigh EG (1977) How does selection reconcile individual advantage with the good of the group? Proc Natl Acad Sci U S A 74:4542–4546

Liao X, Rong S, Queller DC (2015) Relatedness, conflict, and the evolution of eusociality. Plos Biol 13:e1002098

Liebig J (2010) Hydrocarbon profiles indicate fertility and dominancy status in ant, bee, and wasp colonies. In: Insect hydrocarbons: biology biochemistry and chemical ecology. Cambridge University Press, Cambridge, pp 254–281

Liebig J, Peeters C, Oldham NJ, Markstädter C, Hölldobler B (2000) Are variations in cuticular hydrocarbons of queens and workers a reliable signal of fertility in the ant *Harpegnathos saltator*? Proc Natl Acad Sci USA 97:4124–4131. https://doi.org/10.1073/pnas.97.8.4124

Liebig J, Monnin T, Turillazzi S (2005) Direct assessment of queen quality and lack of worker suppression in a paper wasp. Proc R Soc Lond Ser B 272:1339–1344

Loope KJ (2015) Queen killing is linked to high worker-worker relatedness in a social wasp. Curr Biol 25:2976–2979

Loope KJ (2016) Matricide and queen sex allocation in a yellowjacket wasp. Sci Nat 103:57

Martin SJ, Châline N, Oldroyd BP, Jones GR, Ratnieks FL (2004) Egg marking pheromones of anarchistic worker honeybees (*Apis mellifera*). Behav Ecol 15:839–844

Martin SJ, Châline NG, Ratnieks FLW, Jones GR (2005) Searching for the egg-marking signal in honeybees. J Negat Reults 2:1–9

Matthews R (1991) Evolution of social behavior in Sphecid wasps. In: Ross KG, Matthews RW (eds) The social biology of wasps. Cornell University Press, Ithaca, pp 570–602

Metcalf RA (1980) Sex ratios, parent-offspring conflict, and local competition for mates in the social wasps *Polistes metricus* and *Polistes variatus*. Am Nat 116:642–654

Meunier J, West SA, Chapuisat M (2008) Split sex ratios in the social Hymenoptera: a meta-analysis. Behav Ecol 19:382–390

Michener CD (1974) The social behavior of the bees: a comparative study, vol 73. Harvard University Press, Cambridge, MA

Mitra A, Gadagkar R (2011) Can Dufour's gland compounds honestly signal fertility in the primitively eusocial wasp *Ropalidia marginata*? Naturwissenschaften 98:157–161

Mitra A, Saha P, Chaoulideer ME, Bhadra A, Gadagkar R (2011) Chemical communication in *Ropalidia marginata*: Dufour's gland contains queen signal that is perceived across colonies and does not contain colony signal. J Insect Physiol 57:280–284

Monnin T (2006) Chemical recognition of reproductive status in social insects. Ann Zool Fenn 43:515–530

Monnin T, Peeters C (1998) Monogyny and regulation of worker mating in the queenless ant *Dinoponera quadriceps*. Anim Behav 55:299–306

Monnin T, Cini A, Lecat V, Fédérici P, Doums C (2009) No actual conflict over colony inheritance despite high potential conflict in the social wasp *Polistes dominulus*. Proc R Soc B 276:1593–1601

Noll FB, Wenzel JW (2008) Caste in the swarming wasps: 'queenless' societies in highly social insects. Biol J Linn Soc 93:509–522

Nonacs P, Hager R (2011) The past, present and future of reproductive skew theory and experiments. Biol Rev 86:271–298

Noonan KM (1978) Sex Ratio of Parental Investment in Colonies of the Social Wasp *Polistes fuscatus*. Science 199:1354–1356. https://doi.org/10.1126/science.199.4335.1354

O'Donnell S (1998) Reproductive caste determination in eusocial wasps (Hymenoptera: Vespidae). Annu Rev Entomol 43:323–346. https://doi.org/10.1146/annurev.ento.43.1.323

Oi CA, Van Oystaeyen A, Oliveira RC, Millar JG, Verstrepen KJ, van Zweden JS, Wenseleers T (2015a) Dual effect of wasp queen pheromone in regulating insect sociality. Curr Biol 25:1638–1640

Oi CA, van Zweden JS, Oliveira RC, Van Oystaeyen A, Nascimento FS, Wenseleers T (2015b) The origin and evolution of social insect queen pheromones: novel hypotheses and outstanding problems. BioEssays 37:808–821

Oi CA, Millar JG, van Zweden JS, Wenseleers T (2016) Conservation of queen pheromones across two species of vespine wasps. J Chem Ecol 42:1175–1180

Oi CA, Oliveira RC, van Zweden JS, Mateus S, Millar JG, Nascimento FS, Wenseleers T (2019) Do primitively eusocial wasps use queen pheromones to regulate reproduction? A case study of the paper wasp *Polistes satan*. Front Ecol Evol 7:199

Oldroyd BP, Ratnieks FL, Wossler T (2002) Egg-marking pheromones in honey-bees *Apis mellifera*. Behav Ecol Sociobiol 51:590–591

Olejarz JW, Allen B, Veller C, Nowak MA (2015) The evolution of non-reproductive workers in insect colonies with haplodiploid genetics. Elife 4:e08918

Olejarz JW, Allen B, Veller C, Gadagkar R, Nowak MA (2016) Evolution of worker policing. J Theor Biol 399:103–116

Oliveira RC, Oi CA, do Nascimento MM, Vollet-Neto A, Alves DA, Campos MC, Nascimento F, Wenseleers T (2015) The origin and evolution of queen and fertility signals in Corbiculate bees. BMC Evol Biol 15:254

Oliveira RC, Oi CA, Vollet-Neto A, Wenseleers T (2016) Intraspecific worker parasitism in the common wasp, *Vespula vulgaris*. Anim Behav 113:79–85. https://doi.org/10.1016/j.anbehav.2015.12.025

Oliveira RC, Vollet-Neto A, Oi CA, van Zweden JS, Nascimento F, Brent CS, Wenseleers T (2017) Hormonal pleiotropy helps maintain queen signal honesty in a highly eusocial wasp. Sci Rep 7:1654

Pamilo P (1991) Evolution of colony characteristics in social insects. II. Number of reproductive individuals. Am Nat 138:412–433

Patalano S, Vlasova A, Wyatt C, Ewels P, Camara F, Ferreirab PG, Asher CL, Jurkowski TP, Segonds-Pichon A, Bachman M, Gonzalez-Navarrete I, Minoche AE, Krueger F, Lowy E, Marcet-Houben M, Rodriguez-Ales JL, Nascimento FS, Balasubramanian S, Gabaldon T, Tarver JE, Andrews S, Himmelbauer H, Hughes WOH, Guigo R, Reik W, Sumner S (2015) Molecular signatures of plastic phenotypes in two eusocial insect species with simple societies. Proc Natl Acad Sci U S A 112:13970–13975. https://doi.org/10.1073/pnas.1515937112

Peeters C (1993) Monogyny and polygyny in ponerine ants with or without queens. In: Keller L (ed) Queen Number and Sociality in Insects. Oxford University Press, Oxford, pp 234–261

Peeters C, Liebig J (2009) Fertility signaling as a general mechanism of regulating reproductive division of labor in ants. In: Fewell J, Gadau J (eds) Organization of insect societies: from genome to socio-complexity. Harvard University Press, Cambridge, MA, pp 220–242

Piekarski PK, Carpenter JM, Lemmon AR, Moriarty Lemmon E, Sharanowski BJ (2018) Phylogenomic evidence overturns current conceptions of social evolution in wasps (Vespidae). Mol Biol Evol 35:2097–2109

Platt TG, Queller DC, Strassmann JE (2004) Aggression and worker control of caste fate in a multiple-queen wasp, *Parachartergus colobopterus*. Anim Behav 67:1–10

Princen SA, Oliveira RC, Ernst UR, Millar JG, van Zweden JS, Wenseleers T (2019a) Honeybees possess a structurally diverse and functionally redundant set of queen pheromones. Proc R Soc B 286:20190517. https://doi.org/10.1098/rspb.2019.0517

Princen SA, Van Oystaeyen A, Petit C, van Zweden JS, Wenseleers T (2019b) Cross-activity of honeybee queen mandibular pheromone in bumblebees provides evidence for sensory exploitation. Behav Ecol 31:303–310. https://doi.org/10.1093/beheco/arz191

Queller DC, Strassmann JE, Solís CR, Hughes CR, DeLoach DM (1993) A selfish strategy of social insect workers that promotes social cohesion. Nature 365:639

Quiñones AE, Pen I (2017) A unified model of Hymenopteran preadaptations that trigger the evolutionary transition to eusociality. Nat Commun 8:15920

Ratnieks FLW (1988) Reproductive harmony via mutual policing by workers in eusocial Hymenoptera. Am Nat 132:217–236. https://doi.org/10.1086/284846

Ratnieks FL (1991) Facultative sex allocation biasing by workers in social Hymenoptera. Evolution 45:281–292

Ratnieks FL (1995) Evidence for a queen-produced egg-marking pheromone and its use in worker policing in the honey bee. J Apic Res 34:31–37

Ratnieks FLW, Reeve HK (1992) Conflict in single-queen hymenopteran societies: the structure of conflict and processes that reduce conflict in advanced eusocial species. J Theor Biol 158:33–65

Ratnieks FLW, Visscher PK (1989) Worker policing in the honeybee. Nature 342:796–797

Ratnieks FLW, Wenseleers T (2005) Policing Insect Societies. Science 307:54–56. https://doi.org/10.1126/science.1106934

Ratnieks FLW, Wenseleers T (2008) Altruism in insect societies and beyond: voluntary or enforced? Trends Ecol Evol 23:45–52. https://doi.org/10.1016/j.tree.2007.09.013

Ratnieks FLW, Foster KR, Wenseleers T (2006) Conflict resolution in insect societies. Annu Rev Entomol 51:581–608. https://doi.org/10.1146/annurev.ento.51.110104.151003

Ratnieks FL, Foster KR, Wenseleers T (2011) Darwin's special difficulty: the evolution of "neuter insects" and current theory. Behav Ecol Sociobiol 65:481–492

Reeve HK (1991) Polistes. In: Ross KG, Matthews RW (eds) The social biology of wasps. Cornell University Press, Ithaca

Reeve HK, Keller L (2001) Tests of reproductive-skew models in social insects. Annu Rev Entomol 46:347–385. https://doi.org/10.1146/annurev.ento.46.1.347

Reeve HK, Nonacs P (1997) Within-group aggression and the value of group members: theory and a field test with social wasps. Behav Ecol 8:75–82

Reeve HK, Peters JM, Nonacs P, Starks PT (1998) Dispersal of first "workers" in social wasps: causes and implications of an alternative reproductive strategy. Proc Natl Acad Sci 95:13737–13742

Reeve HK, Starks PT, Peters JM, Nonacs P (2000) Genetic support for the evolutionary theory of reproductive transactions in social wasps. Proc R Soc Lond Ser B 267:75–79

Rehan SM, Toth AL (2015) Climbing the social ladder: the molecular evolution of sociality. Trends Ecol Evol 30:426–433. https://doi.org/10.1016/j.tree.2015.05.004

Ross KG, Matthews RW (eds) (1991) The social biology of wasps. Cornell University Press, Ithaca

Saigo T, Tsuchida K (2004) Queen and worker policing in monogynous and monandrous colonies of a primitively eusocial wasp. Proc R Soc Lond Ser B 271:S509–S512

Seger J (1983) Partial bivoltinism may cause alternating sex-ratio biases that favour eusociality. Nature 301:59

Shreeves G, Field J (2002) Group size and direct fitness in social queues. Am Nat 159:81–95. https://doi.org/10.1086/324125

Shreeves G, Cant MA, Bolton A, Field J (2003) Insurance–based advantages for subordinate co–foundresses in a temperate paper wasp. Proc R Soc Lond Ser B 270:1617–1622. https://doi.org/10.1098/rspb.2003.2409

Sledge MF, Boscaro F, Turillazzi S (2001) Cuticular hydrocarbons and reproductive status in the social wasp *Polistes dominulus*. Behav Ecol Sociobiol 49:401–409

Spradbery J (1971) Seasonal changes in the population structure of wasp colonies (Hymenoptera: Vespidae). J Anim Ecol 40:501–523

Starr CK (1984) Sperm competition, kinship, and sociality in the aculeate Hymenoptera. Sperm Compet Evol Anim Mat Syst 428:459

Stelinski LL, Oakleaf R, Rodriguez-Saona C (2007) Oviposition-deterring pheromone deposited on blueberry fruit by the parasitic wasp, *Diachasma alloeum*. Behaviour 144:429–445. https://doi.org/10.1163/156853907780755996

Strassmann JE (1981) Evolutionary implications of early male and satellite nest production in *Polistes exclamans* colony cycles. Behav Ecol Sociobiol 8:55–64

Strassmann JE (1989) Early termination of brood rearing in the social wasp, *Polistes annularis* (Hymenoptera: Vespidae). J Kansas Entomol Soc 62:353–362

Strassmann JE, Klingler CJ, Arevalo E, Zacchi F, Husain A, Williams J, Seppä P, Queller DC (1997) Absence of within–colony kin discrimination in behavioural interactions of swarm–founding wasps. Proc R Soc Lond Ser B 264:1565–1570

Strassmann JE, Sullender BW, Queller DC (2002) Caste totipotency and conflict in a large–colony social insect. Proc R Soc Lond Ser B 269:263–270

Sumner S, Lucas E, Barker J, Isaac N (2007) Radio-tagging technology reveals extreme nest-drifting behavior in a eusocial insect. Curr Biol 17:140–145. https://doi.org/10.1016/j.cub.2006.11.064

Sumner S, Bell E, Taylor D (2018) A molecular concept of caste in insect societies. Curr Opin Insect Sci 25:42–50

Suzuki T (1986) Production schedules of males and reproductive females, investment sex ratios, and worker-queen conflict in paper wasps. Am Nat 128:366–378

Tannure-Nascimento IC, Nascimento FS, Zucchi R (2008) The look of royalty: visual and odour signals of reproductive status in a paper wasp. Proc R Soc Lond Ser B 275:2555

Tarpy DR, Gilley DC, Seeley TD (2004) Levels of selection in a social insect: a review of conflict and cooperation during honey bee (*Apis mellifera*) queen replacement. Behav Ecol Sociobiol 55:513–523

Taylor D, Bentley MA, Sumner S (2018) Social wasps as models to study the major evolutionary transition to superorganismality. Curr Opin Insect Sci 28:26–32

Tibbetts EA (2007) Dispersal decisions and predispersal behavior in *Polistes* paper wasp 'workers'. Behav Ecol Sociobiol 61:1877–1883

Tibbetts EA, Dale J (2004) A socially enforced signal of quality in a paper wasp. Nature 432:218–222. https://doi.org/10.1038/nature02949

Tibbetts EA, Reeve HK (2000) Aggression and resource sharing among foundresses in the social wasp *Polistes dominulus*: testing transactional theories of conflict. Behav Ecol Sociobiol 48:344–352

Tibbetts EA, Sheehan MJ (2011) Facial patterns are a conventional signal of agonistic ability in *Polistes exclamans* paper wasps. Ethology 117:1138–1146

Toth AL, Rehan SM (2017) Molecular Evolution of Insect Sociality: An Eco-Evo-Devo Perspective. Annu Rev Entomol 62:419–442. https://doi.org/10.1146/annurev-ento-031616-035601

Trivers RL, Hare H (1976) Haploidploidy and the evolution of the social insect. Science 191:249–263. https://doi.org/10.1126/science.1108197

Tsuchida K, Saigo T, Nagata N, Tsujita S, Takeuchi K, Miyano S (2003) Queen-worker conflicts over male production and sex allocation in a primitively eusocial wasp. Evolution 57:2365–2373

Turillazzi S (2012) The evolution of social behaviour in insects and the hover wasps. In: The biology of hover wasps. Springer, New York, pp 233–263

Turillazzi S, West-Eberhard MJ (1996) Natural history and evolution of paper-wasps, vol 400. Oxford University Press, Oxford

Turillazzi S, Sledge MF, Dapporto L, Landi M, Fanelli D, Fondelli L, Zanetti P, Dani FR (2004) Epicuticular lipids and fertility in primitively social wasps (Hymenoptera Stenogastrinae). Physiol Entomol 29:464–471

Van Oystaeyen A, Oliveira RC, Holman L, van Zweden JS, Romero C, Oi CA, d'Ettorre P, Khalesi M, Billen J, Wackers F, Millar JG, Wenseleers T (2014) Conserved class of queen pheromones stops social insect workers from reproducing. Science 343:287–290. https://doi.org/10.1126/science.1244899

van Zweden JS, d'Ettorre P (2010) Nestmate recognition in social insects and the role of hydrocarbons. Insect hydrocarb: Biol, Biochem Chem Ecol 11:222–243

van Zweden JS, Bonckaert W, Wenseleers T, d'Ettorre P (2014) Queen Signaling in Social Wasps. Evolution 68:976–986. https://doi.org/10.1111/Evo.12314

Wenseleers T (2007) Nepotism absent in insect societies—or is it? Mol Ecol 16:3063–3065

Wenseleers T, Ratnieks FL (2004) Tragedy of the commons in *Melipona* bees. Proc R Soc Lond Ser B 271:S310–S312

Wenseleers T, Ratnieks FLW (2006a) Comparative analysis of worker reproduction and policing in eusocial Hymenoptera supports relatedness theory. Am Nat 168:E163–E179. https://doi.org/10.1086/508619

Wenseleers T, Ratnieks FLW (2006b) Enforced altruism in insect societies. Nature 444:50–50. https://doi.org/10.1038/444050a

Wenseleers T, van Zweden JS (2017) Sensory and cognitive adaptations to social living in insect societies. Proc Natl Acad Sci U S A 114:6424–6426

Wenseleers T, Ratnieks FL, Billen J (2003) Caste fate conflict in swarm-founding social Hymenoptera: an inclusive fitness analysis. J Evol Biol 16:647–658

Wenseleers T, Hart AG, Ratnieks FLW (2004a) When resistance is useless: Policing and the evolution of reproductive acquiescence in insect societies. Am Nat 164:E154–E167. https://doi.org/10.1086/425223

Wenseleers T, Helanterä H, Hart A, Ratnieks FL (2004b) Worker reproduction and policing in insect societies: an ESS analysis. J Evol Biol 17:1035–1047

Wenseleers T, Badcock N, Erven K, Tofilski A, Nascimento F, Hart A, Burke T, Archer M, Ratnieks F (2005a) A test of worker policing theory in an advanced eusocial wasp, *Vespula rufa*. Evolution 59:1306–1314

Wenseleers T, Tofilski A, Ratnieks FL (2005b) Queen and worker policing in the tree wasp *Dolichovespula sylvestris*. Behav Ecol Sociobiol 58:80–86

Wenseleers T, Gardner A, Foster KR (2010) Social evolution theory: a review of methods and approaches. In: Szekely T, Moore AJ, Komdeur J (eds) Social behaviour: genes, ecology and evolution. Cambridge University Press, Cambridge, pp 132–158

Wenseleers T, Helantera H, Alves DA, Duenez-Guzman E, Pamilo P (2013) Towards greater realism in inclusive fitness models: the case of worker reproduction in insect societies. Biol Lett 9:20130334. https://doi.org/10.1098/Rsbl.2013.0334

West S (2009) Sex allocation, vol 48. Princeton University Press, Princeton

Wilson EO (1971) The insect societies. Harvard University Press, Cambridge, MA

Wossler TC, Crewe RM (1999) Honeybee queen tergal gland secretion affects ovarian development in caged workers. Apidologie 30:311–320. https://doi.org/10.1051/apido:19990407

Zanette LR, Miller SD, Faria CM, Almond EJ, Huggins TJ, Jordan WC, Bourke AF (2012) Reproductive conflict in bumblebees and the evolution of worker policing. Evolution 66:3765–3777

Chapter 9
Post-embryonic Development in Brazilian Social Wasps

Edilberto Giannotti and Agda Alves da Rocha

Abstract This chapter presents a review of morphological studies of the immature stages of social Brazilian wasps, discussing aspects such as oviposition; instar larva, pre-pupa, and pupal stages; as well as duration times of the immature stages. Those themes have been explored in less than 20% of the species recorded for that country. Most of those articles represent only descriptions of the last instar larva, so that there is still much to explore in terms of comparative studies.

Keywords Polistinae · Instar · Morphology · Larvae · Pupae

9.1 Introduction

Social wasps, as well as other representatives of the order Hymenoptera, show holometabolous development, in which all individuals pass through larval and pupa stages after the eclosion of the egg until the adult phase. The larvae of representative of the family Vespidae are very similar in many species, with most of them being of the vermiform type, with soft and segmented bodies without appendices (apodal). The cephalic capsule, with its buccal parts, and the spiracles are generally more heavily sclcrified than the rest of the body cuticle, therefore being measurable structures for studying larval development (growth). Their distinctive characteristics and classifications are based principally on the morphology of the cephalic capsule and the parts of the buccal apparatus (Reid 1942).

Reid (1942) pioneered the study of the larval morphologies of the Vespidae and likewise considered it to be the most important feature in terms of systematic

E. Giannotti
Departamento de Zoologia, Instituto de Biociências, UNESP – Universidade Estadual Paulista, Rio Claro, SP, Brazil
e-mail: edilberto.giannotti@unesp.br

A. A. da Rocha (✉)
Instituto Multidisciplinar em Saúde, UFBA – Universidade Federal da Bahia, Vitória da Conquista, BA, Brazil
e-mail: rocha.agda@ufba.br

F. Prezoto et al. (eds.), *Neotropical Social Wasps*,
https://doi.org/10.1007/978-3-030-53510-0_9

characters. This author comparatively studied 51 species of mature larvae (from the last instar) of both solitary and social wasps and prepared an identification key for the subfamilies. Additional studies relevant to identifying immature social wasps were published in the 1970s. In Brazil, Dias-Filho (1975) described the larval morphologies of 30 species of social wasps. Richards (1978) proposed a key for the identification of the larvae of the American genera of Vespidae. Wheeler and Wheeler (1979) proposed a classification of the larval body forms of Hymenoptera and noted that the "vespoid" type is the most common, except among the larvae of *Mischocyttarus* de Saussure, 1853, which have ventral lobes on their first abdominal segments known as the "mischocyttaroid" type. Additionally, Spradbery (1973), Edwards (1980) and Chao and Hermann (1983) make excellent contribution as the post-embryonic development of social wasps occurs.

The present chapter presents a review of morphological studies of the immature stages of wasps, indicating the time required for egg to adult development among social Brazilian wasps.

9.2 Ovipositioning

The queens of social wasp colonies are females with more developed ovaries, and the fertilization of the eggs occurs when they pass from the ovariole to the common oviduct through a structure called the spermatheca. There, a set of stored spermatozoids are released and penetrate through the micropile – the funnel-shaped microscopic canal in the chorion of the egg (Spradbery 1973).

Before ovipositioning, the female wasp examines each cell with its antenna to confirm that it is empty, as well as to determine the position in which the egg will be fixed. The queen then inserts its abdomen into the cell, moving its body to attain an angle of approximately 45° in relation to the cell wall, and spreads a secretion from an accessory gland on its reproductive organ to fix the egg to the wall (Edwards 1980, Spradbery 1973), and spreads a secretion from an accessory gland on its reproductive organ to fix the egg to the wall. That behavior usually requires a few minutes (mean of 0.9–9.5 min) in the Brazilian species that have so far been studied (Table 9.1).

In a new nest, in a pre-emergence stage, the queen will put one egg in each cell, usually just after the construction of a shallow calyx, with the egg being fixed to the cell corner closest to the central axis of the comb. One reason for the egg being placed in that position is that the external faces of cells under construction are rounded and therefore without reference angles (Fig. 9.1a). In a large nest, with reutilized cells, the queen will affix the egg to any of the angles of the cell, at a depth that can vary from half to two thirds its depth (Fig. 9.1b).

Table 9.1 Mean duration (in minutes) of the ovipositioning behaviors of the females of social wasp species occurring in Brazil

Species	Ovipositioning (min)	References
Mischocyttarus cerberus styx Richards, 1940	6.1 ± 2.9	Giannotti (1999)
Mischocyttarus consimilis Zikán, 1949	3.4 ± 0.8	Torres (2009)
Mischocyttarus labiatus (Fabricius, 1804)	6.9	Litte (1981)
Polistes canadensis (Linnaeus, 1758)	9.5 ± 3.2	Torres et al. (2009)
Polistes lanio (Fabricius, 1775)	4.6 ± 1.8	Giannotti and Machado (1999)
Polistes versicolor (Olivier, 1791)	2.7 ± 1.1	Gobbi (1977)
Agelaia pallipes (Olivier, 1791)	1.8 ± 0.3	Simões (1977)
Parachartergus fraternus (Gribodo, 1891)	0.9	Mateus (2005)
Protopolybia exigua (Saussure, 1854)	1.8 ± 0.6	Simões (1977)

Fig. 9.1 (**a**) Nest of *Polistes actaeon* (Haliday, 1836) in a pre-emergence stage, showing three upper cells containing eggs. (Photograph by Dr. André Sunao Nishiuchi Murakami); (**b**) queen of *Mischocyttarus cerberus styx* (Richards, 1940) laying egg in a recently constructed cell. (Photograph by Dr. Olga Coutinho Togni)

9.3 Morphologies of the Immature Stages

Wasp eggs are narrow at their base (which is fixed to the cell wall), but rounded and wide at the apical extremity, and may be curved (Fig. 9.2a). They are normally white (with an external transparent, thin, and fragile chorion) and of varying sizes, although only a few millimeters long (Table 9.2). In more advanced stages, segmentation can often be observed in the developing embryo.

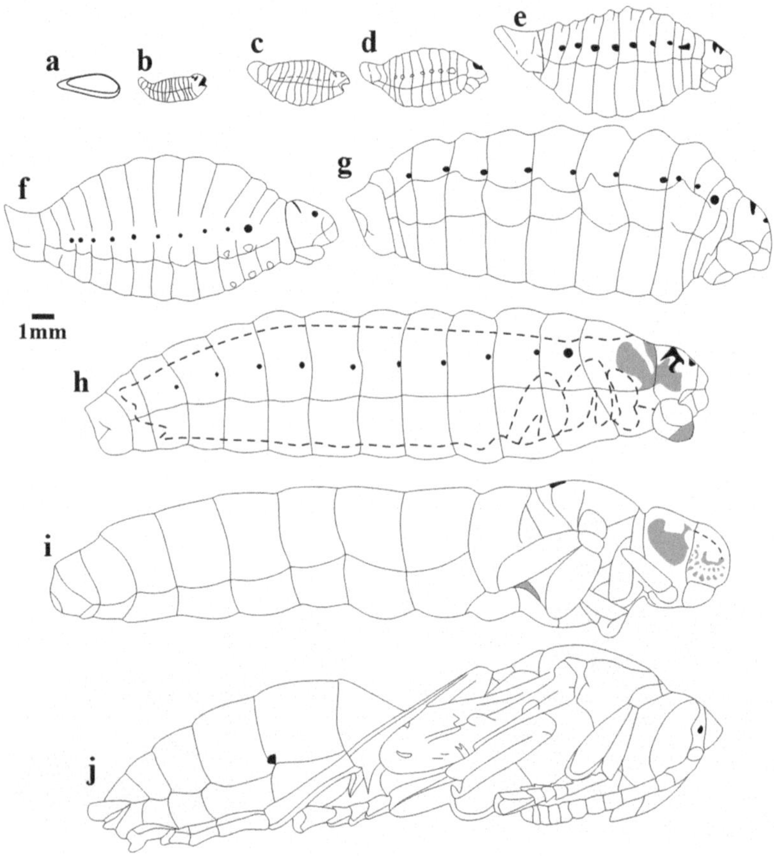

Fig. 9.2 Immature stages of *Polistes lanio* (Fabricius, 1775) in lateral view: (**a**) egg; (**b** and **c**) first larval instar (L1); (**d**) L2; (**e**) L3; (**f**) L4; (**g**) L5; (**h**) pre-pupa; (**i**) pre-pupa at the moment of change to a pupa; (**j**) pupa. (Adapted from Giannotti 1995)

9.3.1 Larval Stage

Social wasps develop through three to five instar stages, with five being more frequent (Fig. 9.2b–g); three instars are only observed in *Angiopolybia pallens* (Lepeletier, 1836) by Cruz (2006) (Table 9.3). Higher instar numbers are generally associated with greater durations of their immature stages. The instar number is determined by measurements of the width of the cephalic capsule (Reid 1942, Dias-Filho 1975, Edwards 1980, Parra and Haddad 1989), which is the hardest part of the larvae (and generally light brown in color).

Other parameters may also be evaluated. Solis et al. (2012), for example, utilized other characters in addition to the cephalic capsule in their study of the external morphology of *Polybia paulista* von Ihering 1896, including the diameter of the spiracles, noting their modifications among the different instar larvae.

Table 9.2 Mean lengths and widths of eggs (in millimeters) observed in social wasp species occurring in Brazil

Species	Length (mm)	Width (mm)	References
Mischocyttarus cassununga (von Ihering, 1903)	1.36 ± 0.07	0.48 ± 0.04	Giannotti and Silva (1993)
Mischocyttarus cerberus styx Richards, 1940	1.31 ± 0.07	0.46 ± 0.03	Giannotti (2006a)
Mischocyttarus drewseni (Saussure, 1857)	1.80 ± 0.15	0.62 ± 0.05	Giannotti and Trevisolli (1993)
Mischocyttarus latior (Fox, 1898)	1.63 ± 0.15	0.55 ± 0.06	Cecílio et al. (2015)
Mischocyttarus nomurae (Richards, 1978)	1.03 ± 0.12	0.40 ± 0.03	Rocha and Giannotti (2016)
Polistes cinerascens Saussure, 1854	1.90 ± 0.20	0.80 ± 0.10	Giannotti (1997)
Polistes lanio lanio (Fabricius, 1775)	2.30 ± 0.16	0.91 ± 0.07	Giannotti (1995)
Polistes simillimus Zikán, 1951	1.18 ± 0.08	0.52 ± 0.05	Prezoto and Gobbi (2005)
Angiopolybia pallens (Lepeletier, 1836)	–	0.52 ± 0.09	Cruz (2006)
Polybia paulista von Ihering, 1896	0.93–1.23	0.31–0.49	Solis et al. (2012)

The growth ratios of larvae vary from 1.30 to 1.64, with most species having values near 1.4 (Table 9.3), and they appear to follow the rule suggested by Dyar, which was initially proposed for Lepidoptera species (Dyar, 1890). That rule, according to Parra and Haddad (1989), states that the cephalic capsule of the larvae will increase its width at a constant ratio with each ecdysis, growing as a geometric progression.

Apodal and soft-bodied larvae have three short thoracic segments and ten larger abdominal segments (Fig. 9.3). They have ten pairs of lateral respiratory openings: the first pair of spiracles opens between the prothorax and the mesothorax. The second pair of spiracles opens between the meso- and metathorax, while the others are located along the abdomen, with the last opening on the eighth abdominal segment (Reid 1942). The first thoracic spiracles of larvae of wasps of the genus *Mischocyttarus* are at least twice the diameter of the other spiracles, as could be seen in *Mischocyttarus nomurae* Richards, 1978, whose first larval spiracles of the fifth instar have diameters of 0.26 mm, and are approximately three times larger than the others (Richards 1978, Rocha and Giannotti 2016).

The head, or cephalic capsule, has an inverted Y-shaped epicranial suture (which is most visible when the cuticle is more heavily sclerified) (Reid 1942). The base of the "Y" forms a median suture, and the bifurcations form frontal sutures. The median suture represents a fracture line in the cuticle at the moment of ecdysis (ecdysial cleavage line). There are two small, deep depressions at the base of the clypeus, called the anterior tentorial pits (the tentorium is an internal supporting structure of the cephalic capsules of larvae and adults) (Fig. 9.4a).

The antennae of the larvae are represented by imaginal discs. A pair of temporal bands or well-defined ocular lines represent the rudiments of compound eyes. The mouthparts consist of the labrum, which is a single plaque that covers the mandibles

Table 9.3 Numbers of larval instars and the growth ratios of various species of social wasps occurring in Brazil

Species	Instar larvae	Growth ratio	References
Mischocyttarus atramentarius Zikán, 1949	4	1.42	Silva (1984)
Mischocyttarus extinctus Zikán, 1935	4	1.46	Raposo-Filho (1981)
Mischocyttarus cassununga (von Ihering, 1903)	5	1.42	Giannotti and Fieri (1991)
Mischocyttarus cerberus styx Richards, 1940	5	1.44	Giannotti (2006a)/Silva et al. (2020)
Mischocyttarus drewseni (Saussure, 1857)	5	1.38	Giannotti and Trevisolli (1993)
Mischocyttarus latior (Fox, 1898)	5	1.37	Cecílio et al. (2015)
Mischocyttarus nomurae Richards, 1978	5	1.48	Rocha and Giannotti (2016)
Polistes canadensis (Linnaeus, 1758)	5	–	Rodrigues (1968)
Polistes carnifex (Fabricius, 1775)	5	–	Rodrigues (1968)
Polistes cinerascens Saussure, 1854	5	1.3	Giannotti (1997)
Polistes lanio lanio (Fabricius, 1775)	5	1.41	Giannotti (1995)
Polistes simillimus Zikán, 1951	5	1.44	Prezoto and Gobbi (2005)
Polistes versicolor (Olivier, 1791)	5	–	Rodrigues (1968)
Agelaia vicina (Saussure, 1854)	5	1.39	Machado and Curado (1996)
Angiopolybia pallens (Lepeletier, 1836)	3	1.64	Cruz (2006)
Apoica pallens (Fabricius, 1804)	5	–	Silva et al. (2006)
Brachygastra lecheguana (Latreille, 1824)	5	1.38	Machado et al. (1988)
Polybia chrysothorax (Lichtenstein, 1796)	5	1.46	Mello and Giannotti (1994)
Polybia fastidiosuscula Saussure, 1854	5	1.4	Tech and Machado (1989)
Polybia occidentalis cinctus (Provancher, 1888)	4	–	Hebling and Machado (1974)
Polybia occidentalis occidentalis (Olivier, 1791)	5	1.3	Machado (1977a)
Polybia paulista von Ihering, 1896	5	1.41	Carvalho and Silva (1975) Machado (1983)
Polybia paulista von Ihering, 1896	5	1.39	Solis et al. (2012)
Protopolybia chartergoides (Gribodo, 1891)	5	–	Silveira (1994)
Protopolybia exigua (Saussure, 1854)	4	1.35	Machado (1974)
Protopolybia sedula (Saussure, 1854)	4	1.35	Machado (1977b)

Adapted from Cruz (2006)

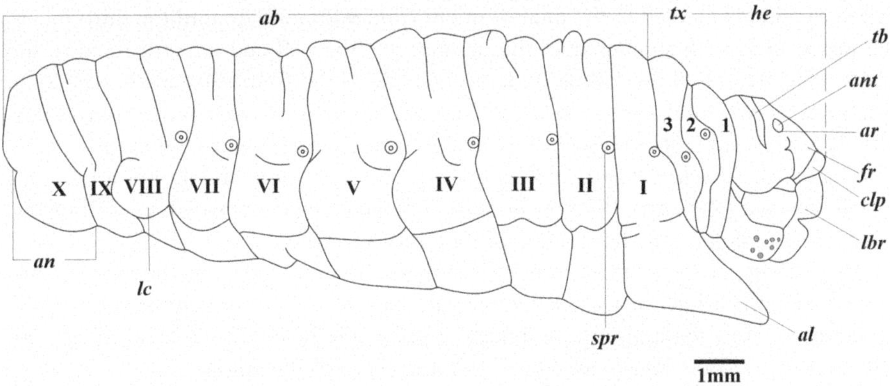

Fig. 9.3 A fifth instar larva (L5) of *Mischocyttarus drewseni* (Saussure, 1857) in lateral view, showing the different regions of the body and its structures (Giannotti & Trevisolli 1993): ab, abdomen; al, abdominal lobe; an, anal segment; ant, antenna; ar, antennal ring; clp, clypeus; fr, frons; he, head; lbr, labrum; lc, lateral corcova; spr, spiracle; tb, temporal band; tx, thorax; 1 to 3, segments of the thorax; I to X, segments of the abdomen

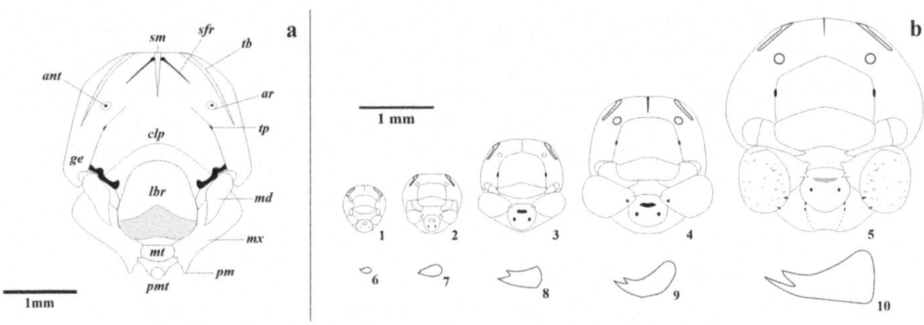

Fig. 9.4 (**a**) Frontal view of the cephalic capsule of a fifth instar larva of *Mischocyttarus drewseni* (Saussure, 1857) (Giannotti & Trevisolli 1993): ant, antenna, ar, antennal ring; clp, clypeus; fs, frontal suture, ge, gena; lbr, labrum; md, mandible; mp, maxillary papilla; ms, median suture; mt, mentum (labium); mx, maxilla; pm, papilla of the mentum; tb, temporal band; tp, anterior tentorial pit; (**b**) frontal view of the cephalic capsules of the first to fifth instar larvae of *Polistes lanio lanio* (Fabricius, 1775) (1 to 5) evidencing the development of the mandible (6 to 10). (Adapted from Giannotti 1995)

(except in the first two instars). The mandibles are more sclerified at their extremities, where one or more teeth may occur, depending on the species (Fig. 9.4b). The teeth, when present, increase with each instar. The maxillas are voluminous and have variable numbers of sensorial papillae at their extremities. The labium also has a pair of sensorial papillae and the openings of the labial glands. Structures such as bristles are generally visible on mature larvae but not visible on first instars (Spradbery 1973).

When eclodes, the larva will have an inverted position in the cell and be fixed to one of the cell walls by the remnants of the chorion and a viscous secretion produced by glands located in the anal segment (Fig. 9.2b–e). Up until the third instar, the larvae are much smaller than the diameter of the cell and are still fixed to the walls. Usually, when the larvae undergo ecdysis to the fourth instar, they become free to move within the interior of the cell, but are held by body projections: laterally located pleural lobes (Fig. 9.2f–g, Fig. 9.3) and dorsal bridges (Giannotti 1995). The larvae of *Mischocyttarus* also show one, two, or three abdominal lobes projected forward (Richards 1978) (Fig. 9.3). The anal segment, which is well-developed and glandular, may have a dorsal lobe in some species .

The fifth instar shows increased body volume, so that it takes up the entire diameter of the cell. Adult wasps will increase the heights of the cells in the nest as the larvae grow, thus maintaining the buccal parts of the larvae at the level of the cell edge to receive food (masticated prey and nectar) from the adults.

Edwards (1980) provided descriptions of the anatomies and physiologies of immature social wasps. The larva have a digestive tube with the posterior intestine atrophied and separated from the median intestine by a thin membrane. As the larvae grow, their feces accumulate at the end of the median intestine and are eliminated in the last ecdysis, before pupation. As such, the larvae only defecate once during their development, forming a layer of meconium that is deposited at the bottom of the cell at each occupation, thus avoiding any pollution of the narrow cells where they develop. Malpighian tubules are present in the larvae but do not have any function. Their labial glands are enormous and produce "larval saliva" (secretions that adults feed on); at the end of the last instar, those glands alter their function and begin to secrete the silk used to weave the cocoon (Edwards 1980).

9.3.2 Weaving the Silk Cocoon

After completing their days of feeding and growth, the mature larvae begin to weave cocoons, using the secretions of their salivary glands that begin to produce silk. The larvae turn about their axis, fixing the silk to the cell edge and weaving a ring/collar (with the central orifice gradually diminishing, until it completely closes off the cell). The larvae then continue their work internally, turning and twisting and covering the walls to the bottom of the cell; they will subsequently return to their original position (Chao and Hermann 1983). The external phase of pupal capsule construction was described by Togni et al. (2010) for the larvae of *Mischocyttarus parallelogrammus* Zikán, 1935 as follows: the larvae begin to weave the cocoon, fixing the silk to the edge of one of the walls of the cell and then attaching it to the other side. That action is subsequently repeated in an irregular fashion. The silk threads cross the cell opening many times, gradually closing it and forming a thick web until it is totally sealed, approximately 2 hours later. The adults will subsequently cover the pupal capsule with wood pulp.

9.3.3 *Pre-pupa or Pharate Pupa*

When the larva returns to its initial position, it defecates, eliminating its fecal mass within the cocoon where it becomes incorporated into the bottom of the cell (the meconium layer) (Chao and Hermann 1983). The larva then begins its metamorphosis. The larva has now attained the pre-pupa or pharate pupa phase, which is not considered an actual stage because there is no alteration of the cephalic capsule width, which remains the same as in the last instar (Giannotti 1995). This phase is marked by the elongation of the pro-, meso-, and metathorax and the first abdominal segment (propodeum). Due to the transparency of the tegument, it is possible to see the pupal structures being formed: legs, eyes, and often antennae (Fig. 9.2h). Figure 9.2i illustrates the moment in which the pre-pupa of *Polistes lanio lanio* (Fabricius, 1775) is undergoing ecdysis to the pupal stage, having legs and wings (although still short) already free from the larval exoskeleton.

9.3.4 *The Pupal Stage and Adult Emergence*

The pupa is an exarate type, with its appendices not adhered to its body (Fig. 9.2j). The wings are not initially distended, being saculiform, short, and voluminous; later, at the end of the pupal phase, it will take on its final form. The sclerotization of the cuticle and the pigmentation of the body will continue until the end of the pupal phase.

Exiting from the cocoon is generally performed without the assistance of adults. The fully formed wasp will cut the pupal capsule with its mandibles and soften the silk with its saliva (which is not, however, digested). Its first emergent behavior is to clean its own body. In addition to the external morphological differences between phases, they also appear to differ in terms of their chemical composition, considering the cuticles of those insects in their egg, larval, pupal, and adult phases (Michelutti et al. 2017). That more recent approach of examining the levels, compositions, and functions of cuticular hydrocarbons among immatures, as discussed by Silva et al. (2020), needs to be more closely studied in Brazil.

9.4 Duration Times of the Immature Stages

Periodic maps are normally made up of the nests in order to study the duration times of the immature stages, recording the numbers of cells and their respective contents: egg, larvae, pupa, or empty. The cells on the map are numbered and that data reproduced in a table so that the duration times of the different immature stages can be calculated. It is also possible to determine the number of times a given cell has been

Table 9.4 Duration time (in days) of the immature stages of various species of social wasps occurring in Brazil

Species	Egg	Larva	Pupa	Total	References
Mischocyttarus atramentarius Zikán, 1949	12.7	25.1	16.8	54.8	Silva (1984)
Mischocyttarus cassununga (von Ihering, 1903)	13.2	32.6	15.6	61.2	Giannotti and Fieri (1991)
Mischocyttarus cerberus styx Richards, 1940	11.7	31.9	19.8	61.7	Giannotti (2006b)
Mischocyttarus drewseni (Saussure, 1857) (PA)	11.1	20.2	14.8	46.1	Jeanne (1972)
Mischocyttarus drewseni (Saussure, 1857) (PR)	11.3	34.7	19.9	65.9	Dantas-de-Araujo (1980)
Mischocyttarus drewseni (Saussure, 1857) (SP)	15.1	26.5	18.9	58.0	Giannotti and Trevisoli (1993)
Mischocyttarus extinctus Zikán, 1935	11.4	20.6	16.8	48.8	Raposo-Filho (1981)
Mischocyttarus labiatus (Fabricius, 1804)	16.1	16.1	16.3	48.5	Litte (1981)
Mischocyttarus latior (Fox, 1898)	14.1	36.9	16.4	67.4	Cecílio et al. (2015)
Polistes cinerascens Saussure, 1854	13.0	23.7	22.2	58.6	Giannotti (1997)
Polistes erythrocephalus Latreille, 1813	17.1	26.6	23.8	67.5	West-Eberhard (1969)
Polistes lanio lanio (Fabricius, 1775)	20.8	40.6	22.6	87.5	Giannotti and Machado (1994)
Polistes versicolor (Olivier, 1791)	10.0	20.9	18.4	49.4	Gobbi (1977)
Polistes simillimus Zikán, 1951	10.2	25.3	18.7	51.9	Giannotti (1994)
Polistes subsericius Saussure, 1854	10.0	32.7	13.0	55.7	Giannotti (1995)
Agelaia pallipes (Olivier, 1791)	6.3	9.1	14.7	30.1	Simões (1977)
Protopolybia sedula (Saussure, 1854)	5.0	8.0	10.0	25.0	Machado (1977b)
Protopolybia exigua (Saussure, 1854) (27 °C)	3–5	18–20	20	41–45	Machado (1974)
Protopolybia exigua (Saussure, 1854) (20 °C)	5	28	28–30	60–63	Machado (1974)

utilized, the productivity of the colony, the immature/adult ratio, and detect cases of oophagy, larviphagy, pupa eating, parasitism and larval predation.

The mean duration times of the immature stages of social wasps are quite variable, with *Protopolybia sedula* (Saussure, 1854) requiring the shortest time (25 days) while *P. lanio* shows the longest development times of its immature stages as compared to other species of social wasps (87.5 days) (Table 9.4). The variations observed in the durations of the immature stages may be related to various factors, such as temperature, relative humidity and local rainfall (see *M. drewseni and P. exigua* in Table 9.4).

Some authors have studied the duration times of the immature stages of *M. cerberus styx*, *M. latior*, and *P. lanio* throughout the different seasons of the year, showing that development times are longer in the winter than in the summer (Fig. 9.5)

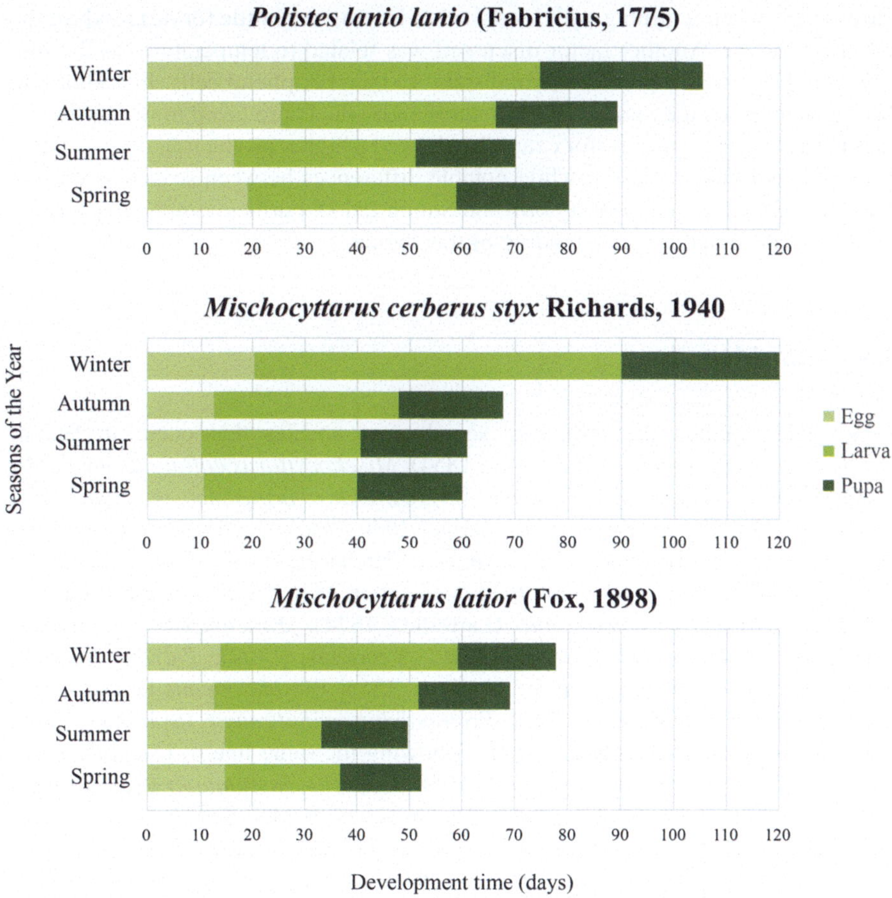

Fig. 9.5 Development times (in days) of three social wasp species during different seasons of the year in Brazil

(Giannotti 2006b and Silva et al. 2020, Cecílio et al. 2015 and Giannotti and Machado 1994, respectively).

Colony phase, nest size and, consequently, the number of wasps caring for the offspring can also influence the development times of the immature stages. Giannotti (2006b), for example, observed that the mean incubation times of the eggs of *M. cerberus* and larval and pupal development were longer in small nests than in large nests.

Examining *M. latior*, Cecílio et al. (2015) observed that the development times of the immature stages can also vary according to their location in the nest, so that immatures held in the first seven cells constructed tend to have larvae with shorter development times than those raised in other cells within the nest. Those authors noted that the larvae located in those central cells received more food resources from the adults, as the center of the nest tends to be the first location where foraging wasps alight when returning with food – often leaving very little (or no) food for the peripheral larvae. Another factor discussed was related to temperature, as the first cells could experience greater thermal stability than peripheral cells. In addition to development time, the numbers of immature individuals produced may vary in different seasons. *M. cerberus*, for example, showed a higher production of immatures in the hot and humid season, while notable differences between seasons were not noted with *M. cassununga* (Biagiotto and Shima 2017). Further comparative studies would provide a better understanding of this theme.

9.5 Considerations

Reid (1942) studied the following social wasp species that occur in Brazil: *Mischocyttarus carbonarius* (Saussure, 1854), *Mischocyttarus collarelus* Richards, 1940, *Mischocyttarus lecointei* (Ducke, 1904), *Mischocyttarus metoecus* (Richards, 1940), *Mischocyttarus superus* Richards, 1940, *Mischocyttarus surinamensis* (Saussure, 1854), *Mischocyttarus synoecus* Richards, 1940, *Polistes carnifex* (Fabricius, 1775), *Angiopolybia pallens* (Lepeletier, 1836), *Apoica pallida* (Olivier, 1791), *Brachygastra lecheguana* (Latreille, 1824), *Brachygastra scutellaris* (Fabricius, 1804), *Metapolybia cingulata* (Fabricius, 1804), *Polybia bistriata* (Fabricius, 1804), *Polybia catillifex* Moebius, 1856, *Polybia micans* Ducke, 1904, *Polybia rejecta* (Fabricius, 1798), *Protopolybia minutissima* (Spinola, 1851), *Protopolybia sedula* (Saussure, 1954). Following the same line of research, Dias-Filho (1975) described the larvae of *Mischocyttarus cassununga* (von Ihering, 1903), *Mischocyttarus cerberus styx* Richards, 1940, *Mischocyttarus drewseni* (Saussure, 1857), *Mischocyttarus latior* (Fox, 1898), *Mischocyttarus paraguayensis* (Zikán, 1935) (= *M. araujoi*), *Mischocyttarus rotundicolis* (Cameron, 1912), *Polistes canadensis* (Linnaeus, 1758), *Polistes cinerascens* (Saussure, 1854), *Polistes simillimus* Zikán, 1951, *Polistes versicolor* (Olivier, 1972), *Agelaia meridionalis* (Ihering, 1904), *Agelaia pallipes* (Olivier, 1791), *Apoica flavissima* Van der Vecht, 1973, *Apoica pallens* (Fabricius, 1804), *Brachygastra augusti* (Saussure, 1854), *Brachygastra lecheguana* (Latreille, 1824), *Parachartergus fraternus* (Gribodo, 1892), *Parachartergus pseudapicalis* (Willink, 1959), *Polybia dimidiata* (Olivier, 1791), *Polybia fastidiosuscula* Saussure, 1854, *Polybia ignobilis* (Haliday, 1836), *Polybia jurinei* Saussure, 1854, *Polybia occidentalis* (Olivier, 1791), *Polybia paulista* von Ihering 1896, *Protonectarina sylveirae* (Saussure, 1854), *Protopolybia exigua* (Saussure, 1854), *Protopolybia sedula* (Saussure, 1854), and *Synoeca cyanea* (Fabricius, 1775), while Giannotti (1998) described the larvae of *Agelaia multipicta* (Haliday, 1836) and Silveira (1994) described the larvae of *Protopolybia chartergoides* (Gribodo, 1891).

Those types of descriptive studies, combined with molecular tools, can make important contributions to phylogenetic reconstructions of the group, as well as to the classification of social wasps. Of the 381 social wasp species recorded for Brazil (see Chap. 16), aspects of post-embryonic development have been studied in approximately 16% (61 species), considering the classic works described above and those that have followed in more than seven decades of research concerning the immatures of Brazilian social wasps. Most of those articles represent only descriptions of the last instar larvae, so that there is still much to explore in terms of comparative studies, their behaviors, descriptions of the other instar larvae and stages, developmental times, and variations related to different climatic and spatial distributions, among other aspects.

References

Biagiotto RH, Shima SN (2017) Comparative Study of the Development of Mischocyttarus cassununga Von Ihering and Mischocyttarus cerberus styx Richards Colonies (Hymenoptera, Vespidae, Mischocyttarini). EntomoBrasilis 10(3):170–177

Carvalho GG, Silva MT (1975) Alguns aspectos do desenvolvimento larval de *Polybia paulista* Richards (Hymenoptera: Vespidae). Studia Entomol 8:555–568

Cecílio DSS, Rocha AA, Giannotti E (2015) Post-embryonic development of *Mischocyttarus latior* (Fox) (Hymenoptera: Vespidae). Sociobiology 62:446–449

Chao JT, Hermann HR (1983) Spinning and external ontogenetic changes in the pupae of *Polistes annularis* (Hymenoptera: Vespidae, Polistinae). Insect Soc 30:496–507

Cruz JD (2006) Aspectos Bioecológicos de *Angiopolybia pallens* (Lepeletier, 1836) (Hymenoptera: Vespidae). Ph.D. thesis, Universidade Estadual Paulista Júlio de Mesquita Filho

Dantas-de-Araújo CZ (1980) Bionomia e comportamento social comparado de *Mischocyttarus drewseni* de Saussure, 1987, nas regiões subtropical (Curitiba, PR) e tropical (Belém, PA) do Brasil (Hymenoptera, Vespidae). Ph.D. thesis, Universidade Federal do Paraná

Dias-Filho MM (1975) Contribuição à Morfologia de Larvas de Vespídeos sociais do Brasil (Hymenoptera, Vespidae). Rev Bras Entomol 19:1–36

Dyar HG (1890) The number of molts of lepidopterous larvae. Psyche 5:420–422

Edwards R (1980) Social wasps, their biology and control. Rentokil, Felcourt

Giannotti E (1994) Notes on the biology of *Polistes simillimus* Zikán (Hymenoptera, Vespidae). Bioikos 8:41–49

Giannotti E (1995) Immature stages of *Polistes lanio* (Fabricius, 1775) (Hymenoptera, Vespidae). Rev Bras Biol 55:527–531

Giannotti E (1997) Biology of the Wasp *Polistes (Epicnemius) cinerascens* Saussure (Hymenoptera: Vespidae). An Soc Entomol Bras 26:61–67

Giannotti E (1998) On the nest of *Agelaia multipicta* (Haliday, 1836) and description of the mature larva (Hymenoptera, Vespidae). Rev Bras Entomol 42:97–99

Giannotti E (1999) Social organization of the eusocial wasp *Mischocyttarus cerberus styx* (Hymenoptera: Vespidae). Sociobiology 33:325–338

Giannotti E (2006a) Caracterização morfológica dos estágios imaturos e número de ínstares larvais da vespa social Mischocyttarus cerberus (Hymenoptera: Vespidae). In: Abstracts of the XXI Congresso Brasileiro de Entomologia, Universidade Federal Rural de Pernambuco, Recife, 06–11 August 2006

Giannotti, E. (2006b). Fatores que afetam o tempo de desenvolvimento dos estágios imaturos da vespa social *Mischocyttarus cerberus* (Hymenoptera: Vespidae). In: Abstracts of the XXI Congresso Brasileiro de Entomologia, Universidade Federal Rural de Pernambuco, Recife, 06–11 August 2006

Giannotti E, Fieri SR (1991) On the brood of *Mischocyttarus (Monocyttarus) cassununga* (Ihering, 1903) (Hymenoptera, Vespidae). Rev Bras Entomol:263–267

Giannotti E, Machado VLL (1994) The seasonal variation of brood stages duration of *Polistes lanio* (Fabricius, 1775) (Hymenoptera, Vespidae). Naturalia 19:97–102

Giannotti E, Machado VLL (1999) Behavioral castes of the social wasp *Polistes lanio* Fabricius (Hymenoptera, Vespidae). Rev Bras Entomol 43:185–190

Giannotti E, Silva CV (1993) *Mischocyttarus cassununga* (Hymenoptera, Vespidae): external morphology of the brood during the post-embryonic development. Rev Bras Entomol 37:309–312

Giannotti E, Trevisolli C (1993) Desenvolvimento pós-embrionário de *Mischocyttarus drewseni* Saussure, 1857 (Hymenoptera - Vespidae). Insecta 2:41–52

Gobbi N (1977) Ecologia de *Polistes versicolor* (Hymenoptera, Vespidae). Tese de Doutorado. Universidade de São Paulo, Ribeirão Preto, São Paulo, p 229f

Hebling NJ, Machado VLL (1974) Análise populacional e biométrica em *Polybia occidentalis* var. *cincta*. Ciência e Cultura 16:340–341

Jeanne RL (1972) Social biology of the neotropical wasp *Mischocyttarus drewseni*. Bull Mus Comp Zool 144:63–150

Litte M (1981) Social biology of the polistine wasp *Mischocyttarus labiatus*: survival in a Colombia rain forest. Smithson Contrib Zool 317:1–26

Machado VLL (1974) Aspectos biológicos de *Protopolybia exigua exigua* (Saussure, 1854) (Hym., Vespidae). Ph.D. thesis, Escola Superior de Agricultura "Luiz de Queiroz"

Machado VLL (1977a) Estudos biológicos de *Polybia occidentalis occidentalis* (Olivier, 1791) (Hym. – Vespidae). An Soc Entomol Bras 6:7–24

Machado VLL (1977b) Aspectos da biologia de *Protopolybia pumila* (Saussure, 1863) (Hym., Vespidae). Rev Bras Biologia 37:771–784

Machado VLL (1983) Análises morfométricas em colônias de *Polybia (Myrapetra) paulista* (Ihering, 1896) (Hymenoptera, Vespidae). Naturalia 8:219–226

Machado VLL, Curado K (1996) Análise Populacional e Morfométrica em uma colônia de *Agelaia vicina (de Saussure, 1854) na fase reprodutiva*. Bioikos 9–10:7–15

Machado VLL, Gravena S, Giannotti E (1988) Análise populacional e morfométrica em uma colônia de *Brachygastra lecheguana* (Latreille, 1824) na fase reprodutiva. An Soc Entomol Bras 17:491–506

Mateus S (2005) Análise dos comportamentos envolvidos na organização social e no processo de enxameio de *Parachartergus fraternus* (Hymenoptera, Polistinae, Epiponini). Ph.D. thesis, Faculdade de Filosofia, Ciências e Letras de Ribeirão Preto – USP

Mello ALR, Giannotti E (1994) Determinação do número de instares larvais da vespa social *Polybia chrysothorax*. In: Proceedings of the VI Congresso de Iniciação científica da UNESP, Universidade Estadual Paulista Júlio de Mesquita Filho, Guaratinguetá

Michelutti KB, Cardoso CAL, Antonialli-Junior WF (2017) Avaliação da Assinatura Química dos Estágios de Desenvolvimento *de Mischocyttarus consimilis* Zikán (Hymenoptera, Vespidae) Empregando Cromatografia Gasosa Acoplada à Espectrometria de Massas. Revista Virtual de Química 9:535–547

Parra JRP, Haddad ML (1989) Determinação do número de ínstares larvais de insetos. FEALQ, Piracicaba

Prezoto F, Gobbi N (2005) Morfometria dos estágios imaturos de *Polistes simillimus* Zikán, 1951 (Hymenoptera, Vespidae). Rev Bras Zoociências 7:47–54

Raposo-Filho JR (1981) Biologia de *Mischocyttarus (Monocyttarus) extinctus* Zikán, 1935 (Polistinae – Vespidae). Dissertation, Universidade Estadual Paulista Júlio de Mesquita Filho

Reid JA (1942) On the classification of the larvae of the Vespidae (Hymenoptera). T Ent Soc London 92:285–331

Richards OW (1978) The social wasps of the Americas excluding the Vespinae. British Museum (Natural History), London

Rocha AA, Giannotti E (2016) External morphology of immatures during the post-embryonic development of *Mischocyttarus nomurae* Richards (Hymenoptera, Vespidae). Sociobiology 63:998–1004

Rodrigues VM (1968) Estudo sobre vespas sociais do Brasil (Hymenoptera, Vespidae): estudo sobre populações de vespas sociais dos gêneros *Polistes* e *Polybia* (Vespidae-Polistinas e Polybiinae). Ph.D. thesis, Faculdade de Filosofia, Ciências e Letras de Rio Claro

Silva MN (1984) Aspectos do desenvolvimento e do comportamento de *Mischocyttarus* (*Kappa*) *atramentarius* Zikan, 1949 (Hymenoptera – Vespidae). Ph.D. thesis, Universidade Estadual Paulista Júlio de Mesquita Filho

Silva CAD, Pires EM, Lacerda MB et al (2006) Immature stages and nest characteristics of *Apoica pallens* (Hymenoptera, Vespidae). Sociobiology 48:365–374

Silva RC, Assis DS, Souza AR, Nascimento FS, Giannotti E (2020) Notes on Brood Morphology and Development of the Neotropical Eusocial Wasp *Mischocyttarus cerberus* (Hymenoptera, Vespidae, Polistinae) Sociobiology 67:301–307

Silveira OT (1994) External morphology of the larva of *Pseudochartergus chartergoides* (Gribodo) (Hym., Vespidae, Polistinae). Entomol Mon Mag 130:71–76

Simões D (1977) Etologia e diferenciação de casta em algumas vespas sociais (Hymenoptera, Vespidae). Ph.D. thesis, Universidade de São Paulo

Solis DR, Dias NB, Fox EGP (2012) External morphology of the Immatures of *Polybia paulista* (Hymenoptera: Vespidae). Flo Entomol 95:890–899

Spradbery JP (1973) Wasps: An account of the biology and natural history of social and solitary wasps. University of Washington Press, Seattle.

Tech GM, Machado VLL (1989) Análise morfométrica em colônias de *Polybia (Myrapetra) fastidiosuscula* Saussure, 1854 (Hymenoptera, Vespidae). Rev Bras Entomol 33:447–454

Togni OC, Locher GA, Giannotti E (2010) De larva à pupa: comportamento larval de *Mischocyttarus (Megacanthopus) parallelogrammus* Zikán (Hymenoptera, Vespidae) na construção do casulo. In: Proceedings of the Encontro Anual de Etologia, Universidade Federal de Alfenas, Alfenas, 12–15 November 2010

Torres VO (2009) Divisão de trabalho, polietismo temporal e aspectos da atividade forrageadora de *Mischocyttarus consimilis* Zikán, 1949 (Hymenoptera: Vespidae). 80 f. Dissertation, Universidade Federal da Grande Dourados

Torres VO, Antonialli-Junior WF, Giannotti E (2009) Divisão de trabalho em colônias da vespa social neotropical *Polistes canadensis canadensis* Linnaeus (Hymenoptera, Vespidae). Rev Bras Entomol 53:593–599

West-Eberhard MJ (1969) The social biology of polistine wasps. Misc Publ Mus Zool Univ Michigan 140:1–101

Wheeler GC, Wheeler J (1979) Larvae of social Hymenoptera. In: Hermann HR (ed) Social insects. Academic Press, New York, pp 287–338

Chapter 10
Evolution and Adaptation of the Wings and Mandibles of Neotropical Social Wasps

Carlos E. Sarmiento and Orlando Tobias Silveira

Abstract Besides the sting, the wings and mandibles of social wasps have attracted the attention of researchers because of their uniqueness to the group and its importance for the various functions of individuals in the colony. In this chapter, we present a summary of the research published in this area and the main characteristics that have been recorded.

Keywords Wing folding · Wing allometry · Denticles · Tooth · Metallic elements · Microtrichia

10.1 The Wings

The ability of insects to fly has been of special interest to humans, and the study of flying has covered many topics (Dudley 2002). In the social wasps, flying is the main means of displacement and, paradoxically, has been considered responsible for the low morphological differentiation of castes in this group of Hymenoptera (Oster and Wilson 1978). In fact, it has been proposed that morphological castes are a consequence of the strong functional specialization in the more complex social insects. In addition to being the basic structures for flight, the wings are used in nest ventilation and in defensive displays (Spradbery 1973).

Nest cooling behavior has been documented in many species of the group such as these of genera *Polistes*, *Mischocyttarus*, and *Polybia*. When environment temperature raises, workers, queens, and occasionally males alight on the nest and fan

C. E. Sarmiento (✉)
Laboratorio de Sistemática y Biología Comparada de Insectos, Instituto de Ciencias Naturales, Universidad Nacional de Colombia, Bogotá, Colombia
e-mail: cesarmientom@unal.edu.co

O. T. Silveira
Coordenação de Zoologia; Museu Paraense Emílio Goeldi, MPEG/MCT, Belém, PA, Brazil
e-mail: orlando@museu-goeldi.br

F. Prezoto et al. (eds.), *Neotropical Social Wasps*,
https://doi.org/10.1007/978-3-030-53510-0_10

it persistently. In addition, some individuals deposit drops of water on the surface of the nest (Jeanne 1972). These strategies reduce the amount of heat accumulated in the nests, keeping the larvae within the proper temperature ranges for their development. In the Vespinae, Sotavalta (1947) determined that the frequency of flutter during cooling is regularly between 117 and 247 movements per second, while it increased to 221 Hz if this flutter was aimed at warding off intruders to the colony (Sadeh et al. 1977; Ishay and Sadeh 1982).

As for colony defense, species such as *Synoeca septentrionalis* Richards, 1978 display highly synchronized episodes of fluttering whose frequency increases as the threat becomes more evident. The sound generated by such movement is quite notorious because of the resonance produced by the nest envelope. In other cases, wasps of the genus *Apoica* regularly keep their wings on the metasoma when in the nest, but when a threat is present, they place them underneath so that the clear light-yellow color of the metasoma signal is displayed and can suddenly exhibit the first levels of warning for those approaching. They also use this display to facilitate foragers returning to the nest at night acting as a lighthouse (Warrant 2008).

The wings are crucial for certain characteristics of social wasps such as the flying range when searching for prey or for material to build the nest. In species such as *Polybia emaciata* Lucas, 1879, *Polybia occidentalis* (Olivier, 1791), *Polybia fastidiosuscula* de Saussure, 1854, *Polistes simillimus* Zikán, 1951, and *Protopolybia exigua* de Saussure, 1854, the flight range is between 50 and 100 meters (Santos et al. 2000; Prezoto and Gobbi 2005; Ribeiro et al. 2008; Hernández et al. 2009; de Souza et al. 2010; López Galé et al. 2015). These distances must take into account the carrying capacity of an individual measured as a percentage of their body weight, which ranges between 39% for *Polybia occidentalis* (Hernández et al. 2008), 30% for *Polybia emaciata* Lucas, 1879 (López Galé et al. 2015), 25.3% for *P. paulista* H. von Ihering, 1896 (Gobbi and Machado 1985), and 13.4% for *P. platycephala* Richards, 1951 (Prezoto et al. 2005).

There are also interesting differences in the maneuverability of the vespids while executing their flights. While Asian stenogastrines of the genera *Parischnogaster*, *Metischnogaster*, and *Eustenogaster* can remain suspended in the air while stealing prey from cobwebs (Turillazzi 2012), in the Neotropical social wasps, these behaviors have not been normally observed. Nevertheless, the difference in the flight style of species of the genera *Agelaia* and *Polybia* compared with that of *Polistes*, *Mischocyttarus*, or *Brachygastra*, among others, is remarkable since the former two oscillate less compared with the latter three (CES personal observations). All these behavioral characteristics depend to a large extent on the wings and on the associated muscular system, so that their detailed study can reveal interesting patterns.

10.1.1 Structure of the Wings

The wings of Neotropical social wasps generally show the characteristics of the Vespoidea: the anterior wings are longer than the posterior wings. They typically have a costal cell, three submarginal cells with variations in the size and shape of the

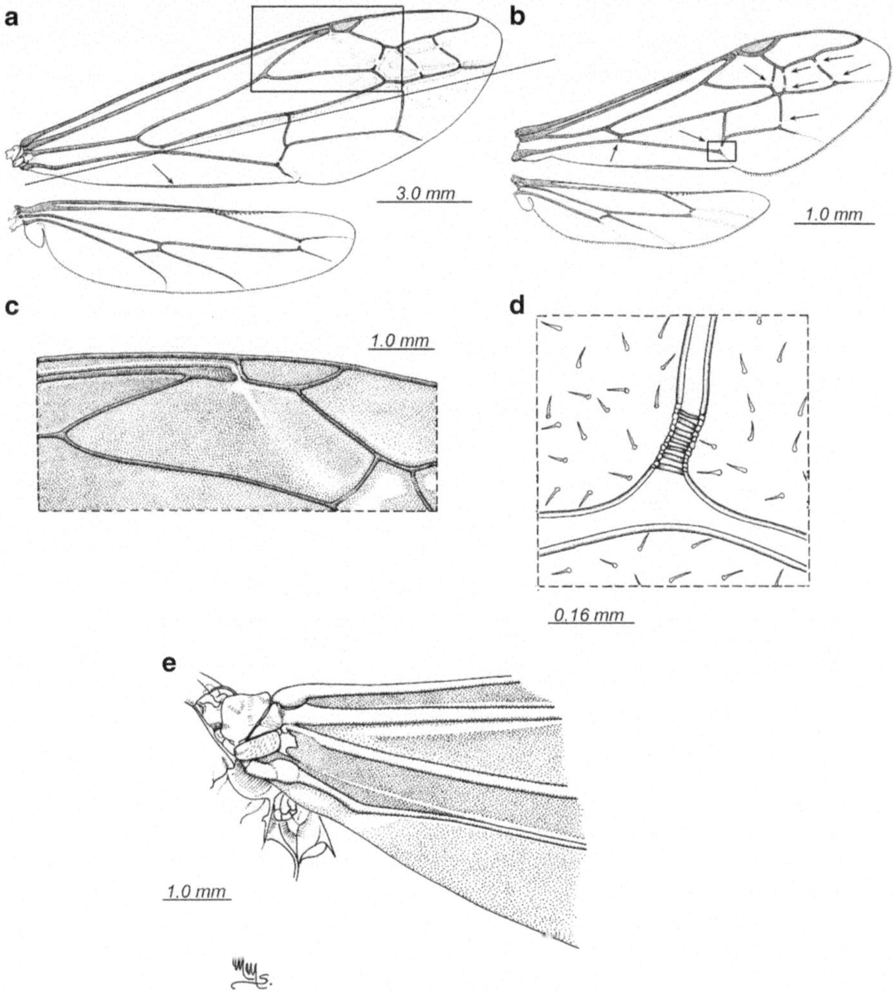

Fig. 10.1 General characteristics of the wings of Neotropical social vespids. (**a**) Wings of *Synoeca septentrionalis*, the box indicates the enlarged region in **c**, the arrow indicates the fold of the hooking system, and the line indicates the sector where the longitudinal fold of the wing occurs; (**b**) wings of *Leipomeles dorsata*, the box indicates the enlarged region in **d**, and the arrows indicate fenestrations or areas of venation less sclerotized where the flexion lines are given; (**c**) detail of the stigma and paraestigma; (**d**) detail of a fenestration in the anterior wing of *Leipomeles dorsata*; (**e**) detail of the basal sclerites in the wing of *Synoeca septentrionalis*, and they are observed in antero-posterior order, humeral plate (subtriangular), medial plate (rectangular), and vanal sclerite (globose)

first cell, a medial cell, and two discoidal cells. The discoidal 1 is very long, synapomorphy of the group, as well as a submedial and a subdiscoidal cell (Fig. 10.1a, b). The posterior wings only have a costal, a medial, and a submedial cell. The stigma is quite developed in the anterior wing, although it is absent in the posterior wing

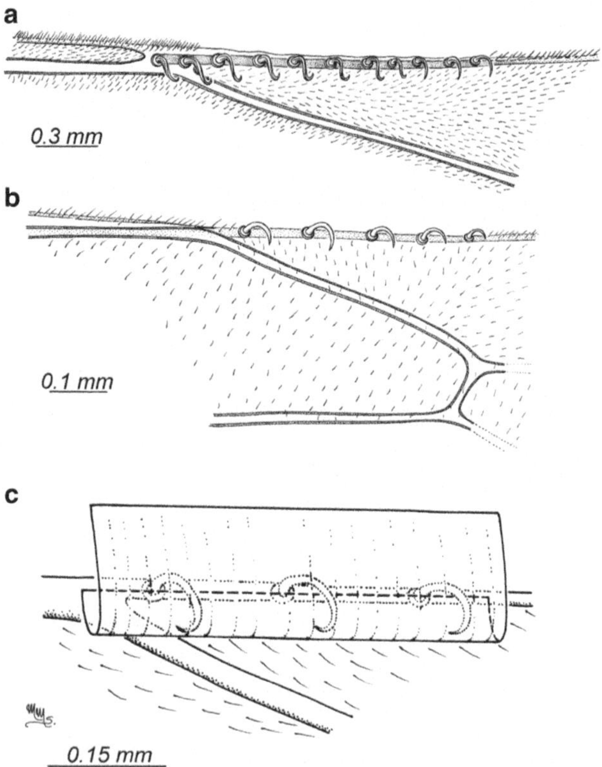

Fig. 10.2 Details of the hooking system on the wings of Neotropical social wasps. (**a**) Dorsal view of the hamuli in *Synoeca septentrionalis* Richards, 1978; (**b**) dorsal view of the hamuli in *Leipomeles dorsata* (Fabricius, 1804); (**c**) detail of the arrangement of the coupling system between the anterior and posterior wing

(Fig. 10.1a, b). Another remarkable structural feature is the set of hooks or hamuli arranged in the anteromedial edge of the posterior wing (Figs. 10.1a, b and 10.2) and a longitudinal fold located in the posteromedial edge of the anterior wing (Duncan 1939) called the frenum (Fig. 10.1a). Hamuli and frenum hook and keep anterior and posterior wings together during flight. The number of hamuli is higher in workers than in queens and males (Spradbery 1973). In addition, the muscular system of the anterior wings is much more developed than that of the posterior wings, because of which these animals are considered functionally anteromotor dipterans (Duncan 1939; Danforth and Michener 1988).

In general, it is recognized that the disposition of the veins plays an important role in maintaining the rigidity of the wings (Combes and Daniel 2003), as well as in the production of specific aerodynamic forces that increase the efficiency during the flight (Young et al. 2009). Evidence has been found for Neotropical social wasps in this regard (García and Sarmiento 2012). However, as properly demonstrated by Perrard et al. (2014) in the Vespinae, the specificities of the spatial changes of the veins in the different groups can point to more complex processes that direct the shape of the wings in different directions for different species. Within the castes, there has been a tendency for the queens to have longer wings than the workers, although there are many cases where this pattern does not occur (Richards 1978).

The wing blade is thin and is whitish and amber or even has a dark coloration; it is covered with small setae. The darkened costal margin is a common feature in several species of the genera such as *Polybia*, *Apoica*, and *Polistes* (Richards 1978). Within the species of the genus *Parachartergus*, a blackish coloration with a wide transparent apical region or milky color can be seen (Richards 1978), a character that has convergently appeared in several other groups of insects that mimic these social wasps.

10.1.2 Wings, Taxonomy, and Systematics

Some characteristics of the wings are important in the recognition of Neotropical social wasps, such as the elongated arrangement of the discoidal cell 1 (Sarmiento and Carpenter 2006) although this characteristic also occurs in several other vespid groups (Carpenter 1982, 1988). On the other hand, Richards (1978) characterizes the presence of the anal lobe of the posterior wing as a property of the Polistinae, and he also highlights the relative position of the veins 1cu1 and cu-a of the posterior wing to recognize genera such as *Brachygastra*, *Epipona*, *Synoeca*, and *Metapolybia*. The parastigma is also analyzed by this author to differentiate species in the genera *Pseudopolybia* and *Agelaia*. Richards (1978) extensively used wing color characteristics for the separation of species in several of the Neotropical genera. In the same sense, but perhaps with less success, this author used the length of the anterior wings for separation species. However, this character is not so easy to use since there are many cases of overlap between the options given by the key proposed by Richards.

In more recent times, studies of social wasp species of the subfamilies Vespinae and Stenogastrinae have employed more explicit mathematical methods such as geometric morphometrics. They have shown that the relative position of the veins and their places of union have much potential for species recognition (Baracchi et al. 2011), and these characteristics can even separate genera and populations (Perrard et al. 2014). The phylogenetic value of the shape of the wings has given diverse results, while in some works they have found interesting coincidences with phylogenetic proposals well established in the Stenogastrinae (Baracchi et al. 2011). Studies of Vespinae, however, have been less conclusive (Perrard et al. 2014).

10.1.3 Arrangement of Wings in Flight and at Rest

During flight the vespids maintain the anterior wings joined to the posterior ones by hooking between the hamuli system of the anteromedial edge of the posterior wing (Fig. 10.2a, b) and the longitudinal fold of the posteromedial edge of the anterior wing (Fig. 10.2c) (Duncan 1939). Although the longitudinal folding of the wings is mentioned in several documents as a key character in the identification of Neotropical species of the Vespidae, and even at times this served as a source of inspiration to speak of Diploptera when referring to the Vespoidea (Spradbery 1973), this property occurs only in the subfamilies Vespinae, Polistinae, and Eumeninae of the six living subfamilies, and it is also convergent in species of Leucospidae, Gasteruptiidae, Figitidae, Pompilidae, and even in some Colletidae (Danforth and Michener 1988). From an evolutionary and functional point of view, it has been proposed that this ability to fold the wings is beneficial for development in the conditions of a narrow cell in the crowded colonies where wasps of the Vespidae family develop (Danforth and Michener 1988).

This wing folding capacity is the result of a complex and diverse array of wing structural modifications. Danforth and Michener (1988) describe two types of modifications of the vespid wings that are also present in other Hymenoptera: the folds and the flexion lines. The flexion lines are associated with changes in the shape of the wings during flight (Wootton 1981). They are dorsal modifications and include the claval line that is located parallel to and anterior to the anal vein, the radial lines that lie between the radial and medial veins, and the medial lines that lie between the medial and cubital veins (Fig. 10.1a, b). The fold is the line along which the bending of the wings occurs. When the flexion lines and the fold line cross a vein, changes are observed in the coloration and disposition of the vein itself, called fenestrations or bullae. The fenestrations of the veins are observed as pale and transversely striated regions (Fig. 10.1a, b, d). These fenestrations are seen in the vein Rs, in the two rm, in the two m-cu, in the cu-v, and at the end of the culB (Fig. 10.1b). They are key in the deformations of the wings during flight. In the Vespidae, such fenestrations are also seen in the distal abscissa of the Cu vein (Danforth and Michener 1988) (Fig. 10.1b).

The forewings in Vespidae fold sharply and longitudinally so that sections of the ventral side come in contact (Danforth and Michener 1988). The fold line begins at the posterior margin of the base of the wing intersecting the 1cu-v vein basally and the vein 2m-cu so that it is subparallel to the medial vein (Fig. 10.1a). This is "type A" folding as described by Danforth and Michener (1988), and it is also present in Gasteruptiidae, Leucospidae, and Figitidae according to these authors; this pattern is different from the type of fold observed in some Pompilidae. In this family, the fold line is located basally costal to the vein M + Cu, and it distally intersects the vein 2m-cu. It does not reach the axillary triangle.

Studies in species of *Polistes* (Danforth and Michener 1988) show that the fold or deployment of the anterior wings results both from the participation of the hind wings and from the structural characteristics and the muscles that affect the wing itself. Thanks to the hamuli, the costal movement of the posterior wing folds the anterior wing while the opposite occurs when the posterior wing moves toward the anal region. However, experiments of posterior wing removal in live animals or the permanence of the folded position in removed forewings suggest that there are intrinsic mechanisms related to the physical disposition of the wing that facilitate its bending (Danforth and Michener 1988). In a very elegant way, Danforth and Michener (1988) conclude that it is likely that there are two stable positions resulting from the tensions generated by the arrangement of veins and wing plates. Specifically, in the place where the fold line crosses the cu-v vein, there is a tension point that generates a double-fold state mechanism or "click" which is in a strongly convex or concave position according to whether the wing is folded or unfolded. This "click" mechanism ensures that the wing remains fully extended during the flight.

Wasps can achieve changes between the folded or unfolded position of their anterior wing by alterations in the curvature of the costal margin. In *Polistes metricus* Say, 1831, the third axillary sclerite also has an important role in the folding process, and when the muscles attached to it contract, it rotates so that its distal end is arranged upward from the plane of the wing and toward the body. Because of this rotation, the vanal sclerite, which is the basal part of the anal vein, moves toward the midline of the body and passes below the posterior margin of the middle sclerite; this change is magnified by a protuberance in the anterior dorsal part of the vanal sclerite. These changes are also in accordance with the alteration of the humeral plate that slopes downward previously resulting in the fold of the two parts of the wing so that they come into contact ventrally (Danforth and Michener 1988). In species like *Synoeca septentrionalis*, a proximal globose thickening of the vanal sclerite can be observed that could fulfill the function described for the equivalent protuberance in *Polistes metricus* (Fig. 10.1e).

10.1.4 Wings and Size

Flight in insects is a subject of great complexity since there are many characteristics of the structures associated with this behavior that define its performance. In the wing, for example, aspects such as the maximum length, the distribution of the area from the base to the apex, the ratio between maximum length and width, the relative position of the veins, the asymmetry in the curvature of the wing in transverse view (camber), and the silhouette and corrugations, among other things, have been shown to be critical to the understanding of the flight performance (Ellington 1984; Danforth 1989; Usherwood and Ellington 2002; Luo and Suo 2005; Tsuyuki et al. 2006).

With the advent of geometric morphometrics, it has been possible to do detailed analysis of the shape of the wing (Johansson et al. 2009; Suárez-Tovar and Sarmiento 2016) since more classical indicators such as the aspect ratio that relates length and maximum width can give similar values despite having large differences in form

(Johansson et al. 2009). In the Neotropical social Vespidae, for example, important changes in shape related to size are observed, so that species with extreme sizes are not simple extensions or reductions of the others but show important alterations in shape and proportion (García and Sarmiento 2012). Small species such as *Leipomeles dorsata* (Fabricius, 1804) have more rounded wings with veins more concentrated towards the proximal region, and with their stigma and their mesosoma being proportionally larger (Fig. 10.1b), while larger species such as *Synoeca septentrionalis* Richards, 1978 have more elongated wings, with more extended distal venation and proportionally smaller stigma and mesosoma (Fig. 10.1a) (García and Sarmiento 2012). Such differences are not associated with forms of social organization or phylogenetic proximity but depend on the change in size (García and Sarmiento 2012).

The comparative study of García and Sarmiento (2012) suggests that species with extreme sizes have higher demands for flight than species of intermediate sizes, and their responses are expressed for different reasons. Small species face greater drag forces that lead to greater optimality in wings with forms that carry a comparatively denser medium (Wootton 1992). More rounded wings increase the lifting force, the movement of recovery, and the maintenance of the angle of attack (Dudley 2002; Walker 2002; Wootton 1992). Likewise, a proportionally larger stigma avoids the bending of the wings and maintains its angle of attack (Danforth 1989). A complementary result is the increase in the size of the mesosoma and the associated musculature, thus achieving greater strength. In large wasps, the changes in the shape of the wing suggest problems of carrying capacity of the body weight, although the results in this sense are less conclusive (García and Sarmiento 2012). These results are not consistent with what was found in the Vespinae, since some larger species may have more rounded wings than smaller ones (Perrard et al. 2014), which means that the complexity of the relationship between wing shape and flight could be greater.

It is worth noting that a proportional increase in both wing and mesosome sizes is also seen at higher altitude where low atmospheric pressure negatively affects their performance (Hodkinson 2005; Dillon et al. 2006). This implies greater investment of resources to be able to fly, even when the resources available to the colony are scarcer leading to a reduced body size (Rodríguez-Jimenez and Sarmiento 2008). In this way, the role played by the wings in the performance and survival of the social wasps is more evident.

Flight has impacted many of the structural characteristics of the social wasps, and as we have seen in this review, the wings exhibit numerous attributes that have affected their functionality. Surely these changes extend beyond the wings and the muscular system that supports them. Future research in these relationships with other parts of the body will be very interesting to better understand the evolution of the shape of the social wasps. The richness of the group in the Neotropics is also another characteristic that invites us to devote increased efforts for understanding this group. Flight has limited the development of morphological castes, which in turn could have influenced the appearance of other forms of specialization within the colonies.

10.2 The Mandibles

The mandibles of social wasps have been of interest for quite a long time. de Saussure (1854) emphasized the structural correspondence of shape and position between the mandibles and the clypeus of the vespids as shown in his commentary below about two "morphological types":

> These two forms (the author referring to different shapes of the clypeal apex) always go hand in hand with that of the mandibles: the first type of one of these organs is found only when the first type of the other is also present, and vice versa. This shows that these forms of clypeus are not mere unimportant accidents, but that they play a real zoological role, which their constancy proves.

The mandibles are vital in the social wasps for a variety of functions such as handling prey, feeding young, collection of material for the construction of nests, and the transport of materials, as well as for self-care, social interactions, and defense. A synthesis of what is known about the organizational structure as well as functional aspects of the mandibles in Neotropical social wasps will be presented.

10.2.1 *Structural Organization of the Mandibles*

In general, the mandibles of social wasps are mobile symmetric structures articulated to the head by two joints (Duncan 1939). Mandibles are wide basally and more flattened at the distal region where four apical teeth are found (Silveira and Santos 2011) (Fig. 10.3a, b). When a wasp is observed in frontal view, the mandibles often appear as subtriangular forms. As seen from the side, one notes the anterior and posterior points of articulation, the latter joint presenting a very conspicuous and darkened round condyle (see Duncan 1939) (Fig. 10.3a, b, d).

The social wasps are "hypognathous," this term referring to those insects in which the mouth parts point downward. The movement of the mandibles is relatively restricted, given their articulation at two fixed points, and they can only perform an "open and close movement" (Duncan 1939; Silveira and Santos 2011). The direction of mandible movement in relation to a plane crossing transversely the head (i.e., the trajectory of the mandibular apex) varies depending on mandible shape. Short and flat mandibles just overlap when closed (i.e., *Polistes*), while elongated mandibles may cross each other with the tips slightly twisted backward (i.e., *Brachygastra*) (Fig. 10.3f) (Silveira and Santos 2011). A regionalization of the mandible's parts of female Polistinae after Silveira and Santos (2011) will be presented below, with some elements added from the pioneering work of Duncan for the Vespinae (1939).

On its external surface, three large areas of the mandible can be defined: (1) the external basal area (Fig. 10.3a, eba) that extends from the basal margin to half the total length of the mandible, excluding the areas near the anterior and posterior

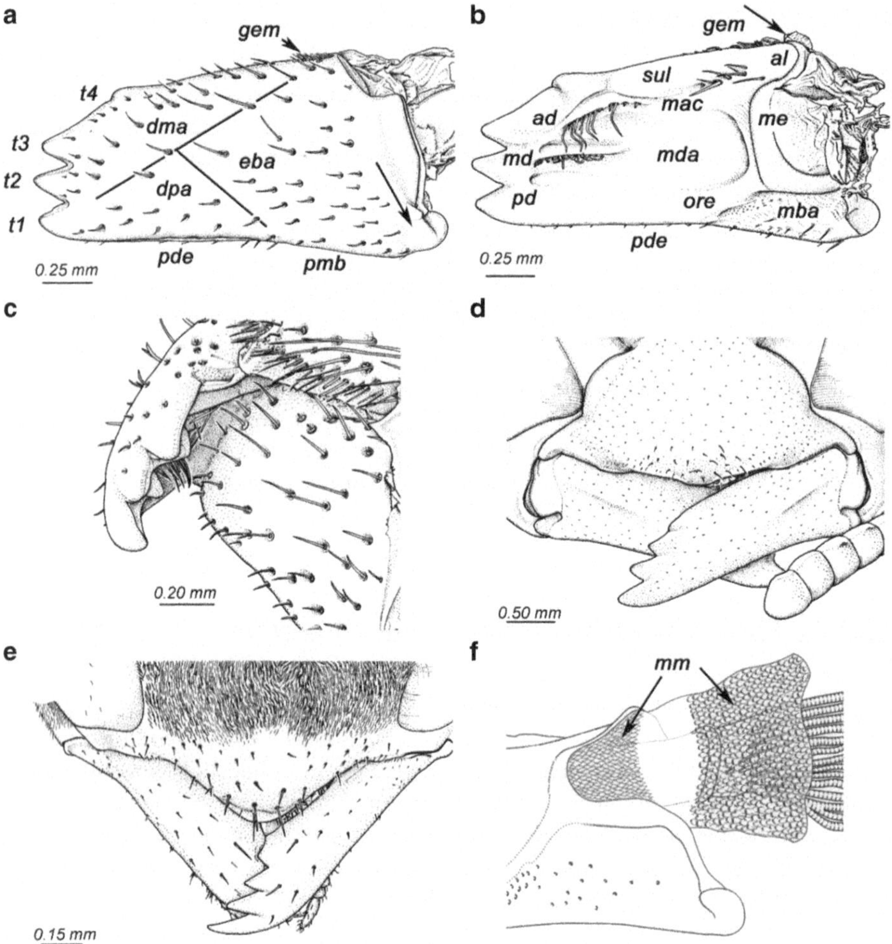

Fig. 10.3 Structural characteristics of the mandibles in Neotropical social vespids. (**a**) External view of the mandible of *Polistes*, the arrow indicates the posterior condyle; (**b**) posterior view of the mandible of *Polistes*, the arrow indicates the anterior condyle; (**c**) external lateral view of the mandible and clypeus in *Polistes*; (**d**) detail of the clypeus and mandibles in *Synoeca*; (**e**) detail of the clypeus and mandibles in *Brachygastra*; (**f**) posterior view of the vespid mandible with microtriquial membrane detail. (**a–e** Redesigned images of Silveira and Santos (2011), **f** redesigned image of López-Cubillos and Sarmiento (2013))

margins so that it resembles an arrowhead; (2) the posterior distal area is quite convex and has continuity with the t1 or posterior apical tooth 1 (Fig. 10.3a, dpa, t1); (3) the median distal area (dma) is adjacent to three of the apical teeth (t2-4) (Fig. 10.3a, dma), and it is variable between species with respect to its convexity. The apical teeth 1 to 3 may be rounded to subacute; tooth 1 is moderately varied in shape and length, and its apex slightly points mesially; the spacing between the teeth 3 and 4 is generally less pronounced, sometimes being noted as a simple notch

(Fig. 10.3a). The anterior margin can be divided into two sections, basal and distal, both having blunt edge; the boundary point between the two sections coincides with the outer extremity of the mesial oblique border (Fig. 10.3a, pmb, pde).

In mesial view (Fig. 10.3b), the mesial oblique border (ore) is a low blunt linear element that starts at the anterior side of the mesial emargination (me) and extends to the posterior margin; this border is the boundary between the mesial basal area and the mesial distal area of the mandible. The mesial emargination is the sector that receives the apodema of adductor muscle, and in its anterior part, there is a semicircular depression called the mesial anterior loop (*al*). On the mesial distal area, there are three subapical denticles. The posterior denticle (*pd*) is the smallest and is sometimes barely visible; the median denticle (*md*) is the most developed, and the anterior denticle (*ad*) is connected to the mesial anterior carina (*mac*). This carina is quite long and subparallel to the anterior margin, and together they define the anterior mesial sulcus (*sul*) (Fig. 10.3).

Rows of well-developed setae are found on the posterior side of the median and anterior denticles. On the external region of the mandible there are sensory setae, sometimes arranged in rows. At the proximal part of the mesial side of the mandible, very close to the mesial anterior loop, is a brush of setae related to the ectal mandibular gland (Fig. 10.3a, b, gem) (Downing and Jeanne 1982).

Silveira and Santos (2011) make a detailed description of the morphological variations of the mandibles in a representative sample of species of all the genera of Polistinae. In *Polistes*, the mandibles are relatively short and flat, and the apexes of the apical teeth (t1-3) line up as a roughly straight line (Figs. 10.3a, b). Basal genera of the Polistinae tend to show this morphological pattern. For example, species of *Agelaia* and *Angiopolybia* have quite robust mandibles (length/width ratio around 2.2), often with the posterior margin slightly converging toward the apex, when seen from the sides. However, tooth 2 is more developed than in *Polistes*, and the apexes of teeth 1 and 2 tend to align perpendicularly to the mandible's base, being more distant from the base than the apex of tooth 3. Tooth 1 is often divergent and directed backward. In mesial view, a very distinctive characteristic of these two genera is the great development of both the anterior denticle and the mesial anterior carina, being the latter high and sharp, indeed higher than the anterior margin, forming with this a wide anterior mesial sulcus. These elements of form jointly constitute a remarkable molar structure in these species of wasps.

Polybia species have slender and more elongated mandibles with a length/width ratio between 2.2 and up to 3.2 in species such as *P. striata* (Fabricius, 1787). More elongated mandibles also tend to show in frontal view the median distal area more twisted medially, although variation is observed between species. In the genera that build astelocyttarous nests, the genera *Asteloeca*, *Metapolybia*, and *Clypearia* show a similar range of shape variation as that observed in *Polybia*. Their length/width ratio ranges from 2.2 to 3.4 in species such as *C. weyrauchi* Richards, 1978, with elongation and twisting of the mandible being associated to modifications in the shape of the mesial denticles and of the anterior mesial carina. However, in *Clypearia* and *Synoeca*, such form changes are strong to the point that, in frontal view, the apical teeth and the median distal area jointly conform a very strong oblique surface roughly parallel to the mandible closing trajectory. Related to these changes, the mesial denticles and the mesial anterior carina are very small structures, almost evanescent.

The species of *Protopolybia* usually have slender mandibles. Their length/width ratio ranges from 2.6 in *P. exigua* (de Saussure, 1854) to 3.7 in *P. fuscatus* (Fox, 1898). The posterior margin is strongly sinuate endin in an enlarged and prominent tooth 1. Tooth 4 is prominent and comparatively large, and the anterior margin of the mandible is straight even at its basal section. In mesial view, the posterior and median denticles are generally short though well raised and sharp, but the anterior denticle and the mesial anterior carina are poorly developed. A torsion of the apical area of the mandible is also characteristic for the genus. Species of the genus *Charterginus* have similar mandibles to those described for *Protopolybia* species such as *P. chartergoides* (Gribodo, 1891) and *P. sedula* (de Saussure, 1854).

10.2.2 Mechanical Interactions Between Structural Parts of the Mandibles

Mandibles of vespids are closed through the action of strong adductor muscles by moving just ventral to the head capsule along a transversal plane, so that the apex of one of the mandibles may pass just below the apex of its opposite pair. Further inward movement allows one mandible to lay over the other in a resting position (see Duncan 1939) (Fig. 10.3c–e). In addition to the direct actions of apical teeth and of the internal elements (mesial denticles and anterior mesial carina) over the diverse substrates, mechanical interactions can occur among the main structural parts along the mandible closing trajectory (Fig. 10.3c–e) such as (a) opposition and crossing of apical teeth at middle of trajectory, (b) cutting by the interaction between apical teeth and mesial denticles of the opposite mandible (occurring after the mandibles cross each other), and (c) cutting by the interaction between the anterior margin of the mandible against the ventral margin of the clypeus (Silveira and Santos 2011).

10.2.3 Metallic Elements in the Mandibles

Studies in progress (Lagos-O and Sarmiento in prep.) have shown that, as reported by other authors (Bronwen et al. 2008; Quicke et al. 1998) for other insects, social wasps show accumulation of metals in the mandibles. This occurrence of metallic elements has been interpreted as a strategy of hardening of the apical teeth of the mandible, which optimize its durability. Lagos-O and Sarmiento (in prep.) have observed that species that use hard materials to build the nest, such as long plant fibers or clay, present greater levels of mandible wear than species using short fibers, easier to remove and manipulate.

10.2.4 The Mandibular Muscle Apparatus

The muscular apparatus that activates the movement of the mandibles in social wasps is a complex structure, and it presents similarities with other groups of social insects such as ants (Gronenberg et al. 1997) and bees (Snodgrass, 1956). Duncan (1939) provided an excellent description of this system for *Vespula pensylvanica* Saussure, 1857, which applies quite well to Neotropical social wasps (López-Cubillos et al. In prep.). The mandibular muscle apparatus consists of a small abductor muscle that opens the mandible and a greatly enlarged adductor muscle that closes it. The abductor muscle originates at the lower part of the gena and inserts on the external part of the mandible base. The adductor muscle is divided into two bundles of fibers: a small one that originates at the vertex and a large one that originates all across the internal dorsolateral surface of the head. An apodema divided into two laminar elements provides connecting surfaces for the fasciae of the adductor muscle, such a connecting system being attached to the distal border of the mesial emargination. Several muscle subgroups can be individualized within the large adductor muscle group that may be responsible for modulating the biting force in these wasps (López-Cubillos et al. In prep.).

At the base of the mandible internally, close to the apodema, there is the microtrichial membrane (*mm*), which may be divided into three areas (Fig. 10.3f): (1) a basal area overlaying the area of the mesial emargination, (2) a second element extending posteriorly to the articulation region, and (3) a membrane area positioned distal to the mandible articulation. The second membrane area is bare, while the first and third membrane regions are covered by microtrichia which, although varying in shape, invariably point toward the apex of the mandible.

This arrangement suggests a passive restraint system similar to a "velcro tape" that can keep the mandibles closed without continuous muscle power. The smooth second membrane region, without microtrichiae, and the opening of the mesal (mesial) gland at this local can jointly facilitate the disengagement of the system (López-Cubillos and Sarmiento 2013). A remarkable aspect of these microtrichiae is that they vary greatly across the social wasp species studied. However, López-Cubillos and Sarmiento (2013) did not find an association between the level of similarity of the microtrichiae and phylogenetic relationships of the species analyzed.

10.2.5 Mandibles and Exocrine Glands

The interior of the mandible houses four exocrine glands: (1) the mandibular ectal gland, (2) the mandibular mesal gland, (3) the intramandibular gland II, and (4) the intramandibular gland I. Except for the intramandibular gland II, these glands have been observed in most Neotropical wasp species studied (Downing 1991; Cely-Ortiz 2011; Penagos-Arévalo et al. 2015; Cely-Ortiz et al. 2017).

The first three glands are of the morphological type called "class 3," in which the flow of the compound produced by the secretory cells is performed by "ducts," formed by specialized elongated cells, discharging the secretion through "pores," typically 0.5 microns, or into internal reservoirs (Noirot and Quennedey 1991). The mandibular ectal gland presents a reservoir connected to the exterior by a set of pores located near the anterior mesial loop, where a brush of setae can be seen associated to the pores. The mesal gland is located at the base of the mandible and releases its products close to the microtrichial membrane. The intramandibular gland II consists of several aggregates of cells within the mandible, especially in the proximity of the mesial denticles (Penagos-Arévalo et al. 2015). The intramandibular gland I is a "class 1" gland, composed of a monolayer of cells covering the entire inner wall of the mandible, although larger cells are seen near the larger setae around the mesial denticles (Penagos-Arévalo et al. 2015).

Regarding functions of these glands, it was initially proposed that the ectal gland has a social function given its location (Downing and Jeanne 1982), but other authors have associated this gland with feeding behavior (Spradbery 1973), defensive behavior (Fortunato et al. 2001; Togni and Giannotti 2007; Turillazzi 2012), or sexual displays (Fortunato and Coster-Longman 2000; Landolt and Akre 1979; Wenzel 1987). García and Noll (2013) relate this gland to age-conditioned activities within the colony of *Polybia paulista* H. von Ihering, 1896, based on changes in size of the gland with age of individuals. For the mesal gland, a role has been suggested in the hierarchical organization of the colony, as well as in the construction of the nest (Downing and Jeanne 1983).

10.2.6 The Shape of the Mandibles and Their Performance

O'Donnell (1995) commented on the adaptive nature of a structure he observed in species of *Agelaia*, *Angiopolybia*, and *Apoica* named by him the "mesial dorsal tooth" and described as an elongated and "bladelike" ridge along the length of the mandible. O'Donnell presented a very precise description of this structure (named as mesial posterior denticle by Silveira and Santos 2011) and suggested that this is a structural modification in species of these genera adapted to collecting of meat in carrion. However, the structure described by O'Donnell (1995) is plesiomorphic for the Polistinae occurring in the basal genera, and considerable variation in shape and sharpness of the denticle occurs within *Agelaia* and *Angiopolybia*. In addition, the structure also occurs in a similar way in other of the more primitive genera in Polistinae. Even in *Chartergus* and *Polybia*, similarly well-prominent and elongated mesial denticles can occur.

On the other hand, a different structural variation on the mesial face of the mandible nearby the anterior margin, which is a synapomorphy for *Agelaia* and *Angiopolybia*, may support a relationship between necrophagy and mandible shape.

As described above for these two genera, part of the anterior mesial denticle and the mesial anterior mesial carina are exceptionally prominent, forming a well-developed "molar surface."

Given the complex behaviors of chewing and handling of materials for the construction of nests, it is expected that the shape of the mandible has a very important role for their performance. This assumption found support in the explicit comparative studies of Sarmiento (2004) and Silveira and Santos (2011) as well as in qualitative analyses (Hansell 1987). Wenzel (1998) categorized the construction material of social wasps' nests into four main groups: mud, plant hairs, short wood fibers, and long wood fibers. Sarmiento (2004) reported a significant correlation between prominence (i.e., height and length) of mesial mandibular denticles and the types of building materials used by polistines. He found that species that use long fibers (more resistant to removal) tend to have more prominent mesial denticles, while those using short fibers have less developed denticles. In this context, it is worth mentioning that Borges et al. (2017) observed that *Charterginus fulvus* (of the morphological type with less development of the mesial elements) almost exclusively uses hairs of *Cecropia* trees in the construction of the nest. These results are generally in agreement with the hypothesis above, of correlation between the prominence and size of the mesial denticles with the greater or lesser "hardness" of the used plant fibers (Sarmiento 2004; Silveira and Santos 2011).

However, after analysis of the morphological variation of a larger group of mandibular characters in a phylogenetic tree of the Polistinae, Silveira and Santos (2011) found that the extreme reduction of mesial denticles observed in some taxa seems partially associated with the torsion of the distal part of the mandible, as well as other morphological characteristics, such as the acute shape of the anterior margin of the mandible. However, such a correlation is far from perfect, and it is possible that the various elements that make up the integral mandible forms observed today may have evolved in response not only to the hardness of the nest building material but to other factors as well.

This brief review shows the great complexity of the mandibles, with its multiple structural components which together with their respective functional counterparts clearly demonstrate the great importance of this organ for the survival of the social wasps. Future research that delves into the three-dimensional organization of this structure and its performance could shed light on the evolution of the use of nest building materials and therefore on the evolution of colony size. Understanding the functions of the mandibular glands is also of great interest, given their likely roles in social interactions and also, once again, in respect of the possible functions of secretions in the preparation of building materials.

Acknowledgment The authors thank their respective institutions (Museu Paraense Emílio Goeldi OTS, National University of Colombia CES) and their colleagues and laboratory students for the support for the development of the research presented in this paper. Authors also thank Marcela Morales for the drawings. Authors thank Fabio Prezoto, Bruno Corrêa Barbosa, Fabio Nascimento, and Alexandre Somavilla for their work with this book.

References

Baracchi D, Dapporto L, Turillazzi S (2011) Relevance of wing morphology in distinguishing and classifying genera and species of Stenogastrinae wasps. Contrib Zool 80:191–199

Borges RC, de Seixas Felizardo SP, dos Santos JNA, Silveira OT (2017) Nest building by a neotropical social wasp using *Cecropia* trichomes as main construction material (Hymenoptera, Vespidae, Polistinae). Insect Soc 64:403–413

Bronwen WC, Stewart A, Huang H, Truss R, Noller B, Rasch R, Zalucki MP (2008) Insect mandibles—comparative mechanical properties and links with metal incorporation. Naturwissenschaften 95:17–23

Carpenter JM (1982) The phylogenetic relationships and natural classification of the Vespoidea (Hymenoptera). Syst Entomol 7:11–38

Carpenter JM (1988) The phylogenetic system of the Gayellini (Hymenoptera: Vespidae: Masarinae). Psyche 95:211–241

Cely-Ortiz C (2011) Morfología de las glándulas de la cabeza de *Polybia emaciata*, *Polybia ignobilis* y *Polybia occidentalis* (Hymenoptera, Vespidae). Trabajo de grado, Departamento de Biología, Universidad Nacional de Colombia – Sede Bogotá, Bogotá, p 23

Cely-Ortiz C, Arévalo AP, Billen J, Sarmiento CE (2017) Description of the Intramandibular Gland II in *Polybia emaciata* Lucas, 1879 (Hymenoptera: Vespidae). Sociobiology 64:347–351

Combes SA, Daniel TL (2003) Flexural stiffness in insect wings I. Scaling and the influence of wing venation. J Exp Biol 206:2979–2987

Danforth B (1989) The evolution of hymenopteran wings: the importance of size. J Zool (Lond) 218:247–276

Danforth B, Michener C (1988) Wing folding in the Hymenoptera. Ann Entomol Soc Am 81:342–349

de Saussure H (1854) Monographie des guepes sociales ou de la tribu des Vespiens. Kessmann, Paris/Geneva, p 256

De Souza A, Rocha M, Netto P, Venâncio D, Prezoto F (2010) Preliminary homing ability study of *Polybia fasidiosuscula* (Hymenoptera: Vespidae) Workers. Sociobiology 56:133–136

Dillon ME, Frazier M, Dudley R (2006) Into thin air: physiology and evolution of alpine insects. Integr Comp Biol 46:1–13

Downing HA (1991) The function and evolution of exocrine glands. In: Ross KG, Matthews RW (eds) The social biology of wasps. Cornell University Press, Ithaca, pp 540–569

Downing HA, Jeanne RL (1982) A description of the ectal mandibular gland in the paper wasp *Polistes fuscatus* (Hymenoptera: Vespidae). Psyche 89:317–320

Downing HA, Jeanne RL (1983) Correlation of season and dominance status with activity of exocrine glands in *Polistes fuscatus* (Hymenoptera: Vespidae). J Kansas Entomol Soc 56:387–397

Dudley R (2002) The biomechanics of insect Flight. Princeton University Press, Princeton, p 496

Duncan C (1939) A contribution to the biology of north American vespine wasps. Stanford University Press, Stanford, p 165

Ellington CP (1984) The aerodynamics of hovering insect Flight. II. Morphological parameters. Philos Trans R Soc Lond B 305:17–40

Fortunato A, Coster-Longman C (2000) Reproductive behavior and relative anatomical structures for the production of pheromones and luminous signal in male *Metischnogaster* (Vespidae: Stenogastrinae). Insects Sociaux Life 3:27–28

Fortunato A, Maile R, Turillazzi S, Morgan ED, Moneti G, Jones G, Pieraccini G (2001) Defensive role of secretion of ectal mandibular glands of the wasp *Polistes dominulus*. J Chem Ecol 27:569–579

García Z, Noll F (2013) Age and morphological changes in the Epiponini wasp *Polybia paulista* Von Ihering (Hymenoptera: Vespidae). Neotrop Entomol 42:293–299

García Z, Sarmiento CE (2012) Comparative analysis reveals changes in wing shape with body size in Neotropical social wasps (Hymenoptera: Vespidae: Polistinae). Zoomorphology 131:25–35

Gobbi N, Machado VL (1985) Material capturado e utilizado na alimentação de *Polybia (Myrapetra) paulista* Ihering, 1896 (Hymenoptera: Vespidae). Anuario Sociedad Entomológica de Brasil 14:189–195
Gronenberg W, Paul J, Just S, Hölldobler B (1997) Mandible muscle fibers in ants: fast or powerful? Cell Tissue Res 289:347–361
Hansell M (1987) Nest building as a facilitating and limiting factor in the evolution of eusociality in the Hymenoptera. In: Harvey P, Partridge L (eds) Oxford surveys in evolutionary biology. Oxford University Press, New York, pp 155–181
Hernández JD, Sarmiento CE, Hernández C (2009) Actividad de forrajeo de *Polybia occidentalis venezuelana* (Hymenoptera, Vespidae). Revista Colombiana de Entomología 35:230–234
Hodkinson ID (2005) Terrestrial insects along elevation gradients: species and community responses to altitude. Biol Rev 80:489–513
Ishay JS, Sadeh D (1982) The sounds of honey bees and social wasps are always composed of a uniform frequency. J Acoust Soc Am 72:671–675
Jeanne RL (1972) Social biology of the neotropical wasp *Mischocyttarus drewseni*. Bull Mus Comp Zool 144:63–150
Johansson F, Söderquist M, Bokma F (2009) Insect wing shape evolution: independent effects of migratory and mate guarding flight on dragonfly wings. Biol J Linn Soc 97:362–372
Landolt P, Akre R (1979) Occurrence and location of exocrine glands in some social Vespidae (Hymenoptera). Ann Entomol Soc Am 72:141–148
López Galé Y, Morinson MV, Caraballo-Gracia P (2015) Amplitud de vuelo y capacidad de carga de la avispa social *Polybia emaciata* Lucas, 1879 (Hymenoptera: Vespidae: Polistinae). Entomotropica 30(11):105–111
López-Cubillos S, Sarmiento CE (2013) A mandible arresting system in neotropical social wasps (Vespidae; Polistinae): structural diversity within homogeneous functionality. Naturwissenschaften 100:429–435
Luo G, Suo M (2005) The effects of corrugation and wing planform on the aerodynamic force production of sweeping model insect wings. Acta Mech Sinica 21:531–541
Noirot C, Quennedey A (1991) Glands, gland cells, glandular units: some comments on terminology and classification. Annales de la Société Entomologique de France 27:123–128
O'Donnell S (1995) Necrophagy by neotropical swarm-founding wasps (Hymenoptera: Vespidae, Epionini). Biotropica 27:133–136
Oster G, Wilson EO (1978) Caste and ecology in the social insects. Princeton University Press, Princeton, p 372
Penagos-Arévalo A, Billen J, Sarmiento CE (2015) Uncovering head gland diversity in neotropical Polistinae wasps (Hymenoptera, Vespidae): comparative analysis and description of new glands. Arthropod Struct Dev 44:415–425
Perrard A, Baylac M, Carpenter JM, Villemant C (2014) Evolution of wing shape in hornets: why is the wing venation efficient for species identification? J Evol Biol 27:2665–2675
Prezoto F, Gobbi N (2005) Flight range extension in *Polistes simillimus* Zikán, 1951 (Hymenoptera, Vespidae). Braz Arch Biol Technol 48:947–950
Prezoto F, Lima M, Machado V (2005) Surveys of prey captured and used by *Polybia platycephala* (Richards) (Hymenoptera: Vespidae). Neotrop Entomol 34:849–851
Quicke DLJ, Wyeth P, Fawke JD, Basibuyuk HH, Vincent JFV (1998) Manganese and zinc in the ovipositors and mandibles of hymenopterous insects. Zool J Linnean Soc 124:387–396
Ribeiro C, Elisei T, Guimaraes D, Prezoto F (2008) Flight range extension in the swarm-founding wasp *Protopolybia exigua* (Sausure, 1854) (Hymenoptera, Vespidae, Epiponini). Sociobiology 51:16–21
Richards OW (1978) The social wasps of the Americas. British Museum (Natural History), London, p 585
Rodríguez-Jimenez A, Sarmiento CE (2008) Altitudinal distribution and body resource allocation in a high mountain social wasp (Hymenoptera: Vespidae). Neotrop Entomol 37:1–8

Sadeh D, Ishay J, Yotam R (1977) Hornet ventilation noise: rhythm and energy content. Experientia 33:335–337

Santos G, Santana-Reis V, Resende J, De Marco P, Filho C (2000) Flying capacity of swarm-founding wasp *Polybia occidentalis occidentalis* (Olivier, 1791) (Hymenoptera, Vespidae). Revista Brasilera de Zoociencias 2:33–39

Sarmiento C (2004) A test of adaptive hypotheses: mandibular traits, nest construction materials, and feeding habits in neotropical social wasps (Vespidae: Polistinae). Insect Soc 51:387–391

Sarmiento C, Carpenter JM (2006) Familia vespidae, 539-556. In: Fernández F, Sharkey MJ (eds) Introducción a los Hymenoptera de la Región Neotropical. Sociedad Colombiana de Entomología y Universidad Nacional de Colombia, Bogotá, p 894

Silveira OT, Santos JN Jr (2011) Comparative morphology of the mandibles of female polistine social wasps (Hymenoptera, Vespidae, Polistinae). Revista Brasileira de Entomologia 55:479–500

Snodgrass RE (1956) Anatomy of the honey bee. Cornell University Press, London, p 162

Sotavalta O (1947) The flight-tone (wing-stroke frequency) of insects. Acta Entomologica Fennica 4:1–117

Spradbery J (1973) Wasps. University of Washington Press, Seattle, p 235

Suárez-Tovar CM, Sarmiento CE (2016) Beyond the wing planform: morphological differentiation between migratory and non-migratory dragonfly species. J Evol Biol 29:690–703

Togni OC, Giannotti E (2007) Nest defense behavior against the attack of ants in colonies of pre-emergent *Mischocyttarus cerberus* (Hymenoptera, Vespidae). Sociobiology 50:675–694

Tsuyuki K, Sudo S, Tani J (2006) Morphology of insect wings and airflow produced by flapping insects. J Intell Mater Syst Struct 17:743–751

Turillazzi S (2012) The biology of hover wasps. Springer-Verlag, Berlin, p 283

Usherwood JR, Ellington CP (2002) The aerodynamics of revolving wings I. Model hawkmoth wing. J Exp Biol 205:1547–1564

Walker JA (2002) Functional morphology and virtual models: physical constraints on the design of oscillating wings, fins, legs and feet at intermediate Reynolds numbers. Integr Comp Biol 42:232–242

Warrant EJ (2008) Seeing in the dark: vision and visual behaviour in nocturnal bees and wasps. J Exp Biol 211:1737–1746

Wenzel JW (1987) Male reproductive behavior and mandibular glands in *Polistes major* (Hymenoptera Vespidae). Insect Soc 34:44–57

Wenzel J (1998) A generic key to the nests of Hornets, yellowjackets and paper wasps worldwide (Vespidae: Vespinae, Polistinae). American Museum Novitates 3224:39

Wootton RJ (1981) Support and deformability in insect wings. J Zool (Lond) 193:447–468

Wootton R (1992) Functional morphology of insect wings. Annu Rev Entomol 37:113–140

Young J, Walker SM, Bomphrey RJ, Taylor GK, Thomas ALR (2009) Details of insect wing design and deformation enhance aerodynamic function and flight efficiency. Science 325:1549–1552

Chapter 11
Sternal Exocrine Glands in Neotropical Social Wasps

Marjorie da Silva, Fernando Barbosa Noll, and Johan Billen

Abstract A great diversity of exocrine glands is found in social insects. The main social function of exocrine glands is the production of pheromones, which are related to the maintenance of social organization of the colony, as interactions of dominance, maintenance of queen status, recruitment for food, search for nesting sites, defense, and reproductive behaviors, among other aspects. The exocrine glands can be classified into two types (class 1 and class 3) according to their structural arrangement. Some of them are common to all social insects, occurring in males and females (e.g., mandibular and salivary glands). Others are characteristic of a *taxon*, as the van der Vecht and Richards glands, found only in vespids of the Vespidae family. An overview about the structure, location, and, whenever possible, function of the exocrine glands of Neotropical social wasps is presented here. Future studies considering large and more detailed morphological analyses, together with information on the chemical aspects of the produced secretions, are desirable for a broader understanding of the function of the pheromones produced by the exocrine glands in the social wasps.

Keywords Behavior · Class 1 · Class 3 · Pheromones · Polistinae · Swarming

The anatomical descriptions found in books generally apply equally to social insects, while the development of the exocrine system distinguishes them clearly from the solitary species (Billen and Morgan 1998). A great diversity of exocrine glands is found in social insects exhibiting, in their majority, functions clearly related to the social organization of the colony (Hölldobler and Wilson 1990). In social wasps, interactions of dominance, maintenance of queen status, recruitment for food, search for nesting sites, defense, and reproductive behaviors are some of

M. da Silva (✉) · F. B. Noll
UNESP, São José do Rio Preto, SP, Brazil
e-mail: fernando.noll@unesp.br

J. Billen
KU Leuven-Bélgica, Leuven, Belgium
e-mail: johan.billen@kuleuven.be

F. Prezoto et al. (eds.), *Neotropical Social Wasps*,
https://doi.org/10.1007/978-3-030-53510-0_11

the aspects mediated by exocrine glands and their secretions (Downing 1991). The main social function of the exocrine glands is the production of pheromones, for which many glands have specialized (Billen and Morgan 1998). However, little is known about the chemical composition or function of glandular secretion in social wasps (Blum 1982; Akre 1982; Downing 1991), especially in Neotropical groups.

Several exocrine glands are common to all social insects and occur in queens, workers, and males, such as the mandibular and salivary glands (Billen and Morgan 1998). Others, however, represent new structures that are characteristic of a family, subfamily, genus, and even a species. As an example, we have the van der Vecht and Richards glands, found only in vespids of the Vespidae family (Billen and Morgan 1998). In this chapter, aspects of the structure, location, and, where possible, function of the exocrine glands present in Neotropical social wasps, with special emphasis on the abdominal sternal glands and the production of pheromones, will be discussed.

11.1 Structure and Classification of Glands

The exocrine glands can be classified into two main types: class 1 and class 3. This arrangement is related to the structural organization of the secretory cells (Billen 1991) and corresponds to the class 1 and class 3 glands proposed by Noirot and Quennedey (1974). Glands with class 1 secretory cells (Fig. 11.1a) are derived directly from the integument of the epidermis, thus presenting an epithelial organization (Billen and Morgan 1998). Secretory cells form a single layer of epithelium as part of the outer tegument through which cells secrete their products directly or as the lining of an internalized reservoir where secretion can be temporarily stored (Billen and Morgan 1998). The apical membrane of the cell is modified in a border with microvilli, resulting in a greater surface and, consequently, a greater secretory capacity (Billen 2011). In the presence of this type of gland, the cuticle presents small pores that allow the transport of the secretion to the external environment (Billen 2011).

The class 3 glands, on the other hand, present a more complex structure (Fig. 11.1b). In this type, the gland is formed by a variable number of bicellular units, which correspond to a secretory cell and a duct (Billen and Morgan 1998). The secretory cell channels and the ductal canal are covered with cuticle, which is continuous with the outer cuticle of the body (Noirot and Quennedey 1974; Delfino et al. 1979). The area of contact between the cell duct and the secretory cell is known as the end apparatus and configures a specialized region to allow the secretion to find its way out of the gland. Glands of this type can open through their cell ducts, directly in the integument or inside a reservoir. The reservoir in these glands is formed by flattened and generally nonsecretory epithelial cells (Billen and Morgan 1998). Glands of this class may be found alone or there may be groups of

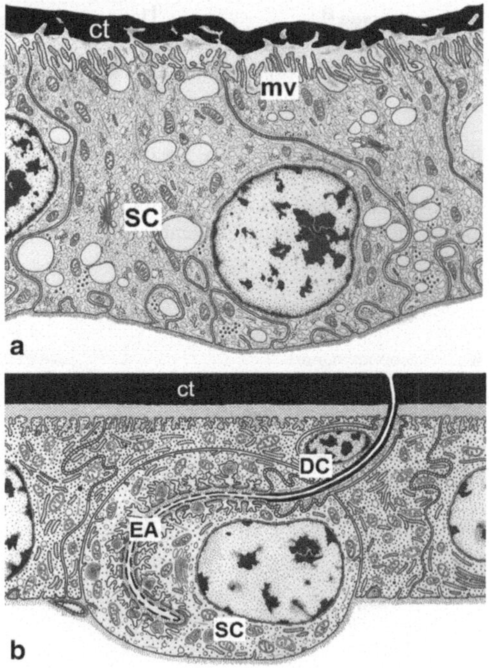

Fig. 11.1 Schematic of the morphology of glandular cells (**a**) epithelial class 1 gland (tibial gland of *Crematogaster* (Hymenoptera: Formicidae)) and (**b**) bicellular class 3 gland (queen tergal gland of *Apis mellifera* Linnaeus, 1758 (Hymenoptera: Apidae)). *CS* secretory cell, *ct* cuticle, *DC* duct, *EA* end apparatus, *mv* microvilli. (Source: Billen 2011)

them. Individual cells scattered in various regions of the body such as around joints and on the head secrete substances that lubricate and somehow maintain the integrity of the cuticle. Groups of class 3 glands produce secretions with different functions. The cuticle associated with these glands is often modified (Downing 1991), and the degree of specialization of these modifications varies greatly (Landolt and Akre 1979; Jeanne et al. 1983; Downing et al. 1985).

The original classification of Noirot and Quennedey (1974) also included the category of glands with class 2 cells. Currently, this nomenclature is no longer used, since these cells were later considered homologous to the oocytes (Noirot and Quennedey 1991).

Both glandular types (class 1 and class 3) may have a location that allows them to release their secretions directly into the external environment or into a reservoir, where secretion can be stored until the release (Billen 2011). Based on these characteristics, Billen (2009a, b, 2011) recently proposed five anatomic types for the exocrine glands, taking into account not only the cellular structure but also the presence or absence of a reservoir (Fig. 11.2): (A) epithelial gland without reservoir, (B) epithelial gland with reservoir, (C) bicellular unit without reservoir, (D) bicellular unit with reservoir, and (E) bicellular unit opening through intersegmental membrane.

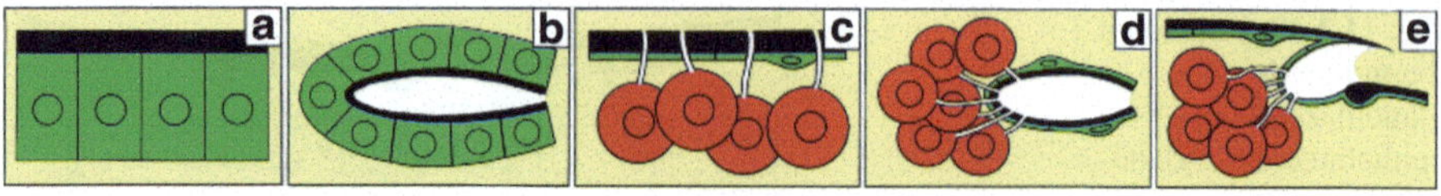

Fig. 11.2 The five anatomical types of exocrine glands found in insects. (**a**) Epithelial gland without a reservoir, (**b**) epithelial gland with the reservoir, (**c**) bicellular glandular unit without a reservoir, (**d**) bicellular glandular unit with the reservoir, and (**e**) bicellular glandular unit with duct opening in the intersegmental membrane. (Source: Billen 2011)

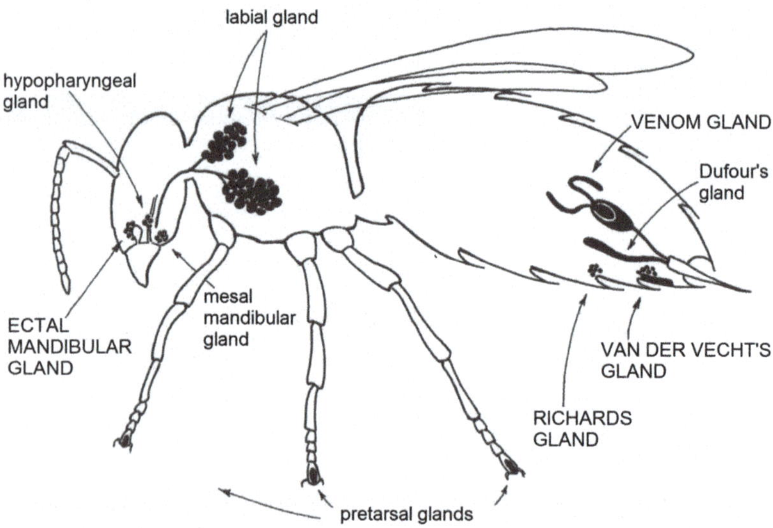

Fig. 11.3 Schematic profile of a wasp showing the location of the main exocrine glands. Names in capital letters indicate the glands that secrete pheromones. (Source: Billen and Morgan 1998)

11.2 The Occurrence of Pheromone-Producing Glands

Glands are specialized structures for the storage, emission, propagation, or evaporation of their secretory substances (Noirot and Quennedey 1991). This means that glandular cells exhibit a specific ability to absorb precursor molecules for subsequent biosynthesis of secretory products and can effectively regulate their release (Billen and Morgan 1998). The pheromone-producing glands are distributed throughout the body of the insect (Fig. 11.3), and their location is directly related to their function (Billen and Morgan 1998). In social wasps, alarm pheromones are produced primarily by the exocrine glands associated with the sting, the defense weapon in these insects.

The evolution of the ovipositor in Hymenoptera to the stinger apparatus resulted in changes in the venom and Dufour glands, whose functions changed from reproductive to defense function (Billen and Morgan 1998). The venom gland, which always releases its secretion through stings, plays an important role in the alarm-defense system. The complex compartmentalized organization of this gland, which comprises a convoluted and internalized gland, prevents autointoxication (Schoeters and Billen 1995a, b). In the wasps, the Dufour gland opening is located ventrally to the sting, and the secretion is released into the oviduct (Billen 1987), which explains why the secretory function of this gland is unknown for these social insects.

Two glands are especially important for Neotropical social wasps, either because they occur more widely within this group or because their secretions play a key role in the performance of some of the activities within these species.

11.2.1 van der Vecht Gland

The independent-founding wasps, such as the *Polistes* and *Mischocyttarus* species, use exocrine secretions to defend their nests against predation by ants. The glandular source of this repellent is the van der Vecht gland, formed by two groups of class 3 cells located laterally in the sixth (last) sternite metasomal, whose ducts open to the external cuticle of this area (van der Vecht 1968; Jeanne 1970, 1996; Noirot and Quennedey 1974; Turillazzi 1979; Turillazzi and Ugolini 1979; Post and Jeanne 1981; Jeanne et al. 1983; Kojima 1983, 1992; Keeping 1990) (Fig. 11.4). This gland was named after van der Vecht's study (Van der Vecht 1965), which observed the presence of a tuft of bristles on a wasp of the genus *Polistes* and later examined and described these cuticle specializations in the terminal sternum of almost all genera of social wasps (van der Vecht 1968).

Polistes, *Mischocyttarus*, *Belonogaster*, and some species of *Ropalidia* and *Parapolybia* construct nests that consist of a single comb suspended by a petiole. To deposit the repellent secretion, females rub the ventral surface of the gastropod against the surface of the petiole (Jeanne 1970; Corn 1972; Hermann and Dirks 1974, 1975; Darchen 1976; Litte 1976, 1979; Gamboa et al. 1978; Turillazzi and Ugolini 1978, 1979; Dani et al. 1992) (Fig. 11.5). The thin petiole increases the effectiveness of the repellent action by concentrating the secretion in a small area (Jeanne 1975). Jeanne (1970) in a study with *Mischocyttarus drewseni* Saussure, 1857, provided evidence that by rubbing the petiole, the wasps established a chemical barrier for the passage of ants that would otherwise discover the nest and the brood. The author associated this chemical barrier to the secretion of the van der Vecht gland. Later, this hypothesis gained support in studies with *Polistes annularis* (Linnaeus) (Hermann and Dirks 1974), *P. gallicus* (Linnaeus) (Turillazzi and Ugolini 1979), and *P. fuscatus* (Fabricius) (Post and Jeanne 1981).

The chemical barrier imposed on the ants, given by the secretion of this gland, is especially useful in the genera of independent foundations (Jeanne et al. 1983). Because in these wasps, nests are left unattended during foraging, since there are no workers until the offspring of the founding queen becomes adult, they are vulnerable to ant predation (Smith et al. 2001). On the other hand, the use of chemical defense to ward off the ants becomes less important in the swarm-founding wasps,

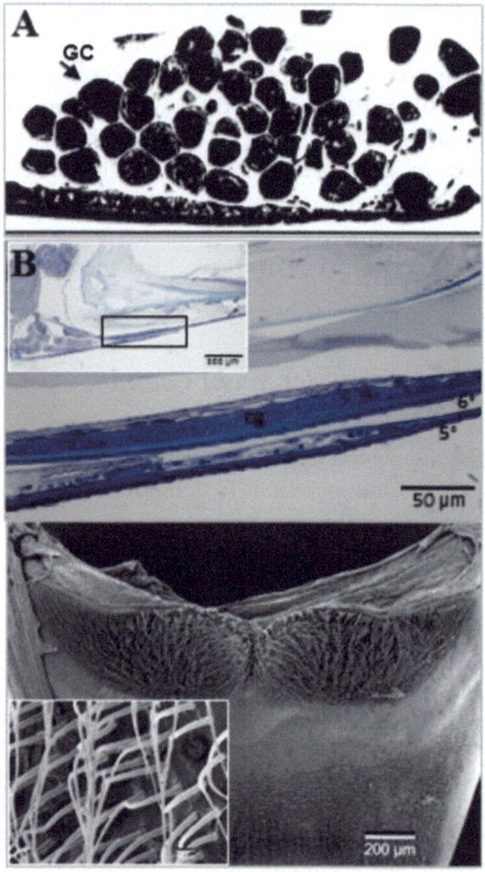

Fig. 11.4 Longitudinal cut of the sixth sternite showing the van der Vecht gland in (**a**) *Dolichovespula maculata* Linnaeus, 1763 (the cuticle of sternite is at the base of the photo and the anterior portion is on the right; increase: 100× (Source: Smith et al. 2001)), and (**b**) *Agelaia pallipes* Olivier, 1791. (**c**) The anterior portion of the sixth sternite of *Apoica pallens* Fabricius, 1804, showing the bristle tuft of the ducts opening (detail to the left). *GC* glandular cell, *5th* fifth sternite, *6th* sixth sternite

since nests are always guarded and the defense against intruders is made in physical form (West et al. O'Donnell and Jeanne 1990; West-Eberhard 1989; Smith et al. 2001).

Smith et al. (2001), in an investigation on the social subfamilies (Stenogastrinae, Vespinae, and Polistinae) within Vespidae, phylogenetically optimized the presence of the van der Vecht gland (Fig. 11.6) and concluded that the independent foundation method and the presence of the gland were present in the ancestral Vespinae + Polistinae. Within this clade, all species presenting independent foundation possess the van der Vecht gland (Table 11.1), and, in all species where the function of this gland was tested, the secretion had a repellent function (Jeanne 1970; Turillazzi and Ugolini 1979; Post and Jeanne 1981; Kojima 1983, 1992; Keeping 1990), and showing this ability was a key adaptation for independent-founding wasps (Smith et al. 2001). However, the repellent function of this gland has not been tested in Vespinae (Matsuura and Yamane 1990).

Since the swarm foundation mode allowed the wasps to protect their nests while others leave for foraging, the production of a secretion to repel ants may have

Fig. 11.5 Newly founded nest of *Mischocyttarus drewseni* (Brazil). A founding female deposits the repellent secretion by rubbing the surface of the sixth sternite against the peduncle of the nest, reducing the chances of the ants discovering the comb containing the brood during the period the queen is foraging. (Source: Jeanne 1970)

acquired secondary importance. According to Jeanne (1991), the greater efficacy in defense against ants is the greater adaptive benefit resulting from the evolution of the mode of foundation by a swarm. For all the species of swarming wasps where the gland of the sixth sternite was analyzed, either the gland or the secretory repellent function was lost (Smith et al. 2001), corroborating the hypothesis raised by Jeanne (1991). Only in 3 of the 21 genera which present swarming foundation the van der Vecht gland is present, and, in two of them (*Agelaia* and *Apoica*), the secretion produced by the gland does not repel the ants (London and Jeanne 2000; Smith et al. 2001). There is no data on the role of this gland for the third genus (*Provespa*) (Smith et al. 2001).

A second function of the secretion of the van der Vecht gland in *Polistes* would be to induce the dominance of the queen by subordinate individuals when applied to the surface of the nest (Dani et al. 2003; London and Jeanne 2000; Dapporto et al. 2007a, 2007b), thus preventing ovarian development in other females. Therefore, the selection in the van der Vecht's organ of the queen may have been different from that which occurred in the female workers. In fact, morphological differences associated with castes have been found in the structure of this gland for some species of this genus, providing evidence of the presence of an incipient morphological dimorphism (Dapporto et al. 2011; Petrocelli and Turillazzi 2013; de Souza et al. 2016). For the genera of swarming wasps that present this gland, although they have lost the repellent function, there are no studies that have analyzed the presence of morphological differences between the castes for this gland nor evidence that the secretion of this one contributes to the maintenance of the hierarchy.

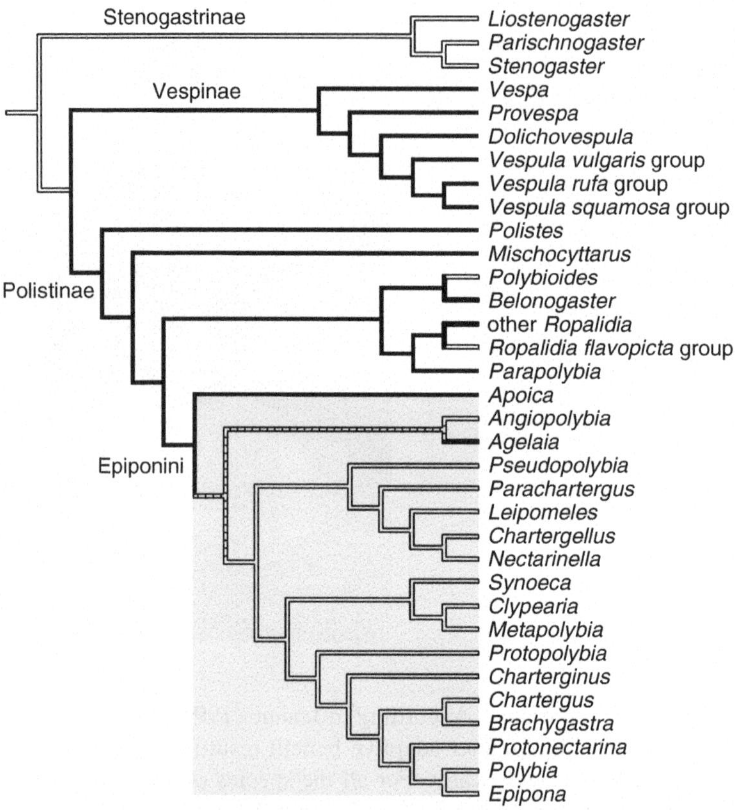

Fig. 11.6 Evolution of the van der Vecht gland and nesting behavior in Vespinae + Polistinae (Wenzel and Carpenter 1994). The black branches represent the presence of the van der Vecht gland; the absence of it is represented by the white branches. Taxa for which the presence of the gland is uncertain are represented by striped branches. The gray shading highlights the swarm-founding lineages. (Source: Smith et al. 2001)

11.2.2 Richards Gland

Several variable organs in structure and complexity, present in the fifth metasomal sternite, have been described for social wasps (Samacá et al. 2013). Heselhaus (1922) was the first to describe a group of cells with their respective ducts present in this region and which was later named the "Richards gland" (Jeanne and Post 1982) since it was Richards (1971) who called attention to the distribution of this gland in Vespidae (Samacá et al. 2013). These class 3 cells secrete their products through the ducts to a reservoir formed by an invagination of the intersegmental membrane between the metasomal sternites 4 and 5 (Samacá et al. 2013). In conjunction with the cells, there are modifications in the exoskeleton where the ducts open externally. Richards (1971) used these modifications as indicative of the distribution of this gland within Neotropical Polistinae. Jeanne and Post (1982) defined

Table 11.1 Glands of the fifth and sixth sternites in females of social wasps (Polistinae)

Species	5°	CC	GE	6°	References
Independent Founding					
Mischocyttarus angulatus Richards, 1945	+	Roughness and pores	+	?	Samacá et al. (2013)
Mischocyttarus atramentarius Zikan, 1949	?	?	?	+	Raposo-Filho et al. (1994)
Mischocyttarus flavitarsis (Saussure, 1854)	+	Pores	+	+	Landolt and Akre (1979), Jeanne et al. (1983)
Mischocyttarus immarginatus Richards, 1940	+	?	+	+	Jeanne et al. (1983), Smith et al. (2002)
Mischocyttarus labiatus (Fabricius, 1804)	+	?	+	+	Smith et al. (2002), Litte (1981)
Mischocyttarus mexicanus (Saussure, 1854)	?	?	?	+	Jeanne et al. (1983)
Polistes aterrimus Saussure, 1853	+	Roughness and pores	+		Samacá et al. (2013)
Polistes canadenses (Linnaeus, 1758)	+	Pores	+	+	Post and Jeanne (1980), Jeanne et al. (1983)
Polistes infuscatus (Fabricius, 1793)	+	Pores	+		Post and Jeanne (1980)
Swarm founding					
Agelaia areata (Say, 1837)	–	Absent	+	+	Jeanne et al. (1983) Smith et al. (2002), Samacá et al. (2013)
Agelaia (near *hamiltoni*)	–	?	?	+	Smith et al. (2002)
Agelaia myrmecophila (Ducke, 1905)	–	?	?	+	Jeanne et al. (1983), Smith et al. (2002)
Agelaia pallipes Olivier, 1792	–	?	–	+	Jeanne (observação pessoal), Observação dos autores
Agelaia panamensis (Cameron, 1906)	?	?	?	+	Jeanne et al. (1983)
Agelaia testacea (Fabricius, 1804)	–	Absent		+	Jeanne et al. (1983) Jeanne (observação pessoal.)
Angiopolybia pallens (Lepeletier, 1836)	–	Absent		–	Jeanne et al. (1983), Smith et al. (2002)
Apoica flavissima (van der Vecht, 1972)	?	?	?	+	Jeanne et al. (1983)
Apoica pallens (Fabricius 1804)	+	Tuft of bristles and ducts grouped in concavities	–	+	Hunt et al. (1995) observação dos autores
Apoica pallida (Olivier, 1792)	+	Tufo de cerdas	–	+	Jeanne et al. (1983)
Brachygastra augusti (Saussure, 1854)	+	Scales	–	–	Jeanne et al. (1983)
Brachygastra bilineolata (Spinosa, 1841)	?	?	?	–	Smith et al. (2001)

(continued)

Table 11.1 (continued)

Species	5°	CC	GE	6°	References
Brachygatra lecheguana (Latreille, 1824)	+	Scales	–	–	Jeanne et al. (1983) Samacá et al. (2013)
Brachygastra smithii Saussure, 1853	?	?	?	–	Jeanne et al. (1983)
Chartergellus communis Richards, 1978	–	Absent	–	–	Jeanne et al. (1983) Observação dos autores
Chartergellus punctatior Richards, 1978	?	?	?	–	Smith et al. (2001)
Charterginus fulvus Fox, 1904	?	?	?	–	Smith et al. (2001)
Chartergus globiventris Saussure, 1854	?	?	?	–	Smith et al. (2001)
Chartergus metanotalis Richards, 1978	?	?	?	–	Smith et al. (2001)
Clypearia apicipennis (Spinosa, 1851)	?	?	?	–	Jeanne et al. (1983)
Clypearia sulcata Saussure, 1853	?	?	?	–	Jeanne et al. (1983)
Epipona guerini (Saussure, 1854)	?	?	?	–	Smith et al. (2001)
Epipona niger (Brethes, 1926)	+	Scales	–	–	Samacá et al. (2013)
Epipona tatua (Civier, 1797)	+	Pores	–	–	Jeanne et al. (1983)
Leipomeles dorsata (Fabricius, 1804)	–	Absent	–	–	Jeanne et al. (1983) Jeanne et al. (1983) Smith et al. 2001, (2002)
Leipomeles spilogastra (Cameron, 1912)	–	Absent	–	–	Samacá et al. (2013)
Metapolybia aztecoides Richards, 1978	+	Scales	–	–	Samacá et al. (2013) Smith et al. (2002)
Metapolybia docilis Richards, 1978	+	Scales	–	–	Jeanne et al. (1983)
Metapolybia cingulata Fabricius, 1804	?	?	?	–	Smith et al. (2001)
Nectarinella championi (Dover, 1925)	–	Absent	–	–	Jeanne et al. (1983) Smith et al. (2001) Jeanne & Keeping (observação pessoal)
Nectarinella xavantinensis (Mateus & Noll, 1997)	?	?	?	–	Smith et al. (2001)
Parachartergus apicalis (Fabricius, 1804)	–	?	–	–	West-Eberhard (1982)
Parachartergus fraternus (Gribodo, 1892)	–	?	–	–	Jeanne et al. (1983) Jeanne & Bouwma (observação pessoal)

(continued)

Table 11.1 (continued)

Species	5°	CC	GE	6°	References
Parachartergus colobopterus (Lichtenstein, 1796)	–	?	–	–	Smith et al. (2001) Jeanne & Keeping (observação pessoal)
Polybia bistriata (Fabricius, 1804)	+	?	–	–	Smith et al. (2002)
Polybia catillifex Moebius, 1856	+	?	–	–	Smith et al. (2002)
Polybia chrysothorax (Lichtenstein, 1796)	?	?	?	–	Jeanne and Post (1982)
Polybia ignobilis (Haliday, 1836)	+	Scales	–	–	Samacá et al. (2013)
Polybia occidentalis (Olivier, 1791)	+	Scales, projections	–	–	Jeanne and Post (1982), Smith et al. (2002) Jeanne (observação pessoal)
Polybia quadricincta Saussure, 1854	?	?	?	–	Jeanne and Post (1982)
Polybia raui Bequard, 1933	+	?	–	–	Smith et al. (2002)
Polybia scrobalis Richards, 1970	+	?	–	–	Smith et al. (2002)
Polybia sericea (Olivier, 1792)	+	?	–	–	Jeanne (1981), Jeanne and Post (1982)
Polybia velutina Ducke, 1907	+	?	–	–	Smith et al. (2002)
Protonectarina sylveirae (Saussure, 1854)	+	Scales	–	–	da Silva et al. (2015), Smith et al. (2001)
Protopolybia amarela Bequaert, 1944	+	Scales	–	–	Samacá et al. (2013)
Protopolybia alvarengai Richards, 1958	?	?	?	–	Jeanne et al. (1983)
Protopolybia exigua (Saussure, 1906)	+	Scales	–	–	Smith et al. (2001), Smith et al. (2002)
Protopolybia scutellaris Bequaert, 1944	?	Scales	–	–	Jeanne et al. (1983)
Pseudopolybia compressa Saussure 1854	?	?	–	–	Jeanne et al. (1983)
Pseudopolybia difficilis (Ducke, 1905)	?	?	?	–	Jeanne et al. (1983)
Synoeca septentrionalis Richards, 1978	+	Scales	–	–	Samacá et al. (2013)
Synoeca surinama (Linnaeus, 1767)	+	Scales	–	–	Jeanne et al. (1983), Smith et al. (2002)
Synoeca virginea (Fabricius, 1804)	+	?	–	–	Jeanne et al. (1983) Jeanne (observação pessoal)

GE glandular epithelium, *CC* cuticular changes, *5 °* class 3 gland in the fifth sternite, *6 °* class 3 gland in the sixth sternite, *?* unknown (there are no studies for the species)

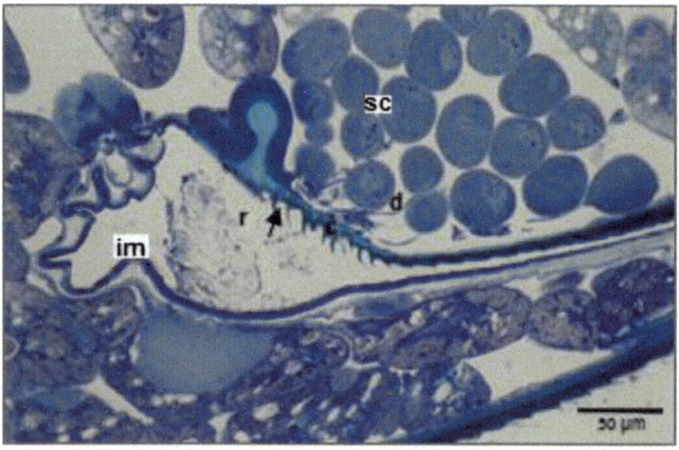

Fig. 11.7 Structure of the Richards gland. Longitudinal section of the posterior region of metasoma of *Metapolybia docilis* Richards 1978 showing the set of secretory cells (cs) and ducts (d) that compound the gland. The arrow indicates the opening of the duct in the cuticle (c), and it is possible to observe the reservoir (r) formed by the invagination of the intersegmental membrane (mi)

the Richards gland as an organ composed of class 3 cells associated with a series of ducts and showing changes in the cuticle at the base of the fifth sternite (where the ducts open externally) (Fig. 11.7).

The fact that, in the Apocrita, the first abdominal segment is fused to the thorax and the rest of the abdomen connected closely with this part, forming the "wasp waist" (Grimaldi and Engel 2005), made the nomenclature of body division different in the hymenopteran of this group. Due to this fact, the literature presents some confusions about the position of the Richards gland in the social wasps, whose correct location is the anterior portion of the fifth metasomal sternite (Fig. 11.8).

Naumann (1975) proposed that this gland is responsible for producing the secretion used for trail marking during the process of founding a new colony, since the scout workers mark the path between the site of the former nest and the place where the new nest will be constructed by rubbing the ventral portion of the metasoma on the vegetation (dragging behavior). Later, Jeanne (1981) experimentally established this relationship for *Polybia sericea* (Olivier, 1792). He noted that during the display of the dragging behavior performed during the swarming process for the marking of the trail, the wasp raises the fifth sternite to expose the Richards gland. The author also comments on the presence of a characteristic odor during this activity, originating from the secretion of the gland. Since then, several publications have recorded the taxonomic distribution and structural variation of this gland within Polistinae (Landolt and Akre 1979; Jeanne and Post 1982; Jeanne et al. 1983; Billen and Noll 2011; Samacá et al. 2013; da Silva et al. 2015).

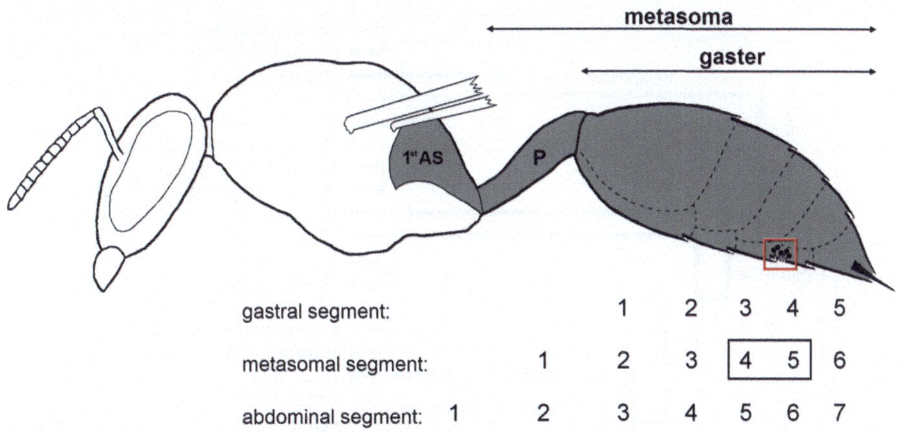

Fig. 11.8 Profile of a wasp (Polistinae) showing the body division and location of the Richards gland. The abdominal segments are colored in gray. *P* petiole, *1st AS* first abdominal segment. Observe the various indications of the position of this gland, depending on the number of "gastral," "metasomal," or "abdominal" sternites. The most commonly used indication, which we adopt in this chapter, is the numbering by the metasomal segments. According to this, the Richards gland is situated between the fourth and fifth sternites (red square)

Despite the experiments performed by Naumann (1975) on *Angiopolybia pallens* (Lepeletier 1836) and Jeanne (1981) in *Polybia sericea* (Olivier 1792) relating the presence of Richards gland to the swarming process, not all Epiponini genera, despite being swarming and presenting the dragging behavior, have glands in the fifth or sixth sternite, as pointed by Richards (1978) and Jeanne et al. (1983). This gland, as well as cuticular modifications, does not occur in the species of *Agelaia*, *Angiopolybia*, *Parachartergus*, *Chartergellus*, *Leipomeles*, and *Nectarinella* (Smith et al. 2002; Samacá et al. 2013). In addition, the records provided by Naumann (1975) for *Agelaia* were later verified as incorrect (Jeanne et al. 1983) (Table 11.1). Smith et al. (2002) have emphatically stated that there is no association between the presence of Richards gland and the dragging behavior performed by the workers during the swarm since this behavior has already been observed for wasp species that do not present this gland. This fact can be observed when the authors optimized the presence of the Richards gland and the founding behavior in the Vespidae phylogeny (Wenzel and Carpenter 1994) (Fig. 11.9).

In a discussion regarding the difference between castes, Richards (1971) describes the presence of different color of a secretion expelled by a structure in the penultimate (fourth) gastral sternite in the queens and workers of *Apoica pallens* (Fabricius, 1804), *Polybia jurinei* Saussure, 1854, and *P. striata* Fabricius, 1787. The author describes the structure as a narrow transverse band with thin pores in the cuticle which may also exhibit distinct coloration. The location of the structure suggests that the pores are the external openings of the Richards gland, and the fact of the presence of the distinct coloration between workers and queens could

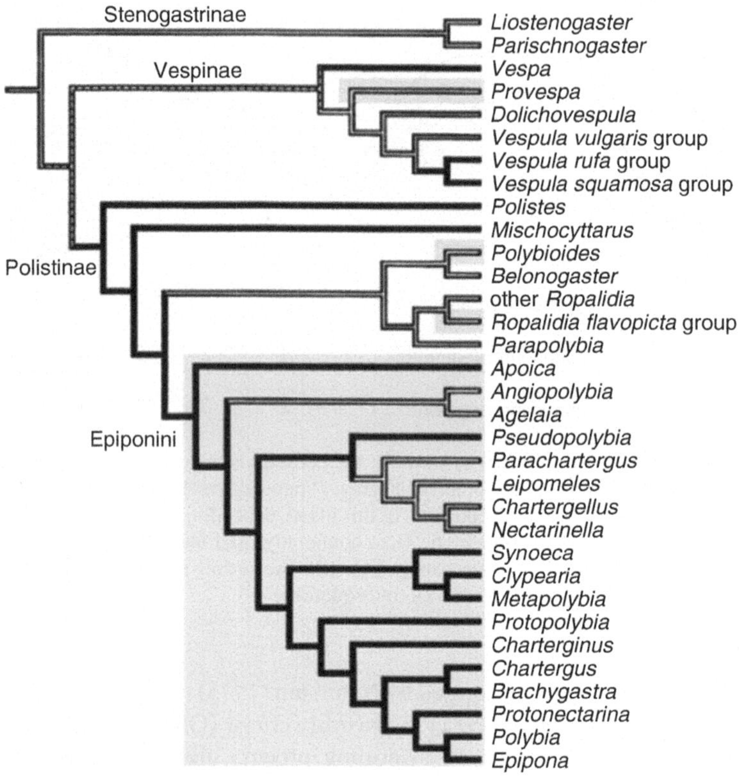

Fig. 11.9 Evolution of the Richards gland and the colony foundation behavior in Vespinae + Polistinae (Wenzel and Carpenter 1994). The black branches represent the presence of the Richards gland; the absence of the gland is represented by the white branches. The striped branches represent taxa for which the presence of the gland is uncertain. Species presenting foundations by swarm are shaded in gray. (Source: Smith et al. 2002)

suggest a role in determining castes or maintaining dominance between them (Richards 1978). Nevertheless, for other species such as *Polybia rejecta* (Fabricius, 1798) and *P. chrysothorax* (Lichtenstein, 1796) there is no difference in color for caste secretion. Moreover, species belonging to the *P. occidentalis* group (Olivier, 1791) and to the genera *Brachygastra* and *Protopolybia* possess the gland, but according to Richards (1971), little or no secretion was observed externally.

In *Mischocyttarus* and *Polistes*, genera of an independent foundation, class 3 cells at the base of the fifth sternite are also present. In the species of these genera, there is also a class 1 gland composed of a layer of secretory cells attached to the inner part of the same sternite (Turillazzi 1979; Post and Jeanne 1980) (Fig. 11.10). In this case, however, no modifications were identified on the outside of the cuticle (Turillazzi 1979; Post and Jeanne 1980; Samacá et al. 2013). Post and Jeanne (1980) and Jeanne et al. (1983) reported the presence of class 1 and class 3 cells in eight species of the same genera (*Mischocyttarus* and *Polistes*) and also found no modification in the cuticle.

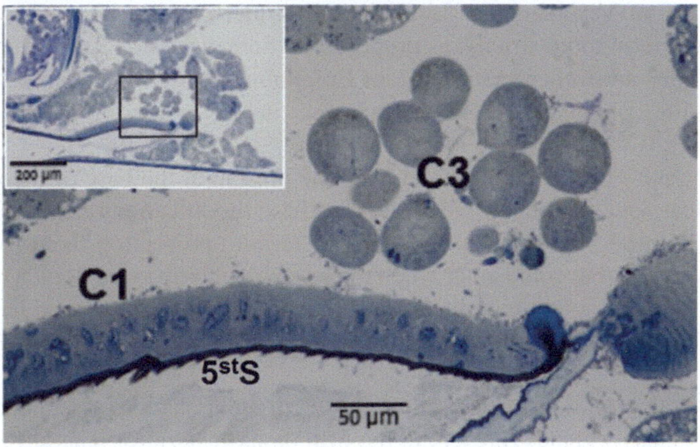

Fig. 11.10 Longitudinal section of the posterior region of the *Mischocyttarus cerberus* metasoma showing the two glandular types present. *C1* class 1 secretory cells, *C3* class 3 secretory cells, *5 ° S* 5 ° sternite. The anterior region is on the right

Samacá et al. (2013) listed five differences between the structure of the fifth sternite gland of independent-founding and swarm-founding wasps:

1. Cuticular changes at the sternite base, where the glandular ducts open externally, occur only in the swarm-founding wasps.
2. In the independent-founding wasps, the proximal part of the posterior area of the anterolateral margin is smooth, but rugged at the distal part, with several pores covering most of the sternite. In the swarm-founding wasps, on the contrary, there is not a rough area and the surface of the sternite is completely smooth after the area where the cuticular modifications are present.
3. Class 1 gland is widely found in independent-founding wasps and it is adjacent to the outer roughened surface of the sternite. Except for *Agelaia areata*, class 1 glands are not present in the fifth sternite in swarm-founding species.
4. The ducts of class 3 cells open only at the posterior part of the antero-costal margin in independent-founding wasps, whereas in swarming species, the ducts go toward the base of the sternite, and their openings are present both before and after anterolateral margin.
5. The second group of class 3 cells is present in the independent-founding species, in the middle of fat cells with their ducts traversing the epithelial gland.

Based on these structural differences, the difficulty of comparisons, as well as the uncertainty about the function of this gland, Samacá et al. (2013) suggest a restriction of the use of the name "Richards gland" only for clusters of class 3 cells whose opening of the ducts is at the base of the anterolateral margin of the fifth sternite, associated with changes in the external cuticle of the same sternite.

Smith et al. (2002), considering as Richards glands any group of glandular cells present in the fifth sternite, performed a study on swarming and communication in social wasps. This concept led the authors to conclude that this gland evolved early in the history of Vespidae, was lost four times, and reappeared once in Vespula. Applying the term more strictly, as proposed by Samacá et al. (2013), the Richards gland becomes an exclusive structure of Epiponini, with two reversals, one in the clade of *Angiopolybia* + *Agelaia* and one in the clade of (*Parachartergus* (*Leipomelles*, (*Chartergellus* + *Nectarinella*))).

The function of the cuticular modifications present at the base of the sternite is not very clear. One possibility would be that these modifications cause the opening of the ducts to be below the surface of the sternite, preventing them from being clogged by debris that could enter when the wasp is rubbing the abdomen during the marking process. Another hypothesis would be that these scales would help in capturing a new layer of secretion from reservoir between applications (Jeanne and Post 1982).

Jeanne et al. (1983) identified four types of cuticular changes associated with the presence of class 3 glands (present in the fifth or sixth sternite). The most common type consists of projections like scales, among which the opening of the ducts of the glandular cells is found. This type of modification had already been described by Jeanne and Post (1982) for some species of *Polybia*. The second type, which had also been identified in *Polybia* (Jeanne and Post 1982), consists of narrow transverse grooves, where the ducts open. The third type of specialization is a tuft of bristles present in the sixth sternite of *Polistes*, *Mischocyttarus*, *Ropalidia*, *Parapolybia*, and *Belonogaster*. In the first three genera, the bristles are long and form a relatively narrow tuft in the midline of the sternite, but in the *Parapolybia* and *Belonogaster*, the bristles cover a wider area. *Apoica* also has hairs on the gland's openings, but it is uncommon to have hairs in a broad band. The fourth type of specialization is known only for *Clypearia sulcata* (Saussure, 1853), where the ducts of the gland cells open into groups in wells between reticulated ridges (Fig. 11.11).

In some genera (*Agelaia*, *Chartergus*, *Synoeca*, and *Epipona*), cuticular modifications are very subtle, with the ducts openings associated with very short setae (*Agelaia*) or slightly recessed in shallow wells (*Synoeca*). However, it is possible to interpret the reticulations in *Clypearia*, the wells in *Synoeca*, and the grooves in some *Polybia* species as different degrees of the same type of modification (opening of the ducts in wells). The scales and the presence of bristles appear to represent completely different types of development (Jeanne et al. 1983).

It is simple to imagine how scale-type modifications derived from an unmodified sclerite – a specialization resulting in exaggeration of the scalelike surface already present. The grooves, on the other hand, involve an invagination of the entire integument. It is not inconceivable that these have evolved from the first, although among the species analyzed the only example of an "intermediate path" is in *Polybia sericea* (Olivier, 1792), where the grooved surface is partly squamous (Jeanne and Post 1982).

Jeanne and Post (1982) analyzed several species of *Polybia* and suggested that the furrows would be a specialization to accommodate larger glands in size, to

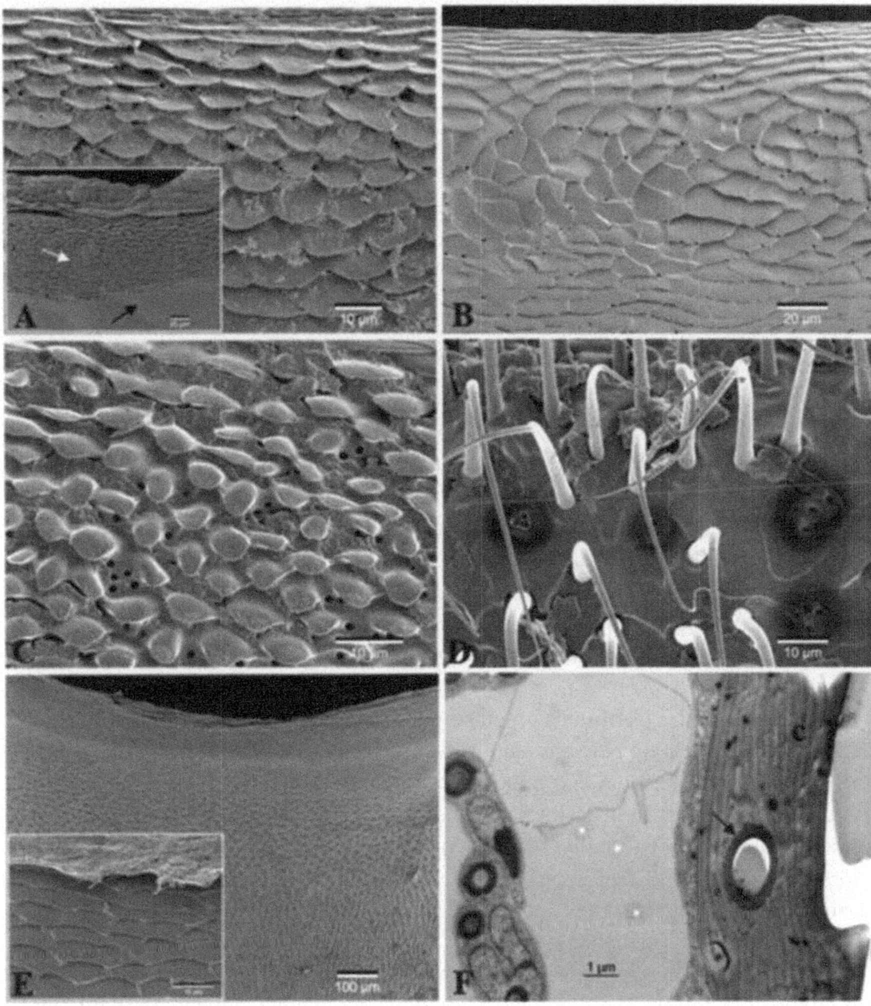

Fig. 11.11 Scanning electron microscopy showing the cuticular changes and external opening of the ducts of the Richards gland, present in the fifth metasomal sternite. (**a**) Difference between the area of the glandular opening (white arrow), detail in a higher magnification on the left, and the remainder of the sternite cuticle (black arrow) in *Protopolybia exigua* (Saussure, 1906); (**b**) *Protonectarin sylveirae* (Saussure, 1854); (**c**) *Polybia occidentalis* (Olivier, 1791), where the scales are most projected; and (**d**) *Apoica pallens* (Fabricius, 1804), where the ducts are grouped into small depressions and many bristles are present. (**e**) Absence of cuticular modifications and pores (detail on the left) in *Chartergellus communis* Richards (1978), evidencing the absence of the Richards gland in this species. (**f**) Electron micrograph showing the external opening of a duct in the cuticle in *P. sylveirae*; *c* cuticle

provide more volume to the reservoir or to enlarge the epidermal surface to accommodate a larger number of glandular cells. In fact, among the studied species, the ones that presented cuticular modifications in the form of grooves had glands with a greater number of cells and larger reservoirs in comparison with species presenting scale-type cuticular modifications. Thus, scales would be more common in species with smaller glands. However, in an analysis with more genera of Polistinae, the largest gland was found in *Pseudopolybia*, which present a scale-type cuticular specialization (Jeanne et al. 1983).

11.2.3 Exocrine Glands in Males

The sternal glands present in males of social wasps have received little attention (Post and Jeanne 1983a). However, structures like those found in females have been discovered in males (Landolt and Akre 1979; Post and Jeanne 1982, Post and Jeanne 1983b) and probably act in the production of pheromones related to reproductive behavior (Post and Jeanne 1982, 1983b, Litte 1979). In a study of species of *Mischocyttarus*, Post and Jeanne (1982) found much intraspecific variation in the development of external glands.

Males of *Mischocyttarus flavitarsis* (Saussure, 1854) Bequaert and *Mischocyttarus drewseni* Saussure, 1857, of Saussure possess many glandular cell ducts in the fifth, sixth, and seventh (terminal) metasomal sternites, whereas the males of *Mischocyttarus mexicanus* (Saussure, 1854) have relatively few glandular cells in the fifth and sixth sternites and none in the seventh. In addition, in *M. flavitarsis*, tufts of bristles are present in the cuticle of the fifth and sixth sternites, where the glandular ducts open externally. In males of *M. flavitarsis*, *M. drewseni*, and *Polistes foederatus* Kohl, 1898 (subgenus *Polistes*), glandular cells are located along the anterior margin of the sternites 3–6 and in the posterior half of the last sternite (seventh) (Jeanne 1982; Turillazzi 1979).

The function of these glands remains unknown. Since males do not rub the abdomen on the nest petiole, the gland probably does not produce an ant repellent. After emergence as adults, males remain only a short time in the nest (approximately 4 days), when they generally stay inactive on the top of the comb (West-Eberhard 1969). When they leave the nest, males spend most of their time looking for females (West-Eberhard 1969). However, the observation that males of *M. flavitar- sis* exhibit the behavior of rubbing the abdomen at sites likely to attract females during mating attempts (Litte 1979) suggests that the secretion of these glands plays a role in reproductive behavior (Post and Jeanne 1983a). This same behavior was observed in males of *Polistes fuscatus* (Fabricius, 1793) suggesting the application of glandular secretion (Post and Jeanne 1983b). Thus, the fact that males rub the gaster on the territorial perches confirms the idea that the products of these glands play some role in the reproduction of these wasps (Post and Jeanne 1983b).

11.3 Conclusions and Perspectives

There is no doubt about the importance of the role of the exocrine glands in insect behavior (Billen and Morgan 1998), including social ones. These glands and their secretions are involved in the most diverse aspects of social wasp life, such as interactions of dominance and maintenance of queen status, foraging, nest search, defense, and breeding behavior (Downing 1991). The swarm-founding way of colony founded and other adaptations to life in the tropics suggest that Epiponini differ significantly from its temperate region relatives to exocrine aspects (Jeanne and Post 1982). Nevertheless, the function of the secretions of these glands remains obscure, especially for Neotropical groups. Although studies on the morphology of the glands do not solve the uncertainty about the function that these structures exert, they make reconstruction of evolutionary history more consistent with thinking in comparative and systematic biology (Samacá et al. 2013). Larger and more detailed morphological analyzes within the group of social wasps, the use of modern techniques with greater accuracy, the integration of this information with chemical analyzes of the produced secretions, and the conduction of bioassays should in the future result in a more detailed understanding of the function of the pheromone-producing exocrine glands in the social wasps.

References

Akre RD (1982) Social Wasps. In: Hermani H (ed) Social insects, Vol. IV. Academic Press, New York, pp 1–105

Billen J (1987) New structural aspects of the Dufour's and venom glands in social insects. Naturwissenschaften 74:340–341

Billen J (1991) Ultrastructural organization of the exocrine glands in ants. Ethol Ecol Evol special issue 1:67–73

Billen J (2009a) Occurrence and structural organization of the exocrine glands in the legs of ants. Arthropod Struct Dev 38:2–15

Billen J (2009b) Diversity and morphology of exocrine glands in ants. Proceedings of the 19th Symposium of the Brazilian Myrmecological Society, Ouro Preto (Brasil), pp 1–6

Billen J (2011) Exocrine glands and their key function in the communication system of social insects. Formosan Entomol 31:75–84

Billen J, Morgan ED (1998) Pheromone communication in social insects – sources and secretions. In: Vander Meer RK, Breed MD, Winston ML, Espelie KE (eds) Pheromone communication in social insects: ants, wasps, bees, and termites. Westview Press, Boulder, pp 3–33

Billen J, Noll FB (2011) Morphology of Richards' gland in the wasp *Metapolybia docilis*. Sociobiology 58:251–255

Blum MS (1982) Pheromonal bases of insect sociality: communication, conundrums and caveats. In: Les Médiateurus Chimiques. INRA Publications, Versailles, pp 149–162

Corn ML (1972) Notes on the biology of *Polistes carnifex* (HymenopteraVespidae) in Costa Rica and Colombia. Psyche 79:150–157

Da Silva M, Noll FB, Billen J (2015) Morphology of Richards' gland in the swarm-founding wasp *Protonectarina sylveirae* (Hymenoptera, Vespidae). Acta Zool 96:530–533

Dani F, Cervo R, Turillazzi S (1992) Abdomen stroking behaviour and its possible functions in *Polistes dominula* (Christ) (Hymenoptera, Vespidae). Behav Process 28:51–58

Dani FR, Jones GR, Morgan DE, Turillazzi S (2003) Reevaluation of the chemical secretion of the sternal glands of *Polistes* social wasps (Hymenoptera Vespidae). Ethol Ecol Evol 15:73–82

Dapporto L, Dani FR, Turillazzi S (2007a) Social dominance molds cuticular and egg chemical blends in a paper wasp. Curr Biol 17:R504

Dapporto L, Santini A, Dani FR, Turillazzi S (2007b) Workers of a *Polistes* paper wasp detect the presence of their queen by chemical cues. Chem Senses 32(8):795–802

Dapporto L, Petrocelli I, Turillazzi S (2011) Incipient morphological castes in Polistes gallicus (Vespidae, Hymenoptera). Zoomorphology 130:197–201

Darchen R (1976) *Ropalidia cincta*, guêpe sociale de la savane du Lampto (Cote d'Ivoire). Annls Soc ent Fr 12:579–601

de Souza AR, Petrocelli I, Lino-Neto J, Santos EF, Noll FB, Turillazzi S (2016) Ontogenic caste differences in the Van der Vecht organ of primitively eusocial Neotropical paper wasps. PLoS One 11(5):e0154521

Delfino G, Marino Piccioli MT, Calloni C (1979) Fine structure of the glands of Van der Vecht's organ in *Polistes gallicus* (L.) (Hymenoptera Vespidae). Mon Zool Ita 13:221–247

Downing HA (1991) The function and evolution of exocrine glands. In: Ross KG, Matthews RW (eds) The social biology of wasps. Cornell University Press, Ithaca, pp 540–569

Downing HA, Post DC, Jeanne RL (1985) Morphology of sternal glands in male polistine wasps (Hymenoptera: Vespidae). Ins Soc 32:186–197

Gamboa GJ, Heacock BD, Wiltjer SL (1978) Division of labor and subordinate longevity in foundress associations of the paper wasp, *Polistes metricus* (Hymenoptera: Vespidae). J Kans Entomol Soc 51:343–352

Grimaldi D, Engel M (2005) Evolution of the insects. Cambridge University Press, New York and Cambridge, 755p

Hermann HR, Dirks TF (1974) Sternal glands in polistine wasps: morphology and associated behavior. J Ga Entomol Soc 9:1–8

Hermann HR, Dirks TF (1975) Biology of *Polistes annularis* (Hymenoptera: Vespidae). I. Spring behavior. Psyche 82:97–108

Heselhaus F (1922) Die Hautdrusen der Apiden und ver-wandter. Formen Zool Jahrb Abt Anat Ontog Tiere 43:369–464

Hölldobler B, Wilson EO (1990) The ants. Harvard University Press, Cambridge, 732p

Hunt JH, Jeanne RL, Keeping MG (1995) Observations on *Apoica pallens*, a nocturnal Neotropical social wasp (Hymenoptera: Vespidae, Polistinae, Epiponini). Ins Soc 42:223–236

Jeanne RL (1970) Chemical defense of brood by a social wasp. Science 168:1465–1466

Jeanne RL (1975) The adaptiveness of social wasp nest architecture. Q Rev Biol 50:267–287

Jeanne RL (1981) Chemical communication during swarm emigration in the social wasp *Polybia sericea* (Olivier). Anim Behav 29:102–113

Jeanne RL (1982) Evidence for an alarm substance in *Polistes canadensis*. Experientia 38:329–330

Jeanne RL (1991) The swarm-founding Polistinae. In: Ross KG, Matthews RW (eds) The social biology of wasps. Cornell University Press, Ithaca, pp 191–231

Jeanne RL (1996) Non-allometric queen-worker dimorphism in *Pseudopolybia difficilis*. J Kansas Entomol Soc 69:370–374

Jeanne RL, Post DC (1982) Richards' gland and associated cuticular modifications in social wasps of the genus *Polybia* (Hymenoptera: Vespidae: Polistinae: Polybiini). Insects Sociaux 29:280–294

Jeanne RL, Downing HA, Post DC (1983) Morphology and function of sternal glands in polistine wasps (Hymenoptera: Vespidae). Zoomorphology 103:149–164

Keeping MG (1990) Rubbing behavior and morphology of van der Vecht's gland in *Belonogaster petiolata* (Hymenoptera: Vespidae). J Insect Behav 3:85–104

Kojima J (1983) Defense of the pre-emergence colony against ants by means of a chemical barrier in *Ropalidia fasciata* (Hymenoptera, VespidaeJ). Jpn J Ecol 33:213–223

Kojima J (1992) The ant repellent function of the rubbing substance in an Old World polistine. *Parapolybia indica* (Hymenoptera Vespidae). Ethol Ecol Evol 4:183–185

Landolt PJ, Akre RD (1979) Occurrence and location of exocrine glands in some social Vespidae (Hymenoptera). Ann Entomol Soc Am 72:141–148

Litte M (1976) Self, kin and social behavior in the polistine wasp *Mischocyttarus mexicanus*. PhD thesis. Cornell Univ., Ithaca, 242 pp

Litte M (1979) *Mischocyttarus flavitarsis* in Arizona: social and nesting biology of a polistine wasp. Z Tierpsychol 50:282–312

Litte M (1981) Social biology of the polistine wasp *Mischocyttarus labiatus*: survival in a Colombian rain forest. Smithson Contrib Zool 327:1–27

London KB, Jeanne RL (2000) The interaction between mode of colony founding, nest architecture and ant defense in polistine wasps. Ethol Ecol Evol 12:13–25

Naumann MG (1975) Swarming behavior: evidence for communication in social wasps. Science 189:642–644

Noirot C, Quennedey A (1974) Fine structure of insect epidermal glands. Annu Rev Entomol 19:61–80

Noirot C, Quennedey A (1991) Glands, gland cells, glandular units: some comments on terminology and classification. Annl Soc Ent Fr 27:123–128

Matsuura M, Yamane S (1990) Biology of the Vespine wasps. Springer-Verlag, Berlin, 330p

O'Donnell S, Jeanne RL (1990) Forager specialization and the control of nest repair in *Polybia occidentalis* Olivier. Behav Ecol Sociobiol 27:359–364

Petrocelli I, Turillazzi S (2013) The morphology of Van der Vecht's organ as a tool to measure caste dimorphism in Polistes paper wasps: a comparative approach. J Zool Syst Evol Res 51:274–278

Post DC, Jeanne RL (1980) Morphology of the sternal glands of *Polistes fuscatus* and *P. canadensis* (Hymenoptera: Vespidae). Psyche 87:49–58

Post DC, Jeanne RL (1981) Colony defense against ants by *Polistes fuscatus* (Hymenoptera: Vespidae) in Wisconsin. J Kans Entomol Soc 54:599–615

Post DC, Jeanne RL (1982) Sternal glands in three species of male social wasps of the genus Mischocyttarurs (Hymenoptera: Vespidae). New York Entomological Society 90:8–15

Post DC, Jeanne RL (1983a) Sternal glands in males of six species of *Polistes (Fuscopolistes)* (Hymenoptera: Vespidae). J Kans Entomol Soc 56:32–39

Post DC, Jeanne RL (1983b) Male reproductive behavior of the social wasp *Polistes fuscatus* (Hymenoptera: Vespidae). Z Tierpsychol 62:157–171

Raposo-Filho JR, Calloni C, Delfino G (1994) Structural and ultrastructural features of the glands in the VI and VII gastral sternites of female *Mischocyttarus (Kappa) atramentarius* Zikán 1949 (Hymenoptera Vespidae). Ethol Ecol Evol 6:47–52

Richards OW (1971) The biology of the social wasps (Hymenoptera: Vespidae). Biol Rev 46:483–528

Richards OW (1978) The social wasps of the Americas excluding the Vespinae. British Museum (Natural History), London, 580p

Samacá E, Billen J, Sarmiento CE (2013) Morphology of the fifth sternal glands of Neotropical social wasps (Hymenoptera, Vespidae, Polistinae). Invertebr Biol 132(2):163–172

Schoeters E, Billen J (1995a) Morphology and ultrastructure of the convoluted gland in the ant *Dinoponera australis* (Hymenoptera: Formicidae). Int J Insect Morphol Embryol 24:323–332

Schoeters E, Billen J (1995b) Morphology and ultrastructure of a secretory region enclosed by the venom reservoir in social wasps (Insecta, Hymenoptera). Zoomorphology 115:63–71

Smith AR, O'Donnell S, Jeanne RL (2001) Correlated evolution of colony defence and social structure: a comparative analysis in eusocial wasps (Hymenoptera: Vespidae). Evol Ecol Res 3:331–344

Smith AR, O'Donnell S, Jeanne RL (2002) Evolution of swarm communication in eusocial wasps (Hymenoptera: Vespidae). J Insect Behav 15:751–764

Turillazzi S (1979) Tegumental glands in the abdomen of some European *Polistes* (Hymenoptera Vespidae). Monit Zool Ital 13:67–70

Turillazzi S, Ugolini A (1978) Nest defense in European Polistes (Hymenoptera Vespidae). Monit Zool Ital 12:72

Turillazzi S, Ugolini A (1979) Rubbing behaviour in some European *Polistes* (Hymenoptera: Vespidae). Monit Zool Ital 16:75–88

Van der Vecht J (1965) The geographical distribution of the social wasps (Hymenoptera: Vespidae). Proc. XII Int. Congr. Ent, London, pp 440–441

Van der Vecht J (1968) The terminal gastral sternite of female and worker social wasps (Hymenoptera, Vespidae). Proc K ned Akad Wet 71:411–422

Wenzel JW, Carpenter JM (1994) Comparing methods: adaptive traits and tests of adaptation. In: Eggleton P, Vane-Wright R (eds) Phylogenetics and ecology. Academic Press, London, pp 79–101

West-Eberhard MJ (1969) The social biology of polistine wasps. Misc Publ Mus Zool Univ Mich 140:1–101

West-Eberhard MJ (1982) The nature and evolution of swarming in tropical social wasps (Vespidae, Polistinae, Polybiini). In: Jaisson P (ed) Social insects in the tropics, vol 1. Universit'e de Paris-Nord, Paris, pp 97–128

West-Eberhard MJ (1989) Scent-trail diversion, a novel defense against ants by tropical social wasps. Biotropica 21:280–281

Chapter 12
Cuticular Hydrocarbon Studies in Neotropical Social Wasps

William Fernando Antonialli-Junior, Viviana de Oliveira Torres, Ivelize Cunha Tannure-Nascimento, and Fábio Santos Nascimento

Abstract Cuticular hydrocarbons are chemical compounds present in all insects and primarily involved in preventing water loss and protecting against pathogenic microorganisms. In social insects, cuticular hydrocarbons evolved in cues that provide multilevel information. In this chapter, we review the studies on cuticular hydrocarbons in Neotropical social wasps. So far, the main research exploring the CHCs of independent and swarm-founding wasps demonstrates their importance in mediating interactions between nestmates and also that these compounds may vary within colony members, differences between sexes, reproductive status, fertility, age, and their relationships with social parasitism.

Keywords Chemical signature · Reproductive signals · Nestmate recognition

12.1 Cuticular Hydrocarbons and Social Insects

The cuticle of the insects consists of several long-chain fatty acids, alcohols, esters, aldehydes, ketones, and hydrocarbons that primarily protect them from desiccation and act as a barrier against microorganisms and other pathogens (Lockey 1988; Gibbs 1998). Secondly, these cuticular compounds, particularly hydrocarbons, also act as mediators in the recognition of nestmates and non-nestmates and thus were

W. F. Antonialli-Junior
Laboratório de Ecologia Comportamental, Universidade Estadual de Mato Grosso do Sul, Dourados, MS, Brazil
e-mail: williamantonialli@yahoo.com.br

V. de Oliveira Torres
Laboratório de Estudo em Vespas Sociais, Universidade Federal da Grande Dourados, Faculdade de Ciências Biológicas e Ambientais, Dourados, MS, Brazil
e-mail: vivianatorres@ufgd.edu.br

I. C. Tannure-Nascimento · F. S. Nascimento (✉)
Departamento de Biologia, Faculdade de Filosofia Ciências e Letras de Ribeirão Preto, Universidade de São Paulo, Ribeirão Preto, SP, Brazil
e-mail: itannure@uol.com.br; fsnascim@usp.br

F. Prezoto et al. (eds.), *Neotropical Social Wasps*,
https://doi.org/10.1007/978-3-030-53510-0_12

considered as short distance pheromones (Howard and Blomquist 2005). These cuticular hydrocarbons (CHCs) are mainly nonvolatile compounds formed by hydrogen and carbon, and although hundreds of these compounds have already been described, it is possible to group them into three major classes: linear alkanes, alkanes with methyl branching, and unsaturated hydrocarbons (Blomquist and Bagnères 2010).

Since the 1980s, studies in literature have highlighted the importance of cuticular hydrocarbons for intraspecific recognition in different species of wasps and other social insects (Lockey 1988; Bonavita–Cougourdan et al. 1987; Singer 1998; Dapporto et al. 2004a; Cotoneschi et al. 2009; Blomquist and Bagnères 2010; Johnson et al. 2011). These CHCs are used as primary pheromones for discrimination between nestmates during brief contact between individuals, whose qualitative and quantitative variations can provide information on species, gender, colony, task, fertility, etc. and thus are considered as a chemical signature or species of individual "chemical identity" (Provost et al. 2008; Blomquist and Bagnères 2010).

The function of these compounds constituting a chemical identity for members of social insect colonies has already been explored, for example, by Liebig et al. (2000) who reported that *Harpegnathos saltator* queen ants (Jerdon, 1851) presented exclusive cuticular hydrocarbons (qualitative difference) and with distinct proportions (quantitative difference) in relation to the workers. The relationship between the CHC profile and the physiological status of the female has already been studied in several species of social wasps (Izzo et al. 2010; Mitra and Gadagkar 2014), in which there may be a relation between the exclusivity of certain compounds and the fertility of females. The specific activity of certain branched alkanes in the queen of the *Dolichovespula saxonica* wasp (Fabricius, 1793) and the *Lasius niger* ant (Linnaeus, 1758) is also indicative of fertility in these species (Holman et al. 2010; van Zweden et al. 2013) In addition, Kather et al. (2011) suggested that in colonies of *Apis mellifera* the proportion of linear alkanes and alkenes is higher in foragers, followed by newly emerged nurse bees and workers, so these cuticular compounds may be associated with the task developed by the female in the colony.

These compounds may also be a signal of the relative age of each member within a colony, as described by Lorenzi et al. (2004) in colonies of *Polistes dominula* (Christ, 1791). In addition, CHCs may be important for gender differentiation of individuals as shown in the studies of Abdalla et al. (2003) with the stingless bee *Melipona bicolor* (Lepeletier, 1836) and Antonialli-Junior et al. (2007) with the ant *Ectatomma vizottoi* (Almeida, 1987).

Since the profile of CHCs can vary according to genetic factors, it can also be used to evaluate interspecific differences, as already demonstrated in termites of the genus *Reticulitermes* (Bagnères et al. 1991), in ants of the genus *Formica* (Martin et al. 2008), in wasps of the subfamily Stenogastrinae (Baracchi et al. 2009), and in male of orchid bees (Santos and Nascimento 2015). Therefore, according to Bagnères and Wicker-Thomas (2010), these compounds may be useful as a complementary taxonomic tool for species distinction, as they are stable, inheritable, and species-dependent metabolic products (Kather and Martin 2012).

Thus, cuticular hydrocarbons may vary depending on both genetic and exogenous factors (Gamboa 1996; Beye et al. 1998; Liang and Silverman 2000; Buczkowski and Silverman 2006; Dronnet et al. 2006). In this sense, geographically closer populations tend to have closer chemical compounds among them (Dapporto et al. 2004b; Bonelli et al. 2015), probably because of the higher level of kinship. On the other hand, colonies that nest in similar environments tend to share more of these compounds among themselves (Wagner et al. 2001). In particular, in the case of the social wasps, the colony's composition of CHCs may be directly related to the material that makes up the substrate of their nests (Singer and Espelie 1992), so if they share the same resources for this purpose, they will also share many of these compounds.

12.2 Social Wasps and Cuticular Hydrocarbons

Relatively a few species had the cuticular hydrocarbon profiles (CHCs) with their biological aspects studied so far (Fig. 12.1). Studies exploring the CHCs of social wasps show their importance in mediating interactions between nestmates (Kudô and Zucchi 2006; Kudô et al. 2007; Kelstrup et al. 2014a), demonstrating that these compounds may vary according to population (Dapporto et al. 2004 a, b), colonies (Layton et al. 1994; Dapporto et al. 2005, Tannure-Nascimento et al. 2007; Antonialli-Junior et al. 2008), sex (Cotoneschi et al. 2009; Neves et al. 2013), reproductive status (Sledge et al. 2001; Kelstrup et al. 2014b; Soares et al. 2014; Torres

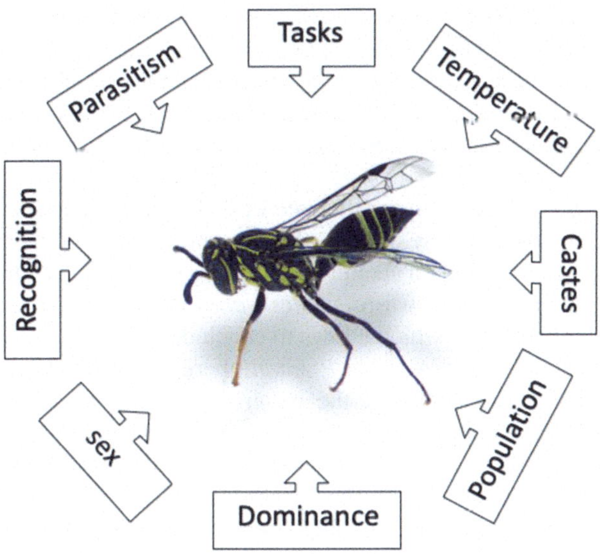

Fig. 12.1 Scheme showing the social and ecological aspects reported on the cuticular chemical profile studies of social wasps. (Photo of *Protopolybia sedula*: André R. de Souza)

et al. 2014), fertility (Monnin 2006; Izzo et al. 2010), age (Neves et al. 2012), and determinants during relationships of social parasitism (Neves et al. 2013).

Neotropical social wasps are represented by the subfamily Polistinae, which comprises 25 genera and more than 900 species, distributed mainly among the tropics and subtropics (Richards 1978), and from these, more than 300 species can be found only in Brazil (Carpenter and Marques 2001). These species of social wasps can independently found their colonies, as it occurs in the Tribes Polistini and Mischocyttarini or by swarming as most genera of the Tribe Epiponini (Jeanne 1980; Carpenter and Marques 2001). Swarm foundations are common in more derived species, in which a swarm with several queens and hundreds or thousands of workers leaves the original colony in search of a suitable place to build a new nest (Jeanne 1991). The independent foundation occurs in less derived species, such as *Polistes*, *Mischocyttarus*, *Belonogaster*, and *Parapolybia* (Gadagkar 1991). In this foundation, a new colony can be initiated by a single inseminated female or haplometrosis, or two or more inseminated females initiate the foundation process by association or pleometrosis (Jeanne 1991).

12.3 Cuticular Hydrocarbons in Neotropical Independent Founding Wasps

Species of Neotropical social wasps of independent foundation that have already had their profile of hydrocarbons studied are *Polistes versicolor* (Olivier, 1791), *Polistes ferreri* (Saussure, 1853), *Polistes satan* (Bequaert, 1940), *Mischocyttarus latior* (Fox, 1898), *Mischocyttarus cassununga* (Von Ihering, 1903), *Mischocyttarus cerberus* (Richards, 1940), *Mischocyttarus consimilis* (Zikán, 1949), and *Mischocyttarus parallelogrammus* (Zikán, 1949) (Tannure-Nascimento et al. 2007; Ferreira et al. 2012; Neves et al. 2012; Neves et al. 2013; Soares et al. 2014; Togni 2014; Torres et al. 2014, 2016; Murakami et al. 2015; Michelutti et al. 2018) (Table 12.1).

The pioneer work to investigate the role of CHCs in a Neotropical species of social wasp was carried out by Tannure-Nascimento et al. (2007), in which they verified that the linear odd chain alkanes and methylated alkanes are the most abundant in the cuticle of the social wasp *Polistes satan*. These compounds vary significantly among their colonies, which allows to conclude that each colony has its own cuticular chemical signature. Also, in *P. satan*, there is a relationship between the chemical cuticular compounds and the pattern of visual signals of the face of females, which in turn may indicate the female hierarchy in the colony (Tannure-Nascimento et al. 2008). Therefore, these signals together could transmit reliable information to the nestmates about their reproductive capacity. In fact, facial patterns may also be good indicators of body size and hierarchical position of females in colonies of temperate climate species *Polistes dominula* (Tibbetts and Dale 2004).

Table 12.1 Summary of the studies on cuticular hydrocarbons made with Neotropical social wasps

Species	Type of study	Type of most relevant compounds	Study	Technique
Independent founding Neotropical wasps				
Polistes versicolor	Reproductive status Effect of temperature on CHCs of individuals and nests	Functional groups related to cuticle and hydrocarbons Increase of linear alkanes and reduction of methyl-branched alkanes	Torres et al. (2014), Michelutti et al. (2018, 2019)	FTIR-PAS GC-MS
Polistes ferreri	Reproductive status Differences among parasitized females and number of parasites	Functional groups: benzene (C-H in the plane and off-plane), -C-CH_3, and -N-H Functional groups: -C-H (CH_2), C-CH_3, C-CH_2, and C-H	Soares et al. (2014), Torres et al. (2016)	FTIR-PAS FTIR-PAS
Polistes satan	Differences among colonies Dominant, founders, substitutes, and subordinates differences	Abundance of alkanes with methyl branching, but alkanes such as C27, C28, and C33 were important in female distinction Long-chain linear alkanes were the most important Synthetic CHC blend did not inhibit ovary activation of subordinates	Tannure-Nascimento et al. (2007, 2008), Oi et al. (2019)	GC-MS
Mischocyttarus latior *M. cerberus* *M. consimilis*	Differences among species	C15, C19, C23, C26, C27, C29, and C30 were important to distinguish species	Ferreira et al. (2012)	GC-MS
Mischocyttarus consimilis	Chemical composition of the cuticle in females with application of juvenile hormone (JH) Females with and without JH application	Functional groups C-H, C-CH_3, CH_3, and N-H were important to distinguish groups of females with and without JH application Linear alkanes, alkanes with methyl branching, and a dimethyl were important in distinguishing the control group from those receiving JH	Neves et al. (2013)	FTIR-PAS GC-MS

(continued)

Table 12.1 (continued)

Species	Type of study	Type of most relevant compounds	Study	Technique
Mischocyttarus cassununga	Differences among colonies	Branched alkanes were the important compounds for differentiation	Murakami et al. (2015)	GC-MS
Mischocyttarus parallelogrammus	Differences among colonies	Branched alkanes (methyl and dimethyl) were important to differentiate the colony-specific profile	Togni (2014)	GC-MS
Neotropical swarm-founding wasps				
Polybia micans	Age, fertility, and caste	Predominance of linear alkanes, but the methyl alkanes were also important to differentiate the groups	Kelstrup et al. (2014b)	GC-MS
Polybia paulista	Recognition trails in male and female nestmates Effect of temperature on CHCs of individuals and nests	Behavioral tests that discuss the possible involvement of cuticular compounds in the processes CHC profile variation of the individuals and nests	Kudô and Zucchi (2006), Kudô et al. (2007), Michelutti et al. (2018, 2019)	None GC-MS
Polybia ignobilis	Effect of temperature on CHCs of individuals and nests	CHC profile variation of the individuals and nests	Michelutti et al. (2018 and 2019)	GC-MS
Synoeca surinama	Reproductive status	Linear alkanes and alkenes; related to JH titers	Kelstrup et al. (2014a)	GC-MS
Synoeca septentrionalis	Caste and sex distinction Differences among colonies	Linear alkanes and alkenes	Santos et al. (2018)	GC-MS
Protopolybia exigua	Differences among colonies Differences in stages of development	Absence of differences based on the proximity of the colonies Significant quantitative differences, with more abundant linear alkanes and decrease according to the development from egg to larvae	Silva et al. (2016)	GC-MS

Cuticular hydrocarbons can be used as indicators of the reproductive status of females in colonies of *P. ferreri* and *P. versicolor*, which have a relationship between the chemical profile and the degree of ovarian development of the female (Soares et al. 2014; Torres et al. 2014). Previous results obtained by Izzo et al. (2010) showed the relationship between the profile of CHCs, reproductive status, and juvenile hormone titers in females of temperate climate species *P. dominula*.

The effect of parasitization by parasites of the genus *Xenos* on *P. ferreri* colonies was described, and it goes beyond alterations in the reproductive potential, as was already pointed out in other species of temperate climate *Polistes* (Hughes et al. 2004; Beani et al. 2011), also altering the CHC profile (Torres et al. 2016).

The first studies on CHCs in *Mischocyttarus* wasps evaluated the effects of the facultative social parasitism of *M. consimilis* on *M. cerberus* and how it changed their chemical signatures (Ferreira et al. 2012; Neves et al. 2013). The first case of facultative social parasitism was described by Montagna et al. 2012. According to these authors, the first contact of the parasite species with the host occurs during the foundation of the colony, and from there, it causes a modification in the chemical composition of the two species, converging a similar profile and allowing the interaction between both species in the same colony.

In fact, studies like Lorenzi and Cervo (1992) and Cervo and Lorenzi (1996) with the temperate climate species *Polistes biglumis bimaculatus* (Geoffroy, 1785) have already reported that the parasite species can acquire part of the chemical signature of the host colony after the first contacts, especially with the substrate of the nests reducing the agonistic behaviors that occur at the beginning of the interactions. Neves et al. (2012) observed that *M. consimilis* females acquire their complete colonial chemical signature by the third or fourth day after their emergence. In this sense, Lorenzi et al. (2004) reported that in temperate climate species *P. dominula*, the cuticle of newly emerged females increases the number of hydrocarbons, more precisely branched alkanes, and that the older ones are probably less prone to changes in hydrocarbons, since they would already have a complete chemical profile. According to Gamboa (1996, 2004), *Polistes* wasps are able to recognize olfactory cues and "learn" the characteristic odor of their nest during the first hours after their emergence.

In *M. parallelogrammus*, more than 80% of the cuticular hydrocarbons identified were methyl- and dimethyl-branched alkanes (Togni 2014), which were also important to distinguish the colony-specific profile during postemergence. In addition, *M. parallelogrammus* females were classified according to the variation of their ovary activation. The cuticular chemical compounds of *M. cassununga* species showed that the branched alkanes were widely predominant and responsible for providing a unique colonial signature (Murakami et al. 2015), which has also been described in other studies with temperate climate paper wasps (Cervo et al. 1996; Dani et al. 2001; Mitra et al. 2014). More recently, combining different approaches as behavioral observations, reproductive physiology, and cuticular hydrocarbons analysis, Silva et al. (2020) demonstrated that in *Mischocyttarus cerberus*, females occupying the highest position in a dominance hierarchy possess a different level of fertility and cuticular chemical profile accordingly.

In another study, Montagna et al. (2015) evaluated the effects of topical application of juvenile hormone on immatures of *M. consimilis*. In this study, the authors evidenced that in primitively eusocial wasps, caste determination may be, at least in part, preimaginal, which resulted in the production of females with more developed ovaries and a profile of CHCs more similar to the queen.

12.4 Cuticular Hydrocarbons in Neotropical Swarm-Founding Social Wasps

As observed for independent founding Neotropical wasps so far, a few papers investigated the role of CHCs in Neotropical swarm-founding wasps. The studies to date were made on *Polybia micans* (Ducke, 1904), *Polybia paulista* (Ihering, 1896), *Synoeca surinama* (Linnaeus, 1767), and *Protopolybia exigua* (Saussure, 1854) (Kudô and Zucchi 2006; Kudô et al. 2007, 2016; Kelstrup et al. 2014a, b; Silva et al. 2016).

Colonies of swarming species are numerous and sometimes polygynous; in this sense, there may exist an efficient mechanism that signals the reproductive status of females. Thus, the studies of Kelstrup et al. (2014a, b) investigated the relationship between physiology and cuticular composition of females with different reproductive status in *S. surinama* and *Pol. micans*, respectively. Kelstrup et al. (2014a) evaluated that JH titers define the reproductive status of *S. surinama* females and consequently their chemical cuticular compounds. However, Kelstrup et al. (2014b) did not find a relationship between JH titers and cuticular chemical compounds in *Pol. micans* females.

Among the studies with Neotropical swarm-founding wasps, Kudô and Zucchi (2006) reported possible odors as recognition cues for male acceptance in *Pol. paulista* colonies. Further, Kudô et al. (2007) studied interactions among females' nestmates in colonies with multiple queens of *Pol. paulista* and observed that even in polygynous colonies, females are able to recognize their nestmates and concluded that each colony has a specific odor. However, none of these studies identified these possible chemical compounds involved in recognition.

Recognition among adult individuals is well documented in social wasps (Gamboa 2004; West-Eberhard and Turillazzi 1996); however, little is known about the recognition of immatures, and in this sense, Kudô et al. (2016) evaluated the ability of *Pol. paulista* workers to recognize larvae of different stages from their own colonies through CHCs. The results showed that *Pol. paulista* workers are capable of recognizing nestmate larvae and that the hydrocarbon profile identified in the cuticle of the larvae allows them to differentiate according to the origin colony (Kudô et al. 2016).

Also, Silva et al. (2016) analyzed the chemical cuticular compounds of brood in *Protopolybia exigua* and concluded that each stage has its own chemical cuticular signature. However, the absence of differences between colonies was attributed to the proximity of the nesting places of the colonies analyzed and the level of kinship among them.

In recent studies, authors showed the effect of temperature on the cuticular hydrocarbons in three distinct wasp species, *P. versicolor*, *Pol. Ignobilis*, and *Pol. paulista*. The latter species did not change its cuticular profile under higher temperatures, but in *P. versicolor* and *Pol. Ignobilis*, there were a significant increase in the proportion of linear alkanes and reduction of methyl-branched alkanes (Michelutti et al. 2018). On the stark contrary, temperature had different effects on the hydrocarbons deposited on the nest envelope of these species (Michelutti et al. 2019), so it is possible that higher temperature may affect insects and their nests in a distinct way. In addition, a significant modification seen in the cuticular chemical profiles associated with the *Pol. occidentalis* workers' age polyethism is conserved in Hymenoptera and provides evidence that these compounds are of importance in the maintenance and cohesion of the colonies.

12.5 Final Considerations

Although literature in general has discussed in recent years the importance of CHCs as mediators of social behavior in wasps and other insects, there are still a few studies about Neotropical species of both independent and swarm founding. It is necessary conducting more research on unsolved questions that only Neotropical social swarm-founding wasps provide, such as the chemical regulation of excess of queens during cyclical oligogyny in Epiponini and how and whether laying workers (intermediates) signal their presence in the colonies. In independent founding Neotropical wasps, aspects involve the variation of cuticular hydrocarbons along the colony cycle and how distinct biomes (climate, altitude gradient, and season) affect the cuticular chemical profiles. These are some examples of questions that we propose for future researchers interested in this area may consider.

Acknowledgment The authors thank Fundação de Amparo a Pesquisa do Estado de São Paulo-Fapesp (2018/10699-0) and Conselho Nacional de Desenvolvimento Científico e Tecnológico (CNPq) for a productivity grant (307702/2018-9 to F.S.N. and 308182/2019-7 to W.F.A-Jr) and Projeto Universal (405082/2018-5).

References

Abdalla FC, Jones GR, Morgan ED, Cruz-Landim C (2003) Comparative study of the cuticular hydrocarbon composition of *Melipona bicolor* Lepeletier, 1836 (Hymenoptera, Meliponini) workers and queens. Genet Mol Res 2(2):191–199

Antonialli-Junior WF, Lima SM, Andrade LHC, Súarez YR (2007) Comparative study of the cuticular hydrocarbon in queens, workers and males of *Ectatomma vizottoi* (Hyminoptera, Formicidae) by Fourier transform-infrared photoacoustic spectroscopy. Genet Mol Res 6:492–499

Antonialli-Junior WF, Andrade LHC, Súarez YR, Lima SM (2008) Intra- and interspecific variation of cuticular hydrocarbon composition in two *Ectatomma* species (Hymenoptera: Formicidae) based on Fourier transform infrared photoacoustic spectroscopy. Genet Mol Res 7(2):559–566

Bagnères AG, Wicker-Thomas C (2010) Chemical taxonomy with hydrocarbons. In: Blomquist GJ, Bagnères AG (eds) Insect hydrocarbons, Biology, biochemistry, and chemical ecology. Cambridge University Press, New York, pp 121–162

Bagnères AG, Killian A, Clement JL, Lange C (1991) Interspecific recognition among termites of the genus *Reticulitermes*: evidence for a role for the cuticular hydrocarbons. J Chem Ecol 17(12):2397–2420

Baracchi D, Dapporto L, Teseo S, Hashim R, Turillazzi S (2009) Medium molecular weight polar substances of the cuticle as tools in the study of the taxonomy, systematics and chemical ecology of tropical hover wasps (Hymenoptera: Stenogastrinae). J Zool Syst Evol 48:109–144

Beani L, Dallai R, Mercati D, Cappa F, Giusti F, Manfredini F (2011) When a parasite breaks all the rules of a colony: morphology and fate of wasps infected by a strepsipteran endoparasite. Anim Behav 82:1305–1312

Beye M, Neumann P, Chapuisat M, Pamilo P, Moritz RFA (1998) Nestmate recognition and the genetic relatedness of nests in the ant *Formica pratensis*. Behav Ecol Sociobiol 43:67–72

Blomquist GJ, Bagnères AG (2010) Insect hydrocarbons: biology, biochemistry, and chemical ecology. Cambridge U.P, New York, 492 p

Bonavita–Cougourdan A, Clément JL, Lange C (1987) Nestmate recognition: the role of cuticular hydrocarbons in the ant *Camponotus vagus* Scop. J Entomol Sci 22(1):1–10

Bonelli M, Lorenzi MC, Christidès JP, Dupont S, Bagnères AG (2015) Population diversity in cuticular hydrocarbons and mtDNA in a mountain social wasp. J Chem Ecol 41:22–31

Buczkowski G, Silverman J (2006) Geographical variation in argentine ant aggression behavior mediated by environmentally derived nestmate recognition cues. Anim Behav 71:327–335

Carpenter JM, Marques OM (2001) Contribuição ao estudo dos vespídeos do Brasil (Insecta, Hymenoptera, Vespoidae, Vespidae). Cruz das Almas, Universidade Federal da Bahia. Publicações Digitais, Vol. II

Cervo R, Dani FR, Turillazzi S (1996) Nestmate recognition in three species of stenogastrinae wasps (Hymenoptera, Vespidae). Behav Ecol Sociobiol 39:311–316

Cotoneschi C, Dani FR, Cervo R, Scala C, Strassmann JE, Queller DC, Turillazzi S (2009) *Polistes dominulus* (Hymenoptera, Vespidae) larvae show different cuticular patterns according to their sex: workers seem not use this chemical information. Chem Senses 34:195–202

Cervo R, Lorenzi MC (1996) Behaviour in usurpers and late joiners of Polistes biglumis bimaculatus (Hymenoptera, Vespidae). Insectes Sociaux 43(3):255–266

Dani FR, Jones GR, Destri S, Spencer SH, Turillazzi S (2001) Deciphering the recognition signature within the cuticular chemical profile of paper wasps. Anim Behav 62:165–171. https://doi.org/10.1006/anbe.2001.1714

Dapporto LP, Theodora C, Spacchini G, Pieraccini, Turillazzi S (2004a) Rank and epicuticular hydrocarbons in different populations of the paper wasp *Polistes dominulus* (Christ) (Hymenoptera, Vespidae). Insect Soc 51:279–286

Dapporto L, Palagi E, Turillazzi S (2004b) Cuticular hydrocarbons of *Polistes dominulus* as a biogeographic tool: a study of populations from the Tuscan archipelago and surrounding areas. J Chem Ecol 30(11):2139–2151

Dapporto L, Sledge FM, Turillazzi S (2005) Dynamics of cuticular chemical profiles of *Polistes dominulus* workers in orphaned nests (Hymenoptera, Vespidae). J Insect Physiol 51(9):969–973

Dronnet S, Lohou C, Christides JP, Bagnères AG (2006) Cuticular hydrocarbon composition reflects genetic relationship among colonies of the introduced termite *Reticulitermes santonensis feytaud*. J Chem Ecol 32:1027–1042. PMID: 16739021

Ferreira AC, Cardoso CAL, Neves EF, Súarez YR, Antonialli-Junior WF (2012) Distinct linear hydrocarbon profiles and chemical strategy of facultative parasitism among *Mischocyttarus* wasps. Genet Mol Res 11:4351–4359

Gadagkar R (1991) *Belonogaster, Mischocyttarus, Parapolybia*, and independent– founding *Ropalidia*. In: Ross KG, Matthews RW (eds) The social biology of wasps. Cornell University Press, Ithaca, pp 149–190
Gamboa GJ (1996) Kin recognition in social wasps. In: Turillazzi S, West-Eberhard MJ (eds) Natural history and evolution of paper-wasps. Oxford University Press, Oxford, pp 161–177
Gamboa GJ (2004) Kin recognition in eusocial wasps. Ann Zool Fenn 41:789–808
Gibbs AG (1998) Water-proofing properties of cuticular lipids. Am Zool 38:471–482
Holman L, Jorgensen CG, Nielsen J, D'Ettorre P (2010) Identification of an ant queen pheromone regulating worker sterility. Proc Royal Soc B 277:3793–3800
Howard RW, Blomquist GJ (2005) Ecological, behavioral, and biochemical aspects of insect hydrocarbons. Annu Rev Entomol 50:371–393
Hughes DP, Kathirithamby J, Beani L (2004) Prevalence of the parasite Strepsiptera in adult *Polistes* wasps: field collections and literature overview. Ethol Ecol Evol 16:363–375
Izzo A, Wells M, Huang Z, Tibbetts E (2010) Cuticular hydrocarbons correlate with fertility, not dominance, in a paper wasp, *Polistes dominulus*. Behav Ecol Sociobiol 64:857–864
Jeanne RL (1980) Evolution of social behavior in the Vespidae. Annu Rev Entomol 25:371–396
Jeanne RL (1991) The swarm-founding Polistinae. In: Ross KG, Matthews RW (eds) The social biology of wasps. Cornell University Press, Ithaca, pp 191–231
Johnson BR, van Wilgenburg E, Tsutsui ND (2011) Nestmate recognition in social insects: overcoming physiological constraints with collective decision making. Behav Ecol Sociobiol 65:935–944
Kather R, Martin SJ (2012) Cuticular hydrocarbon profiles as a taxonomic tool: advantages, limitations and technical aspects. Physiol Entomol 37:25–32
Kather R, Drijfhout FP, Martin SJ (2011) Task group differences in cuticular lipids in the honey bee *Apis mellifera*. J Chem Ecol 37:205–212
Kelstrup HC, Hartfelder K, Nascimento FS, Riddiford LM (2014a) The role of juvenile hormone in dominance behavior, reproduction and cuticular pheromone signaling in the caste-flexible epiponine wasp, *Synoeca surinama*. Front Zool 11. https://doi.org/10.1186/s12983-014-0078-5
Kelstrup HC, Hartfelder K, Nascimento FS, Riddiford LM (2014b) Reproductive status, endocrine physiology and chemical signaling in the Neotropical, swarm-founding eusocial wasp *Polybia micans*. J Exp Biol 217(13):2399–2410
Kudô K, Zucchi R (2006) Effect of a foreign male's age on acceptance into a conspecific colony in a Neotropical wasp. Entomol Sci 9(4):355–357
Kudô K, Tsuchida K, Mateus S, Zucchi R (2007) Nestmate recognition in a neotropical polygynous wasp. Insect Soc 54(1):29–33
Kudô K, Oliveira LA, Mateus S, Zucchi R, Nascimento FS (2016) Nestmate larval discrimination by workers in the swarm-founding wasp *Polybia paulista*. Ethol Ecol Evol. https://doi.org/10.1080/03949370.2015.1129363
Layton JM, Camann MA, Espelie KE (1994) Cuticular lipid profiles of queens, workers, and males of social wasp Polistes metricus say are colony-specific. J Chem Ecol 20(9):2307–2321
Liang D, Silverman J (2000) 'You are what you eat': diet modifies cuticular hydrocarbons and nestmate recognition in the argentine ant, *Linepithema humile*. Naturwissenschaften 87:412–416
Liebig J, Peeters C, Oldham NJ, Markstädter C, Hölldobler B (2000) Are the variations in cuticular hydrocarbons of queens and workers a reliable signal of fertility in the ant *Harpegnathos saltator*? Proc Natl Acad Sci U S A 97(8):4124–4131
Lorenzi MC, Cervo R (1992) Behaviour of Polistes biglumis bimaculatus (Hymenoptera, Vespidae) foundresses on alien conspecific nests. In: Biology and Evolution of Social Insects (Billen J, ed), Leuven University Press, Leuven Belgium, 273–279
Lorenzi MC, Sledge MF, Laiolo P, Sturlini E, Turillazzi S (2004) Cuticular hydrocarbon dynamics in young adult *Polistes dominulus* (Hymenoptera: Vespidae) and the role of linear hydrocarbons in nestmate recognition systems. J Insect Physiol 50(10):935–941
Martin SJ, Helantera H, Drijfhout FP (2008) Evolution of species–specific cuticular hydrocarbon patterns in *Formica* ants. Biol J Linn Soc 95(1):131–140

Michelutti KB, Soares ERP, Sguarizi-Antonio D, Piva RC, Súarez YR, Cardoso CAL, Antonialli-Junior WF (2018) Influence of temperature on survival and cuticular chemical profile of social wasps. J Thermal Biol 71:221–231

Michelutti KB, Piva RC, Lima-Junior SE, Cardoso CAL, Antonialli-Junior WF (2019) Effect of temperature on the chemical profiles of nest materials of social wasps. J Thermal Biol 84:214–220

Mitra A, Gadagkar R (2014) The Dufour's gland and the cuticle in the social wasp *Ropalidia marginata* contain the same hydrocarbons in similar proportions. J Insect Sci 14:9. Available online: http://www.insectscience.org/14.9

Mitra A, Ramachandran A, Gadagkar R (2014) Nestmate discrimination in the social wasp *Ropalidia marginata*: chemical cues and chemosensory mechanism. Anim Behav 88:113–124

Monnin T (2006) Chemical recognition of reproductive status in social insects. Ann Zool Fenn 43(5–6):515–530

Montagna TS, Neves EF, Antonialli-Junior WF (2012) First report of interspecific facultative social parasitism in the paper wasp genus *Mischocyttarus* Saussure (Hymenoptera, Vespidae). Revista Brasileira de Entomologia 56(2):263–265

Montagna TS, Raizer J, Antonialli-Junior WF (2015) Effect of larval topical application of juvenile hormone on caste determination in the independent-founding eusocial wasp *Mischocyttarus consimilis* (Hymenoptera: Vespidae). Open J Anim Sci 5:174–184

Murakami ASN, Nunes TM, Desuó IC, Shima SN, Mateus S (2015) The cuticular hydrocarbons profiles in the colonial recognition of the Neotropical eusocial wasp, *Mischocyttarus cassununga* (Hymenoptera, Vespidae). Sociobiology 62(1):109–115

Neves EF, Andrade LHA, Súarez YR, Lima SM, Antonialli-Junior WF (2012) Age-related changes in the surface pheromones of the wasp *Mischocyttarus consimilis* (Hymenoptera: Vespidae). Genet Mol Res 11(3):1891–1898

Neves EF, Montagna TS, Andrade LHC, Súarez YR, Lima SM, Antonialli-Junior WF (2013) Social parasitism and dynamics of cuticular hydrocarbons in paper wasps of the genus *Mischocyttarus*. J Kansas Entomol Soc 86(1):69–77

Oi CA, Oliveira RC, van Zweden JS, Mateus S, Millar JG, Nascimento FS, Wenseleers T (2019) Do primitively Eusocial wasps use queen pheromones to regulate reproduction? A case study of the paper wasp *Polistes satan*. Front Ecol Evol 7:199

Provost E, Blight O, Tirard A, Renucci M (2008) Hydrocarbons and insects' social physiology. In: Maes RP (ed) Insect physiology: new research. Nova Science Publishers, New York, pp 19–72

Richards OW (1978) The social wasps of the Americas excluding the Vespinae. British Museum (Natural History), London, 580 p

Santos AB, Nascimento FS (2015) Cuticular hydrocarbons of orchid bee males: interspecific and chemotaxonomy variation. PLoS One 10(12):e0145070. https://doi.org/10.1371/journal.pone.0145070

Santos EB, Shemilt S, de Carvalho CAL, Martin SJ (2018) Life history and chemical ecology of the warrior wasp *Synoeca septentrionalis* (Hymenoptera: Vespidae, Epiponini). PLoS One 13(3):e0194689. https://doi.org/10.1371/journal.pone.0194689

Silva ERS, Michelutti KB, Antonialli-Junior WF, Batistote M, Cardoso CAL (2016) Chemical signatures in the developmental stages of *Protopolybia exigua*. Genet Mol Res 15:gmr.15017586

Silva RC, Prato AS, Oi CA, Turatti ICC, Nascimento FS (2020) Dominance hierarchy, ovarian activity and cuticular hydrocarbons in the primitively eusocial wasp *Mischocyttarus cerberus* (Vespidae, Polistinae, Mischocyttarini). J Chem Ecol. https://doi.org/10.1007/s10886-020-01206-1

Singer TL (1998) Roles of hydrocarbons in the recognition systems of insects. Am Zool 38(38):394–405

Singer TL, Espelie KE (1992) Social wasp use nest paper hydrocarbons for nestmate recognition. Anim Behav 44:63–68

Sledge MF, Boscaro F, Turillazzi S (2001) Cuticular hydrocarbons and reproductive status in the social wasp *Polistes dominulus*. Behav Ecol Sociobiol 49:401–409

Soares ERP, Torres VO, Antonialli-Junior WF (2014) Reproductive status of females in the eusocial wasp *Polistes ferreri* Saussure (Hymenoptera:Vespidae). Neotrop Entomol 43:500–508

Tannure-Nascimento IC, Nascimento FS, Turatti IC, Lopes NP, Trigo JR, Zucchi R (2007) Colony membership is reflected by variations in cuticular hydrocarbon profile in a Neotropical paper wasp, *Polistes satan* (Hymenoptera, Vespidae). Genet Mol Res 6(2):390–396

Tannure-Nascimento IC, Nascimento FS, Zucchi R (2008) The look of royalty: visual and odour signals of reproductive status in a paper wasp. Proc Royal Soc B 275:2555–2561

Tibbetts EA, Dale J (2004) A socially enforced signal of quality in a paper wasp. Nature 432:218–222

Togni OC (2014) Biologia e Ecologia Comportamental da Vespa Eussocial Primitiva *Mischocyttarus (Megacanthopus) parallelogrammus* (Hymenoptera, Vespidae). Tese de doutorado em Ciências Biológicas (Zoologia) – Universidade Estadual Paulista, RioClaro, 158 f

Torres VO, Sguarizi-Antonio D, Lima SM, Andrade LHC, Antonialli-Junior WF (2014) Reproductive status of the social wasp *Polistes versicolor* (Hymenoptera, Vespidae). Sociobiology 61(2):218–224

Torres VO, Soares ERP, Lima LD, Lima SM, Andrade LHC, Antonialli-Junior (2016) Morphophysiological and cuticular chemical alterations caused by *Xenos* entomophagus endoparasites in the social wasp *Polistes ferreri* (Hymenoptera, Vespidae). Parasitology 143:1939. https://doi.org/10.1017/S0031182016001529

Van Zweden JS, Bonckaert W, Wenseleers T, d'Etorre P (2013) Queen signaling in social wasps. Evolution 68(4):976–986

Wagner D, Tissot M, Gordon D (2001) Task-related environment alters the cuticular hydrocarbon composition of harvester ants. J Chem Ecol 27(9):1805–1819

West-Eberhard MJ, Turillazzi S (1996) Natural history and evolution of paper wasps. Oxford University Press, New York, 400p

Chapter 13
Biogeographical Hypotheses for the Neotropical Social Wasps

Alexandre Somavilla, Antônio F. Carvalho, and Rodolpho S. T. Menezes

Abstract Biogeographical studies for Neotropical polistines are very recent and with scarce information. For instance, there is a paucity of discussions linking multiple sources of biological data (e.g., morphological, molecular, and physiological) with historical processes. Nevertheless, in this chapter, we reviewed biogeographical hypotheses for some genera of Neotropical social wasps such as *Angiopolybia*, *Apoica*, *Brachygastra*, *Chatergellus*, *Epipona*, *Mischocyttarus*, *Polistes*, *Pseudopolybia*, and *Synoeca*.

Keywords Biogeography · Polistinae · Epiponini · Mischocyttarus · Polistes

13.1 Introduction

Biogeographical hypotheses for Neotropical polistines (Vespidae: Polistinae) are very recent and lack information. Hence, many issues remain unresolved or have never been discussed for the subfamily. This gap in our knowledge is due to the absence of discussions linking multiple sources of biological data (e.g., morphological, molecular, and physiological) with historical processes (Carvalho et al. 2015a). Interesting topics such as colonization routes, population genetics, divergence times, and phylogeography were hitherto little explored in social wasps (for exceptions see Menezes et al. 2015, 2017, 2020), limiting our ability to reach a

A. Somavilla (✉)
Instituto Nacional de Pesquisas da Amazônia, Manaus, Amazonas, Brazil
e-mail: alexandresomavilla@gmail.com

A. F. Carvalho
Instituto Nacional da Mata Atlântica, Santa Teresa, Espírito Santo, Brazil
e-mail: dcarvalhoaf@gmail.com

R. S. T. Menezes
Faculdade de Filosofia, Ciências e Letras, Universidade de São Paulo, Ribeirão Preto, São Paulo, Brazil
e-mail: rstmenezes@gmail.com

F. Prezoto et al. (eds.), *Neotropical Social Wasps*,
https://doi.org/10.1007/978-3-030-53510-0_13

better understanding of the biogeographical patterns and evolutionary processes behind the diversity of the group.

The first biogeographical hypothesis suggested that the social wasps probably originated in the tropics and it could be explained by the high diversity of species in this region (Wheeler 1922; Richards and Richards 1951). According to this hypothesis, the Vespinae/Polistinae ancestral group probably occurred in Southeast Asia, which can be sustained by the occurrence in sympatry in the eastern tropics of the three subfamilies that present social behavior (Stenogastrinae, Vespinae, and Polistinae), by the presence of ancestral forms in the nest architecture (West-Eberhard 1969), and by the less diversity of morphological traces (Van der Vecht 1965) of the species in that region. Such information support Van der Vecht (1965) hypotheses. According to these assumptions, the dispersion of polistines to the New World would have occurred across the Bering Strait during the Middle Tertiary, and the wide distribution of these wasps in the New World would probably have been achieved after the last ice age (Richards 1978).

Contrastingly, Carpenter (1981) criticized the relationship between center of diversity and origin and suggested that the distribution pattern of these wasps is "broadly Gondwanic." Furthermore, Carpenter (1993) suggested that the separation between Africa and South America during the Early Cretaceous (~120–100 million years ago) was an important event in the evolutionary history of the group. Carpenter (1996) also reviewed the two main hypotheses on Polistinae biogeography and additionally performed component analysis for the subgenera of *Polistes* Latreille, 1802. The cladograms obtained in that study corroborated the idea of the Gondwana distribution of the subgenera and rejected the expected dispersal across the Bering Strait (as defended by Van der Vecht 1965; Richards 1973, 1978).

Phylogenetic inferences has provided valuable information regarding the evolutionary history of Neotropical social wasps. *Polistes*, for example, has been the better studied genus in this sense, and it is suggested that all subgenera of the Old World are invariably positioned at the base of the proposed phylogenetic trees, supporting the hypothesis of the center of origin in the Old World, and that the subgenera of the New World are derived (Carpenter 1996; Arévalo et al. 2004; Pickett and Wenzel 2004; Pickett et al. 2006; Pickett and Carpenter 2010; Santos et al. 2015). Thus, taking into account the phylogenetic positioning of the *Polistes* subgenera, this is a strong evidence for a more recent colonization in the western tropics (Carvalho et al. 2015b). In the same way, considering that the Mischocyttarini Carpenter 1993 and Epiponini Lucas, 1867 tribes only occur in the Neotropics and that no fossil of these groups has been found in the Old World, it is believed that both tribes appeared in the New World after the separation of Africa and South America (~120 million years ago) (Carvalho et al. 2015a).

13.2 Biogeographical Hypotheses for *Polistes*

The most recent hypothesis for the colonization process of the New World *Polistes* species suggests that during the Cretaceous, the transoceanic dispersion of Southeast Asia, following a single route, first reaches South America and later North America

(Santos et al. 2015). Such a phenomenon is usually dismissed as unlikely due to the seemingly overwhelming scale of the geographical barrier involved – Pacific Ocean. However, there is evidence for other cases of arthropods that have surpassed oceanic barriers (Sharma and Giribet 2012), as well as a high frequency of natural oceanic dispersion by some species (Sharma and Giribet 2012). Considering the clear dispersal ability of *Polistes*, as evidenced by its cosmopolitan distribution, as well as its high colonization success following human introductions, a scenario of transoceanic dispersal remains plausible (Santos et al. 2015).

In addition, there is a general perception that *Polistes* as a whole may have a Gondwanan origin, a possibility that cannot be ruled out under the results of the current analyzes performed for the group. In the last work (Santos et al. 2015), only one species from the Afrotropics was represented in the phylogenetic analyses. The inclusion of further African species probably could lead to Africa as the ancestral area for the *Polistes* (*Polistes*) clade and for the *Polistes* of the New World (Santos et al. 2015; Carvalho et al. 2015b).

Phylogenetic studies with *Polistes* using morphological, molecular, and behavioral data established the phylogenetic basal positions of the Old World subgenera (Carpenter 1996; Arévalo et al. 2004; Pickett and Wenzel 2004; Pickett et al. 2006; Santos et al. 2015) but have not entirely resolved the relationships among derived New World groups. Pickett et al. (2006) carried out a meta-analysis using both previously available and new data to propose a robust phylogeny for this group. For example, the study placed *Polistes* sensu stricto as the sister group of the New World subgenera, which are arranged into five subgenera disposed in three main clades: (*A phanilopterus* + ((*Palisotius* + *Epicnemius*) + (*Onerarius* + *Fuscopolistes*))). However, Santos et al. (Santos et al. 2015) performed an analysis using a larger number of taxa, molecular and morphological data, and established *Polistes* (*Polistela*) as the sister group of the other New World *Polistes* subgenera, and forming three main clades: (*Epicnemius* + ((*Fuscopolistes* + *Onerarius*) + (*Palisotius* + *Aphanilopterus*))).

While the subgenera *Polistes* (*Onerarius*) and *Polistes* (*Palisotius*) have a wide distribution in the Americas and *Polistes* (*Fuscopolistes*) have a distribution typically Nearctic, *Polistes* (*Epicnemius*) contains species widespread in Central and South America that are not very informative from the perspective of the progression rule principle and are not very informative from a biogeographical point of view. Within the subgenus *Polistes* (*Aphanilopterus*), three clades are separated and exhibit two phylogeny-distribution patterns (Santos et al. 2015); one clade apparently tracked the colonization route from eastern South America toward the Amazon, Central America, and North America; and two clades reveal spatial progression from North America toward eastern South America. Such data strengthens the round-trip hypothesis proposed by Carvalho et al. (2015b) that predicts an early colonization wave from eastern South America toward the west – only seen in *Polistes* (*Aphanilopterus*) – and multiple waves from the Amazon Forest toward the east, a route recorded in most living New World Polistinae.

13.3 Biogeographical Hypotheses for *Mischocyttarus*

Mischocyttarus de Saussure, 1853, is divided into 11 subgenera and grouped into 2 main clades: *Clypeopolybia* and *Mischocyttarus* sensu stricto and a second clade formed by the other species (Silveira 2008). However, *Clypeopolybia* exhibits polytomies and is not very informative from the biogeographical point of view. *Mischocyttarus* sensu stricto contains *Mischocyttarus acunai* Alayo which presents distribution in North and Central America; *Mischocyttarus tomentosus* Zikán and *Mischocyttarus smithii* (de Saussure) both distributed in northwestern Neotropics; and *Mischocyttarus drewseni* de Saussure is widely distributed in the Neotropics.

The second major branch has the subgenus *Monogynoecus* as the most basal. A review of the distribution of species included in this subgenus reveals that eight of the ten species are restricted to the northwestern Neotropics. However, there is no progression of species relationships in the phylogeny to allow the proposal of ancestral or more recent areas for this lineage. In contrast, phylogeny-distribution inferences are evident in this branch for the subgenera *Mischocyttarus* (*Kappa*) and *Mischocyttarus* (*Omega*). All ancient species included in the phylogeny of both subgenera are those from the northwestern and/or central Neotropics, whereas the more derived species *Mischocyttarus* (*Kappa*) *funerulus* Zikán and *Mischocyttarus* (*Omega*) *buyssoni* (Ducke) are only found in the eastern Neotropics and discretely in the central region of Brazil (see Carvalho et al. 2015b).

13.4 Biogeographical Hypotheses for Some Genera of Epiponini

The Epiponini tribe consists of 19 genera with Neotropical distribution, but there are few biogeographical studies for the group. In fact, more extensive biogeographical studies were carried out only for the genera *Brachygastra* Perty, 1883 (Silva and Noll 2014), and *Synoeca* de Saussure, 1852 (Menezes et al. 2015, 2017). In addition, for some genera of Epiponini, distribution and phylogeny information were combined to assist in the elaboration of a proposal for possible dispersal and colonization routes for Neotropical social wasps (Carvalho et al. 2015b). Recently, Menezes et al. (2020) performed a phylogenomic and biogeographic study for Epiponini. This study indicates Amazonian as the major source of Neotropical swarm-founding social wasp diversity.

Brachygastra is widely distributed, occurring from the Southern United States to southern Brazil. Silva and Noll (Silva and Noll 2014) combined phylogenetic information and geographic distribution data of *Brachygastra* species, and they suggested a possible influence of a terrestrial bridge between the northern and southern hemispheres, as well as a probable origin of the genus in northern South America.

The first molecular phylogeny for a genus of Epiponini was carried out to investigate the phylogenetic relationships and biogeographical history of the *Synoeca* species (Menezes et al. 2015). In this study, based on analyzes of divergence time and historical biogeography, the authors proposed an Amazonian origin and three main dispersion events for the group. The oldest dispersal route probably occurred in southern South America between the Amazon and Atlantic forest. A second route occurred from the Amazon toward Central America via the Isthmus of Panama. Finally, the last and most recent colonization route occurred toward the Brazilian northeast between the Amazonian and Atlantic forests.

In addition, the first phylogeographic study on Neotropical social wasps was performed by Menezes et al. (2017) for *Synoeca* species, specifically *Synoeca cyanea* (Fabricius) and *Synoeca ilheensis* Lopes and Menezes (treated in the work as *Synoeca* aff. *septentrionalis*). Based on multiple sources of data (multilocus DNA sequences, climatic niche models, and chromosome features) and analytical methods, the authors proposed a contrasting pattern of historical colonization, south-north (*Synoeca cyanea*) and north-south (*Synoeca ilheensis*), and clinal chromosomal variation along the Brazilian Atlantic Forest.

Other Epiponini genera that had previous phylogenies and geographic distribution data available were analyzed in a biogeographical context by Carvalho et al. (2015b) with the aim of proposing dispersion events carried out by Neotropical social wasps.

In the genus *Angiopolybia* Araujo, 1946, *Angiopolybia pallens* (Lepeletier) presents the widest distribution and occurs from Panama to the east of South America. This species has a disjoint distribution between the Amazon and the Atlantic Forest, where it was suggested that the western lineages are more derived in relation to the Atlantic populations (Carvalho et al. 2014). The other three species (*Angiopolybia obidensis* (Ducke), *Angiopolybia paraensis* (Spinola), and *Angiopolybia zischkai* (Richards)) are exclusive in the northwest of the Neotropics.

Apoica Lepeletier, 1836, is widely distributed in the Neotropics and consists of ten species. The only available phylogeny for the genus was proposed by Pickett and Wenzel (2007), but did not completely resolve the relationships among species. However, *Apoica arborea* de Saussure is sister of the other species of the group, and it occurs in the northwest of the Neotropics and central Brazil. Some species such as *Apoica flavissima* van der Vecht, *Apoica gellida* van der Vecht, *Apoica pallida* (Olivier), and *Apoica strigata* Richards are widely distributed in the Neotropical region, occurring both in dry areas of the Cerrado and Caatinga and in wetlands of the Amazon and Atlantic Forest. In addition, *Apoica pallens* (Fabricius) and *Apoica thoracica* du Buysson are widely distributed throughout the Atlantic Forest.

For *Chatergellus* Bequaert, 1938, which has a distribution from Costa Rica to Southeastern Brazil, two species with apomorphic characteristics of the genus *Chatergellus zonatus* (Spinola) and *Chatergellus sanctus* Richards occur in the Amazon and Atlantic Forest and eastern portion of South America, respectively. *Chatergellus communis* Richards and *Chatergellus atectus* Richards, with plesiomorphic characteristics, occur discretely in central Brazil and North and Central America, respectively. All other species are restricted to the northwest of the Neotropics.

Epipona Latreille, 1802, has an Amazonian distribution; most of species are endemic to the northwest of the Neotropics. However, *Epipona tatua* (Cuvier) and *Epipona media* Cooper are also found discreetly in the Atlantic Forest in the states of Bahia, Espírito Santo, and São Paulo.

Pseudopolybia de Saussure, 1863, presents a distribution in the Amazon region, Atlantic Forest, and central Brazil. *Pseudopolybia langi* Bequaert and *Pseudopolybia difficilis* (Ducke) are endemic to the northwest of the Neotropics, while *Pseudopolybia vespiceps* (de Saussure) and *Pseudopolybia compressa* (de Saussure) are the only species that also colonized all of South America, being considered the two species of the genus with plesiomorphic characteristics.

These genera were discussed biogeographically based on current phylogeny and distribution; however, such conclusions may not be considered conclusive. Additional assessments using phylogeographic approaches, mainly within diverse genera and groups of species within social wasps, are recommended to establish more robust conclusions. In addition, particular analyzes of widely distributed species may clarify important questions about the colonization and routes used by Polistinae in the Neotropical region.

Acknowledgment AS thanks Fundação de Amparo à Pesquisa do Estado do Amazonas for the postdoctoral scholarship (FAPEAM – FIXAM, process number 062.01427/2018). AFC thanks Fundação de Amparo à Pesquisa do Estado de São Paulo (FAPESP 2011/13391-2) for the PhD scholarship, Conselho Nacional de Desenvolvimento Científico e Tecnológico (CNPq: PCI-INMA 302376/2020-8) and Coordenação de Aperfeiçoamento de Pessoal de Nível Superior (CAPES) for two postdoctoral fellowships. RSTM is grateful to FAPESP for postdoctoral fellowships and grants (2015/02432-0 and 2016/21098-7).

References

Arévalo E, Zhu Y, Carpenter JM, Strassmann JE (2004) The phylogeny of the social wasp subfamily Polistinae: evidence from microsatellite flanking sequences, mitochondrial COI sequence, and morphological characters. BioMed Central Evolutionary Biology 4:8. https://doi.org/10.1186/1471-2148-4-8

Carpenter JM (1981) The phylogenetic relationships and natural classification of the Vespoidea (Hymenoptera). Syst Entomol 7:11–38

Carpenter JM (1993) Biogeographic patterns in Vespidae (Hymenoptera): two views of Africa and South America. p. 139-154. In: P Goldblatt (ed) Biological relationship between Africa and South America. New Haven, Yale University Press.

Carpenter JM (1996) Phylogeny and biogeography of *Polistes*. In: Turillazzi S, West-Eberhard MJ (eds) Natural history and evolution of paper-wasps, vol Xiv. Oxford University Press, Oxford, New York, Tokyo, pp 18–57. 400p

Carvalho AF, Santos GMM, Menezes RST, Costa MA (2014) Genetic diversity of *Angiopolybia pallens* (Lepeletier) (Hymenoptera, Vespidae, Polistinae) explained by the disjunction of South American rainforests. Genet Mol Res 13:89–94

Carvalho AF, Menezes RST, Somavilla A, Costa MA, Del Lama MA (2015a) Polistinae biogeography in the Neotropics: history and prospects. J Hymenopt Res 42:93–105

Carvalho AF, Menezes RST, Somavilla A, Costa MA, Del Lama MA (2015b) Neotropical Polistinae (Vespidae) and the progression rule principle: the round-trip hypothesis. Neotrop Entomol 44:01–08

Menezes RST, Brady SG, Carvalho AF, Del Lama MA, Costa MA (2015) Molecular phylogeny and historical biogeography of the Neotropical swarm-founding social wasp genus *Synoeca* (Hymenoptera: Vespidae). PLoS One 10(3):e0119151

Menezes RST, Brady SG, Carvalho AF, Del Lama MA, Costa MA (2017) The roles of barriers, refugia, and chromosomal clines underlying diversification in Atlantic Forest social wasps. Sci Rep 7(1):1–6

Menezes RST, Lloyd MW, Brady SG (2020) Phylogenomics indicates Amazonia as the major source of Neotropical swarm-founding social wasp diversity. Proc R Soc B Biol Sci 287(1928):20200480

Pickett KM, Carpenter JM (2010) Simultaneous analysis and the origin of eusociality in the Vespidae (Insecta: Hymenoptera). Arthropod Syst Phylogeny 68(1):3–33

Pickett KM, Wenzel J (2004) Phylogenetic analysis of the New World *Polistes* (Hymenoptera: Vespidae: Polistinae) using morphology and molecules. J Kansas Entomol Soc 77(4):742–760

Pickett KM, Wenzel JW (2007) Revision and cladistic analysis of the nocturnal social wasp genus, *Apoica* Lepeletier (Hymenoptera: Vespidae: Polistinae: Epiponini). Am Mus Novit 3562:1–30

Pickett KM, Carpenter JM, Wheleer WC (2006) Systematics of *Polistes* (Hymenoptera: Vespidae), with a phylogenetic consideration of Hamilton's haplodiploidy hypothesis. Ann Zool Fennici 43:390–406

Richards OW (1973) The subgenera of *Polistes* Latreille (Hymenoptera, Vespidae). Revista Brasileira de Entomologia 17:85–104

Richards OW (1978) The social wasps of the Americas (excluding the Vespinae). British Museum of Natural History, London, p 580

Richards OW, Richards MJ (1951) Observations on the social wasps of South America (Hymenoptera: Vespidae). Transactions of the Royal Entomological Society of London, p 169

Santos BF, Payne A, Pickett KM, Carpenter JM (2015) Phylogeny and historical biogeography of the paper wasp genus *Polistes* (Hymenoptera: Vespidae): implications for the overwintering hypothesis of social evolution. Cladistics 31:535–549

Sharma PP, Giribet G (2012) Out of the Neotropics: late cretaceous colonization of Australasia by American arthropods. Proc R Soc B 279:3501–3509

Silva M, Noll FB (2014) Biogeography of the social wasp genus *Brachygastra* (Hymenoptera: Vespidade: Polistinae). J Biogeogr:1–10

Silveira OT (2008) Phylogeny of wasps of the genus *Mischocyttarus* de Saussure (Hymenoptera, Vespidae, Polistinae). Revista Brasileira de Entomologia 52(4):510–549

Van der Vecht J (1965) The geographical distribution of the social wasps (Hymenoptera, Vespidae). Proceedings of the XII International Congress of Entomology, p 440–441

West-Eberhard MJ (1969) The social biology of Polistine wasps. Misc Pub Mus Zool Univ Mich 140:1–101

Wheeler WM (1922) Social life among the insects. Lecture II. Part 2. Wasps solitary and social. Sci Mon 15:119–131

Chapter 14
Chromosome Diversity and Evolution in Neotropical Social Wasps

Rodolpho S. T. Menezes

Abstract Karyological features are useful characters for evolutionary and taxonomic studies. Moreover, cytogenetic analyses are important tools for understanding the genomic organization of any species. However, within Vespidae, there is a paucity of cytogenetic studies, especially for Neotropical social wasps as a whole. Despite that, the group exhibits a high chromosomal diversity within a chromosomal interspecific variation from $n = 5$ to $n = 34$. The karyological features of Neotropical social wasps are reviewed and future directions for cytogenetic research are given.

Keywords Chromosome evolution · Cytogenetics · Polistinae · Social wasps · Vespidae

14.1 Introduction

Chromosomal number and morphology are relevant aspects of genomic organization, and their variations may reflect interspecific divergence (Goodisman et al. 2008). Moreover, they are useful characters for evolutionary and taxonomic studies, since closely related species may have more similar karyotypes than among more distant species (Sumner 2003). Thus, several studies were conducted with the objective of understanding the effects that the structure and chromosome number can have on a taxon. For example, Kandul et al. (2007) analyzing the high karyotype diversity in the genus of blue butterflies, *Agrodiaetus*, showed evidence of a role of chromosomal changes in the speciation process in the group. In addition, the use of

R. S. T. Menezes (✉)
Departamento de Biologia, Faculdade de Filosofia, Ciências e Letras - Universidade de São Paulo (FFCLRP/USP), Ribeirão Preto, SP, Brazil
e-mail: rstmenezes@gmail.com

F. Prezoto et al. (eds.), *Neotropical Social Wasps*,
https://doi.org/10.1007/978-3-030-53510-0_14

chromosomal data may contribute to elucidate taxonomic problems. Cristiano et al. (2013) combining cytogenetic and molecular data suggest that the ant *Acromyrmex striatus* be allocated to a new genus.

Another relevant aspect in studies involving chromosomes is the high chromosome diversity presented by some particular groups of taxa. Some hypotheses have been proposed to interpret this chromosomal diversity: the fusion hypothesis (White 1954, 1973), which predicts that chromosomal evolution proceeds from high to low chromosomal number driven by chromosomal fusion; the fission hypothesis (Todd 1970, 1975), which is an alternative to the first hypothesis with an increase in chromosomal number by chromosomal fission; and the modal hypothesis (Matthey 1973), in which the chromosomal number changes from a modal number by chromosomal fusion and fission. However, analyses of a large volume of chromosomal data on ants (Hymenoptera, Formicidae) made possible the proposal of the minimum interaction theory (Imai et al. 1994). This theory is similar to the fission hypothesis and suggests that chromosomal evolution evolves toward increasing chromosomal number by chromosomal fission, enabling the reduction of genetic risks associated with deleterious translocations.

Extreme intraspecific variation in chromosome number among closely related animal species has been reported for several groups. In mammals, a curious case is between the Chinese deer *Muntiacus reevesi*, with $n = 23$ chromosomes, and the Indian deer, *Muntiacus muntjak*, with $n = 3$ chromosomes (Wurster and Benirschke 1970; Yang et al. 1997). However, the highest ranges of within-genus karyotype variation are registered in insects. For example, gall-inducing scale insects of the genus *Apiomorpha* have an interspecific chromosome number range from $n = 2$ to $n = 96$ (Cook 2000), and despite morphological similarity and very recent time of species divergence, *Agrodiaetus* blue butterflies exhibit karyotypic diversity ranging from $n = 10$ to $n = 134$ chromosomes (Lukhtanov et al. 2005). Interestingly, in these butterflies, differences in chromosomal number accumulate faster between sympatric than parapatric species (Lukhtanov et al. 2005; Kandul et al. 2007).

In Hymenoptera, several studies involving chromosomes were performed. The most studied group is Formicidae, with more than 750 species with known chromosome number (Lorite and Palomeque 2010). In addition, about 370 species of parasitoid wasps were analyzed using cytogenetic techniques, allowing inferences in the process of chromosomal evolution, as well as knowledge of the diversity about the chromosomal number of the group ($n = 2$ to 23) (Gokhman 2006). However, information about the karyotype of Neotropical social wasps is scarce with only 35 karyotyped species to date, and this group has a high chromosomal diversity represented by the variation in haploid chromosome number of $n = 5$–34 (Table 14.1) (Pompolo and Takahashi 1987, 1990a, 1990b; Hoshiba et al. 1989; Menezes et al. 2013, 2014, 2017, 2019).

Table 14.1 Chromosome number described for Neotropical social wasps. Brazilian states: AM, Amazônia; BA, Bahia; DF, Distrito Federal; ES, Espírito Santo; MA, Maranhão; MS, Mato Grosso do Sul; PE, Pernambuco; PR, Paraná; RS, Rio Grande do Sul; SP, São Paulo

Genus	Species	n	2n	Locality	References
Tribe Polistini					
Polistes	*canadensis*	16		Brazil	Kerr (1952), quoted from Pompolo and Takahashi (1990b)
	cinerascens	27	54	Ribeirão Preto (SP)	Pompolo and Takahashi (1990b)
	simillimus	28		Ribeirão Preto (SP)	Pompolo and Takahashi (1986)
	versicolor	31	62	Ribeirão Preto (SP)	Pompolo and Takahashi (1986); Pompolo and Takahashi (1990b)
Tribe Mischocyttarini					
Mischocyttarus	*cassununga*	32		Ribeirão Preto (SP)	Pompolo and Takahashi (1990a)
	consimilis	33	66	Dourados, Mundo Novo, and Ponta Porã (MS); Guaíra, Dr. Oliveira Castro, and Santa Cruz de Monte Castelo (PR)	Cunha et al. (2017)
	sp.	34		Ribeirão Preto (SP)	Pompolo and Takahashi (1990a)
Tribe Epiponini					
Agelaia	*multipicta*		64	Ribeirão Preto (SP)	Pompolo and Takahashi (1987)
	pallipes	32	64	Ribeirão Preto (SP)	Pompolo and Takahashi (1987)
Brachygastra	*lecheguana*		56	Ribeirão Preto (SP)	Pompolo and Takahashi (1987)
Clypearia	*weyrauchi*	21		Iranduba (AM)	Menezes et al. (2014)
Epipona	*media*		66	Santa Teresinha (BA)	Menezes et al. (2014)

(continued)

Table 14.1 (continued)

Genus	Species	n	2n	Locality	References
Metapolybia	*docilis*		38	São Luiz (MA)	Menezes et al. (2014)
	decorata	17, 18	34, 35, 36	Igrapiúna, Itacaré, Ilhéus, Una, Camacan, Arataca, and Santa Teresinha (BA); Linhares and Colatina (ES)	Menezes et al. (2013)
	sp.	19		Ribeirão Preto (SP)	Pompolo and Takahashi (1990a)
	cingulata		32	Iranduba (AM)	Menezes et al. (2014)
Parachartergus	*smithii*		54	Ribeirão Preto (SP)	Pompolo and Takahashi (1990a)
	fraternus		54	Brasília (DF)	Menezes et al. (2014)
	pseudapicalis		52	Itamaraju (BA)	Menezes et al. (2014)
Polybia	sp.1	16		Ribeirão Preto (SP)	Pompolo and Takahashi (1990a)
	sp.2		34	Ribeirão Preto (SP)	Pompolo and Takahashi (1990a)
(*Myrapetra*)	*occidentalis*		34	Ribeirão Preto (SP)	Pompolo and Takahashi (1990a); Menezes et al. (2014)
(*Myrapetra*)	*paulista*	17	34	Ribeirão Preto (SP)	Pompolo and Takahashi (1987)
(*Myrapetra*)	*scutellaris*		34	Franca (SP)	Pompolo and Takahashi (1987)
(*Trichinothorax*)	*sericea*	27		Ribeirão Preto (SP)	Pompolo and Takahashi (1990a)
(*Formicicola*)	*rejecta*		30	Ilhéus (BA)	Menezes et al. (2014)
(*Apopolybia*)	*jurinei*		10	Manaus (AM)	Menezes et al. (2014)
Protonectarina	*sylveirae*		58	Ribeirão Preto (SP)	Pompolo and Takahashi (1987)
Protopolybia	*exigua*	31		Ribeirão Preto (SP)	Pompolo and Takahashi (1990a)
	sedula		42	Ribeirão Preto (SP)	Pompolo and Takahashi (1990a)
Pseudopolybia	*vespiceps*	8	16	Jardinópolis (SP) and Manaus (AM)	Pompolo and Takahashi (1990a); Menezes et al. (2014)

(continued)

Table 14.1 (continued)

Genus	Species	n	2n	Locality	References
Synoeca	*cyanea*		38, 40, 42	Carlos Barbosa (RS); Santa Teresa (ES); and Senhor do Bonfim (BA)	Menezes et al. (2017)
	ilheensis		38, 40, 41	Ilhéus, Itacaré, Itabuna and Santa Teresinha (BA); Moreno (PE); and Vila Regência (ES)	Menezes et al. (2017)
Synoeca	*surinama*		64	Ilhéus and Itajuípe (BA)	Menezes et al. (2019)

14.2 History of Chromosomal Studies in Neotropical Social Wasps

Cytogenetic studies involving social wasps are scarce and began in the 1940s. Two Palearctic social wasps, *Polistes dominula* (Christ, 1791) ($n = 21$ chromosomes) and *Polistes gallicus* (Linnaeus, 1767) ($n = 14$), were the first species of Polistinae with chromosomal numbers determined (Pardi, 1942). However, *Polistes canadensis* (Linnaeus, 1758) ($n = 16$) was the first social wasp with geographical distribution in the Neotropical region to have its chromosome number determined (Kerr, 1952). After 30 years, other Neotropical social wasps were analyzed using cytogenetic techniques, but only involving chromosomal number description (see Table 14.1). In 1990, two Neotropical species of *Polistes* had their chromosomes studied by the C-band technique (it shows the chromosomal distribution of heterochromatin).

In recent years, cytogenetic studies in Neotropical social wasps have advanced from simple determinations of chromosome number to studies involving comparisons of distinct populations and molecular cytogenetic techniques [Fluorescence in situ Hybridization (FISH)]. For example, Menezes et al. (2013) using classical and molecular cytogenetic techniques in populations of *Metapolybia decorata* Gribodo 1896 described a numeric-structural intraspecific chromosomal variation produced by telomeric fusion and elimination of a probable inactive centromere in this species. Moreover, Menezes et al. (2014) found a frequent involvement of guanine-cytosine (GC) base pair-rich regions in some types of chromosomal rearrangements in social wasps representing the Epiponini tribe. Additionally, Menezes et al. (2017) showed an interesting pattern of clinal chromosomal variation, involving decreased latitudinally oriented GC content along the Atlantic Forest in two armadillo wasps, *Synoeca cyanea* (Fabricius, 1775) and *S. ilheensis* Lopes and Menezes, 2017.

In a recent work, Menezes et al. (2019) investigated whether the evolutionary dynamics of heterochromatin and DNA ribosomal (rDNA) clusters played a role in the genomic reshuffling and chromosomal evolution of *Synoeca* using three closely related species: *Synoeca cyanea*, *S. ilheensis*, and *S. surinama*. Interestingly, *S. surinama* was karyotyped for the first time and showed several microchromosomes (usual in bird karyotypes, but not in the order Hymenoptera), and to our knowledge, this represents an unprecedented number of this chromosomal type for Hymenoptera.

The authors suggested an intense evolutionary process of fission of regular-sized chromosomes from an ancestor-like *S. surinama* as the putative evolutionary mechanism for the origin of these microchromosomes in this species. The chromosomal mapping of rDNA clusters showed a striking variability in the number and chromosomal position of this repetitive DNA family among the three species as well as intraspecific variation, revealing an intense evolutionary dynamics of this type of DNA sequences, and highlights rapid chromosomal evolution in these insects.

14.3 Chromosomal Morphology

Despite the significant variation in the chromosome number observed in Neotropical social wasps, there is a prevalence of metacentric and submetacentric chromosomes (see Table 14.2). The small number of acrocentric chromosomes may be related to the process of chromosomal evolution in social wasps. For example, Menezes et al. (2014) used the comparative method proposed by Mayrose et al. (2010) combining chromosome number and phylogenetic tree data to test different models of chromosomal evolution, and therefore, the authors suggested a major role for chromosomal fusions during the chromosomal evolution process in Epiponini. Furthermore, based on such analyses, the authors suggested a karyotype with a high number of chromosomes (probably $n = 33$) as the putative ancestral Epiponini karyotype.

It is noteworthy that the low quality of metaphase photos from early cytogenetic works and the absence of refined chromosomal banding techniques prevent a more accurate information about the morphology of social wasp chromosomes.

14.4 Prospects

It is undeniable that chromosomal studies on social wasps are scarce and restricted only to the conventional staining technique. So far only six species of social wasps have been studied using FISH, which allows chromosomal mapping of specific DNA sequences. Repetitive ribosomal DNA clusters were located on the first chromosomal pair of the three *Metapolybia* species analyzed [*M. decorata*, *Metapolybia cingulata* (Fabricius, 1804), and *Metapolybia docilis* Richard, 1978)] suggesting a conservation in the location of these clusters in the genus. In an opposite way, *Synoeca* species showed inter- and intraspecific variation in the chromosomal number and position of rDNA clusters. Despite this contrasting pattern about the chromosomal position and number of rDNA clusters showed by these two closely related genera (for phylogenetic information, see Menezes et al. 2020), nothing is known about the chromosomal location of other repetitive DNA families in social wasps as a whole.

Studies combining morphological, molecular, and chromosomal data may be crucial to determine accurately the diversity of social wasps as well as their

Table 14.2 Chromosomal morphology described for Neotropical social wasps

Genus	Species	Chromosomal morphology (n)	References
Polistes	*cinerascens*	Metacentrics, submetacentrics, and acrocentrics	Pompolo and Takahashi (1990b)
	simillimus	Metacentrics and submetacentrics	Pompolo and Takahashi (1986)
	versicolor	Metacentrics and submetacentrics	Pompolo and Takahashi (1986); Pompolo and Takahashi (1990b)
Mischocyttarus	*cassununga*	Metacentrics and submetacentrics	Pompolo and Takahashi (1990a)
	consimilis	7 metacentrics, 13 submetacentrics, and 13 acrocentrics	Cunha et al. (2017)
	sp.	Metacentrics, submetacentrics, and 2 acrocentrics	Pompolo and Takahashi (1990a)
Agelaia	*pallipes*	Submetacentrics	Pompolo and Takahashi (1987)
	multipicta	Submetacentrics	Pompolo and Takahashi (1987)
Brachygastra	*lecheguana*	Metacentrics, submetacentrics, and acrocentrics	Pompolo and Takahashi (1987)
Clypearia	*weyrauchi*	5 metacentrics, 8 submetacentrics, and 8 acrocentrics	Menezes et al. (2014)
Epipona	*media*	33 acrocentrics	Menezes et al. (2014)
Metapolybia	*docilis*	6 metacentrics, 8 submetacentrics, and 5 acrocentrics	Menezes et al. (2014)
	decorata	11 metacentrics, 1 acrocentrics, and 6 pseudoacrocentrics	Menezes et al. (2013)
	sp.	14 metacentrics/submetacentrics and 5 acrocentrics	Pompolo and Takahashi (1990a)
	cingulata	7 metacentrics, 6 submetacentrics, and 3 acrocentrics	Menezes et al. (2014)
Parachartergus	*smithii*	Metacentrics, submetacentrics, and acrocentrics	Pompolo and Takahashi (1990a)
	fraternus	4 metacentrics, 5 submetacentrics, and 18 acrocentrics	Menezes et al. (2014)
	pseudapicalis	5 metacentrics, 4 submetacentrics, and 17 acrocentrics	Menezes et al. (2014)

(continued)

Table 14.2 (continued)

Genus	Species	Chromosomal morphology (n)	References
Polybia	sp.1	15 metacentrics/ submetacentrics and 1 acrocentrics	Pompolo and Takahashi (1990a)
	sp.2	Submetacentrics	Pompolo and Takahashi (1990a)
	(*Myrapetra*) *occidentalis*	16 metacentrics/ submetacentrics and 1 acrocentric	Pompolo and Takahashi (1990a); Menezes et al. (2014)
	(*Myrapetra*) *paulista*	Metacentrics and submetacentrics	Pompolo and Takahashi (1987)
	(*Myrapetra*) *scutellaris*	Metacentrics and submetacentrics	Pompolo and Takahashi (1987)
	(*Trichinothorax*) *sericea*	24 metacentrics/ submetacentrics and 3 acrocentrics	Pompolo and Takahashi (1990a)
	(*Formicicola*) *rejecta*	8 metacentrics, 6 submetacentrics, and 1 acrocentrics	Menezes et al. (2014)
	(*Apopolybia*) *jurinei*	5 submetacentrics	Menezes et al. (2014)
Protonectarina	*sylveirae*	Submetacentrics and acrocentrics	Pompolo and Takahashi (1987)
Protopolybia	*exigua*	30 metacentrics/ submetacentrics and 1 acrocentrics	Pompolo and Takahashi (1990a)
	sedula	20 metacentrics/ submetacentrics and 1 acrocentrics	Pompolo and Takahashi (1990a)
Pseudopolybia	*vespiceps*	Submetacentrics	Pompolo and Takahashi (1990a); Menezes et al. (2014)
Synoeca	*cyanea*	BA: 14 metacentrics, 4 submetacentrics, and 2 acrocentrics ES: 11 metacentrics and 8 submetacentrics RS: 10 metacentrics and 9 submetacentrics	Menezes et al. (2017)
	ilheensis	BA: 10 metacentrics, 10 submetacentrics, and 1 acrocentric ES: 8 metacentrics, 9 submetacentrics, and 2 acrocentrics PE: 10 metacentrics, 9 submetacentrics, and 1 acrocentric	Menezes et al. (2017)
Synoeca	*surinama*	17 metacentrics and 15 microchromosomes	Menezes et al. (2019)

evolutionary history. In addition, exciting topics in genetics such as molecular cytogenetics and cytogenomics should be explored in order to better understand the pattern of chromosomal evolution in the Neotropical social wasps.

Acknowledgment I am thankful to Fundação de Amparo à Pesquisa do Estado de São Paulo (FAPESP) for postdoctoral fellowships and grants 2015/02432-0 and 2016/21098-7.

References

Cook LG (2000) Extraordinary and extensive karyotypic variation: a 48-fold range in chromosome number in the gall-inducing scale insect *Apiomorpha* (Hemiptera: Coccoidea: Eriococcidae). Genome 2:255–263

Cristiano MP, Cardoso DC, Fernandes-Salomão TM (2013) Cytogenetic and molecular analyses reveal a divergence between *Acromyrmex striatus* (Roger, 1863) and other congeneric species: taxonomic implications. PLoS One 8(3):1–9

Cunha DA, Menezes RST, Costa MA, Lima SM, Andrade LH, Antonialli WF (2017) Integrated analyses of Cuticular hydrocarbons, chromosome and mtDNA in the Neotropical social wasp *Mischocyttarus consimilis* Zikán (Hymenoptera, Vespidae). Neotrop Entomol 2:1–7

Gokhman VE (2006) Karyotypes of parasitic Hymenoptera: diversity, evolution and taxonomic significance. Insect Science 13(4):237–241

Goodisman MAD, Kovacs JL, Hunt BG (2008) Functional genetics and genomics in ants (Hymenoptera: Formicidae): the interplay of genes and social life. Myrmecol News 11:107–117

Hoshiba H, Matsuura M, Imai HT (1989) Karyotype evolution in the social wasps (Hymenoptera, Vespidae). Jpn J Genet 64:209–222

Imai HT, Taylor RW, Crozier RH (1994) Experimental bases for the minimum interaction theory. I. Chromosome evolution in ants of the *Myrmecia pilosula* species complex (Hymenoptera: Formicidae: Myrmeciinae). Jpn J Genet 69:137–182

Kandul NP, Lukhtanov VA, Pierce NE (2007) Karyotypic diversity and speciation in *Agrodiaetus* butterflies. Evolution 61(3):546–559

Kerr WE (1952) A variação do número de cromossomos na evolução dos Hymenoptera. Sci Genet 4:182–190

Lorite P, Palomeque T (2010) Karyotype evolution in ants (Hymenoptera: Formicidae), with a review of the known ant chromosome numbers. Myrmecological News 13:89–102

Lukhtanov VA, Kandul NP, Plotkin JB, Dantchenko AV (2005) Reinforcement of pre-zygotic isolation and karyotype evolution in *Agrodiaetus* butterflies. Nature 436(7049):385

Mayrose I, Backer MS, Otto SP (2010) Probabilistic models of chromosome number evolution and the inference of polyploidy. Syst Biol 59:132–144

Matthey R (1973) The chromosome formulae of eutherian mammals. In: Chiarelli AB, Capanna E (eds) Cytotaxonomy and vertebrate evolution. Academic Press, London, pp 531–616

Menezes RST, Gazoni T, Costa MA (2019) Cytogenetics of warrior wasps (Vespidae: *Synoeca*) reveals intense evolutionary dynamics of ribosomal DNA clusters and an unprecedented number of microchromosomes in Hymenoptera. Biol J Linn Soc 126(4):925–935

Menezes RST, Brady SG, Carvalho AF, Del Lama MA, Costa MA (2017) The roles of barriers, refugia, and chromosomal clines underlying diversification in Atlantic Forest social wasps. Sci Rep

Menezes RST, Carvalho AF, Correia JP, Silva TS, Somavilla A, Costa MA (2014) Evolutionary trends in the chromosome numbers of swarm-founding social wasps. Insect Soc 61:385–393

Menezes RST, Lloyd MW, Brady SG (2020) Phylogenomics indicates Amazonia as the major source of Neotropical swarm-founding social wasp diversity. Proc R Soc B Biol Sci 287(1928):20200480

Menezes RST, Silva TM, Carvalho AF, Andrade-Souza V, Silva JG, Costa MA (2013) Numerical and structural chromosome variation in the swarm-founding wasp *Metapolybia decorata* Gribodo 1896 (Hymenoptera, Vespidae). Genetica 141:273–280
Pardi L (1942) Ricerche sui Polistini. 8: La spermatogenesi di Polistes gallicus (L.) e di P.(Leptopolistes) omissus (Weyrauch). Sci Genet 3:14–22
Pompolo SG, Takahashi CS (1986) Karyotype of the species of wasps of the genus *Polistes* (Polistinae: Hymenoptera). Insect Soc 33:142–148
Pompolo SG, Takahashi CS (1987) Cytogenetics of Brazilian Polybiini wasps (Hymenoptera, Vespidae, Polistinae). Rev Brasil Genet 10:483–496
Pompolo SG, Takahashi CS (1990a) Karyotype of ten species of social wasps (Hymenoptera, Polistinae, Polybiini). Rev Brasil Genet 13:469–477
Pompolo SG, Takahashi CS (1990b) Chromosome numbers and C banding in two species of the genus *Polistes* (Hymenoptera, Polistinae, Polistinii). Insect Soc 37:251–257
Sumner AT (2003) Chromosomes: organization and function. Blackwell Publishing Co, North Berwick, United Kingdom
Todd NB (1970) Karyotypic fissioning and canid phylogeny. J Theor Biol 26:445–480
Todd NB (1975) Chromosomal mechanisms in the evolution of artiodactyls. Paleobiology 1:175–188
White MJD (1954) Animal cytology and evolution, 2nd edn. Cambridge University Press, London
White MJD (1973) Animal cytology and evolution, 3rd edn. Cambridge University Press, London
Wurster DH, Benirschke K (1970) Indian muntjac, *Muntiacus muntjak*: a deer with a low diploid chromosome number. Science 168:1364–1366
Yang F, O'Brien PC, Wienberg J, Neitzel H, Lin CC, Ferguson-Smith MA (1997) Chromosomal evolution of the Chinese muntjac (*Muntiacus reevesi*). Chromosoma 1:37–43

Chapter 15
Phylogeny and Classification of the Neotropical Social Wasps

Orlando T. Silveira, Sergio R. Andena, Alexandre Somavilla, and James M. Carpenter

Abstract Social wasps are common elements of the Neotropical fauna, known to most people living in this region because of their large nests and the aggressive defensive behavior of many species. Polistinae is a widely distributed taxon in the Neotropics, with more than 300 species described within 21 genera, with 20 of these genera endemic. In this chapter, we review the history and current state of taxonomic knowledge of this group of insects, since the pioneering works of Latreille, de Saussure, and Ducke, through Zikán and Richards, until the phylogeny-based classifications of Carpenter and other authors. In the last two decades, phylogenetic investigation in the Vespidae received the important input of DNA sequences, and analyses made with these new data matrices have produced phylogenetic hypotheses differing in several respects from those based on morphology and behavior alone. However, at higher levels, the relationships of some groups are not yet established such as relationship of Stenogastrinae to other vespids (which impacts on questions regarding the origin of the eusocial behavior in the Vespidae) and the possible paraphyly of the subfamilies Masarinae and Eumeninae (for the latter, a crucial issue is the possibility of a closer relationship of zethines to the Vespinae and Polistinae). At less inclusive levels, the relationships among the polistine tribes are still uncertain, in particular the position of the genus *Polistes* traditionally considered as the basalmost within the subfamily. Furthermore, phylogeny within some large taxa such as the genera *Polistes* and *Mischocyttarus* and the tribe Epiponini remain as major fields of investigation for the systematists dedicated to this fascinating group of hymenopterans.

O. T. Silveira (✉)
Museu Paraense Emílio Goeldi/MPEG, Belém, Brazil
e-mail: orlando@museu-goeldi.br

S. R. Andena
Universidade Estadual de Feira de Santana/UEFS, Feira de Santana, Brazil

A. Somavilla
Instituto Nacional de Pesquisas da Amazônia, Manaus, Brazil

J. M. Carpenter
American Museum of Natural History/AMNH, New York City, New York, USA

F. Prezoto et al. (eds.), *Neotropical Social Wasps*,
https://doi.org/10.1007/978-3-030-53510-0_15

Keywords Vespidae · Phylogeny · Classification · Taxonomic history · Neotropical wasps

15.1 Introduction

Social wasps are common elements of the Neotropical fauna, being known by most people living in this region because of their large nests and the aggressive defensive behavior of many species. In some regions of Latin America, wasps play central roles in rituals, legends, sympathies, and superstitions among the natives (von Ihering 1968; Lenko and Papavero 1996). Natives from Mexico used to eat the honey produced by some species of the genus *Brachygastra*, as well as the larvae and pupa present inside these wasps' nests (Lenko and Papavero 1996).

The social behavior of some species of Hymenoptera has fascinated entomologists around the world along the centuries. One of the first studies on the behavior of wasps was carried out by René de Réamur (1683–1757). Later, Charles Darwin in his most important book *On the Origin of Species* discussed the evolution of sociality in insects in the light of his theory of natural selection. In the twentieth century, William Hamilton, who studied wasps of the genus *Polistes*, created the concept of "inclusive fitness" (Hamilton 1964). In the 1970s and 1980s, publications of E. O. Wilson introduced the term "sociobiology" to deal with the whole context of the life of social organisms and their features and behaviors, like those related to reproductive division of labor, cooperative brood care, social defensive behavior, etc. (Wilson 1971).

Wasps are included in one of the most diverse orders of insects, the Hymenoptera, with about 125,000 described species, which also includes the bees and ants (Fernández and Sharkey 2006). Despite many people having the perception that all bees, ants, and wasps are social, actually only a fraction of the aculeate Hymenoptera (those that have a sting) are truly social, i.e., according to Wilson's definition of "eusociality," which is characterized by the presence of cooperative brood care, overlapping generations of adults within a colony, and a labor division into reproductive and nonreproductive groups. Within the Aculeata (Fig. 15.1) only four groups have eusocial representatives: (a) the family Crabronidae with the genus *Microstigmus* Ducke, 1907 presenting social species; (b) the bees, with several eusocial lineages (both cases "a" and "b" included in the superfamily Apoidea); (c) the family Formicidae, with all ants being eusocial; and d) the Vespidae, with three subfamilies of eusocial wasps (the latter two taxa of ants and wasps are included in the superfamily Vespoidea; see Brothers and Carpenter 1993, Brothers 1999; Sharkey et al. 2012) (Fig. 15.2).

Microstigmus is a Neotropical genus with approximately 30 species. Reports of the sociobiology of some species can be found in Matthews (1991) and Lucas et al.

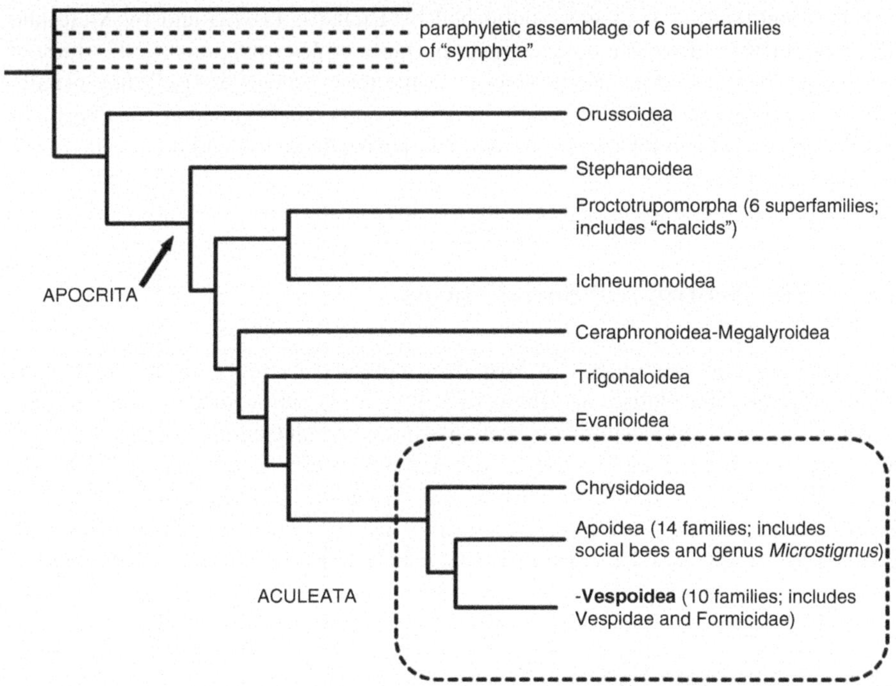

Fig. 15.1 Simplified cladogram showing phylogenetic relationships between major Hymenoptera lineages. (Adapted from Sharkey et al. 2012)

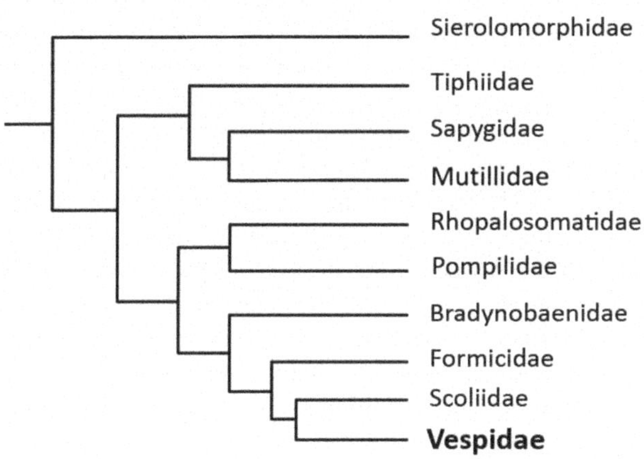

Fig. 15.2 Cladogram showing phylogenetic relationships between families of Vespoidea. (After Brothers 1999)

(2011). Identification keys were published by Richards (1972) and by Melo and Matthews (1997). However, despite the importance of *Microstigmus* for the study of the evolution of sociality, when people use the term "social wasps," they usually refer to wasps of the family Vespidae, due to its great taxonomic diversity, conspicuousness in several habitat types, exuberant defensive behavior, and the beauty of the architecture of nests.

15.2 The Neotropical Social Wasps

Among the social subfamilies of Vespidae, natural occurrence in the Neotropics mainly concerns the Polistinae (Table 15.1, Fig. 15.3), while species of the mainly Holarctic genus *Vespula* Thomson 1869 (Vespinae) reach Central America (Kimsey and Carpenter 2012; Landolt et al. 2010), and the species *V. germanica* and *V. vulgaris* have been introduced into Chile and Argentina (Masciocchi et al. 2010). Besides that, also *Vespa orientalis* has recently been reported from Chile (Ríos et al. 2020). Most of the social species of Vespidae belong to the subfamily Polistinae.

15.3 Taxonomic History

The studies on the higher classification of the vespids started in a context of the writing of general entomological monographs during the eighteenth and nineteenth centuries. Latreille (1802) treated the taxon as the family "Vespariae," which he subdivided into three major sections, one of these including the social vespids and containing the genera *Vespa*, *Polistes*, and *Epipona* (the last two being created in that work). *Epipona* was the first generic name proposed for a taxon representative of the Neotropical fauna, specifically for the species *E. nidulans* (Fabricius, 1793) and *E. tatua* (Cuvier, 1797), the first being a junior synonym of *Chartergus artifex* (Christ, 1791). Lepeletier (1836), a colleague and disciple of Latreille (see Dupuis 1974), also dealt with the higher classification of vespids, but in the scope of a more specialized "hymenopterological" work. However, in view of the method and criteria used (i.e., priority given to socio-behavioral characteristics), social wasps were separated from other vespids of solitary habits. He named the social vespids the "Polistides," with eight genera, including the new *Polybia*, *Agelaia*, *Apoica*, and *Chartergus*, which presently comprise only Neotropical species.

Carpenter (1982) approached the taxonomic history of the vespids, highlighting the works of Henri de Saussure (1829–1905), Joseph Bequaert (1886–1982), and Owain Richards (1901–1984), of which the first and the last produced more comprehensive works with greater impact on the knowledge of the Neotropical fauna. De Saussure described around 50 genus-level taxa (genera and subgenera; van der Vecht and Carpenter 1990), and in his most important work *Études sur La famille des Vespides* (published in three volumes from 1852 to 1858), he divided the

Table 15.1 Characters and distribution of Vespidae subfamilies and tribes. For Polistinae, the constituent genera and the numbers of described extant species of those with distribution in the New World are reported

Subfamily	Tribe	Genera (no. of species in Americas)	Distribution
Polistinae Eusocial, nests of carton, insect predators; longitudinally folded anterior wing; simple tarsal claws; hind coxa without dorsal carina; hind wing almost always with small anal lobe (except Apoica) *Colonies started by one or a few related females (independent founders: IND) or by swarms of queens and several or many workers (swarm founders: SWA)	**Ropalidiini** (IND; SWA)	*Belonogaster*, *Parapolybia*, *Polybioides*, *Ropalidia*	Africa, Asia, and Australia
	Larval meconium extracted from the cell by adult females; nests with variable architecture		
	Polistini (IND)	*Polistes* (93)	Cosmopolitan (except New Zealand)
	Metasoma not petiolate, 1st. Segment cone shaped; nest with naked single comb		
	Mischocyttarini (IND)	*Mischocyttarus* (252)	Americas (mostly Neotropical)
	Tarsal articles with asymmetrical lobes; larva with ventral processes on abdominal sternum 1; nests normally with naked single comb		
	Epiponini (SWA) Male antenna without "tiloyds"; nests more often with multiple combs, covered by an envelope	*Apoica* (10), *Agelaia* (31), *Angiopolybia* (4), *Pseudopolybia* (4), *Parachartergus* (17), *Chartergellus* (13), *Nectarinella* (3), *Leipomeles* (4), *Protopolybia* (38), *Charterginus* (6), *Protonectarina* (1), *Chartergus* (3), *Brachygastra* (17), *Polybia* (59), *Synoeca* (6), *Epipona* (5), *Asteloeca* (3), *Clypearia* (8), *Metapolybia* (18)	Americas (mostly Neotropical)

(continued)

Table 15.1 (continued)

Subfamily	Tribe	Genera (no. of species in Americas)	Distribution
	Palaeopolistes, *Protopolistes*, *Palaeopolybia* (extinct: Palaeogene) Tribal designation dubious		Europe – France, UK
Vespinae			Holarctic and Oriental
Eusocial, nests of carton, insect predators; longitudinally folded anterior wing; simple tarsal claws; hind wing often without anal lobe; metasoma sessile (without petiole)			
Stenogastrinae			Oriental and New Guinea
Eusocial, insect predators; wing not folded longitudinally; secretions of the Dufour's gland used in breeding offspring and colony defense (anti-ant protectors); labium without acroglossal buttons; pronotal lobe very distant from tegula			
Eumeninae			Cosmopolitan
Solitary or subsocial, nests in galleries in the ground or in plants, or constructed with mud or plant materials; insect predators; longitudinally folded anterior wing; mesoscutum with posterior lateral process (parategula); usually bifid tarsal claws			
Masarinae			Americas, Southern Europe, Central Asia,
Solitary, foraging pollen and nectar, nests in the soil or in mud cells juxtaposed to rocks or plants; anterior wing more often not folded longitudinally at rest, marginal cell with apex rounded and away from margin; femur of mid leg without a ringed groove at the base			Africa, and Australia
Euparagiinae (recent-Cretaceous)			Western North America (recent) and Asia and Africa (Cretaceous)
Solitary; predators of curculionid beetles, nests in galleries in the ground; anterior wing not longitudinally folded, second discal cell with apical digitiform projection; posterior wing with very long anal lobe			
Priorvespinae (extinct)			Asia (Cretaceous)
Protovespinae (extinct)			Asia (Cretaceous)

Vespidae into three tribes – Masariens, Euméniens, and Vespiens – based on the level of sociality, which, according to the author, would be enough to separate the groups. Masariens and Euméniens were solitary, while the Vespiens were social. Besides taxonomic work, de Saussure presented extensive biological data, as well as detailed studies on nest architecture. Some terms created by him to describe general architectural features, like "stelocyttarous" and "phragmocyttarous," were widely employed by Richards (1971, 1978) and are still in use.

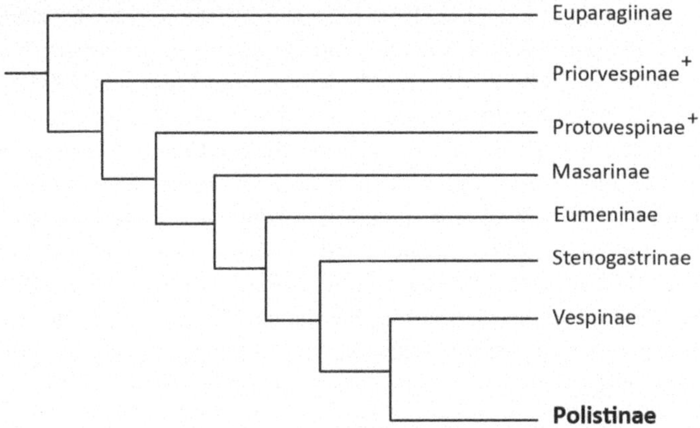

Fig. 15.3 Cladogram showing phylogenetic relationships between subfamilies of Vespidae. (Based on Perrard et al. 2017)

Richards described 15 genus-level taxa (genera and subgenera) and around 150 species and subspecies of Polistinae (van der Vecht and Carpenter 1990) and made the first revision of the genus *Mischocyttarus* de Saussure (Richards 1945). His most prominent book *The Social Wasps of the Americas, Excluding the Vespinae*, despite being out of date in many aspects, is still a precious source of data about taxonomy and biology for species of all genera of Neotropical social wasps.

Carpenter (1982) made the first cladistic analysis of the group, diverging from Richards (1962), who treated the vespids as a superfamily Vespoidea, including three families: Masaridae, Eumenidae, and Vespidae. Carpenter demonstrated non-monophyly for several taxa at different levels in Richards' classification. Following the hierarchical relations in the obtained cladogram, Carpenter (1982) treated the whole group as a single family, Vespidae, with six subfamilies: Euparagiinae, Masarinae, Eumeninae, Stenogastrinae, Vespinae, and Polistinae. Two additional extinct subfamilies were later included in Vespidae: Priorvespinae, by Carpenter and Rasnitsyn (1990; for the fossil genus *Priorvespa* Carpenter and Rasnitsyn 1990, with five species described from Russia and Mongolia), and, more recently, Protovespinae in Perrard et al. (2017); for *Protovespa haxairei* Perrard and Carpenter, 2017). Among these subfamilies, Stenogastrinae, Vespinae, and Polistinae are constituted of eusocial species, with considerable diversification as to the types of organization of the societies, architecture of the nests, mechanisms of foundation of colonies, and reproductive dominance (Jeanne 1975, 1980).

In subsequent works, with collaborators, Carpenter progressively deepened the systematic knowledge "within" the Vespidae subfamilies, thus initiating a detailed analysis of the phylogenetic relationships among genera (Carpenter and Cummings 1985; Carpenter 1987, 1988a, 1988b; Carpenter and Kimsey 2009), reviewing the validity and synonymy of generic names (van der Vecht and Carpenter 1990; Carpenter and Day 1988; Carpenter 1996a, 1996b, 1996c), and initiating the publication of catalogs and checklists (Carpenter 1996a, 1996b, 1996c, 2001; Carpenter and Kojima 1996; Carpenter and Kojima 1997; Carpenter and Madl 2009; Carpenter et al. 2009, 2010a, 2010b; Kojima and Carpenter 1997).

Especially important for the classification of Neotropical social wasps were two papers (Carpenter 1991, 1993), respectively, on the phylogeny of genera of the subfamily Polistinae and on biogeographic patterns in Vespidae, from which resulted a revised version of tribe-level classification for polistines, formed by Polistini, Ropalidiini, Epiponini, and the new tribe Mischocyttarini containing only *Mischocyttarus* de Saussure, 1853 (see Table 15.1). Later, Carpenter (1997b) clarified nomenclatural questions concerning the names of tribes, especially on the precedence and correct authorship of the name "Epiponini Lucas," as opposed to "Polybiini Bequaert" previously used by Richards (1962).

15.4 Genus-Level Classifications

Excepting the Ropalidiini, the other three tribes of Polistinae occur in the Neotropics, including the American species of Polistini/*Polistes*, Mischocyttarini/*Mischocyttarus*, and the 19 genera of Epiponini, the last two tribes being restricted to Americas and mostly Neotropical. In Table 15.1, we present the genera of Polistinae, including their distribution and number of species (for those present in the Neotropics). Somavilla and Carpenter (Chapter 18, this volume) present an identification key to the genera of the subfamily Polistinae.

A major part of the genus-level classification of the Neotropical social wasps was worked out by A. Ducke in several articles published in the first two decades of the twentieth century (Ducke 1904, 1905a, 1905b, 1907, 1910, 1913, 1914, 1918). Ducke created nine new genera of social wasps – most used until very recently and three of them still being valid nowadays. With these new taxa, Ducke promoted the recognition of considerable diversity of form, social organization, and architecture in natural groups traditionally treated as parts of *Polybia*. So it was with *Metapolybia*, *Protopolybia*, and *Synoecoides* (in Ducke 1905a) and *Protonectarina* and *Stelopolybia* (in Ducke 1910), the latter soon being divided with the creation of *Gymnopolybia* (in Ducke 1914). Not less important, with the new genera *Megacanthopus* (in Ducke 1904) and *Monacanthocnemis* (in Ducke 1905a), Ducke also rescued a considerable number of species previously classified in *Polybia* but which had very different morphological and behavioral characteristics compared to the more typical species of this genus, being in most aspects more similar to *Mischocyttarus*, created by de Saussure in 1853. However, in a 1913 paper, Ducke himself synonymized his new generic names with *Mischocyttarus*, an older name that should keep nomenclatural priority. With the exception of *Gymnopolybia*, all other generic names created by Ducke were used by Richards (1978), as genera or as subgenera of *Mischocyttarus*. However, some of them are now treated as synonyms, such as *Stelopolybia* (= *Agelaia* Lepeletier) and *Pseudochartergus* (=

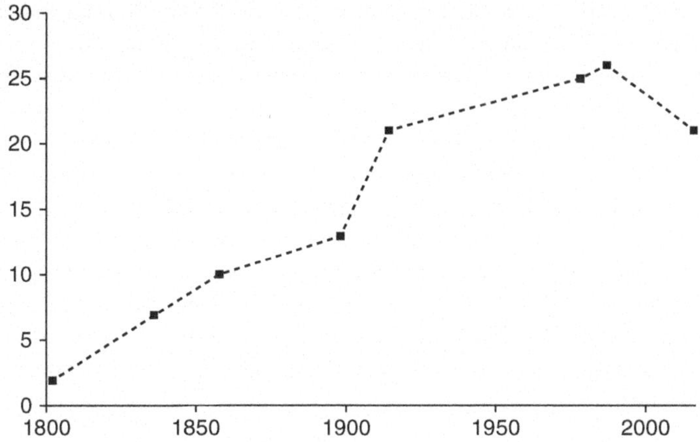

Fig. 15.4 Variation over time in the number of valid names of genera of Polistinae

Protopolybia) in Epiponini and *Monacanthocnemis* (= *Omega* of de Saussure) in Mischocyttarini.

Figure 15.4 summarizes the history of the generic classification for the Neotropical Polistinae from 1802 (i.e., Latreille) to the present, showed as a result of the balance between the creation of new names and "sinking" of synonyms. One may see the continuous growth of the number of names up to the year 1989, when a process of reduction of problematic taxa was initiated, based on the use of more strict criteria regarding the feasibility of reciprocal diagnosis between some very similar genera (*Pseudochartergus* = *Protopolybia*; Carpenter and Wenzel 1989) (*Marimbonda* = *Leipomeles*; Carpenter 2004) and also the impediment of paraphyletic groups (*Occipitalia* = *Clypearia*; Carpenter et al. 1996a, 1996) (*Synoecoides* = *Polybia*; Carpenter et al., 2000).

15.5 Fossils of Neotropical Wasps

In Vespidae, fossils have been recorded in Priorvespinae, Protovespinae, Euparagiinae, Eumeninae, Vespinae, and Polistinae, the oldest ones being of the Cretaceous period of the Mesozoic era (Carpenter and Rasnitsyn 1990; Brothers and Rasnitsyn 2008; Perrard et al. 2015, 2017; Poinar Jr. 2005). Martill et al. (2007) tentatively attributed to the subfamily Eumeninae a fossil from the Crato Formation (Cretaceous), in Ceará, Brazil, based on the form of the metasomal petiole. However, the authors do not point out, in the first place, any unequivocal characters that would diagnose the specimen as Vespidae. A positively confirmed ancient fossil of a eumenine species is *Symmorphus senex,* Carpenter 2000, from New Jersey Cretaceous amber (Carpenter 2000).

Perrard et al. (2017) presented several recent additions of Cretaceous fossils of vespids assigned to the extant solitary subfamily Euparagiinae (two species of the genus *Curiosivespa* Rasnitsyn, 1975) and to a presumed solitary extinct subfamily Protovespinae. However, the oldest record of a social wasp is represented by the fossil remains of a nest of the *ichno* genus and species *Brownichnus favosites* (Brown, 1941) from about 66 million years ago, on the Cretaceous-Paleocene transition (see Wenzel 1990; Genise 2000). Perrard et al. (2015) presented information on nine fossil species of Polistinae, all coming from amber deposits of Paleocene to Miocene ages. Most of these species belong to *Polistes* and were found in European localities. Only one of these fossils is from the Neotropics, *Agelaia electra* (Carpenter and Grimaldi 1997), found in amber from the Miocene of the Dominican Republic, on the island of Santo Domingo (Hispaniola), in the Caribbean.

15.6 Taxa Created for Social Parasites

Social parasitism in Vespidae is known only in Vespinae, in two genera, *Vespula* and *Dolichovespula*, and in four European species of *Polistes* (Carpenter and Perera 2006; Cervo 2006; Schmid-Egger et al. 2017). The latter constitute a monophyletic group and were treated as the subgenus *Sulcopolistes* Bluthgen, 1938; Weyrauch 1937 even erected for them the new genus *Pseudopolistes*. Carpenter (1997a) in his phylogenetic analysis pointed out that the group caused the paraphyly of *Polistes* s. str. Concerning the Neotropical Polistinae, no social parasites are known. Despite this, J. F. Zikán, in his studies on the *Mischocyttarus* fauna of the southeastern region of Brazil, especially of Itatiaia National Park (states of Rio de Janeiro and Minas Gerais), described several species that he supposed to be social parasites. Zikán was strongly influenced by the work of Weyrauch (1937) on parasitic species in Europe and believed to see in *Mischocyttarus* the same patterns of morphological differentiation observed in European parasitic wasps, in those characters "destined to subjugate the host queen of a nest" (Zikán 1949: 233). So, Zikán based his "social parasites" exclusively on the morphological differences he could see, without the necessary behavioral evidence. Richards (1978) discussed the subject in detail and synonymized virtually all of the parasite's names created by Zikán who typically added the suffix "oides" to the name of the presumed host species.

15.7 Phylogenetic Classifications

Recently, published phylogenetic studies in Vespidae have been frequent, using different sources of characters, i.e., nest architecture, behavior, morphology, and molecular characters (see below). For the social vespids, if at suprageneric levels the phylogeny has already been reasonably worked out, within the genera (especially in the Neotropical fauna) the existence of more detailed phylogenetic results is still limited mainly to the smaller genera (see below). In taxa with large numbers of species, such as *Polistes*, *Mischocyttarus*, and *Polybia*, the available phylogenetic hypotheses are still, in general, more fully informative about the relationships between subgenera or species groups. The latter, moreover, being simply informal groupings of species (which do not require a type designation) represent a convenient tool for referring to supraspecific taxa that one does not want to name formally, sometimes because of their small size, sometimes because its monophyly (or validity, in a phylogenetic sense) is not guaranteed (see below work in progress on *Polistes*).

The classification of subfamilies of Vespidae proposed by Carpenter (1982), and later modified by inclusion of extinct fossil subfamilies (Carpenter and Rasnitsyn 1990; Perrard et al. 2017) (see Fig. 15.3), has been challenged mainly by phylogenetic studies of DNA sequences. Results from Schmitz and Moritz (1998) and Hines et al. (2007) based on data obtained by "Sanger sequencing" methods (for relatively few loci) of samples with low representation of the taxonomic diversity in the subfamilies (low number of genera and species), and based on the exclusive use of molecular data, indicated different arrangements for the relationships between subfamilies, with emphasis on the basal positioning of Stenogastrinae and even paraphyly of Eumeninae in relation to the clade formed by Vespinae + Polistinae. Carpenter (2003) and Pickett and Carpenter (2010) criticized these two studies under various methodological aspects, especially for ignoring an extensive available base of morphological characters, as well as some behavioral ones. A reanalysis of these molecular data, combined with morphological and behavioral characters, published by Pickett and Carpenter (2010) supported the relationships between subfamilies as proposed by Carpenter (1982). More recently two papers returned to the problem of the relationships among subfamilies, based on large genomic (and transcriptomic) datasets (Bank et al. 2017; Piekarski et al. 2018). Both studies retrieved Stenogastrinae as sister group to other vespids and a paraphyletic Eumeninae with zethines more closely related to Vespinae + Polistinae (as in Hines et al. 2007). The paper by Piekarski et al. (2018) goes further in pointing out a paraphyletic Masarinae, based on a sister group relationship between *Gayella eumenoides* and two species of Euparagiinae. Especially this second paper takes the obtained results too strongly as evidence to propose changes in vespid classification, i.e., with the raising of tribes Zethini and Gayellini to subfamily rank.

If the sample of Bank et al. (2017) was still strongly deficient in regard to taxonomic diversity within Vespidae (2 Polistinae, 2 Vespinae, 1 Stenogastrine; mostly based on the eumenine fauna of Central Europe), the paper by Piekarski et al. (2018) shows much better representation of the more diversified tropical taxa (9 Ropalidiini, 19 Epiponini, 5 *Mischocyttarus*, 9 *Polistes*, 9 Stenogastrinae). However, both studies contradict recent morphology-based studies on eumenines (Hermes et al. 2014; Perrard et al. 2017; Lopes and Noll 2017), as well as the mixed morphological-molecular analysis of the Vespidae by Pickett and Carpenter (2010). We think that, despite the molecular basis underlying such classificatory proposals, it is premature to adopt them without properly assessing the evidence provided by morphological and behavioral characters. Particularly, as to the questioning about the validity of

Masarinae, it is important to note that a single Gayellini species was studied by Piekarski et al. (2018), and this is not sufficient to test the monophyly of Gayellini in the first place, much less to propose it as a higher-ranked taxon sister to the Euparagiinae. We recognize the significance and innovative character of these two works, inasmuch as the subject of subfamily relationships in Vespidae is of utmost importance for the study of the evolution of social behavior.

In the context of Neotropical social taxa, the study by Pickett and Carpenter (2010) showed for the first time an alternative arrangement of the phylogenetic relationships among the Polistinae tribes, with Ropalidiini being the basalmost lineage and *Polistes* as sister group to the Epiponini. This frame of a relatively more recent origin for the Polistini tribe contrasts with the previous hypothesis (Carpenter 1991, 1993) with the tribe as sister group to the remaining Polistinae, which was consistent with the traditional interpretation of *Polistes* as a relatively older cosmopolitan lineage that retains plesiomorphic characters in Polistinae (see Ross and Matthews 1991; Turillazzi and West-Eberhard 1996). However, subsequent papers based on molecular data corroborated such a pattern of relationships for polistine tribes (Piekarski 2014; Piekarski et al. 2018).

Among the tribes with Neotropical components, phylogenetic studies began in the 1990s on the genus-level phylogeny of Polistinae, with major contributions to knowledge of the relationships within Epiponini (Carpenter 1991) and within the genus *Polistes* (Carpenter 1996). Since then many works have been published on the whole subfamily Polistinae (Arévalo et al. 2004) and on *Polistes* (Pickett and Wenzel 2004; Pickett et al. 2006; Santos et al. 2015), as well as a considerable number of studies on Neotropical genera by Andena et al. (2007a, 2007b, 2009a, 2009b, 2009c; Andena and Carpenter 2012), Carpenter et al. (2000), Menezes et al. (2015), Pickett and Wenzel (2007), Santos Jr. et al. (2015), and Silveira (2008). The main results of these studies are presented below.

15.8 Phylogeny of Polistinae

Richards (1978), to elucidate the relationships among the genera of Polistinae, presented a dendrogram almost completely resolved based on only 11 characters, without a formal cladistic analysis. Carpenter (1991) correctly pointed out that the data presented were not enough to support that degree of resolution and, furthermore, the branches were mostly supported only by primitive characters or had no support at all. In that work, Carpenter proposed the first phylogeny for the Polistinae (Fig. 15.5), and upon the resultant cladogram he tested a hypothesis on the origin of eusociality in wasps proposed by West-Eberhard (1978).

In the cladogram of Carpenter (1991), *Polistes* is the basalmost lineage, being sister to a polytomy that includes *Mischocyttarus*, a clade of the old world ropalidiines ((*Polybioides*, *Belonogaster*) + (*Ropalidia*, *Parapolybia*)), and also the epiponines. The latter, with *Apoica* as the basalmost genus, had as a prominent synapomorphy the foundation of colonies by swarming, a feature shared by all other

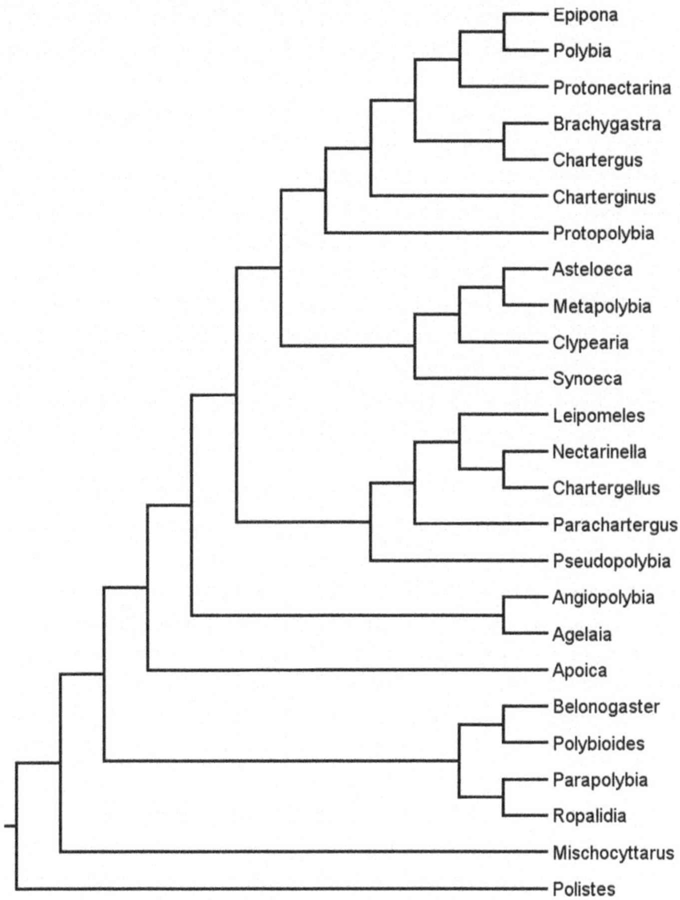

Fig. 15.5 Genus-level phylogeny of Polistinae (Based on Wenzel and Carpenter 1994)

Epiponini but which arose independently in Ropalidiini (and in *Provespa*, Vespinae) (Fig. 15.5).

Despite the great importance of the work by Carpenter (1991), many polytomies were still pending resolution in Epiponini, such as that involving the genera *Agelaia* and *Angiopolybia* (both being basal branches in the tree) and another one referring to a major clade here named "*Polybia*-component," where the relationships among the lineages were not fully resolved, i.e., *Protonectarina,* (*Protopolybia* + *Pseudoc hartergus*), *Synoecoides*, *Polybia,* (*Chartergus* + *Brachygastra*). Additionally, *Clypearia*, *Asteloeca*, *Metapolybia*, and *Occipitalia* were all grouped into a clade, but without resolution of the relationships.

It is worth noting in that study the presence of the genus *Pseudochartergus*, although it had already been synonymized with *Protopolybia* by Carpenter and Wenzel (1989). Some other generic names in that work also would be synonymized in subsequent works (see below). As pointed out by Carpenter (1991), that phylogeny was part of a work in progress, and preliminary data were employed. Despite the lack of resolution of some clades, this monumental work served as base for many others (see below).

Wenzel (1993) presented a new phylogeny for the Polistinae based solely on nest characters. Other authors had already employed nest characters for classification (see Ducke 1914); however, their characters were based on generalizations such as "astelocyttarus nest" (with cells built directly on the substrate). The publication by Wenzel (1993) brought a new perspective to the use of nest architecture in phylogenetic analyses, no longer using general aspects but decomposing the architecture into elementary characters. Unfortunately, the work of Wenzel (1993) resulted in a less resolved tree than that proposed by Carpenter (1991). A major problem in the study by Wenzel (1993) was the basal positioning of *Apoica*, composing a polytomy along with *Mischocyttarus* and *Polistes*. In addition, as in Carpenter (1991), the basal epiponine genera *Agelaia* and *Angiopolybia* remained unresolved as to their relationships.

Wenzel and Carpenter (1994) added to these previously published data on morphology and nest architecture (Carpenter 1991; Wenzel 1993) new larval morphological characters and proposed a new phylogenetic hypothesis for the Polistinae. It is important to note that the combined data resolved the polytomies present in Carpenter (1991) and Wenzel (1993). In Wenzel and Carpenter (1994), *Apoica* was sister to the remaining Epiponini. At the base of the latter, a clade formed by *Agelaia* + *Angiopolybia* was now sister group of the remaining genera grouped into a component with completely resolved relations. In addition, the data of Wenzel and Carpenter (1994) showed evidence (later confirmed) of synonyms in some pairs of genera (already mentioned above).

In the 1990s and early 2000s, molecular genetic techniques began to become popular and, as a result, began to be used in cladistic works. Discussions about the reliability and quality of molecular data were common at that time, with the dispute of "molecular vs. morphology" promoting great debates until today. Arévalo et al. (2004) published the first phylogeny of Polistinae based on "total evidence," a term coined to define the joint use of molecular and morphological data. The authors pointed out "The best estimator for a species phylogeny might therefore not necessarily depend on the sole use of one set of characters (morphological or molecular), but on a combination of different characters that can prove to be a better estimator of species trees," citing Hillis and Wiens (2000), Moore (1995), Nixon and Carpenter (1996), Wiens and Lingluff (2001), and Gatesy et al. (2003). Arévalo et al. (2004) proposed a phylogeny that resolved the relations of the subgroups within Polistini (see below) and also focused on the relations among the other tribes of Polistinae (Epiponini, Ropalidiini, and Mischocyttarini). Despite the low number of species (only 36) used in the study, the three tribes were supported as monophyletic (either for morphological characters alone or when combined with the molecular ones), but the relationships among tribes were not fully resolved. Regarding the relationships between the genera of Epiponini, the combined analysis of Arévalo et al. (2004) was congruent with that proposed by Carpenter (1991), nevertheless managing to resolve the sister group relationship between *Agelaia* and *Angiopolybia*.

As seen above, in the study by Pickett and Carpenter (2010), the three tribes of Polistinae with Neotropical species (Polistini, Mischocyttarini, and Epiponini) resulted as monophyletic, but their relationships are different from those previously proposed in other studies, where Polistini was usually basal relative to Ropalidiini, Mischocyttarini, and Epiponini. As already mentioned above, recent phylogenomic studies have corroborated this pattern of relationships (Piekarski et al. 2018).

Concerning the relationships within the Epiponini, in the study of Pickett and Carpenter (2010), ideas on the relations of the basal genera experienced few changes, but their sample is clearly insufficient in respect of taxonomic diversity, with most genera being underrepresented and several being absent (i.e., *Angiopolybia*, *Parachartergus*, *Leipomeles*, *Chartergus*, *Synoeca*, *Clypearia*). *Apoica* formed the basalmost lineage in relation to the remaining genera, having as sister group a large clade formed by *Agelaia* and, at successively less inclusive levels, the clade *Pseudopolybia* + *Chartergellus* and then a large component previously referred to as "*Polybia*-component." In fact, one of the most important results in this work is the paraphyly of *Polybia* in relation to a clade encompassing *Epipona*, *Asteloeca*, and *Metapolybia*. As pointed out by the authors, the relationships between genera are generally quite different from previous studies. However, the sister group relationship between *Metapolybia* and *Asteloeca* is maintained.

In the phylogenomic study of Piekarski et al. (2018), the Epiponini sample is poorer in species (but richer in genera) than in the study by Pickett and Carpenter (2010), with 19 species and 18 genera (compared to 34 species and 12 genera in Pickett and Carpenter 2010). Several results in this paper disagree with those of previous studies, the most unexpected being a basal position for *Angiopolybia* within the tribe component and a sister group relation between *Apoica* and *Agelaia*. In addition, contrary to other previous studies (e.g., Wenzel and Carpenter 1994), *Chartergus* does not appear as sister group to *Brachygastra* (but as more closely related to *Protopolybia* and *Charterginus*), and *Metapolybia* appears more closely related to *Clypearia* (rather than to *Asteloeca*). Equally worthy of mention is a sister group relationship between *Epipona* and *Synoeca*, which had not appeared before in large-scale phylogenetic studies, although two previous morphological studies had pointed to similarities between these genera in the form of some elements of the sting apparatus and in the mandibles of the females (Silveira and Silveira 1994; Silveira and Santos Jr. 2011). The same criticisms presented above with respect to the subfamily-level relationships obtained in the phylogenomic studies of Bank et al. (2017) and Piekarski et al. (2018) apply to relations within lower-level clades, as the tribe Epiponini, i.e., sampling is too restricted and important pieces of (available) evidence are discarded (morphology and behavior). The evolutionary history of all 19 genera of Epiponini was also studied by Menezes et al. (2020), employing phylogenomic data, and presenting similar results to those of Pierkarski et al. (2018).

15.9 Phylogeny of Polistini

Polistes is the only genus of the tribe Polistini, described by Latreille (1802), with the type species *P. gallicus* (= *Vespa gallica* Linnaeus, 1767) and having as main diagnostic character the conical shape of Tergum I. *Polistes* is one of the most familiar, species-rich, and widespread taxa of social wasps, with 240 described species, and found in all habitable continents (Richards 1978; Carpenter 1996a, 1996b, 1996c; Santos et al. 2015). Despite the cosmopolitan distribution, the highest diversity has been reported in tropical forests (Richards 1973, 1978; Carpenter 1993, 1996a, 1996b; Nguyen et al. 2017). For the New World, which includes the Neotropical and Nearctic regions, 93 species have been recorded (Richards 1978; Carpenter 1996a; Buck et al. 2012). Forty-one *Polistes* species are known in Brazil.

Richards (1973) proposed the first formal division for the genus *Polistes*, subdividing it into 12 subgenera, with 5 distributed in the New World (*Aphanilopterus* Meunier, 1888; *Epicnemius* Richards, 1973; *Fuscopolistes* Richards, 1973; *Onerarius* Richards, 1973; *Palisotius* Richards, 1973) and 7 in the Old World (*Gyrostoma* Kirby and Spence, 1828; *Megapolistes* van der Vecht, 1968; *Nygmopolistes* Richards, 1973; *Polistella* Ashmead, 1904; *Stenopolistes* van der Vecht, 1972; *Sulcopolistes* and *Polistes* sensu stricto Latreille, 1802). However, Carpenter et al. (1996) in the first cladistic study for the genus, based on morphological characters and species from all biogeographical regions, found that some of these subgenera were not monophyletic. As a result, he proposed the synonymy of eight of these subgenera and a new classification, with only *Polistes* (*Aphanilopterus*) for all New World species and three other subgenera for Old World species: *Polistes* (*Gyrostoma*) (East Asia and Indo-Australian), *Polistes* (*Polistella*) (Austral-Asian region), and *Polistes* (*Polistes*) (Eurasia and African continent) (Fig. 15.6).

Polistes is one of the most studied genera with respect to phylogeny, and several works were carried out after Carpenter et al. (1996), using combined morphological and molecular characters, which largely corroborated this author's classification, despite some incongruences (Arévalo et al. 2004; Pickett and Wenzel 2004; Pickett et al. 2006; Pickett and Carpenter 2010; Santos et al. 2015). However, due to the relatively low number of species used and considering questions related to the support of branches in the analyzes, these studies had no effect on classification.

Somavilla (2016), working with 90 of the 93 New World species and analyzing morphological data (females, males, and larvae), nests architecture, and molecular characters of six gene regions, recovered all subgenera described by Richards (1973) suggesting the possible revalidation of them. According to the author, the species of the New World should be classified as follows: *Polistes* (*Aphanilopterus*), 52 species; *Polistes* (*Epicnemius*), 24 species; *Polistes* (*Fuscopolistes*), 13 species; *Polistes* (*Palisotius*), 3 species; and *Polistes* (*Onerarius*), 1 species (Fig. 15.6).

Although the species of the Old World were not the focus of Somavilla (2016), the author suggests in his phylogeny that the classification should remain as proposed by Carpenter et al. (1996): *Polistes* (*Gyrostoma*) with 23 valid described species, *Polistes* (*Polistella*) with 94 valid species – the most diverse subgenus – and

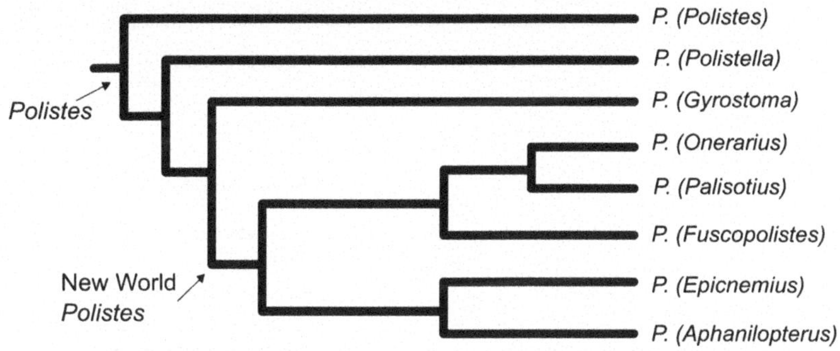

Fig. 15.6 Subgenus-level phylogeny of *Polistes*/Polistini. (Based on Somavilla et al., in preparation)

Polistes (*Polistes*) with 26 described species. In this same work, Carpenter suggests as phylogenetic hypothesis for these subgenera the following: *Polistes* (*Gyrostoma*) + *Polistes* (*Polistella*) + (*Polistes* (*Polistes*) + *Polistes* (*Aphanilopterus*)). In addition, Somavilla et al. (2018) presented an illustrated atlas for the male genitalia of the species of Neotropical *Polistes*.

15.10 Phylogeny of Mischocyttarini

Mischocyttarus is the single genus in Mischocyttarini, being one of the largest among the social vespid genera, with about 250 species described in 11 subgenera: *Mischocyttarus* s. str. de Saussure, 1853; *Clypeopolybia* Brèthes, 1923; *Monogynoecus* Richards, 1941; *Scytokeraia* Cooper, 1997; *Phi* de Saussure, 1854; *Kappa* de Saussure, 1854; *Megacanthopus* Ducke, 1904; *Haplometrobius* Richards, 1978; *Artifex* Silveira, 2008; *Cerberus* Silveira, 2008; and *Omega* de Saussure, 1854. The distribution is essentially Neotropical with a few species reported from north of Mexico.

Old classifications, such as those of Richards (1945, 1978) and even Cooper (1996a, 1996b, 1997a, 1997b, 1998a, 1998b), were made with traditional group recognition techniques, by arbitrary or informal means, based on similarities. Silveira (2008) was the first cladistic study on *Mischocyttarus* (Fig. 15.7), dealing with the problem of the monophyly of the subgenera and the phylogenetic relationships between groups of species. Monophyly of the genus was corroborated in the analyses, as well as of the majority of previously recognized subgenera. However, *Haplometrobius* (with more than 70 species) as conceived by Richards (1978) was not monophyletic, but some of its species groups were. The groups of *M. artifex* and *M. cerberus* were raised to subgenera (*Artifex* Silveira and *Cerberus* Silveira, respectively), and a more restricted concept of *Haplometrobius* was adopted, circumscribed to the group of *M. iheringi* (the type species of this subgenus). The concept of the subgenus *Omega* was expanded to include the groups of species of

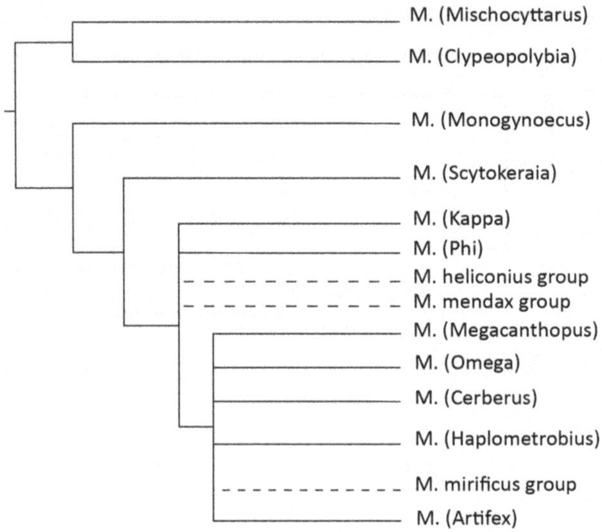

Fig. 15.7 Subgenus-level phylogeny of *Mischocyttarus*/Mischocyttarini. (Based on Silveira 2008)

M. surinamensis and *M. prominulus*, formerly considered parts of *Haplometrobius*. In addition to a new proposal in the subgeneric classification, this study reviewed the limits and diagnoses of all groups of species of the subgenera *Phi* and *Hamplometrobius* sensu Richards.

Regarding subgeneric relationships, a major basal pattern was found in all analyses, consisting of the division of the genus into two monophyletic groups, one of them formed by the clade *Mischocyttarus* s. str. + *Clypeopolybia* and the other forming a large component constituted by the other species groups of the genus. These two large clades, bearing considerable resemblance in content to the original taxa *Mischocyttarus* de Saussure and *Megacanthopus* Ducke, were informally named as "Saussurea" and "Duckea," in honor of those authors. Other relationships between subgenera varied to some degree according to different character weighing schemes (unweighted vs. implied weighting). Especially the relations of the subgenera *Monogynoecus* and *Omega* with the other groups remained uncertain and of considerable importance for the general understanding of the phylogeny in the genus.

In respect to the use of molecular data, the most recent work, and with greater taxonomic representation, is that of Pickett and Carpenter (2010), with sequences for 17 species of 8 subgenera. Unfortunately, in this work, most of the morphological characters used are not informative on the relationships within *Mischocyttarus*. The groupings found largely corroborate the monophyly of represented subgenera, except for a flagrant polyphyly of the subgenus *Phi*. However, as already mentioned above, this study was designed to investigate relationships between subfamilies rather than relationships within genera, presenting sampling deficiencies at this level. The authors themselves, regarding some results differing from those obtained by Silveira (2008), recognized the convenience of postponing the discussion about phylogeny in *Mischocyttarus*.

Since the phylogenetic study of Silveira (2008), efforts were directed by the author to the so-called "alpha" taxonomy (i.e., on identity and nomenclature of specific taxa) within subgenera and groups of species. Revisions were published for the species groups of *M. alfkenii* and *M. paraguayensis* of the subgenus *Phi* (Silveira 2013) and for the groups of *M. punctatus* and of *M. filiformis* of the subgenus *Omega* (Silveira et al. 2015; Borges and Silveira 2019). Other revisional studies as well as a reevaluation of the phylogeny in *Mischocyttarus* (with the addition of molecular characters) are underway.

15.11 Phylogeny of the Epiponini

At the genus-level, Epiponini is the most diverse group of Polistinae, with 19 genera and around 245 species described. However, the phylogeny of the species of each genus is a recent topic (Noll 2013). Since the publication of Richards (1978), several species have been described, as well as data about the biology, nests, and males. Besides, Richards (1978) had described several "forms" or "varieties," which need revision. In the 2000s, Carpenter, Pickett, Andena and Noll, and other collaborators began the study of each genus of Epiponini. Andena et al. (2007a, 2007b) published the phylogeny of the Epiponini genera *Angiopolybia* and *Pseudopolybia*. In the same year, Pickett and Wenzel (2007) published the first combined (morphological and molecular) cladistic analysis for the species of *Apoica*. Later, cladistic analysis, new identification keys, and other data have been published as follows: *Synoeca* (Andena et al. 2009a), *Charterginus* (Andena et al. 2009b), *Epipona* (Andena et al. 2009c), and *Brachygastra* (Andena and Carpenter, 2012). *Synoeca*, a small genus with only six species described, has been subject of debate due to its phylogeny. Cely and Sarmiento (2011) disagreed with the work of Andena et al. (2009a, 2009b, 2009c) and proposed a new phylogeny for the genus. As discussed by Carpenter et al. (2013), the paper of Cely and Sarmiento (2011) had many mistakes. Menezes et al. (2015) presented a different tree based only on molecules and Bayesian analysis. Santos-Junior et al. (2015) published a preliminary phylogeny for the genus *Protopolybia*. Andena and Carpenter and collaborators are still working on phylogeny of *Clypearia*, *Agelaia*, *Metapolybia*, and *Chartergellus*.

15.12 Final Considerations

As discussed above, the classification of social wasps has been troublesome. However, since the beginning of the last century, researchers have been committed to the studies of the group. Here, we highlighted the monumental work of Henri de

Saussure and Adolfo Ducke and later, at 1970s and 1980s until presently, Richards and Carpenter, who have stimulated the studies on social wasps with many publications.

Unfortunately, nowadays, despite the advances on the knowledge of the social wasps, only a few researchers are devoted to the taxonomy of the group, considering its diversity and the number of species to be discovered. It is especially true in the tropics, where the diversity is high and much of the original habitats has been devastated, even before we have a framework of its richness.

We hope that this chapter may have given you an idea of the state of the art on social wasps. Also we strongly hope to stimulate new works in the group.

References

Andena SR, Carpenter JM (2012) A phylogenetic analysis of the social wasp genus *Brachygastra* Perty, 1833, and description of a new species (Hymenoptera: Vespidae: Epiponini). Am Mus Novit 3753:1–38

Andena SR, Noll FB, Carpenter JM, Zucchi R (2007a) Phylogenetic analysis of the Neotropical *Pseudopolybia* de Saussure, 1863, with description of the male genitalia of *Pseudopolybia vespiceps* (Hymenoptera: Vespidae, Epiponini). Am Mus Novit 3586:1–11

Andena SR, Noll FB, Carpenter JM (2007b) Phylogenetic analysis of the Neotropical social wasps of the genus *Angiopolybia* Araujo, 1946 (Hymenoptera, Vespidae, Epiponini). Zootaxa 1427:57–64

Andena SR, Carpenter JM, Pickett KM (2009a) Phylogenetic analysis of species of the neotropical social wasp *Epipona* Latreille, 1802 (Hymenoptera, Vespidae, Polistinae, Epiponini). ZooKeys 20:385–398

Andena SR, Noll FB, Carpenter JM (2009b) A phylogenetic analysis of *Synoeca* de Saussure, 1852, a neotropical genus of social wasps (Hymenoptera: Vespidae: Epiponini). Entomol Am 115:81–89

Andena SR, Noll FB, Carpenter JM (2009c) Cladistic analysis of *Charterginus* fox, 1898 (Hymenoptera, Vespidae, Epiponini), a Neotropical genus of social wasps. J Nat Hist 43:1183–1193

Arévalo E, Zhu Y, Carpenter JM, Strassmann JE (2004) The phylogeny of the social wasp subfamily Polistinae: evidence from microsatellite flanking sequences, mitochondrial COI sequence, and morphological characters. BMC Evol Biol 4:8

Bank S, Sann S, Mayer C, Meusemann K, Donath A, Podsiadlowski L, Kozlov A, Petersen M, Krogmann L, Meier R, Rosa P, Schmitt T, Wurdack M, Liu S, Zhou X, Misof B, Peters RS, Niehuis O (2017) Transcriptome and target DNA enrichment sequence data provide new insights into the phylogeny of vespid wasps (Hymenoptera: Aculeata: Vespidae). Mol Phylogenet Evol 116:213–226

Brothers DJ (1999) Phylogeny and evolution of wasps, ants and bees (Hymenoptera, Chrysidoidea, Vespoidea and Apoidea). Zool Scr 28:223–249

Brothers DJ, Carpenter JM (1993) Phylogeny of Aculeata: Chrysidoidea and Vespoidea (Hymenoptera). J Hymenopt Res 2:227–304

Brothers DJ, Rasnitsyn AP (2008) A new genus and species of Euparagiinae from the Late Cretaceous of southern Africa (Hymenoptera: Vespidae). Alavesia 2:73–76

Borges RC, Silveira OT (2019) Revision of the species-group of *Mischocyttarus* (*Omega*) *filiformis* (de Saussure 1854), with description of three new species (Hymenoptera, Vespidae). Zootaxa 4657:545–564

Buck M, Cobb TP, Stahlhut JK, Hanner RH (2012) Unravelling cryptic species diversity in eastern Nearctic paper wasps, *Polistes* (*Fuscopolistes*), using male genitalia, morphometrics and DNA barcoding, with descriptions of two new species (Hymenoptera: Vespidae). Zootaxa 3502:1–48
Carpenter JM (1982) The phylogenetic relationships and natural classification of the Vespoidea (Hymenoptera). Syst Entomol 7:11–38
Carpenter JM (1987) Phylogenetic relationships and classification of the Vespinae (Hymenoptera: Vespidae). Syst Entomol 12:413–431
Carpenter JM (1988a) The phylogenetic system of the Stenogastrinae (Hymenoptera: Vespidae). J NY Entomol Soc 96:140–175
Carpenter JM (1988b) The phylogenetic system of the Gayellini (Hymenoptera: Vespidae; Masarinae). Psyche 95:211–241
Carpenter JM (1991) Phylogenetic relationships and the origin of social behavior in the Vespidae. In: Ross KG, Matthews RW (eds) The social biology of wasps. Cornell University Press, Ithaca, pp 7–32. 678
Carpenter JM (1993) Biogeographic patterns in the Vespidae (Hymenoptera): two views of Africa and South America. In: Goldblatt P (ed) Biological relationships between Africa and South America. Yale University Press, New Haven, pp 139–155
Carpenter JM (1996a) Distributional checklist of the species of the genus *Polistes* (Hymenoptera: Vespidae; Polistinae, Polistini). Am Mus Novit 3188:1–39
Carpenter JM (1996b) Phylogeny and biogeography of *Polistes*. In: Turillazzi S, West-Eberhard MJ (eds) Natural history and evolution of paper-wasps. Oxford University Press, Oxford, pp 18–57
Carpenter JM (1996c) Generic classification of the Australian pollen wasps (Hymenoptera: Vespidae; Masarinae). J Kansas Entomol Soc 69:384–400
Carpenter JM (1997a) Phylogenetic relationships among European *Polistes* and the evolution of social parasitism (Hymenoptera: Vespidae, Polistinae). Memoir Mus Natl Hist 173:135–161
Carpenter JM (1997b) A note on the name of paper wasp tribes (Insecta: Hymenoptera: Vespidae). Nat Hist Bulletin Univer Ibaraki 1:15–16
Carpenter JM (2000) A vespid wasp from New Jersey cretaceous amber. In: Grimaldi D (ed) Studies on fossils in Amber, with particular reference to the cretaceous of New Jersey. Backhuys Publishers, Leiden, pp 333–337
Carpenter JM (2001) Checklist of species of the subfamily Masarinae (Hymenoptera: Vespidae). Am Mus Novit 3325:1–39
Carpenter JM (2003) On "Molecular Phylogeny of Vespidae (Hymenoptera) and the Evolution of Sociality in Wasps.". Am Mus Novit 3389:1–20
Carpenter JM (2004) Synonymy of the genus *Marimbonda* Richards, 1978, with *Leipomeles* Möbius, 1856 (Hymenoptera: Vespidae; Polistinae), and a new key to the genera of paper wasps of the New World. Am Mus Novit 3465:1–16
Carpenter JM, Cummings JM (1985) A character analysis of the north American potter wasps (Hymenoptera: Vespidae; Eumeninae). J Nat Hist 19:877–916
Carpenter JM, Day MC (1988) Nomenclatural notes on Polistinae (Hymenoptera: Vespidae). Proc Entomol Soc Wash 90:323–328
Carpenter JM, Grimaldi DA (1997) Social wasps in amber. Am Mus Novit 3203:1–7
Carpenter JM, Kimsey LS (2009) The genus *Euparagia* Cresson (Hymenoptera: Vespidae; Euparagiinae). Am Mus Novit 3643:1–11
Carpenter JM, Kojima J (1997) Checklist of the species in the subfamily Vespinae (Insecta: Hymenoptera: Vespidae). Nat Hist Bulletin Ibaraki Univer 1:51–92
Carpenter JM, Kojima J, Wenzel JW (2000) *Polybia*, paraphyly and polistine phylogeny. Am Mus Novit 3298:1–24
Carpenter JM, Madl M (2009) A catalogue of the Vespidae of the Malagasy subregion (Insecta, Hymenoptera). Linz Biol Beitr 41:1871–1935
Carpenter JM, Rasnitsyn AP (1990) Mesozoic Vespidae. Psyche 97:1–20
Carpenter JM, Wenzel JW (1989) Synonymy of the genera *Protopolybia* and *Pseudochartergus* (Hymenoptera, Vespidae, Polistinae). Psyche 96:177–186

Carpenter JM, Wenzel JW, Kojima J (1996a) Synonymy of the genus *Occipitalia* Richards, 1978, with *Clypearia* Saussure, 1854 (Hymenoptera: Vespidae; Polistinae, Epiponini). J Hymenopt Res 5:157–165

Carpenter JM, Kojima J (1996) Checklist of the subfamily Stenogastrinae (Hymenoptera: Vespidae). J NY Entomol Soc 104(1–2):21–36

Carpenter JM, Perera EP (2006) Phylogenetic relationships among yellowjackets and the evolution of social parasitism. Am Mus Novit 3507:1–19

Carpenter JM, Garcete-Barrett BR, Hermes MG (2006) Catalog of the Neotropical Masarinae. Rev Bras Entomol 50:335–340

Carpenter JM, Gusenleitner J, Madl M (2009) A catalogue of the Eumeninae (Hymenoptera: Vespidae) of the Ethiopian region excluding Malagasy subregion. Part I: introduction, key to genera, genera *Aethiopicodynerus* GUSENLEITNER 1997 to *Cyrtolabulus* van der VECHT 1969. Linz Biol Beitr 41:513–638

Carpenter JM, Gusenleitner J, Madl M (2010a) A catalogue of the Eumeninae (Hymenoptera: Vespidae) of the Ethiopian region excluding Malagasy subregion. Part II: Genera *Delta* de Saussure 1885 to *Zethus* Fabricius 1804 and species incertae sedis. Linz Biol Beitr 42:95–315

Carpenter JM, Gusenleitner J, Madl M (2010b) A catalogue of the Eumeninae (Hymenoptera: Vespidae) of the Ethiopian region excluding Malagasy subregion. Part III: classification, additions, corrections and index. Linz Biol Beitr 42:919–1004

Carpenter JM, Andena SR, Noll FB, Wenzel JW (2013) Well, what about intraspecific variation? Taxonomic and phylogenetic characters in the genus *Synoeca* de Saussure (Hymenoptera, Vespidae). Zootaxa 3682:421–431

Cely CC, Sarmiento CE (2011) What about intraspecific variation? Reassessment of taxonomic and phylogenetic characters in the genus *Synoeca* de Saussure (Hymenoptera: Vespidae: Polistinae). Zootaxa 2899:43–59

Cervo R (2006) *Polistes* wasps and their social parasites: an overview. Ann Zool Fenn 43:531–549

Cooper M (1996a) The subgenus *Monogynoecus* Richards of *Mischocyttarus* de Saussure (Hym., Vespidae, Polistinae) with descriptions of two new species. Entomol.'s Mon Mag 132:23–28

Cooper M (1996b) The *mendax* group of *Mischocyttarus* (Hym., Vespidae, Polistinae) with descriptions of new species. Entomol.'s Mon Mag 132:273–280

Cooper M (1997a) A new subgenus of *Mischocyttarus* de Saussure (Hym., Vespidae). Entomol.'s Mon Mag 133:117–129

Cooper M (1997b) The subgenus *Megacanthopus* Ducke of *Mischocyttarus* de Saussure (Hym., Vespidae), with a key and three new species. Entomol.'s Mon Mag 133:217–223

Cooper M (1998a) Two new species of *Mischocyttarus* (Hym., Vespidae) with notes on some members of the *iheringi* group. Entomol.'s Mon Mag 132:89–93

Cooper M (1998b) New species of the *artifex* group of *Mischocyttarus* de Saussure (Hym., Vespidae) with a partial key. Entomol.'s Mon Mag 134:293–306

Ducke A (1904) Sobre as vespidas sociaes do Pará. Bol Mus Para Emílio Goeldi 4:317–374

Ducke A (1914) Über Phylogenie und Klassifikation der socialen Vespiden. Zool Jb Syst 36:303–330

Ducke A (1905a) Nouvelles contributions à la connaissance des vespides sociales de l'Amérique du Sud. Revue d'Entomologie (Caen) 24:5–24

Ducke A (1905b) Sobre as vespidas sociaes do Pará. (1o Supplemento). Bol Mus Para Emílio Goeldi 4:652–698

Ducke A (1907) Novas contribuições para o conhecimento das vespas (Vespidae sociales) da região neotropical. Bol Mus Para Emílio Goeldi 5:152–199

Ducke A (1910) Révision des guêpes sociales polygames d'Amérique. Ann Hist Nat Mus Nat Hung 8:449–544

Ducke A (1913) Zur Synonymie einiger Hymenopteren. Deut Entomol Z 1913:330–333

Ducke A (1918) Catalogo das vespas do Brazil. Rev Mus Paul 10:314–374.

Dupuis C (1974) Pierre André Latreille (1762–1833): The Foremost Entomologist of his Time. Annu Rev Entomol 19:1–14

Fernández F, Sharkey MJ (2006) Introducción a los Hymenoptera de la Región Neotropical, Serie Entomología Colombiana, Sociedad Colombiana de Entomología, Bogotá D.C., Colombia, p 896

Gatesy J, Amato G, Norell M, Desalle R, Hayashi C (2003) Combined support for wholesale taxic atavism in gavialine crocodylians. Syst Biol 52:403–422

Genise JF (2000) The ichnofamily Celliformidae for *Celliforma* and allied ichcnogenera. Ichnos 7:267–282

Hamilton WD (1964) The Genetical evolution of social behaviour. I J Theoretical Biology 7:1–16

Hermes MG, Melo GR, Carpenter JM (2014) The higher-level phylogenetic relationships of the Eumeninae (Insecta, Hymenoptera, Vespidae), with emphasis on *Eumenes* sensu lato. Cladistics 30:453–484

Hillis DM, Wiens JJ (2000) Molecules versus morphology in systematics: conflicts, artifacts, and misconceptions. In: Wiens JJ (ed) Phylogenetic analysis of morphological data. Smithsonian Institution Press, Washington, pp 1–19

Hines HM, Hunt JH, O'Connor TK, Gillespie JJ, Cameron SA (2007) Multigene phylogeny reveals eusociality evolved twice in vespid wasps. Proc Natl Acad Sci USA 104:3295–3299

von Ihering R (1968) Dicionario dos animais do Brasil. Editora da Universidade de Brasilia, São Paulo, p 790

Jeanne RL (1975) The adaptiveness of social wasp Nest architecture. Q Rev Biol 50:267–287

Jeanne RL (1980) Evolution of social behavior in the Vespidae. Annu Rev Entomol 25:371–396

Kimsey LS, Carpenter JM (2012) The Vespinae of North America (Vespidae, Hymenoptera). J Hymenopt Res 28:37–65

Kojima J, Carpenter JM (1997) Catalog of species in the polistine tribe Ropalidiini (Hymenoptera: Vespidae). Am Mus Novit 3199:1–96

Landolt P, Sierra JM, Unruh TR, Zack RS (2010) A new species of *Vespula*, and first record of *Vespa crabro* L. (Hymenoptera: Vespidae) from Guatemala, Central America. Zootaxa 2629:61–68

Latreille PA (1802) Histoire naturelle, générale et particulière des crustacés et insectes. Tome 3

Lenko K, Papavero N (1996) Insetos no Folclore, 2a. rev. ampl. edn. FAPESP, São Paulo, Plêiade, p 468

Lepeletier A (1836) Histoire naturelle des insectes - Hyménoptères. In: Tome 1. Librairie Encyclopedique de Roret, Paris

Lopes RB, Noll FB (2017) Cladistic analysis of the *Zethus* (*Zethoides*) Fox, 1899 (Hymenoptera, Vespidae, Eumeninae) species groups with insights on the current subgeneric classification of *Zethus* Fabricius, 1804. Insect Systematics (online) p 103–129

Lucas ER, Martins RP, Zanette LRS, Field J (2011) Social and genetic structure in colonies of the social wasp *Microstigmus nigrophthalmus*. Insect Soc 58:107–114

Martill DM, Bechly G, Loveridge R (2007) The Crato fossil beds of Brazil. Window into an ancient world. Cambridge University Press, Cambridge, UK, p 625

Masciocchi M, Beggs JR, Carpenter JM, Corley JC (2010) Primer registro de *Vespula vulgaris* (Hymenoptera: Vespidae) en la Argentina. Rev Soc Entomol Argent 69:267–270

Matthews RW (1991) Evolution of social behavior in sphecid wasps. In: Ross KG, Matthews RW (eds) The social biology of wasps. Cornell University Press, Ithaca and London, pp 570–602

Melo GAR, Matthews RW (1997) Six new species of *Microstigmus* wasps (Hymenoptera: Sphecidae), with notes on biology. J Nat Hist 31:421–437

Menezes RST, Brady SG, Carvalho AF, Del Lama MA, Costa MA (2015) Molecular Phylogeny and Historical Biogeography of the Neotropical Swarm-Founding Social Wasp Genus *Synoeca* (Hymenoptera: Vespidae). Plos One. https://doi.org/10.1371/journal.pone.0119151

Menezes RST, Lloyd MW, Brady SG (2020) Phylogenomics indicates Amazonia as the major source of Neotropical swarm-founding social wasp diversity. Proc R Soc B 287:20200480

Moore WS (1995) Inferring phylogenies from mtDNA variation: mitochondrial-gene trees versus nuclear-gene trees. Evolution 49:718–726

Nixon KC, Carpenter JM (1996) On simultaneous analysis. Cladistics 12(3):221–241

Noll FB (2013) "Marimbondos": a review on the Neotropical swarm-founding polistines. Sociobiology 60:347–354

Nguyen LTP, Vu TT, Lee JXQ, Carpenter JM (2017) Taxonomic notes on the *Polistes stigma* group (Hymenoptera, Vespidae: Polistinae) from continental Southeast Asia, with descriptions of three new species and a key to species. Raffles B Zool 65:269–279

Perrard A, Lopez-Osorio F, Carpenter JM (2015) Phylogeny, landmark analysis and the use of wing venation to study the evolution of social wasps (Hymenoptera: Vespidae: Vespinae). Cladistics (online):1–20

Perrard AF, Grimaldi D, Carpenter JM (2017) Early lineages of Vespidae (Hymenoptera) in cretaceous amber. Syst Entomol 42:379–386

Pickett KM, Carpenter JM (2010) Simultaneous analysis and the origin of eusociality in the Vespidae (Insecta: Hymenoptera). Arthropod Systemat Phyl 68:3–33

Pickett KM, Wenzel JW (2004) Phylogenetic analysis of the New World *Polistes* (Hymenoptera: Vespidae, Polistinae) using morphology and molecules. J Kansas Entomol Soc 77:742–760

Pickett KM, Wenzel JW (2007) Revision and Cladistic analysis of the nocturnal social wasp genus *Apoica* Lepeletier (Hymenoptera: Vespidae; Polistinae, Epiponini). Am Mus Novit 3562:1–30

Pickett KM, Carpenter JM, Wheeler WC (2006) Systematics of *Polistes* (Hymenoptera: Vespidae), with a phylogenetic consideration of Hamilton's haplodiploidy hypothesis. Ann Zool Fenn 43:390–406

Piekarski PK (2014) Monophyly of eusocial wasps (Hymenoptera: Vespidae): molecules and morphology tell opposing histories. J Undergraduate Res 4:11–14

Piekarski PK, Carpenter JM, Lemmon AR, Lemmon EM, Sharanowuski BJ (2018) Phylogenomic evidence overturns current conceptions of social evolution in wasps (Vespidae). Mol Biol Evol 35:2097–2109

Poinar G Jr (2005) Fossil Trigonalidae and Vespidae (Hymenoptera) in Baltic amber. Proc Entomol Soc Wash 107:55–63

Richards OW (1945) A revision of the genus *Mischocyttarus* de Saussure (Hymen., Vespidae). Transact Royal Entomol Soc, London 95:295–462

Richards OW (1962) A Revisional study of the Masarid wasps (Hymenoptera, Vespoidea). – British museum (natural history), London

Richards OW (1971) The Biology of the Social Wasps (Hymenoptera, Vespidae). Biol Rev 46:483–528

Richards OW (1972) Species of south-American wasps of genus *Microstigmus*-Ducke (Hymenoptera: Sphecoidea, Pemphredoninae). Transact Royal Entomol Soc London 124:123–148

Richards OW (1973) The subgenera of *Polistes* Latreille (Hymenoptera, Vespidae). Rev Bras Entomol 17:85–103

Richards OW (1978) The social wasps of the Americas excluding the Vespinae. British Museum (Natural History), London, p 580

Ríos M, Barrera-Medina R, Contreras JM (2020) Primer reporte del género *Vespa* Linnaeus (Hymenoptera: Vespidae: Vespinae) en Chile. Rev Chil Entomol 46:237–242

Ross KG, Matthews RW (1991) The social biology of wasps, vol xvii. Cornell University Press, Ithaca, New York, p 678

Santos JN Jr, Silveira OT, Carpenter JM (2015) Phylogeny of *Protopolybia* Ducke, 1905 and taxonomic revision of the *Protopolybia exigua* species-group (Hymenoptera: Vespidae, Polistinae), with description of four new species. Zootaxa 3956:151–182

Santos BF, Payne A, Pickett KM, Carpenter KM (2015) Phylogeny and historical biogeography of the paper wasp genus *Polistes* (Hymenoptera: Vespidae): implications for the overwintering hypothesis of social evolution. Cladistics 31:535–549

Saussure HF de (1853–1858) Études sur la familie des Vespides. 2. Monographie des guêpes sociales, ou de la tribu dês Vespiens. Paris: Masson, and Geneva: J. Cherbuliez

Schmid-Egger C, van Achterberg K, Neumeyer R, Morinière J, Schmidt S (2017) Revision of the West Palaearctic *Polistes* Latreille, with the descriptions of two species – an integrative approach using morphology and DNA barcodes (Hymenoptera, Vespidae). Zookeys 713:53–112

Schmitz J, Moritz RFA (1998) Molecular phylogeny of Vespidae (Hymenoptera) and the evolution of sociality in wasps. Mol Phylogenet Evol 9:183–191

Sharkey MJ, Carpenter JM, Vilhelmsen L, Heraty J, Liljeblad J, Dowling APG, Schulmeister S, Murrays D, Deans AR, Ronquist F, Krogmann L, Wheeler WC (2012) Phylogenetic relationships among superfamilies of Hymenoptera. Cladistics 27:1–33

Silveira OT (2008) Phylogeny of wasps of the genus *Mischocyttarus* de Saussure (Hymenoptera, Vespidae, Polistinae). Rev Bras Entomol 52:510–549

Silveira OT (2013) Social wasp species of *Mischocyttarus* (*Phi*) related to *M. alfkenii* (Ducke) and *M. paraguayensis* Zikán (Hymenoptera, Vespidae, Polistinae). Rev Bras Entomol 57:173–196

Silveira OT, Silveira AT (1994) Comparative morphology of skeletal parts of the sting apparatus in neotropical polistine social wasps. Sociobiology 25:295–327

Silveira OT, Silva SS, Felizardo SPS (2015) Notes on social wasps of the group of *Mischocyttarus* (*Omega*) *punctatus* (Ducke), with description of six new species (Hymenoptera, Vespidae, Polistinae). Rev Bras Entomol 59:154–168

Somavilla A (2016) Evidência total das espécies de *Polistes* do Novo Mundo (Hymenoptera: Polistinae): uma abordagem cladística. Instituto Nacional de Pesquisas da Amazônia, Tese de Doutorado

Somavilla A, Oliveira ML, Andena SR, Carpenter JM (2018) An illustrated atlas for male genitalia of the New World *Polistes* Latreille, 1802 (Vespidae: Polistinae). Zootaxa 4504 (3):301-344

Turillazzi S, West-Eberhard MJ (1996) Natural history and evolution of paper wasps. Oxford Univ Press, p 416

van der Vecht J, Carpenter JM (1990) A catalogue of the genera of the Vespidae (Hymenoptera). ZoologischeVerhandelingen, Leiden 260:3–62

Wenzel JW (1990) A social wasp's nest from the cretaceous period, Utah, USA, and its biogeographical significance. Psyche 97:21–29

Wenzel JW (1993) Application of the biogenetic law to behavioral ontogeny: a test using nest architecture in paper wasps. J Evol Biol 6:229–247

Wenzel JW, Carpenter JM (1994) Comparing methods: adaptive traits and tests of adaptation. In: Eggleton P, Vane-Wright RI (eds) Phylogenetics and ecology. Academic Press, London, pp 79–101

West-Eberhard MJ (1978) Polygyny and the evolution of social behavior in wasps. J Kansas Entomol Soc 51:832–856

Weyrauch W (1937) Zur Systematik und Biologie der Kuckuckswespen *Pseudovespa*, *Pseudovespula* und *Pseudopolistes* Zool Jahr Abt Syst Ökol 70:243–290

Wiens JJ, Lingluff SJL (2001) How lizards turn into snakes: a phylogenetic analysis of body-form evolution in anguid lizards. Evolution 55:2303–2318

Wilson EO (1971) The insect societies. Belknap Press, Cambridge, MA, p 548

Zikán JF (1949) O gênero *Mischocyttarus* Saussure (Hymenoptera, Vespidae), com a descrição de 82 espécies novas. Bol Parq Nac Itatiaia 1:1–125

Chapter 16
List of Species of Social Wasps from Brazil

Alexandre Somavilla, Bruno Corrêa Barbosa, Marcos Magalhães de Souza, and Fábio Prezoto

Abstract In this chapter, we list the 381 species of social wasps from Brazil and inform the Brazilian state distribution: *Agelaia* (*n* = 21 species), *Angiopolybia* (*n* = 4), *Apoica* (*n* = 9), *Asteloeca* (*n* = 3), *Brachygastra* (*n* = 13), *Chartergellus* (*n* = 10), *Charterginus* (*n* = 2), *Chartergus* (*n* = 3), *Clypearia* (*n* = 7), *Epipona* (*n* = 2), *Leipomeles* (*n* = 4), *Metapolybia* (*n* = 16), *Mischocyttarus* (*n* = 144), *Nectarinella* (*n* = 2), *Parachartergus* (*n* = 13), *Polistes* (*n* = 43), *Polybia* (*n* = 51), *Protonectarina* (*n* = 1), *Protopolybia* (*n* = 24), *Pseudopolybia* (*n* = 4), and *Synoeca* (*n* = 5).

Keywords Vespidae · Classification · Taxonomic history · Neotropical wasps · Polistinae

In order to update the list of social wasp species and occurrence states in Brazil originally published by Owain Westmacott Richards (1978) entitled *The Social Wasps of the Americas Excluding the Vespinae*, new data were compiled from records of biological species; papers published in journals, books, and course completion papers such as theses, dissertations, and monographs; as well as unpublished data from 1978 to 2019. We also used the Brazilian Fauna Taxonomic Catalog (Hermes et al. 2020).

A. Somavilla (✉)
Instituto Nacional de Pesquisas da Amazônia, Manaus, Amazonas, Brazil
e-mail: alexandresomavilla@gmail.com

B. C. Barbosa · F. Prezoto
Laboratório de Ecologia Comportamental e Bioacústica (LABEC), Programa de Pós-graduação em Biodiversidade e Conservação da Natureza, Universidade Federal de Juiz de Fora, Juiz de Fora, Minas Gerais, Brazil
e-mail: fabio.prezoto@ufjf.edu.br

M. M. de Souza
Instituto Federal de Educação, Ciência e Tecnologia do Sul de Minas Gerais, Inconfidentes, Minas Gerais, Brazil

F. Prezoto et al. (eds.), *Neotropical Social Wasps*,
https://doi.org/10.1007/978-3-030-53510-0_16

In our compilation, 21 genera were identified – *Agelaia* (n = 21), *Angiopolybia* (*n* = 4), *Apoica* (*n* = 9), *Asteloeca* (*n* = 3), *Brachygastra* (*n* = 13), *Chartergellus* (*n* = 10), *Charterginus* (*n* = 2), *Chartergus* (*n* = 3), *Clypearia* (*n* = 7), *Epipona* (*n* = 2), *Leipomeles* (*n* = 4), *Metapolybia* (*n* = 16), *Mischocyttarus* (*n* = 144), *Nectarinella* (*n* = 2), *Parachartergus* (*n* = 13), *Polistes* (*n* = 43), *Polybia* (*n* = 51), *Protonectarina* (*n* = 1), *Protopolybia* (*n* = 24), *Pseudopolybia* (*n* = 4), *Synoeca* (*n* = 5) – a total of 381 social wasps species are found in Brazil, see Table 16.1.

Table 16.1 List of 382 registered social wasp species in Brazil and their states of occurrence

Genus	Subgenus	Species	Author	Year	Occurrence in Brazil
Agelaia		*acreana*	Silveira and Carpenter	1996	AC, AM
Agelaia		*angulata*	(Fabricius)	1804	AC, AP, AM, BA, ES, MA, MT, MG, PA, PR, RJ, RO, RR, SC, SP, RS
Agelaia		*angulicollis*	(Spinola)	1851	AM, MT, PA
Agelaia		*brevistigma*	(Richards)	1978	AM, AP
Agelaia		*cajennensis*	(Fabricius)	1798	AC, AP, AM, BA, MT, PA, RO, SP
Agelaia		*centralis*	(Cameron)	1907	AM, BA, ES, MA, MT, MG, PA, RJ, SP
Agelaia		*constructor*	(de Saussure)	1854	AP, AM, PA
Agelaia		*flavipennis*	(Ducke)	1905	AM, GO, MA, MT, PA, RO
Agelaia		*fulvofasciata*	(DeGeer)	1773	AC, AP, AM, MA, MT, PA, RO, RR
Agelaia		*hamiltoni*	(Richards)	1978	AC, AM, PA, RO
Agelaia		*lobipleura*	(Richards)	1978	AM, MT, RO
Agelaia		*melanopyga*	Cooper	2000	RO

(continued)

Table 16.1 (continued)

Genus	Subgenus	Species	Author	Year	Occurrence in Brazil
Agelaia		*multipicta*	(Haliday)	1836	AP, AM, DF, ES, GO, MT, MG, MS, PA, PR, RJ, RS, RO, RR, SC, SP
Agelaia		*myrmecophila*	(Ducke)	1905	AM, AP, BA, MA, MG, MT, PA, RO
Agelaia		*ornata*	(Ducke)	1905	AM, RR, RO
Agelaia		*pallidiventris*	(Richards)	1978	AM, RO
Agelaia		*pallipes*	(Olivier)	1792	AM, AC, CE, DF, GO, MA, MT, MG, MS, PA, PR, RN, RJ, RO, RS, RR, SP
Agelaia		*pleuralis*	Cooper	2000	AC
Agelaia		*testacea*	(Fabricius)	1804	AC, AP, AM, GO, MA, MT, PA, RO, RR
Agelaia		*timida*	Cooper	2000	AM, RO
Agelaia		*vicina*	(de Saussure)	1854	AL, BA, CE, ES, GO, MA, MT, MG, PR, RS, RJ, SC, SP
Angiopolybia		*obidensis*	(Ducke)	1904	AP, AM, MT, PA
Angiopolybia		*pallens*	(Lepeletier)	1836	AC, AP, AM, BA, ES, MA, MT, PA, PE, RR, SC, SP

(continued)

Table 16.1 (continued)

Genus	Subgenus	Species	Author	Year	Occurrence in Brazil
Angiopolybia		*paraensis*	(Spinola)	1854	AC, AP, AM, BA, MT, MA, PA, RO
Angiopolybia		*zischkai*	Richards	1978	RO
Apoica	*(Apoica)*	*albimacula*	(Fabricius)	1804	AM, PA
Apoica	*(Apoica)*	*ambracarina*	Pickett	2003	AC, MT
Apoica	*(Deuterapoica)*	*arborea*	de Saussure	1854	AC, AM, GO, MA, MT, PA, RO, TO
Apoica	*(Apoica)*	*flavissima*	Van der Vecht	1972	AC, AL, AM, AP, ES, GO, MA, MT, MG, MS, PA, PB, RJ, RO, RR, SP
Apoica	*(Apoica)*	*gelida*	Van der Vecht	1972	AM, DF, GO, MT, MG, PA, PE, RN, RJ, RO, SP
Apoica	*(Apoica)*	*pallens*	(Fabricius)	1804	AP, AM, BA, DF, GO, MT, MG, MA, MS, PA, PR, RO, RS, RJ, SC, SP
Apoica	*(Apoica)*	*pallida*	(Olivier)	1792	AC, AP, AM, BA, CE, GO, MA, MT, PA, RO, RR, SP, TO
Apoica	*(Apoica)*	*strigata*	Richards	1978	AC, AM, MA, MT, PA, TO
Apoica	*(Apoica)*	*thoracica*	duBuysson	1906	AC, AP, AM, DF, ES, GO, MT, MG, PA, PR, RO, RR, SP, TO

(continued)

Table 16.1 (continued)

Genus	Subgenus	Species	Author	Year	Occurrence in Brazil
Asteloeca		*lutea*	Carpenter	2004	AC
Asteloeca		*traili*	(Cameron)	1906	AC, AM, MA, PA
Asteloeca		*ujhelyii*	(Ducke)	1909	AC, AM
Brachygastra		*albula*	Richards	1978	AM, MT, RO
Brachygastra		*augusti*	(de Saussure)	1854	AP, AM, DF, GO, MA, MT, MG, MS, PA, PR, RO, RS, SP
Brachygastra		*bilineolata*	Spinola	1841	AM, AP, GO, MT, PA, RS, RO, SC, SP
Brachygastra		*buyssoni*	(Ducke)	1905	AM, MT
Brachygastra		*cooperi*	(Richards)	1978	RO
Brachygastra		*fistulosa*	Naumann	1968	MS, RJ, SP, SC
Brachygastra		*lecheguana*	(Latreille)	1804	AM, AP, BA, CE, DF, ES, GO, MA, MT, MG, MS, PA, PR, PB, PE, PI, RN, RS, RJ, RO, RR, SC, SP
Brachygastra		*moebiana*	(de Saussure)	1867	AM, DF, GO, MT, MS, RJ, RO, SP, MG
Brachygastra		*mouleae*	Richards	1978	MG, MS, SP
Brachygastra		*myersi*	Bequaert	1943	RJ
Brachygastra		*propodealis*	Bequaert	1943	RO
Brachygastra		*scutellaris*	(Fabricius)	1804	AC, AP, AM, MA, MT, PA, PE, PI, RJ, RO

(continued)

Table 16.1 (continued)

Genus	Subgenus	Species	Author	Year	Occurrence in Brazil
Brachygastra		*smithii*	(de Saussure)	1854	AC, AM, MT, PA, RO, RR, SP
Chartergellus		*amazonicus*	Richards	1978	AM, RO
Chartergellus		*atectus*	Richards	1978	?
Chartergellus		*communis*	Richards	1978	AM, DF, BA, ES, GO, MA, MT, MG, PA, RO
Chartergellus		*flavoscutellatus*	Somavilla	2019	AC
Chartergellus		*jeannei*	Andena and Soleman	2015	AM
Chartergellus		*nigerrimus*	Richards	1978	AC, AM, AP
Chartergellus		*punctatior*	Richards	1978	PA
Chartergellus		*sanctus*	Richards	1978	ES
Chartergellus		*zonatus*	(Spinola)	1851	PA, RO
Chartergellus		*zucchii*	Mateus and Andena	2015	AC
Charterginus		*fulvus*	Fox	1898	AC, AM, MA, MT, PA, RO
Charterginus		*xanthura*	(de Saussure)	1854	AP
Chartergus		*artifex*	(Christ)	1791	AC, AM, GO, MT, PA, RO, RR, SC
Chartergus		*globiventris*	de Saussure	1854	AM, BA, GO, MA, MT, MG, PA
Chartergus		*metanotalis*	Richards	1978	AC, AM, GO, MT, PA
Clypearia		*angustior*	Ducke	1906	BA, MG, PA, RJ, SP
Clypearia		*apicipennis*	(Spinola)	1851	AM, PA
Clypearia		*duckei*	Richards	1978	AC, AM, AP, PA
Clypearia		*humeralis*	Richards	1978	MT
Clypearia		*nigrior*	Richards	1978	AC

(continued)

Table 16.1 (continued)

Genus	Subgenus	Species	Author	Year	Occurrence in Brazil
Clypearia		*sulcata*	(de Saussure)	1854	AC, AM, AP, PA, RO
Clypearia		*weyrauchi*	Richards	1978	AM, PA
Epipona		*media*	Cooper	2002	AM, PA, GO
Epipona		*tatua*	(Cuvier)	1797	AC, AP, AM, DF, ES, GO, MG, MA, MT, PA, RO, SP
Leipomeles		*albogrisea*	(Richards)	1978	MT
Leipomeles		*dorsata*	(Fabricius)	1804	AP, AM, BA, ES, MT, PA
Leipomeles		*pusilla*	(Ducke)	1904	AP, AM, PA
Leipomeles		*spilogastra*	(Cameron)	1912	AP, AM
Metapolybia		*acincta*	Richards	1978	PA, RO
Metapolybia		*alfkenii*	(Ducke)	1904	AM
Metapolybia		*araujoi*	Somavilla and Andena	2018	PA
Metapolybia		*aztecoides*	Richards	1978	MT, PA
Metapolybia		*bromelicola*	Araujo	1945	RJ
Metapolybia		*cingulata*	(Fabricius)	1804	AC, AP, AM, BA, CE, GO, MA, MG, MT, PA, PE, SP
Metapolybia		*decorata*	(Gribodo)	1896	AP, AM, BA, PA, RJ, SC
Metapolybia		*docilis*	Richards	1978	MG, MT, RJ, SP
Metapolybia		*encantata*	Cooper	1999	AM
Metapolybia		*fraudator*	Carpenter and Andena	2019	AC, AM, PA
Metapolybia		*miltoni*	Andena and Carpenter	2012	MA, SP
Metapolybia		*nigra*	Richards	1978	AM, PA
Metapolybia		*rufata*	Richards	1978	AM, AP
Metapolybia		*suffusa*	(Fox)	1898	MA, MT, MS, SP

(continued)

Table 16.1 (continued)

Genus	Subgenus	Species	Author	Year	Occurrence in Brazil
Metapolybia		*unilineata*	(R. von Ihering)	1904	AM, RR, SP
Metapolybia		*servilis*	Cooper	1999	CE
Mischocyttarus	*(Kappa)*	*acreanus*	Silveira	2006	AC
Mischocyttarus	*(Clypeopolybia)*	*adjectus*	Zikán	1935	RJ
Mischocyttarus	*(Kappa)*	*adolphi*	Zikán	1907	AP, AM, PA
Mischocyttarus	*(Omega)*	*alboniger*	Richards	1978	RR
Mischocyttarus	*(Phi)*	*alfkenii*	(Ducke)	1904	AC, AP, AM, PA
Mischocyttarus	*(Phi)*	*alternatus*	Zikán	1949	RJ
Mischocyttarus	*(Omega)*	*annulatus*	Richards	1978	MT
Mischocyttarus	*(Kappa)*	*anthracinus*	Richards	1945	MG
Mischocyttarus	*(Phi)*	*aracatubaensis*	Zikán	1949	SP
Mischocyttarus	*(Phi)*	*araujoi*	Zikán	1949	MG, RJ, SC, SP
Mischocyttarus	*(Mischocyttarus)*	*aripuanaensis*	Silveira	1998	MT
Mischocyttarus	*(Artifex)*	*artifex*	Ducke	1914	AM, AP, MG, PA
Mischocyttarus	*(Omega)*	*asahi*	Borges and Silveira	2019	PA
Mischocyttarus	*(Phi)*	*bahiae*	Richards	1945	MG, BA, PE
Mischocyttarus	*(Omega)*	*bahiaensis*	Zikán	1949	MG, BA, SP
Mischocyttarus	*(Kappa)*	*bertonii*	(Ducke)	1908	MS, RS, SP, PA, MG
Mischocyttarus	*(Phi)*	*brackmanni*	Zikán	1949	RJ
Mischocyttarus	*(Omega)*	*buyssoni*	(Ducke)	1906	MG, RJ
Mischocyttarus	*(Phi)*	*cabauna*	Zikán	1949	RJ
Mischocyttarus	*(Phi)*	*campestris*	Raw	1985	DF
Mischocyttarus	*(Artifex)*	*capichaba*	Zikán	1949	ES
Mischocyttarus	*(Clypeopolybia)*	*carbonarius*	(de Saussure)	1854	AM, ES, PA, MA, RR
Mischocyttarus	*(Monogynoecus)*	*carinulatus*	Zikán	1949	BA, ES, MG, SP
Mischocyttarus	*(Phi)*	*cassununga*	(R. von Ihering)	1903	BA, DF, ES, MG, RJ, RS, SC, SP, PA, PE
Mischocyttarus	*(Phi)*	*catharinaensis*	Zikán	1949	SC
Mischocyttarus	*(Omega)*	*caxiuana*	Silveira	2015	PA

(continued)

Table 16.1 (continued)

Genus	Subgenus	Species	Author	Year	Occurrence in Brazil
Mischocyttarus	*(Phi)*	*cearensis*	Zikán	1945	CE, BA, PB
Mischocyttarus	*(Cerberus)*	*cerberus*	Ducke	1898	BA, DF, MA, MT, MG, MS, PA, GO, SP
Mischocyttarus	*(Phi)*	*chapadae*	(Fox)	1898	MT
Mischocyttarus	*(Omega)*	*claretianus*	Zikán	1949	PR
Mischocyttarus	*(Clypeopolybia)*	*clypeatus*	Zikán	1935	RJ
Mischocyttarus	*(Megacanthopus)*	*collarellus*	Richards	1940	AM, PA
Mischocyttarus	*(Megacanthopus)*	*collaris*	(Ducke)	1904	AM
Mischocyttarus	*(Phi)*	*confirmatus*	Zikán	1935	RJ
Mischocyttarus	*(Omega)*	*confusoides*	Zikán	1949	RJ
Mischocyttarus	*(Omega)*	*confusus*	Zikán	1935	RJ, MG
Mischocyttarus	*(Phi)*	*consimilis*	Zikán	1941	MT, SP, MG
Mischocyttarus	*(Phi)*	*costalimai*	Zikán	1949	RJ
Mischocyttarus	*(Phi)*	*crypticus*	Zikán	1949	RJ
Mischocyttarus	*(Phi)*	*cryptobius*	Zikán	1949	RJ
Mischocyttarus	*(Haplometrobius)*	*curitybanus*	Zikán	1949	RJ, PR
Mischocyttarus	*(Omega)*	*decimus*	Richards	1978	PA
Mischocyttarus	*(Phi)*	*declaratus*	Zikán	1949	RJ
Mischocyttarus	*(Mischocyttarus)*	*drewseni*	de Saussure	1857	AP, AM, BA, DF, ES, GO, MT, MG, MS, PA, PR, RJ, RS, SC, SP
Mischocyttarus	*(Clypeopolybia)*	*duckei*	(du Buysson)	1908	AM, PA
Mischocyttarus	*(Kappa)*	*efferus*	(Silveira)	1898	MA, MG, PA, RJ
Mischocyttarus	*(Phi)*	*extinctus*	Zikán	1935	MG, PR, PA, RJ, SP
Mischocyttarus	*(Kappa)*	*fidus*	Silveira	2006	PA
Mischocyttarus	*(Omega)*	*filiformis*	(de Saussure)	1854	PA
Mischocyttarus	*(Clypeopolybia)*	*flavicans*	(Fabricius)	1804	AM, PA, RO
Mischocyttarus	*(Phi)*	*flavicornis*	Zikán	1935	BA, GO, MT, MA, PA, TO
Mischocyttarus	*(Kappa)*	*flavoscutellatus*	Zikán	1935	MG, RJ, SP

(continued)

Table 16.1 (continued)

Genus	Subgenus	Species	Author	Year	Occurrence in Brazil
Mischocyttarus	*(Phi)*	*fluminensis*	Zikán	1949	MG, RJ
Mischocyttarus	*(Monogynoecus)*	*foveatus*	Richards	1941	AM, AP, MT, PA
Mischocyttarus	*(Kappa)*	*frontalis*	(Fox)	1898	GO, MT, MG, MS
Mischocyttarus	*(Kappa)*	*funerulus*	Zikán	1949	MG, RJ
Mischocyttarus	*(Omega)*	*garbei*	Zikán	1935	ES, RJ
Mischocyttarus	*(Artifex)*	*giffordi*	Raw	1985	DF, MG, BA
Mischocyttarus	*(Phi)*	*gilvus*	Zikán	1949	SP
Mischocyttarus	*(Kappa)*	*gomesi*	Silveira	2013	RO
Mischocyttarus	*(Mischocyttarus)*	*gynandromorphus*	Richards	1945	MT
Mischocyttarus	*(Omega)*	*hoffmanni*	Zikán	1949	RS, SC, SP
Mischocyttarus	*(Omega)*	*ignotus*	Zikán	1949	RS, RJ, SP, MG
Mischocyttarus	*(Haplometrobius)*	*iheringi*	Zikán	1935	RJ, MG
Mischocyttarus	*(Kappa)*	*imitator*	(Ducke)	1792	AP, AM, DF, GO, MA, MT, MG, PA, RO,
Mischocyttarus	*(Phi)*	*infrastrigatus*	Zikán	1949	RJ, SP
Mischocyttarus	*(Kappa)*	*injucundus*	(de Saussure)	1854	AP, AM, BA, MA, PA, RR, SP
Mischocyttarus	*(Monogynoecus)*	*insolitus*	Zikán	1949	AM
Mischocyttarus	*(Artifex)*	*interjectus*	Zikán	1935	RJ
Mischocyttarus	*(Artifex)*	*interruptus*	Richards	1978	AM, RO
Mischocyttarus	*(Phi)*	*itatiayaensis*	Zikán	1935	RJ
Mischocyttarus	*(Omega)*	*kallindusfloren*	Borges and Silveira	2019	PA
Mischocyttarus	*(Mischocyttarus)*	*labiatus*	(Fabricius)	1804	AM, BA, DF, GO, MT, MG, PA, RJ, RR, RO, SP
Mischocyttarus	*(Phi)*	*lanei*	Zikán	1949	BA, SP
Mischocyttarus	*(Kappa)*	*latior*	(Fox)	1898	GO, MT, MG, SP, MS
Mischocyttarus	*(Kappa)*	*latissimus*	Richards	1978	AM
Mischocyttarus	*(Monogynoecus)*	*lecointei*	(Ducke)	1918	AP, AM, PA, RO
Mischocyttarus	*(Megacanthopus)*	*malaris*	Richards	1978	AM

(continued)

Table 16.1 (continued)

Genus	Subgenus	Species	Author	Year	Occurrence in Brazil
Mischocyttarus	*(Phi)*	*mamirauae*	Raw	1998	AM
Mischocyttarus	*(Phi)*	*maracaensis*	Raw	1992	RR
Mischocyttarus	*(Phi)*	*marginatus*	(Fox)	1898	BA, DF, GO, MT, MG, RJ, SP
Mischocyttarus	*(Mischocyttarus)*	*mattogrossoensis*	Zikán	1935	BA, MT, MG, DF, PA, SP
Mischocyttarus	*(Monogynoecus)*	*melanops*	Cooper	1996	PA
Mischocyttarus	*(Omega)*	*melanoxanthus*	Richards	1978	GO, MT
Mischocyttarus	*(Kappa)*	*metathoracicus*	(de Saussure)	1854	AP, AM, MT, PA, RJ, RR
Mischocyttarus	*(Monogynoecus)*	*metoecus*	Richards	1940	AM, PA
Mischocyttarus	*(Phi)*	*mimicus*	Zikán	1949	RJ
Mischocyttarus	*(Artifex)*	*mirificus*	Zikán	1935	MG, RJ
Mischocyttarus	*(Kappa)*	*mirus*	Silveira	2006	MG
Mischocyttarus	*(Kappa)*	*mocsaryi*	(Ducke)	1909	PA
Mischocyttarus	*(Monogynoecus)*	*montei*	Zikán	1949	BA, SP, MG
Mischocyttarus	*(Phi)*	*mourei*	Zikán	1949	PR, SC, SP, MG
Mischocyttarus	*(Phi)*	*mutator*	Zikán	1949	RJ
Mischocyttarus	*(Omega)*	*nazgul*	Borges and Silveira	2019	PA
Mischocyttarus	*(Cerberus)*	*nomurae*	Richards	1978	CE, BA, MG
Mischocyttarus	*(Artifex)*	*oecothrix*	Richards	1940	AM, PA
Mischocyttarus	*(Omega)*	*omicron*	Richards	1978	MT
Mischocyttarus	*(Phi)*	*pallidipes*	Richards	1945	SC
Mischocyttarus	*(Omega)*	*pallidus*	Zikán	1935	AM
Mischocyttarus	*(Phi)*	*paraguayensis*	Zikán	1935	RS, SC, MG
Mischocyttarus	*(Megacanthopus)*	*parallellogrammus*	Zikán	1935	MG, SP
Mischocyttarus	*(Omega)*	*paulistanus*	Zikán	1935	RJ, SP
Mischocyttarus	*(Kappa)*	*pertinax*	Silveira	2006	PA
Mischocyttarus	*(Kappa)*	*picturatus*	(Bequaert)	1898	MT
Mischocyttarus	*(Clypeopolybia)*	*piger*	Richards	1945	?
Mischocyttarus	*(Phi)*	*plaumanni*	Zikán	1949	PR, SC, SP
Mischocyttarus	*(Kappa)*	*porantin*	Silveira	2006	PA
Mischocyttarus	*(Omega)*	*prominulus*	Richards	1941	AM, PA, RR
Mischocyttarus	*(Phi)*	*proximus*	Zikán	1949	SP

(continued)

Table 16.1 (continued)

Genus	Subgenus	Species	Author	Year	Occurrence in Brazil
Mischocyttarus	*(Kappa)*	*pseudomimeticus*	(Schulz)	1904	AM
Mischocyttarus	*(Omega)*	*punctatus*	(Ducke)	1904	MG, MA, PA, RS
Mischocyttarus	*(Clypeopolybia)*	*richardsi*	Zikán	1949	SC, PR
Mischocyttarus	*(Phi)*	*riograndensis*	Richards	1978	RS
Mischocyttarus	*(Omega)*	*rodriguesi*	Silveira	1978	PA
Mischocyttarus	*(Mischocyttarus)*	*rotundicollis*	(Cameron)	2015	AM, AL, BA, DF, ES, GO, MT, MG, PA, PR, RJ, RO, RS, SC, SP
Mischocyttarus	*(Omega)*	*ryani*	Silveira	2015	BA, MG
Mischocyttarus	*(Kappa)*	*santacruzi*	Raw	2000	BA
Mischocyttarus	*(Megacanthopus)*	*saturatus*	Zikán	1949	AM, PA
Mischocyttarus	*(Haplometrobius)*	*saussurei*	Zikán	1949	MG, SP
Mischocyttarus	*(Artifex)*	*schadei*	Zikán	1949	MG
Mischocyttarus	*(Phi)*	*scitulus*	Zikán	1949	RJ
Mischocyttarus	*(Phi)*	*similaris*	Zikán	1949	SP
Mischocyttarus	*(Phi)*	*similatus*	Zikán	1949	RJ
Mischocyttarus	*(Mischocyttarus)*	*smithii*	de Saussure	1853	AP, AM
Mischocyttarus	*(Kappa)*	*socialis*	(de Saussure)	1854	AM, MG, RJ, SP, PR, SC
Mischocyttarus	*(Omega)*	*souzalopesi*	Zikán	1949	SP
Mischocyttarus	*(Omega)*	*surinamensis*	de Saussure	1854	AC, AM, BA, CE, MA, MT, PA, RJ, RO, RR
Mischocyttarus	*(Artifex)*	*sylvestris*	Richards	1945	AM, PA, PB
Mischocyttarus	*(Artifex)*	*synoecus*	(Richards)	1898	AM, PA, AP
Mischocyttarus	*(Kappa)*	*tapuya*	(Schulz)	1905	PA
Mischocyttarus	*(Haplometrobius)*	*tectus*	Cooper	1996	PA
Mischocyttarus	*(Omega)*	*tenuis*	Richards	1945	PA
Mischocyttarus	*(Haplometrobius)*	*tertius*	Richards	1978	MT
Mischocyttarus	*(Kappa)*	*timbira*	Silveira	2006	MA, SP
Mischocyttarus	*(Mischocyttarus)*	*tomentosus*	Zikán	1935	AC, AM, BA, ES, MT, PA, RO
Mischocyttarus	*(Haplometrobius)*	*travassosi*	Zikán	1949	RJ

(continued)

Table 16.1 (continued)

Genus	Subgenus	Species	Author	Year	Occurrence in Brazil
Mischocyttarus	*(Omega)*	*tricolor*	Richards	1945	GO, MT, SP, MG
Mischocyttarus	*(Artifex)*	*undulatus*	Ducke	1905	AP, AM, MT
Mischocyttarus	*(Omega)*	*vaqueroi*	Zikán	1943	PA
Mischocyttarus	*(Omega)*	*verissimoi*	Silveira	2015	AM
Mischocyttarus	*(Phi)*	*wagneri*	(du Buysson)	1908	MG, RJ, SP
Mischocyttarus	*(Clypeopolybia)*	*wygodzinskyi*	Zikán	1949	RJ, SP
Mischocyttarus	*(Artifex)*	*ypiranguensis*	da Fonseca	1926	SP, MG
Mischocyttarus	*(Kappa)*	*zikaninus*	Richards	1918	AM, PA
Nectarinella		*manauara*	Silveira and Santos	2016	AM
Nectarinella		*xavantinensis*	Mateus and Noll	1998	MT
Parachartergus		*amazonensis*	Ducke	1905	AM, PA
Parachartergus		*colobopterus*	(Lichtenstein)	1796	PA, RO
Parachartergus		*fasciipennis*	Ducke	1905	AM
Parachartergus		*flavofasciatus*	(Cameron)	1906	AP, AM, RO
Parachartergus		*fraternus*	(Gribodo)	1892	AM, AP, DF, GO, MA, MT, MG, PA, RO, SP
Parachartergus		*fulgidipennis*	(de Saussure)	1854	AM, AP, PA
Parachartergus		*griseus*	(Fox)	1898	AM, PA
Parachartergus		*lenkoi*	Richards	1978	AM, MT, RO
Parachartergus		*pseudapicalis*	Willink	1959	BA, GO, MT, MG, PE, RJ, RO, SC, SP
Parachartergus		*richardsi*	Willink	1959	AM, PA
Parachartergus		*smithii*	(de Saussure)	1854	AM, AP, GO, MT, MG, MS, PA, RO, SP, MG
Parachartergus		*wagneri*	duBuysson	1904	MG, RJ, SP

(continued)

Table 16.1 (continued)

Genus	Subgenus	Species	Author	Year	Occurrence in Brazil
Parachartergus		*weyrauchi*	Willink	1959	AC, AM, RO
Polistes	*(Aphanilopterus)*	*actaeon*	Haliday	1836	ES, MG, RJ, RS, SC, SP
Polistes	*(Aphanilopterus)*	*angulinus*	Richards	1951	RO
Polistes	*(Aphanilopterus)*	*aterrimus*	de Saussure	1853	?
Polistes	*(Aphanilopterus)*	*atrox*	Richards	1978	PA
Polistes	*(Aphanilopterus)*	*bequaertianus*	Willink	1954	?
Polistes	*(Aphanilopterus)*	*bicolor*	Lepeletier	1836	AC, AM, AP, MT, MG, PA
Polistes	*(Aphanilopterus)*	*biguttatus*	Haliday	1836	RJ, SP, PR, SC, RO, RS
Polistes	*(Aphanilopterus)*	*billardieri*	Fabricius	1804	AM, AP, BA, GO, MT, MG, MS, PA, PB, PE, PR, RN, RR, RS, SP
Polistes	*(Aphanilopterus)*	*binotatus*	de Saussure	1853	RJ
Polistes	*(Aphanilopterus)*	*brevifissus*	Richards	1978	AP, BA, ES, GO, MT, MS, MG, PA, PB, PR, RJ, RS, RR, SP
Polistes	*(Aphanilopterus)*	*canadensis*	(Linnaeus)	1758	AM, AP, BA, CE, GO, MA, MT, MG, MS, PA, PB, PE, PI, RO, RN, RR, SP, RS
Polistes	*(Aphanilopterus)*	*carnifex*	(Fabricius)	1775	AM, BA, ES, GO, MA, MT, MG, PA, PR, PE, RN, RJ, RS, SP

(continued)

Table 16.1 (continued)

Genus	Subgenus	Species	Author	Year	Occurrence in Brazil
Polistes	*(Aphanilopterus)*	*cavapyta*	de Saussure	1853	AM, RS
Polistes	*(Aphanilopterus)*	*cavapytiformis*	Richards	1978	PR, SC, RS, MG
Polistes	*(Aphanilopterus)*	*cinerascens*	de Saussure	1854	AL, BA, CE, ES, GO, MT, MG, PA, PR, RJ, RS, SC, SP
Polistes	*(Aphanilopterus)*	*claripennis*	Ducke	1904	AM, PA
Polistes	*(Aphanilopterus)*	*consobrinus*	de Saussure	1858	PR, RS, SC, SP
Polistes	*(Aphanilopterus)*	*davillae*	Richards	1978	AM, MT, MG
Polistes	*(Aphanilopterus)*	*deceptor*	Schulz	1905	PA, SP
Polistes	*(Aphanilopterus)*	*erythrocephalus*	Latreille	1813	AC, AM, GO, MT, MG, PA, PR, RJ, SP
Polistes	*(Aphanilopterus)*	*ferreri*	de Saussure	1853	BA, MT, MG, MS, PA, PR, RS, SC, SP
Polistes	*(Aphanilopterus)*	*geminatus*	Fox	1898	BA, GO, MT, MG, MS, PA, SP
Polistes	*(Aphanilopterus)*	*goeldii*	Ducke	1904	AM, MG, MT, PA
Polistes	*(Aphanilopterus)*	*infuscatus*	Lepeletier	1836	PA
Polistes	*(Aphanilopterus)*	*lanio*	(Fabricius)	1775	AC, AP, AM, BA, ES, GO, MT, MG, PA, PR, RJ, RS, RO, SC, SP
Polistes	*(Aphanilopterus)*	*major*	Palisot de Beauvois	1818	AM, PA, ES, SP, MT,
Polistes	*(Aphanilopterus)*	*melanosoma*	de Saussure	1853	AM, BA, ES, GO, MG, RJ, RS, SC, SP
Polistes	*(Aphanilopterus)*	*niger*	Brèthes	1903	MT, MG, SP

(continued)

Table 16.1 (continued)

Genus	Subgenus	Species	Author	Year	Occurrence in Brazil
Polistes	*(Aphanilopterus)*	*notatipes*	Richards	1978	RS, SP
Polistes	*(Aphanilopterus)*	*obscurus*	de Saussure	1863	PR
Polistes	*(Aphanilopterus)*	*occipitalis*	Ducke	1904	AP, AM, ES, MA, MG, PA, RJ, RO, SP
Polistes	*(Aphanilopterus)*	*occultus*	Silveira	1994	PA
Polistes	*(Aphanilopterus)*	*pacificus*	Fabricius	1804	AP, AM, BA, ES, MA, MG, MT, PA, RJ, RS, SP
Polistes	*(Aphanilopterus)*	*paraguayensis*	Bertoni	1921	MT
Polistes	*(Aphanilopterus)*	*ridleyi*	Kirby	1890	PE (Fernando de Noronha)
Polistes	*(Aphanilopterus)*	*rufiventris*	Ducke	1904	AM, PA, RO
Polistes	*(Aphanilopterus)*	*satan*	Bequaert	1940	DF, GO, MG, RJ, SP
Polistes	*(Aphanilopterus)*	*simillimus*	Zikán	1948	BA, MG, MS, PB, PR, RJ, RS, SC, SP
Polistes	*(Aphanilopterus)*	*subsericeus*	de Saussure	1854	AP, AM, BA, MT, MG, PA, RJ, SP, MA
Polistes	*(Aphanilopterus)*	*testaceicolor*	(Bequaert)	1798	AC, AP, AM, MA, MT, PA, RO
Polistes	*(Aphanilopterus)*	*thoracicus*	Fox	1898	GO, MT, PA
Polistes	*(Aphanilopterus)*	*torresae*	Silveira	1994	PA, RO
Polistes	*(Aphanilopterus)*	*versicolor*	(Olivier)	1792	AC, AP, AM, BA, ES, GO, MA, MT, MS, MG, PA, PR, RN, RJ, RS, RO, RR, SC, SP

(continued)

Table 16.1 (continued)

Genus	Subgenus	Species	Author	Year	Occurrence in Brazil
Polybia	*(Trichinothorax)*	*affinis*	DuBuysson	1908	AP, AM, MT, PA, RO
Polybia	*(Trichinothorax)*	*batesi*	Richards	1978	AM
Polybia	*(Myrapetra)*	*belemensis*	Richards	1970	AM, AP, PA, RO
Polybia	*(Myrapetra)*	*bicyttarella*	Richards	1951	AM, AP, PA, SP
Polybia	*(Alpha)*	*bifasciata*	de Saussure	1854	AC, AL, AM, BA, ES, GO, MT, MG, PA, PR, RJ, RO, SP
Polybia	*(Myrapetra)*	*bistriata*	(Fabricius)	1804	AC, AL, AP, AM, BA, ES, MA, MT, MG, PA, RO, SP
Polybia	*(Pedothoeca)*	*brunnea*	(Curtis)	1844	MG, PA
Polybia	*(Myrapetra)*	*catillifex*	Möbius	1856	AM, BA, RJ, PA, RO, SP
Polybia	*(Trichinothorax)*	*chrysothorax*	(Lichtenstein)	1796	AM, BA, CE, ES, GO, MA, MT, MG, PA, PR, RJ, RN, SP
Polybia	*(Synoecoides)*	*depressa*	(Ducke)	1905	AM, BA, MA, PI, RO
Polybia	*(Myrapetra)*	*diguetana*	Buysson	1905	AM, RO
Polybia	*(Cylindroeca)*	*dimidiata*	(Olivier)	1792	AP, AM, BA, DF, ES, GO, MA, MT, MG, PA, RJ, RS, RO, RR, SP
Polybia	*(Myrapetra)*	*dimorpha*	Richards	1978	AM, PA, RR
Polybia	*(Platypolybia)*	*dubitata*	Ducke	1910	AM

(continued)

Table 16.1 (continued)

Genus	Subgenus	Species	Author	Year	Occurrence in Brazil
Polybia	*(Trichinothorax)*	*eberhardae*	Cooper	1993	AM, MT, RO
Polybia	*(Pedothoeca)*	*emaciata*	Lucas	1879	AC, AP, AM, DF, GO, MT, MG, PA, RJ, RO, SP
Polybia	*(Myrapetra)*	*erythrothorax*	Richards	1978	GO, MG, MT, PA, RJ, SP
Polybia	*(Myrapetra)*	*fastidiosuscula*	de Saussure	1854	AM, BA, DF, ES, GO, MA, MT, MG, PA, RS, RJ, SC, SP
Polybia	*(Myrapetra)*	*flavifrons*	Smith	1857	BA, DF, ES, GO, MA, MT, MG, PA, RS, RJ, SC, SP
Polybia	*(Trichinothorax)*	*flavitincta*	Fox	1898	AM, BA
Polybia	*(Furnariana)*	*furnaria*	R. von Ihering	1904	AC, AM, PA, RO
Polybia	*(Trichinothorax)*	*gorytoides*	Fox	1898	AM, AP, MT, PA, RO
Polybia	*(Trichinothorax)*	*ignobilis*	(Haliday)	1836	AM, BA, CE, DF, ES, GO, MA, MT, MG, MS, PA, PB, PI, PR, PE, RN, RS, RJ, RO, RR, SC, SP

(continued)

Table 16.1 (continued)

Genus	Subgenus	Species	Author	Year	Occurrence in Brazil
Polybia	*(Platypolybia)*	*incerta*	Ducke	1907	AM, PA
Polybia	*(Apopolybia)*	*jurinei*	de Saussure	1854	AC, AP, AM, BA, CE, DF, ES, GO, MA, MT, MG, MS, PA, RJ, RO, SP
Polybia	*(Myrapetra)*	*juruana*	R. von Ihering	1904	AC, AM, MT
Polybia	*(Polybia)*	*liliacea*	(Fabricius)	1804	AC, AP, AM, GO, MA, MG, MT, MS, PA, RO, RR, SP
Polybia	*(Trichinothorax)*	*lugubris*	de Saussure	1854	ES, MG, PR, RS, RJ, SC, SP
Polybia	*(Trichinothorax)*	*micans*	Ducke	1904	AP, AM, BA, MA, MT, PA, RO
Polybia	*(Trichinothorax)*	*minarum*	Ducke	1906	BA, MG, PA, PR, RS, RJ, SC, SP
Polybia	*(Myrapetra)*	*occidentalis*	(Olivier)	1792	AM, AP, BA, CE, DF, ES, GO, MA, MT, MG, MS, PA, PB, PI, PR, PE, RN, RS, RJ, RO, RR, SC, SP

(continued)

Table 16.1 (continued)

Genus	Subgenus	Species	Author	Year	Occurrence in Brazil
Polybia	*(Myrapetra)*	*parvulina*	Richards	1970	AP, AM, PA, RO
Polybia	*(Myrapetra)*	*paulista*	H. von Ihering	1896	BA, DF, GO, MT, MG, MS, PA, PI, RJ, SP
Polybia	*(Myrapetra)*	*platycephala*	Richards	1978	AC, AP, AM, BA, GO, MT, MG, PA, RJ, RO, RS, SP
Polybia	*(Platypolybia)*	*procellosa*	Ducke	1907	AM, BA, PA, RO
Polybia	*(Trichinothorax)*	*punctata*	DuBuysson	1908	BA, ES, MG, PR, RS, RJ, SC, SP
Polybia	*(Alpha)*	*quadricincta*	de Saussure	1854	AC, AP, AM, MT, MG, PA, RJ, RO, SP
Polybia	*(Formicicola)*	*rejecta*	(Fabricius)	1798	AC, AP, AM, BA, ES, GO, MA, MT, MG, PA, PE, RN, RJ, RO, RR, SP
Polybia	*(Myrapetra)*	*roraimae*	Raw	1998	RR
Polybia	*(Myrapetra)*	*ruficeps*	Schrottky	1902	BA, CE, DF, GO, MA, MT, MG, MS, PI, RN, SC
Polybia	*(Trichinothorax)*	*rufitarsis*	Ducke	1904	AP, AM, PA, RO
Polybia	*(Myrapetra)*	*scrobalis*	Richards	1970	AP, AC, AM, DF, MA, MT, PA, RO
Polybia	*(Myrapetra)*	*scutellaris*	(White)	1841	BA, DF, ES, GO, MT, MG, PA, RS, RJ, SC, SP

(continued)

Table 16.1 (continued)

Genus	Subgenus	Species	Author	Year	Occurrence in Brazil
Polybia	*(Trichinothorax)*	*sericea*	(Olivier)	1792	AP, AM, BA, CE, DF, ES, GO, MA, MT, MG, MS, PA, PB, PI, PR, RN, RS, RJ, RO, RR, SC, SP
Polybia	*(Alpha)*	*signata*	Ducke	1910	AP, AM, BA, ES, MG
Polybia	*(Apopolybia)*	*similima*	Smith	1862	AC
Polybia	*(Pedothoeca)*	*singularis*	Ducke	1905	AM, AP, GO, AM, MT, PA, RO
Polybia	*(Pedothoeca)*	*spinifex*	Richards	1978	RO
Polybia	*(Polybia)*	*striata*	(Fabricius)	1787	AC, AM, AP, DF, ES, GO, MA, MG, MT, MS, PA, RJ, RO, SP, TO
Polybia	*(Trichinothorax)*	*tinctipennis*	Fox	1898	AM, ES, GO, MT, MG, PA, RJ, RO, SP
Polybia	*(Trichinothorax)*	*velutina*	Ducke	1905	AC, AM, BA
Protonectarina		*sylveirae*	(de Saussure)	1854	BA, CE, DF, ES, GO, MT, MG, MS, PR, PI, RS, RJ, SC, SP
Protopolybia		*acutiscutis*	(Cameron)	1906	AM, MA, MT, PA, RO
Protopolybia		*aliciae*	Silveira	2004	PA
Protopolybia		*alvarengai*	Richards	1978	RO
Protopolybia		*amarella*	Bequaert	1944	AC, AM, PA

(continued)

Table 16.1 (continued)

Genus	Subgenus	Species	Author	Year	Occurrence in Brazil
Protopolybia		*bella*	(R. von Ihering)	1903	AP, AM
Protopolybia		*bituberculata*	Silveira and Carpenter	1996	AM, MA, PA
Protopolybia		*chanchamayensis*	Bequart	1944	AC
Protopolybia		*chartergoides*	(Gribodo)	1892	AC, AP, AM, MA, MT, PA, RO, SP
Protopolybia		*clypeata*	Santos, Silveira, and Carpenter	2015	AC
Protopolybia		*diligens*	(Smith)	1857	AM, MA, PA
Protopolybia		*djaneteae*	Santos and Silveira	2020	AM
Protopolybia		*duckei*	(du Buysson)	1905	BA
Protopolybia		*duckeianus*	(Richards)	1978	AP, AM
Protopolybia		*eldinaris*	Santos and Silveira	2020	MT, PA
Protopolybia		*emortualis*	(de Saussure)	1855	AC, AM, PA
Protopolybia		*exigua*	(de Saussure)	1854	AC, AM, BA, CE, DF, ES, GO, MA, MT, MG, MS, PA, PB, PE, PI, RJ, RO, RR, SC, SP
Protopolybia		*fuscatus*	(Fox)	1898	AM, PA
Protopolybia		*holoxantha*	(Ducke)	1904	AP, AM, PA
Protopolybia		*lidiae*	Santos and Silveira	2017	PA
Protopolybia		*minutissima*	(Spinola)	1851	AM, MG; MT, PA, RO
Protopolybia		*nitida*	(Ducke)	1904	AC, AP, AM, PA
Protopolybia		*potiguara*	Santos, Silveira, and Carpenter	2015	PB
Protopolybia		*rotundata*	(Ducke)	1910	AM, MT

(continued)

Table 16.1 (continued)

Genus	Subgenus	Species	Author	Year	Occurrence in Brazil
Protopolybia		*sedula*	(de Saussure)	1854	AC, AM, AP, BA, CE, DF, ES, GO, MA, MT, MG, MS, PA, PR, RJ, SC, SP
Pseudopolybia		*compressa*	(de Saussure)	1854	AP, AM, BA, ES, GO, MT, PA, RJ, RO
Pseudopolybia		*difficilis*	(Ducke)	1905	AP, AM, MT, PA, RO
Pseudopolybia		*langi*	Bequaert	1944	AP, AM
Pseudopolybia		*vespiceps*	(de Saussure)	1863	AP, AM, BA, DF, ES, GO, MA, MT, MG, MS, PA, PE, RJ, RO, RR, SC, SP
Synoeca		*chalibea*	de Saussure	1852	AC, AM, PA, MT, RO
Synoeca		*cyanea*	(Fabricius)	1775	BA, DF, ES, MG, MS, PA, PE, PR, RS, RJ, SC, SP
Synoeca		*ilheensis*	Lopes and Menezes	2017	BA, ES, PE, SP
Synoeca		*surinama*	(Linnaeus)	1767	AM, AP, BA, DF, GO, MA, MT, MG, MS, PA, PB, PE, RJ, RO, RR, TO, SP

(continued)

Table 16.1 (continued)

Genus	Subgenus	Species	Author	Year	Occurrence in Brazil
Synoeca		*virginea*	(Fabricius)	1804	AC, AM, AP, MA, MT, PA, PI, RO, RR

AC Acre, *AL* Alagoas, *AP* Amapá, *AM* Amazonas, *BA* Bahia, *CE* Ceará, *DF* Distrito Federal, *ES* Espírito Santo, *GO* Goiás, *MA* Maranhão, *MT* Mato Grosso, *MS* Mato Grosso do Sul, *MG* Minas Gerais, *PA* Pará, *PB* Paraíba, *PR* Paraná, *PE* Pernambuco, *PI* Piauí, *RJ* Rio de Janeiro, *RN* Rio Grande do Norte, *RS* Rio Grande do Sul, *RO* Rondônia, *RR* Roraima, *SC* Santa Catarina, *SP* São Paulo, *SE* Sergipe, *TO* Tocantins

References

Hermes MG, Somavilla A, Andena SR (2020) Catálogo Taxonômico da Fauna do Brasil: Vespidae In: https://fauna.jbrj.gov.br/

Richards OW (1978) The social wasps of the Americas (excluding the Vespinae). British Museum of Natural History, London, p 580

Chapter 17
The Old World Versus New World Social Wasps: Similarities, Differences, and Threats

Stephen J. Martin

Abstract The Vespinae subfamily is compromised of the largest (in size) and most studied groups of social wasps in the world. These include the large hornets (*Vespa*), the widespread yellow-jackets (*Vespula* and *Dolichovespula*), and three species of tropical nocturnal wasps (*Provespa*). The Vespinae evolved out of the mountainous regions of Central China and spread throughout the Northern Hemisphere. The *Vespa* (and *Provespa*) spread throughout the tropical regions of Asia; however, they are not naturally present in the Neotropics; instead, this area is dominated by the much lesser studied Epiponini, a highly diverse family of swarm-founding social wasps that belong to the other major branch of social wasps, the Polistinae. The life cycle of Vespinae (Old World) wasps will be presented and compared against a Neotropical (New World) species of social wasp. The diverse nest structures and various traits including polygyny and swarm-founding found in Neotropical social wasps will no doubt challenge the traditional theories of the evolution of sociality, which is currently largely based on the single queen, independently founding species. Furthermore, some members of the Vespinae have become very successful in establishing viable and expanding populations throughout the world, well beyond their natural range, which will also be discussed.

Keywords *Vespa* · *Provespa* · *Vespula* · *Dolichovespula* · Hornets · Yellow-jackets · Wasp invasions

17.1 Introduction

There are over 1000 described species of social wasps (Vespidae) which fall into two subfamilies: the Vespinae and Polistinae. Generally, the Vespinae dominate the temperate regions, while the Polistinae dominate the tropical and subtropical regions of the world. Although all species of the Vespidae are social, in that they all build

S. J. Martin (✉)
School of Environment and Life Sciences, The University of Salford, Manchester, UK
e-mail: s.j.martin@salford.ac.uk

F. Prezoto et al. (eds.), *Neotropical Social Wasps*,
https://doi.org/10.1007/978-3-030-53510-0_17

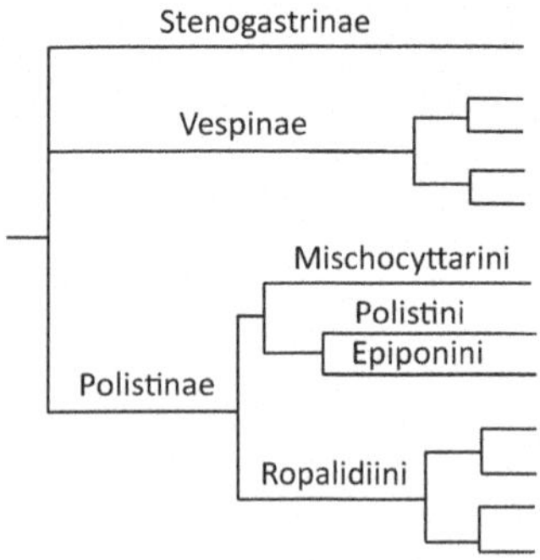

	No. of Species	No. of papers	Region	Nest foundation
7 genera	58	95	O (Tropics)	Ind
Vespa	23	986	O (Temperate/Tropics)	Ind
Provespa	3	6	O (tropics)	Swarm
Vespula	19	778	O/N (Temperate)	Ind
Dolichovespula	14	120	O /N (Temperate)	Ind
Mischocyttarus	200+	143	N(tropics)	Ind
Polistes	300+	1525	N/O (sub/tropics)	Ind
19 genera	234	157	N (tropics)	Swarm
Belonogaster	65	35	O (sub/tropics)	Ind
Polybioides	6	14	O (tropics)	Swarm
Ropalidia	112	194	O (sub/tropics)	Ind
Parapolybia	5	13.	O (tropics)	Ind

Fig. 17.1 The different families of social wasps, including the pre-social Stenogastrinae, indicating the approximate number of genera/species along with the number of published papers on Web of Science, the region the wasps occur (O, Old World; N, New World), and mode of nest foundation (Ind, independent; Swarm, swarming)

complex nests and have communal brood care and overlapping generations, two distinctly different ways of colony foundation have evolved. The more basal method is the independent foundation of colonies, where one or a small number of queens establish a colony after a period of hibernation. The more recently evolved swarm-founding method is where the reproductive queens together with a number of workers "bud off" from the original colony to start a new one. This behavior is restricted to the tropical regions both in the Old World (*Polybioides*) and the New World (Epiponini).

Despite their often conspicuous nature, global distribution, and large diversity of nest types, behaviors, and communication systems, there is a strong publication bias among social wasps toward Vespinae species (Fig. 17.1), which consist of a small number of genera that are species poor when compared to the Polistinae. This bias is further compounded by the fact that even within the underrepresented Polistinae, the vast majority of the 1525 papers study only a small number of the 300+ species of *Polistes*. Again, in the extremely diverse Epiponini, which contains more genera than all other genera of social wasps combined, the majority of the 157 papers all study members of a single genus, the *Polybia*. Although the independent-founding species outnumber the swarm founders by roughly three to one, the swarm-founding species are the more ecologically dominant group in the tropics, especially in the Neotropics, but remain the most understudied but potentially fascinating group of social wasps (Hunt 2007).

17.2 Hornets and Yellow-Jackets (Old World Social Wasps)

Worldwide there are around 22 species of hornets (*Vespa*), 23 species of *Vespula*, and 19 species of *Dolichovespula*, with the latter two groups often called yellow-jackets (Archer 2012). Both the yellow-jackets and hornets evolved out of the

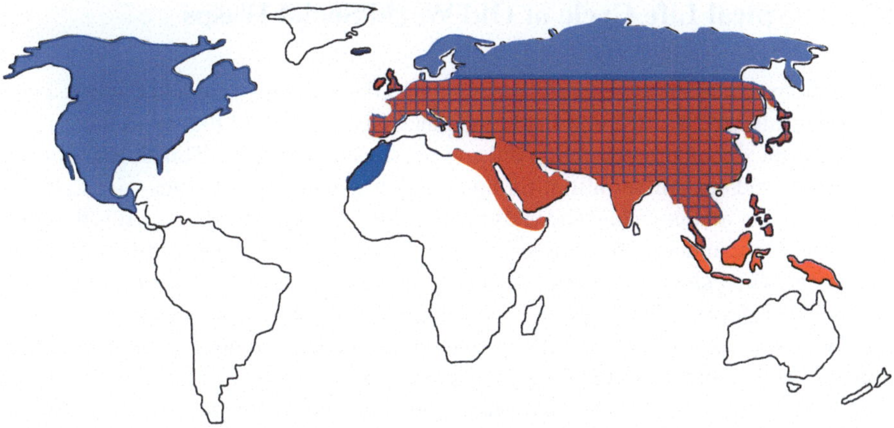

Fig. 17.2 Natural world distribution of Vespinae, with the hornets (*Vespa*) in red, the yellow-jackets (*Vespula* and *Dolichovespula*) in blue, and the hatched area indicating where both groups occur. The *Provespa* (not shown) are restricted to the Oriental regions, i.e., from India (East Himalayas), Southern China, to Vietnam and via the Malaysian Peninsula to Sumatra, West Java, and Borneo

mountains of Central China and so are well adapted to temperate environments, i.e., those that have a strong seasonality, which is reflected in their global distribution (Fig. 17.2). As you travel west from China, the number of hornet species drops from 11 to only 1 species that naturally reached Northern Europe. Unlike the larger hornets, the smaller yellow-jackets continued spreading west reaching the New World, since only the smaller yellow-jackets were able to cross the Bering Sea land bridge connecting Europe to the Americas up to 15,000 years ago. Thus, there are no true hornets (*Vespa*) in the New World. The lack of insect prey during the winter months forces all yellow-jacket and hornet queens to undergo a period of hibernation prior to nest initiation in the following spring by a mated queen, and so all species have an annual colony life cycle (see below). This cycle is even maintained in tropical regions, although colonies here can be started at any time of the year. Furthermore, the reproductive queen and nonreproductive workers are often strikingly divergent morphological between the castes, again a trait attributed to winter survival. In tropical climates, yellow-jackets are probably outcompeted by the very abundant Polistinae paper wasps since they both are generalist insect predators, whereas the larger hornets have more successfully expanded into tropical regions of Asia and have even evolved to exploit the nocturnal niche in the form of the three *Provespa* species. The *Provespa* is the only group of swarm-founding species within the Vespinae which are restricted to the tropical regions of Asia (Matsuura and Yamane 1984). Otherwise, their nest structure is similar to that of all other Vespinae.

17.3 Typical Life Cycle of Old World Social Wasps

Each colony is started by a single queen, mated the previous year and having spent the winter hibernating in a dry, enclosed space, such as an old log. With the arrival of spring, the queen emerges and feeds on nectar and tree resin. This helps activate her ovaries and sustain her initial foraging and nest-building activities, since all the fat stores in her abdomen have been used up. Prior to nest establishment, it is thought that queens can undergo migrations of up to 100 km, although very little data actually exists. Then she establishes her embryo nest that consists of a vertical petiole (stalk) which forms a single point of attachment of the nest to a surface. The lower end forms the initial comb, and the queen applies a chemical ant repellent from her Van der Vetch glands onto the stalk, which protects the brood from ants when she is away foraging for food and wood pulp. The entire structure is surrounded by an envelope that initially consists of a single layer of paper (wood pulp and saliva), to which more layers are added later. During the development of the embryo nest, the queen builds around one new brood cell each day. At night, the queen is often found curled around the petiole on top of the comb in order to speed up the development of her brood by using her body to produce heat. It takes several weeks to rear the first batch of adult workers (Fig. 17.3). These first few workers are also the smallest adults produced, as it's more efficient. With the appearance of the first workers, the queen never leaves the safety of her nest. Now both the physical size of the colony and its worker population increase. Each additional comb maintains a regular round pattern, since new cells are constructed around the edge of the existing comb. The comb construction pattern is mirrored in the brood pattern with the eggs, larvae, and sealed brood, appearing in a series of concentric circles (Fig. 17.3). This circular pattern persists as cells can be reused up to three times in the early combs. As the colony grows rapidly, the internal nest temperature starts to stabilize out at a steady 30 °C, which along with an increased food supply helps speed up the development of the brood. The length of time to develop from an egg to an adult worker in a medium-sized hornet (e.g., *Vespa velutina*) drops from around 50 days in the embryo nest to just 29 days in a mature colony. As the colony reaches its maximum size, the ratio of larvae in the colony to adult workers drops below 2, i.e., two larvae for every worker. This point is associated with the start of sexual production in hornets and yellow-jackets, so the colony enters the reproductive phase.

As the production of male and then queen brood increases, a corresponding decline in worker production occurs. The male and queen destine eggs are laid in some of largest cells built, usually found on the lower combs. After they emerge, the new queens remain in the nest for 1–2 weeks, during which their weight increases by 40% due to the fat stores in the abdomen. They leave the nest and mate away from it, but little is known about their mating behavior. Mating within colony has been seen in the Japanese giant hornet (*Vespa mandarinia*), where males from other colonies gather at the nest entrance and wait for the new queens to emerge, although this behavior appears unique to this species. The main reproductive strategy of the Vespinae is to produce a large number of queens, since only a very small fraction

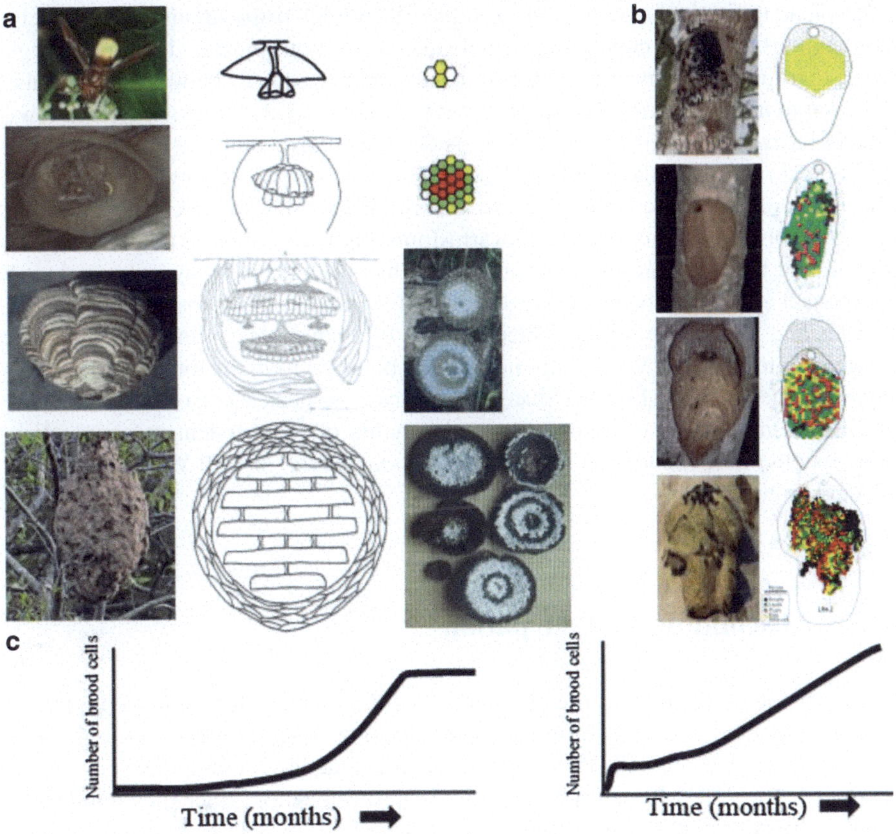

Fig. 17.3 Comparison of colony cycles in (**a**) an independently founding species such as the hornet (*Vespa velutina/simillima*) and (**b**) a swarm-founding species such as the warrior wasp *Synoeca septentrionalis*; (**c**) compares the different patterns of colony growth of the two types of nest founding, with a long period of very slow growth typical of independent-founding colonies, since a single individual (queen) is required to do all the jobs to raise the first batch of brood (which is followed by a rapid period of growth), whereas in the swarm-founding species, the colony grows rapidly since a large number of cells are built within days, so avoiding any time lag

($\ll$1%) of overwintering mated queens will establish a new colony the next year. The entire colony cycle takes between 5 and 10 months depending on the size of the colony and species (Edwards 1980; Spradbery 1973).

17.4 Polygyny in the Vespinae

Among the Polistines, polygyny (presence of several reproductive adults) is a common trait in members of the *Polybioides*, *Ropalidia*, and *Epiponini*. Although polygyny is extremely rare in Vespinae living in temperate environments, it is also found in Vespinae in tropical regions. For example, the hornet *Vespa affinis* is exclusively monogynous (colonies containing a single queen) in subtropical regions, but polygynous in tropical regions, where no dominance behavior occurs between the queens, and even in mature colonies all queens had fully developed ovaries. Where yellow-jackets have established populations outside their native range or expanded into subtropical regions in the Southern United States and Mexico, polygyny and even perennial colonies have been observed. This indicates that the cyclical oligogyny that has evolved in some species of swarm-founding wasps appears not to be occurring among the Vespinae. As in swarm-founding species, polygyny in hornets and yellow-jackets offers a better survival strategy for small nests against predators such as birds and ants in the tropics, since the loss of a single queen in a monogynous colony would be fatal. Monogamy however is believed to maintain kin relatedness among the nestmates, which is reduced in polygynous species; one of the best model study systems investigating the costs and benefits of independent or cooperative (polygynous) nest foundation is in *Ropalidia marginata*, an Old World Polistinae (Gadagkar 2001).

17.5 Absconding vs. Usurpation

In swarm-founding species, the absconding of the entire colony due to predation pressure from animals (e.g., birds) and changes in the local environment (e.g., lack of food) is not uncommon. However, in the Vespinae absconding has very rarely been recorded. A few hornet species undergo nest relocation, since they build their small vulnerable embryo nest within an enclosed space only to later establish a new colony in an open area. However, unlike absconding, both nests are maintained over a period of weeks until all the emerging workers and eventually the queen have moved to the new colony. If an embryo nest is destroyed or predated, the queen does not rebuild a new nest but instead tries to invade another nest and attempt a takeover by killing the resident queen, even when a few workers may be present. This is known as "queen usurpation" and is a common but unseen behavior that may be a very important population control behavior in hornets and yellow-jackets. Fights between the queens often lead to injuries, and even the victor may be damaged sufficiently to either cause the colony to subsequently fail or become smaller than usual. It is possible that any late emerging queens may never attempt to establish their own colony and just attack other ones, which could lead to a population of late emerging "parasitic" queens, as found in the social bumblebees (*Bombus*).

17.6 Invasive Yellow-Jacket and Hornet Species

The accidental introduction of social wasps to areas outside their natural range has become all too frequent. Currently 25 species of social wasps have been introduced around the world, with 16 species belonging to the Polistinae and 9 to the Vespinae (Table 17.1). The four most invasive species are all Vespinae and include two yellow-jackets (*Vespula vulgaris* and *V. germanica*) and two hornets (*Vespa crabro* and *Vespa velutina*), which have all become particularly widespread and abundant with a range of impacts on biodiversity and ecosystem function. Hornets prey on other smaller species of social insects, with *Vespa velutina* having a big impact on beekeepers managing honeybees (*Apis mellifera*) in Europe (Martin 2018).

Table 17.1 List of social wasps known to be accidentally introduced into foreign countries and whether the introduction failed or became established

	Failed introduction	Successful introduction
Vespinae		
Oriental hornet *Vespa orientalis*	Madagascar, Mexico, Fujian Province of China, Belgium, UK	
Yellow hornet *Vespa simillima*	British Columbia, Canada	
Vespa affinis	California New Zealand	
Giant hornet *Vespa mandarinia*		Canada (Vancouver) & USA (Bellingham)
Vespa crabro		USA Guatemala
Yellow-legged hornet *Vespa velutina*		France, Spain, Portugal, Belgium, UK, Germany, South Korea, Japan
Common yellow-jacket *Vespula vulgaris*		South America, Iceland, St. Helen, New Zealand
German yellow-jacket *Vespula germanica*		North and South America, South Africa, New Zealand, Ascension Is., Madeira, Canary Is., Iceland
Vespula pensylvanica		Hawaii
Polistinae		
Mischocyttarus flavitarsis idahoensis *Ropalidia marginata*		Hawaii
Polistes (15 species)		Pacific Islands, Ascension Is., Australia, New Zealand, California.

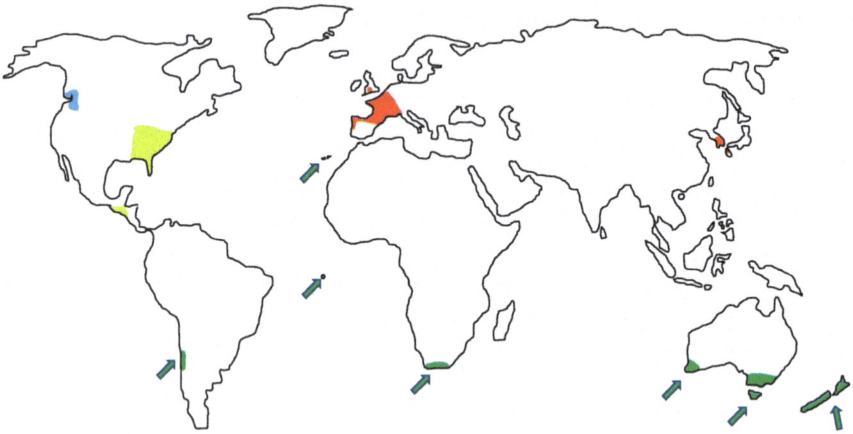

Fig. 17.4 Accidental introductions of the hornets *Vespa mandarinia* (blue), *Vespa velutina* (red), *Vespa crabro* (yellow), and three *Vespula* species (green) that successfully established viable thriving populations (see Table 17.1 for details)

Every new invasive population is always started by the accidental movement of a mated "hibernating" queen via timber or horticultural pots or within building materials, to name a few known examples. The ever-increasing global movement of goods, commodities, and people has aided the introduction of wasps into many new countries around the world (Fig. 17.4). Most introductions fail to establish a new viable population, and these normally go unnoticed, but a small proportion of hornet invasion does get recorded due their large size (Table 17.1).

Despite several accidental introductions of hornets into various countries, only *Vespa velutina* and *Vespa crabro* populations persisted, both following a similar pattern. For example, a single introduction of *Vespa crabro* into North America via New York occurred around 1840. Likewise, a single or very small number of *Vespa velutina* queens were introduced into France in 2003/2004 and South Korea in 2003, probably from the same location in Eastern China. In the United States, over the next 150 years, the distribution of *Vespa crabro* spread west to the Dakotas, south to Florida and Louisiana, and more recently north into the Canadian states of Ontario and Quebec and then to Guatemala, whereas *Vespa velutina* spread through South Korea at 30 km per year and 100 km per year in Europe. From these two populations, large jumps followed to other nearby countries, e.g., Japan from Korea as well as Portugal and the United Kingdom from France. The establishment of *Vespa mandarinia* into North America (Vancouver area) is worrying due to this species infamous ability to conduct mass attacks on honeybee colonies (Matsuura and Yamane 1984). Although the introduction is only a few years old colonies already exist over 160 km apart.

Unlike the hornets, the yellow-jackets have established major populations in several oceanic islands including Hawaii, Ascension Island, and the Canaries, as well as in the Southern Hemisphere (New Zealand, Australia, Tasmania, Chile, and South Africa) (Fig. 17.4). In all cases, these populations became well established and damaging to the local biodiversity, especially on islands where the yellow-jackets outcompete many local species, even birds (honeydew specialists) in New Zealand. This is because social insects (ants, termites, wasps, and bees) are among some of the most successful invasive groups on our planet owing to their high reproductive rates, high dispersal abilities, general diet, behavioral flexibility, and superior competitive abilities.

The unique combination of traits that allow hornets and yellow-jackets to get established outside their range appears to depend on several factors: (1) the ability to establish an entire colony from a single mated female, which swarm-founding species normally cannot do, (2) a hibernation period during which time accidental transportation can occur, and (3) polyandry, which is the behavior of mating with several males.

The vast majority of social insect species (ants, termites, bees, and wasps) just mate with a single male (monogynous). For example, the average mating frequency for most hornets is just above 1, since most queens mate with a single male (a low number mate with two or even three males, but these are rare events). However, there are exceptions to this general rule, such as *Vespa velutina*, *Vespula vulgaris*, *Vespula germanica*, and *Vespula pensylvanica*, which all become major pests outside their native range. All these species have an average mating frequency of 2 or above, with some queens being mated by up to seven males. This helps limit the genetic bottleneck all invasive species go through, since the greater number of fathers increases the genetic diversity of the offspring within the colony. This may be very important during the early stages of establishment when the mating pool is very limited. In the case of *Vespa crabro* getting established in the United States, a species which normally mates with a single male, mating with up to three males has been detected. Therefore, maybe only promiscuous queens have any chance establishing a new viable population.

17.7 Summary

The Vespinae are the most frequently studied of all social wasps, due to the smallish number of species, their accessibility, and pest status in many countries, yet their nests and life cycles are all quite similar. However, the diverse nest structures and traits, such as caste flexibility and polygyny, found in the Neotropical social wasps when studied will no doubt challenge the traditional theories of the evolution of sociality, which are currently largely based on the single queen, independently founding species.

References

Archer ME (2012) Vespine wasps of the world; behaviour, ecology & taxonomy of the vespinae. Siri Scientific Press, Manchester

Edwards R (1980) Social wasps: their biology & control. The Rentokil library. Rentokil, East Grinstead

Gadagkar R (2001) The social biology of Ropalidia marginata. Harvard University Press, Cambridge. ISBN 978-0-674-00611-9

Hunt JH (2007) The evolution of social wasps. Oxford University Press, Oxford

Martin SJ (2018) The Asian hornet (*Vespa velutina*) threats, biology & expansion. Northern Bee Books, Bristol. ISBN: 978-0-86098-281-4

Matsuura M, Yamane S (1984) Biology of the vespine wasps. Springer-Verlag, Berlin

Spradbery JP (1973) Wasps: an account of the biology and natural history of solitary and social wasps. Sidgwick & Jackson, London. ISBN 0-906527-15-5

Chapter 18
Key to the Genera of Social Wasps (Polistinae) Occurring in Neotropics

Alexandre Somavilla and James M. Carpenter

Abstract Polistinae are 25 genera and more than 1,000 described species, and in Brazil 21 genera and nearly 372 species have been recorded. Here, we propose an identification key to the genera of social wasps (Polistinae) occurring in the Neotropics.

Keywords Vespidae · Polistinae · Identification key · Neotropical wasps · Classification

Polistines occur throughout the world but are most diverse in tropical regions, especially the Neotropics. Worldwide there are 25 genera and more than 1,000 species, and in Brazil 21 genera and nearly 380 species have been recorded (Hermes et al. 2020). The Neotropical fauna belongs to three tribes (Carpenter 1993): Polistini (*Polistes*), Mischocyttarini (*Mischocyttarus*), and Epiponini (19 genera).

For the elaboration of this genera key, we used only those species with records of occurrence in the Neotropics and characteristics of adult females. Images are from Carpenter and Marques (2001) and Carpenter (2004). All scale bars are 1 mm.

1. Metasomal segment I subsessile, evenly conical in dorsal view (Fig. 18.1a); propodeum with orifice dorsally acute (Fig. 18.1a); pronotum with a carina posterior to the lateral fovea (Fig. 18.2a)..***Polistes***

1′. Metasomal segment I differently shaped, in dorsal view basally petiolate to long petiolate or sessile with width greater than length (Fig. 18.1b); propodeum with orifice often broadly rounded dorsally (Fig. 18.1b); pronotum with a carina anterior to the lateral fovea, if the latter is present (Fig. 18.2b)................................**2**

A. Somavilla (✉)
Instituto Nacional de Pesquisas da Amazônia, Manaus, Amazonas, Brazil
e-mail: alexandresomavilla@gmail.com

J. M. Carpenter
American Museum of Natural History (AMNH), New York, NY, USA
e-mail: carpente@amnh.org

F. Prezoto et al. (eds.), *Neotropical Social Wasps*,
https://doi.org/10.1007/978-3-030-53510-0_18

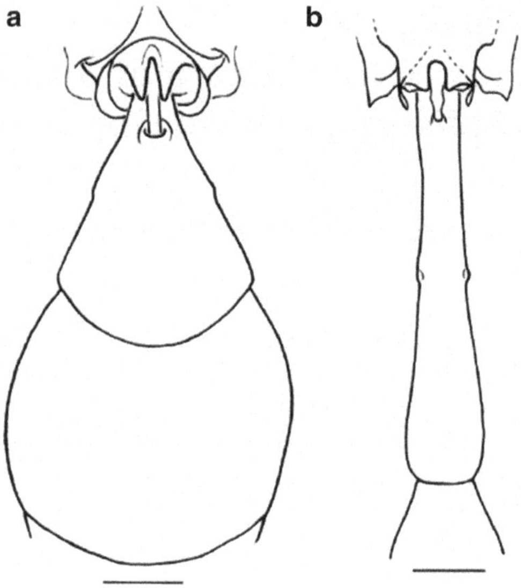

Fig. 18.1 Metasomal segments I and II in dorsal view of *Polistes consobrinus* (**a**) and *Mischocyttarus rotundicollis* (**b**)

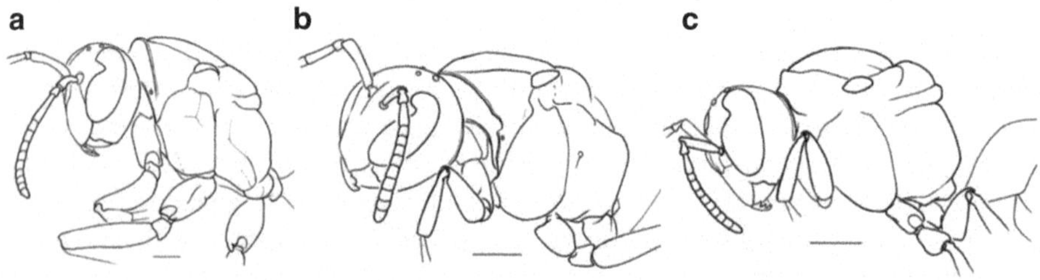

Fig. 18.2 Head and mesosoma in lateral view of *Polistes consobrinus* (**a**), *Parachartergus fraternus* (**b**), and *Chartergus artifex* (**c**)

2. Mid- and hindtarsi with third and fourth segments asymmetrical, inner lobe longer than outer lobe (Fig. 18.3a); forecoxa not dorsolaterally produced (Fig. 18.4a); metasomal segment I petiolate (Fig. 18.1b), in dorsal view with width half or less that of II, and at least twice as long as wide................................***Mischocyttarus***

2′. Mid- and hindtarsi with third and fourth segments symmetrical (Fig. 18.3b); forecoxa usually dorsolaterally produced (Fig. 18.4b); metasomal segment I variously shaped...**3**

3. Pronotum without lateral fovea (Fig. 18.2c)...**4**

3′. Pronotum with lateral fovea (Fig. 18.2a, b), sometimes very shallow....................**10**

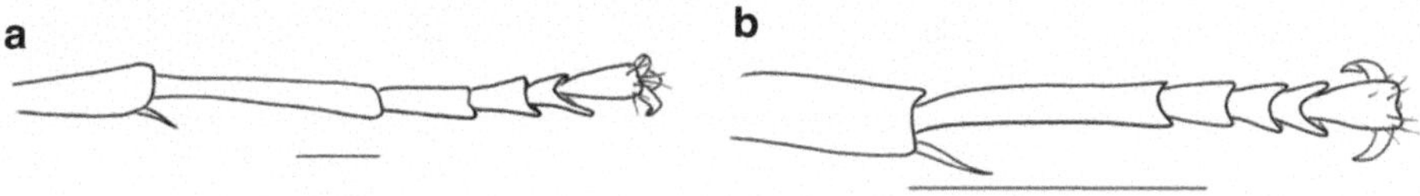

Fig. 18.3 Hindtarsus of *Mischocyttarus rotundicollis* (**a**) and *Brachygastra augusti* (**b**)

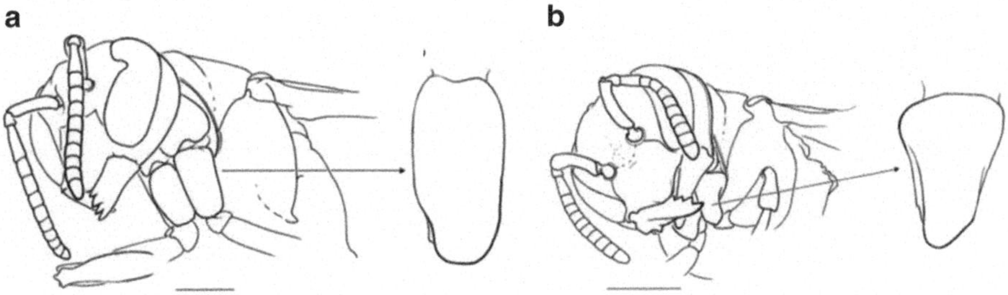

Fig. 18.4 Head and prothorax in oblique frontal view and, in evidence, procoxa of *Mischocyttarus rotundicollis* (**a**) and *Brachygastra augusti* (**b**)

4. Scutellum angled in profile (Fig. 18.5a); metanotum compressed, vertical (Fig. 18.5b); head with occipital carina present dorsolaterally (Fig. 18.5a); pronotum without pretegular carina (Fig. 18.5b); metasomal segment I very compressed (Fig. 18.5b)..***Brachygastra***

4′. Scutellum rounded in profile (Fig. 18.2c); metanotum not compressed or vertical (Fig. 18.5c); head with occipital carina present or absent; pronotum with pretegular carina (Fig. 18.5c), metasomal segment I compressed or usually developed (Fig. 18.5c)..**5**

5. Propodeum with dorsolateral carinae (Fig. 18.6a); metanotum with medial tubercle (Fig 18.6a); head with occipital carina present dorsolaterally (Fig. 18.2c), metasomal segment I compressed..***Chartergus***

5′. Propodeum without carinae (Fig. 18.6b); metanotum without tubercle (Fig. 18.6b); head without occipital carina (Fig. 18.6b); metasomal segment I not compressed...**6**

6. Forecoxa dorsolaterally produced (Fig. 18.4b)..**7**

6′. Forecoxa not dorsolaterally produced (Fig. 18.4a)...**9**

7. Clypeus apically emarginate (Fig. 18.7b); propleuron without lateral, crenate groove (Fig. 18.8b); cuticle black, bare and shining...............................***Epipona***

7′. Clypeus apically bluntly pointed (Fig. 18.7a); propleuron with lateral, crenate groove (Fig. 18.8a); cuticle not shining..**8**

8. Metasomal segment I in dorsal view with maximum width 3× basal width (Fig. 18.6b)...***Asteloeca***

8′. Metasomal segment I in dorsal view with maximum width 2× basal width...***Metapolybia***

a b c

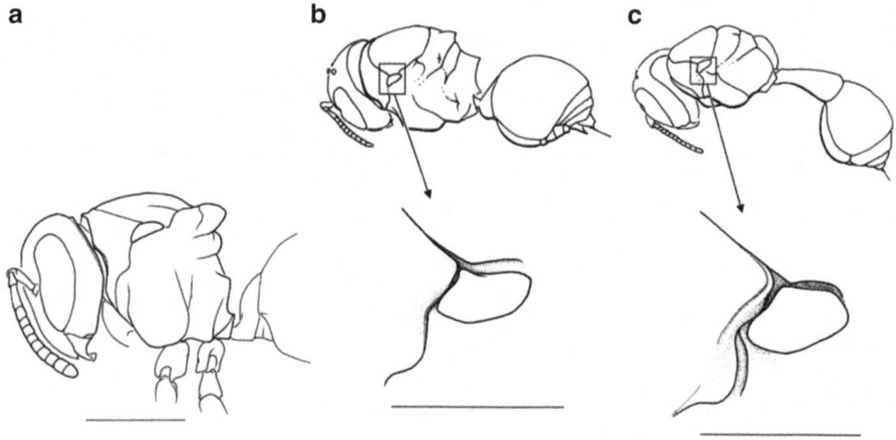

Fig. 18.5 Head and mesosoma in lateral view of *Brachygastra augusti* (**a**), posterolateral view, in evidence, pronotum and tegula of *Brachygastra augusti* (**b**), and *Epipona guerini* (**c**)

a b

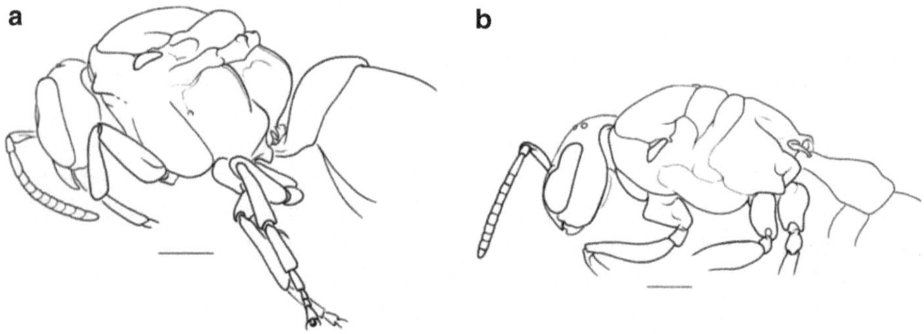

Fig. 18.6 Head and mesosoma in oblique posterolateral view of *Chartergus artifex* (**a**) and *Clypearia sulcata* (**b**)

a b

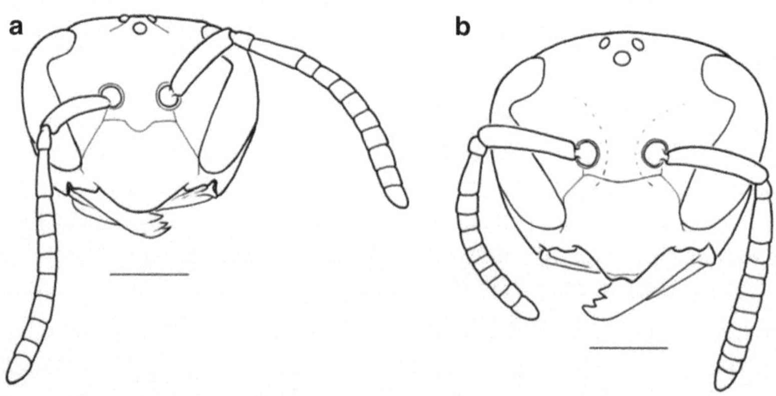

Fig. 18.7 Head in frontal view of *Asteloeca traili* (**a**) and *Epipona guerini* (**b**)

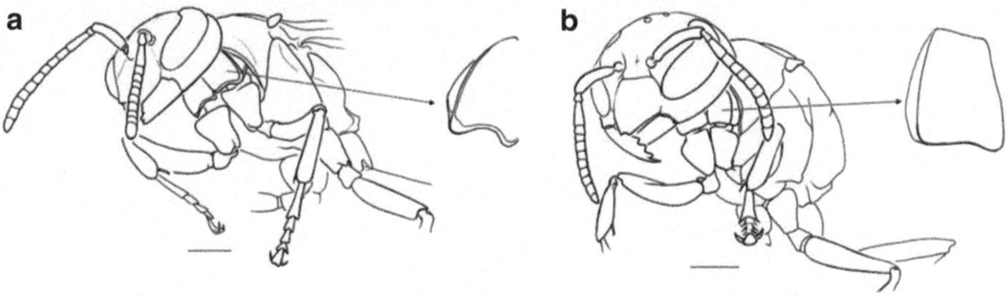

Fig. 18.8 Head and prothorax in oblique frontal view, in evidence, propleuron of *Asteloeca traili* (**a**) and *Epipona guerini* (**b**)

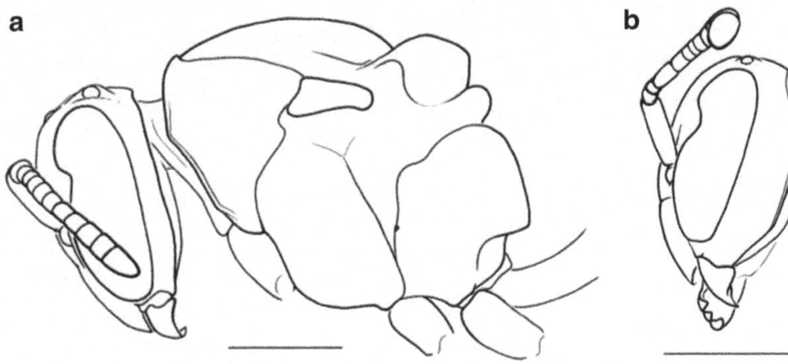

Fig. 18.9 Head and mesosoma in lateral view of *Charterginus fulvus* (**a**) and head of *Protonectarina sylveirae* (**b**)

9. Propodeum posteromedially concave (Fig. 18.6b); body with some pale maculations, not metallic; head in lateral view with tempora narrower than eye at ocular sinus (Fig. 18.9a); clypeus longer than wide.. ***Clypearia***

9′. Propodeum posteromedially convex (Fig. 18.6a); body without pale maculations, cuticle partly to entirely bluish metallic or yellowish with some bluish highlights; head in lateral view with tempora as wide or wider than eye at ocular sinus (Fig. 18.9b); clypeus at most as long as wide................................. ***Synoeca***

10. Pronotum with dorsal carina sinuous in lateral view (Fig. 18.10a); forewing with prestigma about as long as pterostigma (Fig. 18.11a)..................... ***Parachartergus***

10′. Pronotum without sinuous carina (Fig. 18.10b); forewing with prestigma shorter than pterostigma (Fig. 18.11b)... **11**

11. Scutum with posterolateral lamella absent anteriorly, not adjoining tegula (Fig. 18.12a)... **12**

11′. Scutum with posterolateral lamella present adjoining tegula (Fig. 18.12b).......... **16**

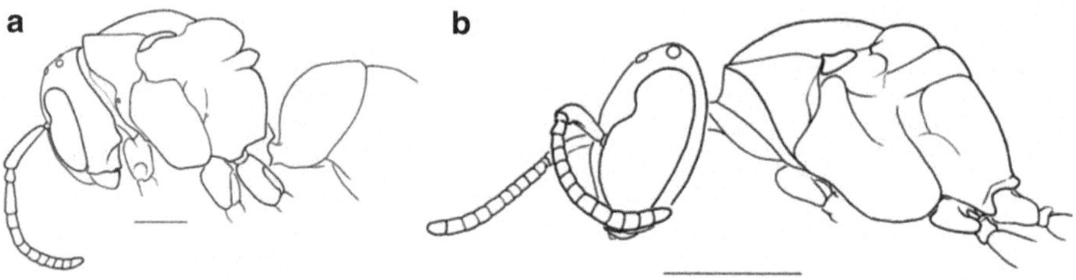

Fig. 18.10 Posterolateral view of *Parachartergus fraternus* (**a**) and *Protopolybia sedula* (**b**)

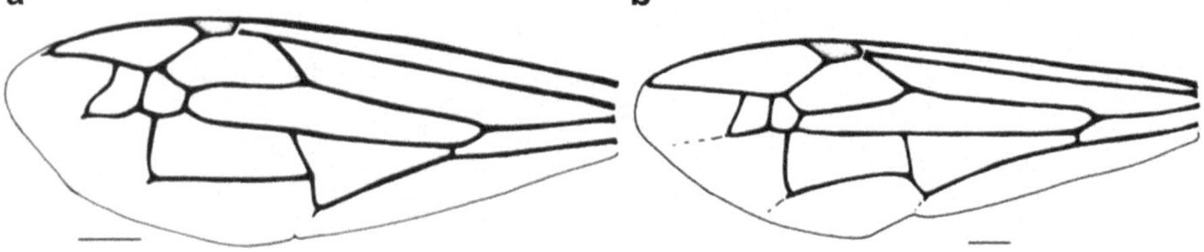

Fig. 18.11 Forewing of *Parachartergus fraternus* (**a**) and *Angiopolybia pallens* (**b**)

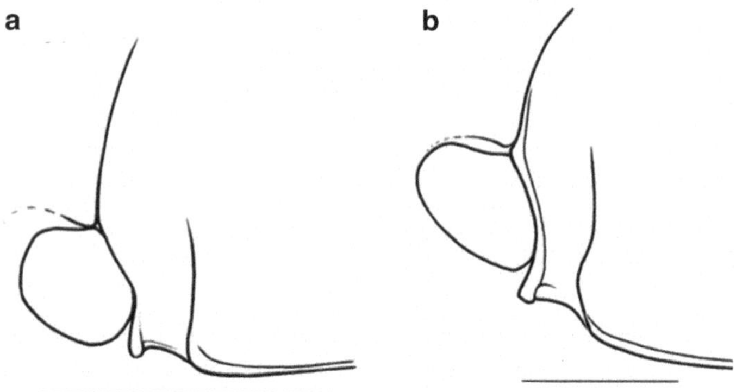

Fig. 18.12 Scutum and tegula in dorsal view of *Polybia bistriata* (**a**) and *Apoica pallens* (**b**)

12. Mesepisternum with dorsal groove (Fig. 18.13a); clypeus with square lateral lobes and sharply pointed apex (Fig. 18.14a)................................***Angiopolybia***

12′. Mesepisternum without dorsal groove (Fig. 18.10b); clypeus with lateral lobes rounded and apex bluntly pointed (Fig. 18.14b) or rounded (Fig. 18.14c) or truncate (Fig. 18.14d)..**13**

13. Metanotum produced posteromedially into elongate lobe overlapping the propodeum (Fig. 18.15a)..***Protopolybia***

13′. Metanotum without posterior lobe...**14**

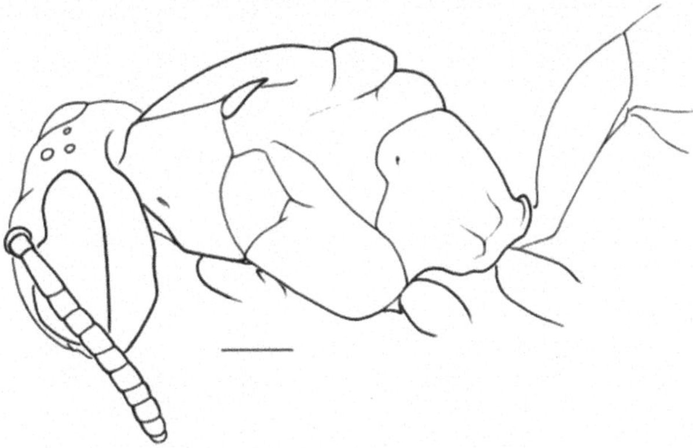

Fig. 18.13 Posterolateral view of *Angiopolybia obidensis* (**a**)

14. Metasomal segment I in dorsal view with short petiole, apically abruptly broadened, wider than half the width of segment II (Fig. 18.15b); pronotum with anteromedial fovea..***Charterginus***
14′. Metasomal segment I in dorsal view differently shaped, narrowly petiolate to sessile; pronotum without anteromedial fovea..**15**

15. Lateral ocelli almost as far apart as distance from each to eyes (Fig. 18.16a); metasomal segment I not petiolate in dorsal view; eyes bristled (Fig. 18.16a)...... .... ***Protonectarina***
15′. Lateral ocelli closer to each other than either is to eyes (Fig. 18.14b); metasomal segment I usually petiolate in dorsal view; eyes usually not bristled...***Polybia***

16. Ocelli enlarged, separated from eyes by less than an ocellar diameter (Fig. 18.16b); hindwing with jugal lobe reduced (Fig. 18.17a)..................***Apoica***
16′. Ocelli normal, separated from eyes by more than an ocellar diameter (Figs. 18.16a, 18.18a); hindwing with jugal lobe normal, not reduced (Fig. 18.17b)...**17**

17. Head in dorsal view with occipital carina present (Fig. 18.18a) **18**
17′. Head in dorsal view with occipital carina absent (Fig 18.19a) **19**

18. Mesepisternum with dorsal groove (Fig. 18.18b) .... ***Agelaia***
18′. Mesepisternum without dorsal groove (Fig. 18.19b) ....***Leipomeles***

19. Metanotum vertical; scutellum more or less angled in profile (Fig 18.19b).. ***Nectarinella***
19′. Metanotum not vertical; scutellum rounded in profile (Fig. 18.19a) **20**

20. Mesepisternum without dorsal groove (Fig. 18.19b); palpal formula 5:3........... .... ***Chartergellus***
20′. Mesepisternum with dorsal groove present, at least as anterior trace (Fig. 18.19a); palpal formula 6:4.. ***Pseudopolybia***

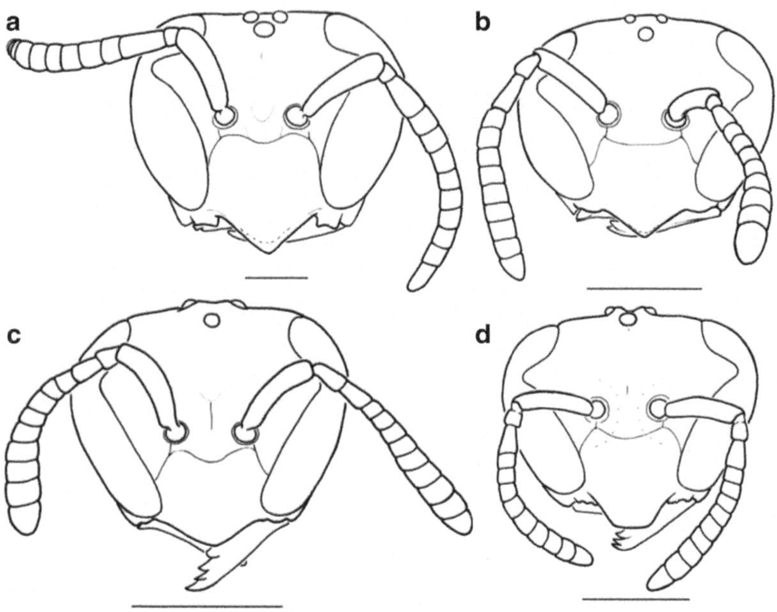

Fig. 18.14 Head in frontal view *Angiopolybia obidensis* (**a**), *Polybia bistriata* (**b**), *Protopolybia sedula* (**c**), and *Charterginus fulvus* (**d**)

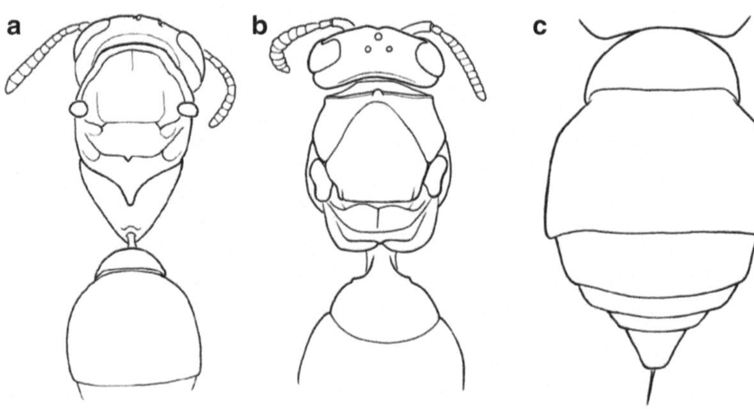

Fig. 18.15 Dorsal view of *Protopolybia sedula* (**a**), *Charterginus fulvus* (**b**), and *Polybia depressa* (**c**)

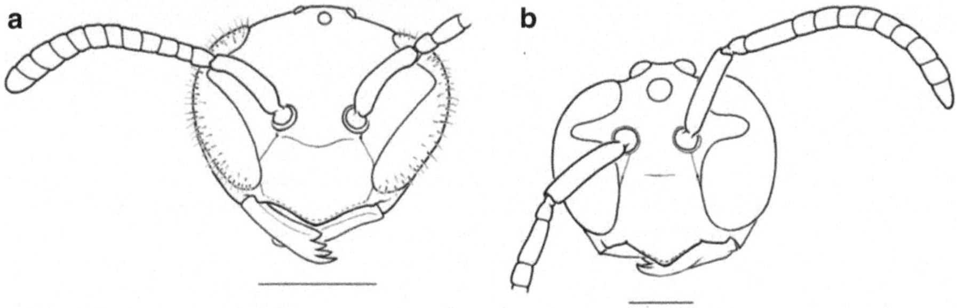

Fig. 18.16 Head in frontal view of *Protonectarina sylveirae* (**a**) and *Apoica pallens* (**b**)

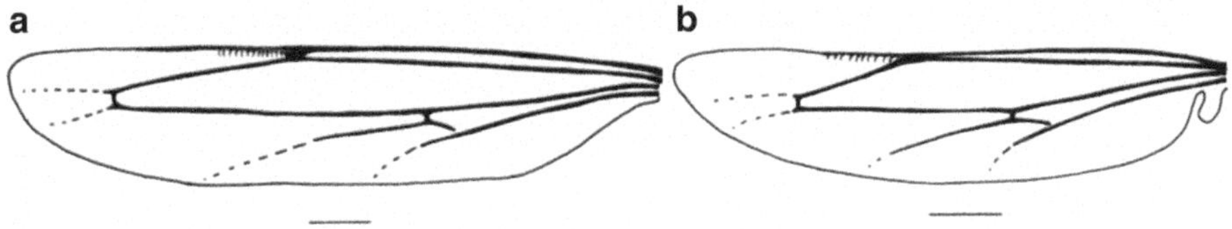

Fig. 18.17 Hindwing of *Apoica pallens* (**a**) and *Agelaia multipicta* (**b**)

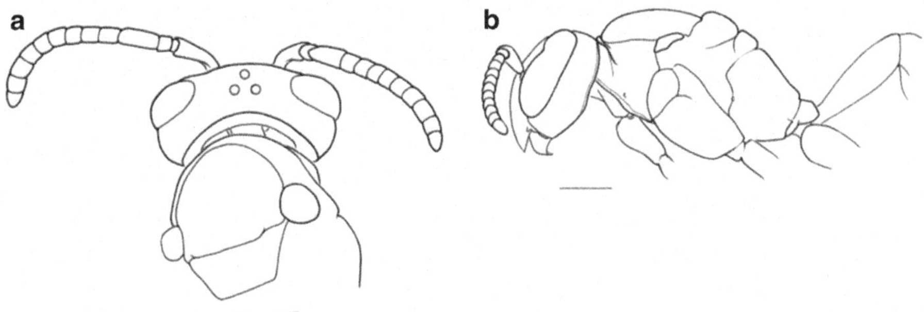

Fig. 18.18 Head and mesosoma in oblique posterior view (**a**) and lateral view (**b**) of *Agelaia multipicta*

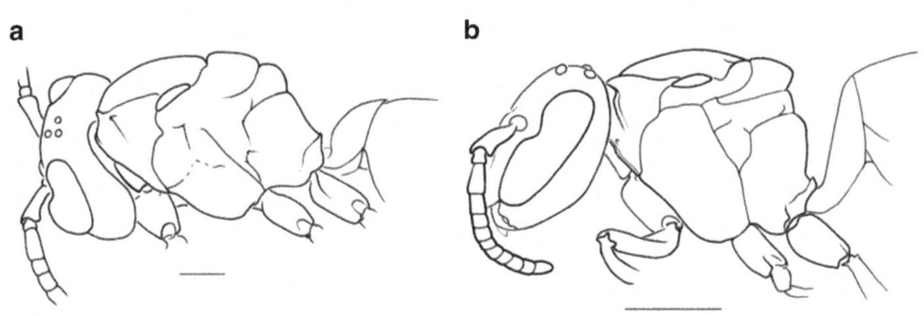

Fig. 18.19 Head and mesosoma in lateral view of *Pseudopolybia vespiceps* (**a**) and *Nectarinella championi* (**b**)

References

Carpenter JM (1993) Biogeographic patterns in Vespidae (Hymenoptera): two views of Africa and South America. p. 139-154. In: Goldblatt P (ed) Biological relationship between Africa and South America. Yale University Press, New Haven

Carpenter JM (2004) Synonymy of the Genus *Marimbonda* Richards, 1978, with *Leipomeles* Möbius, 1856 (Hymenoptera: Vespidae; Polistinae), and a New Key to the Genera of Paper Wasps of the New World. American Museum Novitates 3465:1–16

Carpenter JM, Marques OM (2001) Contribuição ao Estudo dos Vespídeos do Brasil. Universidade Federal da Bahia, Departamento de Fitotecnia. Série Publicações Digitais, v. 3, CD-ROM

Hermes MG, Somavilla A, Andena SR (2020) Catálogo Taxonômico da Fauna do Brasil: Vespidae. In: http://fauna.jbrj.gov.br.

Chapter 19
Research Techniques Used in the Study of Social Wasps

Robert L. Jeanne and Benjamin J. Taylor

Abstract This overview describes and evaluates research methods used in the study of Neotropical social wasps. The emphasis is on field techniques; laboratory and museum procedures and detailed analytical methods can be found elsewhere. The first part of the chapter addresses techniques for working with collected materials. Topics include studies of biodiversity (wasp species richness and abundance), locating colonies in the field, collecting and preserving individuals and whole colonies, and a look at data obtainable from collected colonies. The second part covers procedures for working with live individuals and colonies in the field, including anesthetization, marking adults for individual recognition, uses of video recording, how to study foraging behavior and nesting behavior, and the bioassaying of pheromones. A section follows on manipulating active colonies for special purposes, which covers how to measure age polyethism, colony productivity, egg-to-adult development time, suggestions for working with epiponine swarms, and a look at the pros and cons of working with artificially housed colonies. The final section is a brief assessment of what to expect in the coming years by way of improvements in hardware and software relevant to the field.

Keywords Technology · Methods · Observation · Vespidae

19.1 Introduction

In many fields of science, major advances are often enabled by the application of new methods or approaches. This may come about either via new technological advances (e.g., the development of the electron microscope led to major advances in

R. L. Jeanne (✉)
Department of Entomology, University of Wisconsin, Madison, WI, USA
e-mail: rljeanne@wisc.edu

B. J. Taylor
Department of Natural Sciences, CUNY-LaGuardia Community College, Long Island City, NY, USA
e-mail: btaylor@lagcc.cuny.edu

F. Prezoto et al. (eds.), *Neotropical Social Wasps*,
https://doi.org/10.1007/978-3-030-53510-0_19

cell biology) or in the application of theory developed in one field to facilitate gains in another (e.g., application of economic theory to ecology). Although perhaps more modest in scope and scale than these examples, the same can be said for the study of the behavior and ecology of the social wasps. Technological innovations of many kinds have led to advances in our understanding of this group. The use of video recording and digital photography, for example, has enabled analysis of behavior at levels of accuracy and detail that are impossible with direct observation. The size and cost of camcorders continue to decline, while the quality of output has advanced to the point where high-definition recording is now the norm.

It is with this in mind that we believe a review of methods and technology used by wasp researchers will be of use, especially to students just entering the field. It is often the case that an innovation developed and applied by an author may remain obscure, despite having the potential to facilitate the work of others. Our aim here is to assemble in an organized way the techniques that have been found useful by those working with Neotropical social wasps. Some of these are high-tech, but others are simply shortcuts or practical solutions that we, and others, have found useful. Our hope is that such a compilation will not only serve as a useful reference, but will stimulate a more conscious attention to methods, which may in itself lead to new and useful techniques.

Although some earlier authors have described methods for taking nests (Rau 1933; Richards and Richards 1951), to our knowledge there has been no comprehensive review of research techniques useful in the study of tropical wasps. Spradbery (1973) and Edwards (1980) provide detailed methods applicable to the study of vespine biology, some of which are adaptable for use with polistines. Here we limit our coverage to methods useful in the study of tropical social wasps. Techniques used in studies of temperate-zone species are included if they may be of use for tropical species. Our focus is on field techniques; specialized laboratory methodology, such as chemical analyses and genomics and proteomics, which can be found elsewhere, are not included. The same goes for methods of controlling wasps as pests. We rely heavily on our own experience with what we have found to work, and readily acknowledge that others would likely put together a different account. Thus, we apologize in advance for any expectation that this compilation is complete.

The account below starts with conducting surveys, moves on to the collection of individuals and whole colonies, then to what can be learned from collected colonies. We next focus on how to work with active colonies in the field, including experimentation as well as observation. Finally, we discuss the manipulation of colonies to achieve special aims.

19.2 Survey Methods

19.2.1 Biodiversity Studies

Studies on social wasp diversity and abundance have appeared at an accelerating rate over recent decades, with 78 cited for Brazil alone from 1982 to 2015 (see Barbosa et al. 2016 for a meta analysis; see also Jacques et al. 2018). When the primary purpose is to determine the species occurring in a region or habitat, a combination of methods will lead to the most complete inventory (De Souza et al. 2011). These may include line censuses—visual searches for individuals and nests along forest edges, roadsides, trails, streams, or an artificial transect (Corbara et al. 2009), malaise trapping (Silveira 2002), light trapping (Aragão and Andena 2016), and sampling of foragers feeding at flowers (Silva-Pereira and Santos 2006), ripe fruit (Dvořák and Landolt 2006), or carrion (Silveira et al. 2005). Of the 11 methods used in such surveys (Barbosa et al. 2016), intensive active searching for individuals in all microhabitats typically yields the greatest number of species, with traps baited with sugar solutions coming in second (De Souza et al. 2011; Jacques et al. 2018). In some studies, malaise and bait traps have taken species not obtained by the other methods used (Barbosa et al. 2016; Jacques et al. 2018). Baited traps tend to be more effective at sampling swarm-founding than independent-founding species (Jacques et al. 2018). Sugary baits, especially sugarcane molasses, are more effective than fruit juices and much more effective than protein baits, with the latter attracting primarily the carrion-feeding species (Jacques et al. 2018). Baiting by spraying a solution of sugar and salt on vegetation and checking those sites frequently is a relatively easy means of enhancing the numbers caught by active searching (Noll and Gomes 2009). The nocturnal genus *Apoica* is best sampled with light traps (Aragão and Andena 2016). For as complete a list of species as possible, a combination of sampling techniques, including active search for nests, sweep netting of foragers, baiting and trapping (De Souza et al. 2011), and canopy sampling by fogging with a knockdown insecticide (Blüthgen and Stork 2007) is recommended. Not surprisingly, the number of species taken by sampling, as a fraction of the total species occurring in the sampled area, rises with the number of person-hours spent searching and the number of months of the year sampled (Somavilla et al. 2014; Barbosa et al. 2016).

A usual final step in reporting species-richness surveys is to apply some estimate of how close the sample comes to being a complete list of all the species occurring in the sampled area. The software EstimateS (Colwell 2013) has been used in a number of recent surveys of Neotropical wasp species diversity (e.g., Corbara et al. 2009; Barbosa et al. 2016). However, EstimateS is no longer being updated and the most recent version (9) does not run fully under current versions of Mac and Windows operating systems (Colwell 2013). More recently developed tools for estimating species richness include SpadeR and others that use the powerful R language (Chao and Colwell 2017). These are available for download from Anne Chao's website (Chao 2019). Other estimators abound, and have been used by other authors (e.g., Silva-Pereira and Santos 2006; Gomes and Noll 2009; Melo et al. 2015; Jacques et al. 2018).

19.2.2 Avoiding Bias

If the goal is more than just a species list, but an accurate estimate of the relative abundance of each wasp species in a habitat or region, the potential for bias using each of the above methods must be carefully considered. Care must be taken to sample the range of habitats occurring in the region (e.g., ground level, understory, canopy, undisturbed vs. disturbed, primary vs. secondary forest, urban vs. natural). Sampling should be done over a wide area. Active searches for nests are likely to have a higher success rate with species whose nests are large and conspicuous than with those with small and/or cryptic nests. Searches and bait trapping close to large colonies may bias numbers in favor of those species.

19.3 Finding Colonies

Other than surveys, most studies of social wasps begin with locating colonies. If the goal is simply to find enough colonies of a target species to work with, several techniques are appropriate, varying from the haphazard to highly structured/organized searches. Active visual search in the species' preferred habitat is the most practical method and can work for most species. *Polybia occidentalis* nests are readily found in low shrubs and trees in pastures and along fencerows and roadsides (O'Donnell and Jeanne 1992; Schueller et al. 2010). For species that utilize narrowly specific nesting substrates or sites, searches can focus on such places. For example, *Angiopolybia pallens* frequently nests under the broad leaves of low plants in the forest understory, and *Polybia rejecta*, *Synoeca virginea*, and *Agelaia myrmecophila* typically nest in close association with the large carton nests of the ant *Azteca* spp. (Jeanne 1991). For species that are specialists nesting on myrmecophilous plants, the most efficient approach is to search for those plants. Other species are synanthropic, nesting on human constructions (e.g., Jeanne 1972). *Metapolybia*, for example, nests on flat surfaces and can often be found on buildings.

For species that form large colonies but are rare, or nest in cavities or high in trees, the lining technique may be useful. Lining has long been used to locate colonies of honey bees (Visscher and Seeley 1989). To locate a colony using this technique, one or more bees foraging at flowers are first captured in a box and allowed to feed on a rich, anise-scented sugar syrup inside the box. As the bee exits, the bearing of its departure toward the nest—the "beeline"—is noted. Repeating this step after moving along the beeline can eventually lead to the nest. Triangulation—plotting (e.g., on a Google Earth map) the intersection of bee lines taken from two or more capture points—can help pinpoint the nest location. Applied to wasps, this technique can take advantage of the attraction of many species to fermenting fruit and to carrion (Fig. 19.1). If foragers feeding on a fruit patch depart to the north, some of the fruits, with foragers on them, can be moved some distance in that direction, perhaps in a box. After they depart, the bearing is adjusted based on the new flight path and the process is repeated. With luck, one or more such steps will bring the observer within sight of the nest. If the surroundings are in low light (e.g., forest understory), chilling the wasp on ice, then tying a few centimeters of white thread around the wasp's petiole can extend the distance of visual contact. As the wasp regains mobility and flies toward the nest, the trailing thread provides a visual cue that can be seen for a greater distance than the wasp alone. It also reduces the wasp's

Fig. 19.1 *Synoeca septentrionalis* foraging on guava fruit. (Photo: R. L. Jeanne. Costa Rica)

flight speed so that the observer can run along with it for a distance, thereby getting closer to the location of the colony. A modified version of this technique is traditionally used in Japan to locate vespine wasp nests (Saga 2019). Instead of flagging the wasp itself, thread is tied to pieces of carrion small enough to be carried by a forager. When a wasp picks up the flagged carrion, the thread provides the extra visual cue, as above. To our knowledge, this technique has not been tried for tropical wasps, but it could be useful for *Agelaia*, *Angiopolybia*, and *Polybia* that feed on carrion (Jeanne and Taylor 2009).

A useful strategy for locating colonies in urban areas is to place advertisements in local newspapers or online classifieds offering free removal of colonies from homeowners' properties (Hastings et al. 1998). If a particular species is being sought, including a picture and description of the wasp and nest structure with the advertisement can improve success. However, the researcher must be ready to deal with a potentially large volume of responses and to take on the time and travel needed to follow up on each. It is good practice to provide the homeowners with an information sheet describing the methods they can use to remove a nest, in the event that their wasps are not the species needed.

19.4 Collecting Wasps

19.4.1 Collecting Individuals from the Nest

If only one or a few individuals from a nest are required for identification, foragers can be captured as they leave or approach the nest, provided enough nearby open space is available to swing a net. If not, the rim of the net can be placed gently against the nest until a wasp crawls onto it to inspect it. The net is then slowly pulled away from the nest and into an open area, and when the wasp eventually flies to return to the nest, it is swept into the net. It can be moved to the tip of the net by exploiting the tendency of a trapped wasp to move upward and toward light. Once there, it can be captured by forcing it into an open vial pushed up inside the net. When it is knocked into the vial, the vial is stoppered.

19.4.2 Collecting Whole Colonies

If the goal is to capture the entire adult population, the nest must be taken at night, when most or all of the foragers are inside (except, of course, for the nocturnal genus *Apoica*).

For species nesting on leaves or twigs, it is often possible to collect the entire nest and its adults by enclosing it in a sturdy transparent plastic bag. If dense vegetation surrounds the nest, this can be clipped away a few hours in advance of collection to provide unhindered access. Without disturbing the colony, the opening of the plastic bag is brought up around the nest and sealed tightly around the supporting twig proximal to the nest with a wire tie. The twig is then clipped. Following collection, care must be taken to prevent the formation of water droplets due to condensation inside the bag, lest wasps be lost to drowning. A few sheets of crumpled paper toweling inside the bag will help. If the adults need to be killed, a wad of cotton soaked in ethyl acetate or chloroform can be dropped into the bag. Colonies so captured can be stored in a refrigerator and the adults will remain fresh enough for dissection for about a day. If the wasps need only be anesthetized, a plastic snap-cap vial with holes in the top and containing ether-soaked cotton can be placed in the bag (wasps are killed by direct contact with ether-soaked cotton).

Species nesting on solid substrates present a different challenge. If the nest carton is sturdy (*Synoeca*, *Chartergus*, *Epipona*), the entrance can be plugged with a tight wad of cotton, which is then soaked with chloroform or diethyl ether. After the buzzing stops, the bottom of the carton can be carefully opened and the anesthetized adults dropped into a plastic bag. For more fragile nests (*Metapolybia*, *Clypearia*), the opening of a plastic bag can be tacked tightly around the nest at night. A small amount of anesthetic is then injected via hypodermic through the bag and the nest carton to subdue the adults. (Extreme care must be taken when storing and working with these chemicals. Ether is flammable and the fumes are explosive. Inhaling chloroform can cause fatigue, dizziness, and headache; elevated doses may damage the liver and kidneys.)

Colonies nesting in cavities can be anesthetized by stuffing a wad of cotton tightly into the entrance at night, then soaking it with diethyl ether or chloroform. After activity ceases, the nest can be excavated.

Collecting nests high in trees requires ingenuity. One of us collected a large nest of *Agelaia areata* from a height of ~20 m in a tree in Mexico (Jeanne 1975b). The large branch bearing the nest was carefully cut during the day and left suspended from a rope. By late afternoon, the adults had settled down and were on or inside the nest. After dark, the branch was slowly lowered on the rope to a point 1 m above the ground, where the nest was collected in a plastic bag.

One of us used a specially rigged net to collect a nest of *Chartergus artifex* and its adults from a height of about 7 m (R. L. Jeanne, unpublished). The metal rim of a standard insect net was removed and replaced with an elastic band (a chain of sturdy rubber bands) that constricted the net opening to about 10 cm; 3-cm loops of string were tied around the elastic at four points around the net, and each was passed through a hole in the side of a 50-cm square frame made of light wood. Each loop was held in place by a nail, stretching the net open. At night, the frame, mounted on a bamboo pole, was brought up around the nest and the nails were simultaneously pulled via a lanyard, releasing the string loops and allowing the net to snap tight around the middle of the nest. The branch bearing the nest was cut the next morning and lowered to the ground.

19.4.3 Preservation

The preservation of collected material falls into several categories, depending on its end use. Wasps collected for museum specimens or taxonomic studies, such as those used to study morphological variation within a colony, can be collected and preserved according to standard entomological techniques (i.e., dry-pinning, or preservation in Kahle's or Dietrich's solution and storage in 70% alcohol). These methods are described in a variety of entomological texts (e.g., Arnett Jr. 2000; Johnson and Triplehorn 2004).

Preservation for molecular analyses requires special procedures in order to maintain the integrity of the molecules to be analyzed. Generally, this means storage at −20°C or lower for DNA analysis and −80°C for RNA and protein work. Roskens et al. (2010) studied the proteins and enzymatic activity of adult and larval wasp saliva and found that placing samples on dry ice and storing them at −80°C protected the integrity of their material. They also developed desiccation methods wherein the samples were dried via silica beads or Drierite (W.A. Hammond Drierite, Xenia, OH), with SDS or protease inhibitors added. They were able to successfully preserve the proteins and enzymatic activity of these desiccated samples for up to 3 weeks at 37°C (Roskens et al. 2010). These techniques facilitate the study of proteomics in non-model species, which typically are not kept in labs and must be collected in the field.

19.5 Working with Collected Colonies

In a series of landmark monographs, O. W. Richards (Fig. 19.2) showed how much could be learned about the biology of Neotropical wasps by quantitative analyses of the adults, brood, and nests of collected colonies (Richards 1945; Richards and Richards 1951; Richards 1978). These works generated questions that stimulated a great deal of new work over subsequent decades. They remain useful references today.

Fig. 19.2 O. W. Richards with his wife, Maude J. (Norris) Richards, holding a nest of *Chartergus artifex* on the grounds of the Museu Paraense Emílio Goeldi, Belém, Pará, Brazil, 1968. (Photo: R. L. Jeanne)

19.5.1 Adults

The most basic information that can be taken from a collected colony is the size of the adult population, broken down into males and females. If the colony is collected at night with no escapees, the totals are especially valuable. Dissection of all the females (or a large random sample) can yield a wide range of data, including the number of queens, their degree of ovary development, and the presence of endoparasites.

Dissection to detect insemination, a marker of queens, is most quickly done with fresh material. The terminal gastral segment is grasped with watchmaker's forceps and pulled out, bringing the gut, ovaries, sting apparatus, and spermatheca with it. Care must be taken not to confuse the spermatheca with the venom sac; the spermathecal duct opens dorsally into the common oviduct, whereas the larger and visibly muscular venom sac ducts to the base of the sting. Under low magnification (dissection microscope or strong hand lens), an opaque, pearly mass in the center of the otherwise clear spermatheca indicates a sperm load; an empty spermatheca is clear throughout (Murakami et al. 2009). In preserved material, the tissues of the spermatheca become opaque and the sac must be crushed on a microscope slide and examined with a compound microscope to detect the presence of sperm, a more time-consuming process (Reed et al. 1988).

Degree of ovary development can be assessed with the same dissection. For the epiponines, if the colony is in an early stage of development, it is likely to have many queens, each with slight ovary development. Presence of one or more mature eggs (i.e., at or close to the size of a laid egg) is good evidence that the female is an active egg-layer. A number of authors have quantified the degree of ovary development among females; excellent recent examples include Murakami et al. (2009) and Desuo et al. (2011).

Unless the species is one with clear queen–worker dimorphism, queens will be reliably recognizable only through dissection to check for insemination and developed ovaries. In later stages of colony development, when the number of queens has been reduced, their greater ovary development may be recognizable by a gaster that is visibly swollen and, in at least some species (e.g., *Polybia occidentalis, P. sericea*), chocolate-brown in color as opposed to the jet black of workers and young queens. A caveat: a heavy load of gregarine gametocysts can swell the gaster of a worker to a deceptive, queen-like size (Bouwma et al. 2005); only dissection will determine whether queens can be reliably recognized externally.

Presence of endoparasites can also be revealed via dissection of the gaster. Gregarine parasites in the gametocyst stage are sometimes found in the gasters of infected adults, where they appear as one or more (sometimes >100) white spheres in the body cavity. Heavy infestations can reduce the productivity of a colony (Bouwma et al. 2005). The range of polistine wasp species subject to infection by these protozoan parasites is poorly known. Nothing is known about how the wasps become infected or how the parasites disperse from their hosts.

Body size is a descriptive species statistic of considerable value to ecological and sociobiological studies of social wasps, but has unfortunately been underappreciated. Among the measures used are full body length, various measures of the thorax, and length of the forewing or part of it. In the interest of facilitating cross-species comparisons, we advocate adoption of a standard measure. For the following reasons, we strongly recommend the full length of the forewing: (1) it is unambiguously measured, unlike full body length that can vary depending on the degree of flexure of the gaster in preserved specimens; (2) because it is a two-dimensional structure it is not subject to potential errors introduced when the two ends of a three-dimensional structure are in different planes of focus under a dissection microscope; (3) it is the longest nonarticulated part of the body, thereby minimizing measurement imprecision; (4) it is a part of the thorax, the tagma that varies least among female castes (Jeanne 2003); (5) it is the most widely used measure of body size in the social wasp literature (see chapter 2, this volume). Richards (1978), for example, reported full wing length, measured from the center of the tegula to the tip of the wing, for 86 Neotropical polistine species. In contrast, in a number of other studies, some subdivision of the wing is measured. We can think of no advantage of this over the full wing. While it is true that in older workers the wingtip may be frayed, in our experience this is a rare occurrence.

A number of morphometric studies of caste differences have appeared in recent years (e.g., Noll and Zucchi 2002). Most of these make use of ten or so measurements of the head, thorax, and abdomen, but often differ in the details. Again, for the sake of enabling cross-species comparisons, it would make sense to standardize these measures.

Body weight, fresh or dried, is an alternative measure of body size that is more useful in certain kinds of studies, such as measures of colony productivity (e.g., Bouwma et al. 2006).

19.5.2 Brood

To a greater extent than the adults, the brood in the nest embodies a record of the colony's developmental history. The age of the colony in number of generations of brood, measured as the egg-to-adult development time in days, can be estimated from the distribution of brood stages in the combs (Richards and Richards 1951). In a recently founded nest, if the oldest comb has a central mass of pupae, with one or two central (oldest) cells having recently produced an adult, the nest is just over one developmental period (or "generation") old. If the central mass of pupae is surrounded by open cells of eggs and larvae, and beyond these by a ring of pupae, the nest is just over two periods old (Fig. 19.3). By examining in this way the distribution pattern of eggs, larvae, and pupae across all the combs of the nest, it is possible to estimate age for up to about three periods. Beyond that, the distribution pattern becomes increasingly irregular, and therefore unreliable as an indicator of age, due to the accumulating effects of brood mortality and interindividual variation in development time. For such older colonies, careful inspection of the oldest cells in the nest can reveal how many times those cells have been used to rear an immature to the pupal stage. Prepupae spin a silken cocoon before they pupate, then void their hindgut contents (meconia) into the bottom of the cell. In some species the silk network at the bottom of the cell is substantial enough that longitudinal sections through the cell bottom will reveal, under magnification, how many layers of meconium-on-silk layers there are. Because meconia are often pasted off-center at the bottom of the cell and thus some may be missed by the cut through the cell, it is critical that a sample of cells be sectioned. An alternative approach is to peel away the cell's carton from around the stacked meconia, then carefully tease them apart to determine their number. If ten cells are sampled and found to contain 6, 6, 6, 6, 6, 6, 6, 5, 5, 4 meconia, the nest is at least six brood generations old. Cells with fewer than six housed one or more brood that failed to survive to pupation.

The number of larval instars characterizing a species can be determined by measuring the maximum width of the head capsule of a sample of larvae (e.g., Rocha and Giannotti 2016). For epiponines it is most convenient to do this under a dissection microscope with fresh-killed larvae still in their cells. If the sample is large enough, the number of peaks in the plotted size-frequency distribution will clearly indicate the number of instars.

For some studies it is useful to report the population of brood in a nest. The amount of brood can be quantified either by counts (numbers of eggs, larvae, pupae) or by biomass. For colony productivity studies, total dry mass of brood is the most direct measure of total colony output. All brood are removed from their cells, stored in a preservative, then dried and weighed (Jeanne and Nordheim 1996).

Fig. 19.3 *Polistes instabilis*. The nest grows downward from its attachment at the top, so the age of the brood decreases downward. Thus, the cells in the larger mass of pupae and below contain brood of the first generation (i.e., first use of those cells for rearing brood). Cells above that contain brood of the second generation (second use). The few opens cells at the very top contain young brood of the third generation. Thus, this nest is slightly greater than two egg-to-adult developmental periods old. (On the twig to the left of the nest is a katydid (*Ancistrocercus* sp.), with the antennae of a second one visible top right. These tettigoniids presumably gain some protection from predators by roosting near wasp nests during the day.) (Photo: R. L. Jeanne. Costa Rica)

19.6 Working with Active Colonies

It is our experience that techniques developed for working with active colonies of one species often do not work well for others, even congeners, and have to be modified. It is our hope that the procedures described below will nevertheless provide useful starting points for those wishing to apply them to little-studied species.

19.6.1 Gaining Access

Other than mating behavior and work with foragers at food sources, investigations into the social behavior of wasps require close access to active colonies. If the nest is in an inconvenient location—too high, over water, etc.—it can often be moved. Over many years of working with *Polybia occidentalis*, we have routinely transplanted nests to shaded positions in low vegetation about 1.5 m above the ground, where they can be observed close-up at eye level by an observer sitting in a field

Fig. 19.4 R. L. Jeanne observing behavior of *Polybia occidentalis* at the nest (visible just under the black spring-clip holding the sunshade). Paint pens are visible in the toolbox next to the chair. Guanacaste, Costa Rica, 2006. (Photo: K. J. Loope)

chair (Fig. 19.4). The nest is prepared for moving by clipping vegetation around it, then the twig bearing the nest is clipped at night and carried (typically over several tens of meters) carefully to its new location and wired, zip-tied, or spring-clipped into place on a twig previously prepared to receive it. We have found for this species that enclosure in a plastic bag is unnecessary during the move, as long as care is taken not to excessively jar the nest or bash it against passing vegetation during transport (exception: if the nest is moved a long distance or in a vehicle, bagging it is prudent). Even gentle jarring of the nest may cause some adults to emerge onto the envelope and the supporting twig, but they will not fly at night (Schueller et al. 2010). However, because wasps will fly up a flashlight beam in the dark, artificial white light must be kept away from the nest until it is at its new site. Wasps do not see into the red portion of the spectrum (Peitsch et al. 1992), so a flashlight or headlamp that produces red light allows the collector to see, yet prevents the wasps from flying. After the nest is attached to the twig, Tanglefoot® can be applied to the base of the twig to prevent predation by ants (Schueller et al. 2010).

For nests that cannot be moved—e.g., *Synoeca* and others built high and/or on solid surfaces—the observer must be moved to the nest. Naumann's use of freestanding scaffolding platforms allowed him to reach the high nests of *Protopolybia acutiscutis* that typically were constructed at the tips of flexible branches or palm fronds (Naumann 1970). One of us (RLJ) used the same system to reach similarly situated nests of *Ropalidia romandi* in Queensland, Australia.

Once close access to the nest is in place, care must be taken to avoid disturbances that might alarm the colony. Three stimuli cause alarm: mechanical disturbance, human breath, and alarm pheromone (a component of venom). Alarmed workers will fly at and attempt to sting nearby moving objects, especially dark-colored ones (Jeanne 1981a). Jarring the nest can be avoided by taking care to eliminate any contact by the observer with the substrate bearing the nest. The risk of the second is reduced by placing oneself downwind of the nest, or otherwise being careful not to exhale on it. The third can be a problem if one is working with alarm pheromones or handling wasps (e.g., for marking—see below) in ways that cause them to release venom. When a mishandled wasp returns to the nest, venom residue on its gaster may trigger an alarm response in nearby individuals. Depending on the aggressiveness of the species, protective clothing—a beekeeper's suit with helmet and veil—may be prudent. In our experience, such protective gear is usually unnecessary when working with the relatively docile *Polybia occidentalis*. Nitrile gloves are effective in preventing the sting from penetrating to the skin. It goes without saying that use of mosquito repellent by the observer must be avoided.

19.6.2 Anesthetization

Anesthetization may be necessary in some cases in order to mark individuals or to collect whole, active colonies. Three methods are typically used to anesthetize: chilling on ice or in a refrigerator, exposure to diethyl ether, and exposure to carbon dioxide (CO_2). Of these, chilling typically requires the greatest amount of time to make wasps immobile, and wasps typically recover most rapidly from it provided the ambient temperature is sufficient. It is likely that it is also the most benign of the three methods, but some studies on honey bees suggest that immobilization by chilling affects hoarding behavior (Mardan and Rinderer 1980). Chilling may also be used as a means of temporary immobilization of an entire colony before transferring it to a new container or observation nest box (see Sect. 19.8 below). Colonies of *Vespula germanica* have successfully been kept overnight in a ~4°C refrigerator with very little loss of individuals (Taylor and Jeanne 2018).

Diethyl ether is a useful anesthetic for handling whole, active colonies. Sealing the nest or other container holding the wasps maximizes its effectiveness as an anesthetizing agent. The duration of anesthesia is lessened when wasps are exposed to sunlight or other forms of UV radiation (Ishay et al. 1994). Repeated uses over a short time period also seem to lessen its effects (B. J. Taylor pers. obs.). Because ether fumes are heavier than air, anesthetization will be more effective if it is placed at a level above where most wasps are located. Care must also be taken when using ether, as prolonged exposure can be lethal to insects. In studies on *Vespula germanica*, Taylor (pers. obs.) found that the queen seemed especially susceptible to its effects. Diethyl ether's effects on other aspects of the biology of wasps are not well-known. Plath (1924) found that ether anesthetization of bumble bees did not cause the bees to lose their memory of nest location. However, nest location may be a particularly well-learned memory and not as readily subject to loss as other memories, such as a recently learned food odor (Alloway 1972).

Carbon dioxide is a commonly used anesthetic for entomological studies. Unlike diethyl ether, the risk of wasp mortality when using CO_2 seems to be low. Indeed, Owen (1962) found that CO_2 never caused death when used to anesthetize *Polistes fuscatus*. Its effects on behavior and physiology have long been reported to be minor and short-lived (Nicolas and Sillans 1989). However, CO_2-induced anesthesia has been found to accelerate age polyethism in *Apis mellifera* (Nicolas and Sillans 1989), reduce pollen gathering (Ribbands 1950), and modify hoarding behavior (Mardan and Rinderer 1980). Prolonged exposure also affects memory and foraging behavior (Nicolas and Sillans 1989). Even more extreme are reports that it increases brood cannibalism in ants (Sorensen et al. 1983) and ejection of larvae in bumble bees (Pomeroy and Plowright 1979). Warwick Kerr once informed the senior author that ether has less effect on the behavior of stingless bees than does CO_2 (Jeanne 1972). Finally, CO_2 has the added disadvantage of requiring that a portable CO_2 container be carried into the field.

So, which method is best? All have their drawbacks, but it largely depends on the study being conducted. When rapid anesthetization is required, CO_2 and ether are the best bet. Ether is easier to use if rapid anesthetization is required in the field, especially when collecting whole colonies. If rapid anesthetization is not required, chilling can be done with little fear that it will cause harm to the wasps. In contrast, ether has the potential to kill if exposure is too long. In addition, as mentioned under "Collecting whole colonies" (below), ether fumes are highly combustible. Ashburner and Thompson Jr (1978) reviewed all three methods and recommended both CO_2 and chilling over ether.

19.6.3 Marking Individuals

Most studies of behavior require the marking of at least some of the adults of a colony for individual recognition. For small colonies, one can get away with a small number of unique color combinations. West-Eberhard was able to mark *Polistes* without capture or anesthetization by daubing one or more colors of paint haphazardly onto the wings and bodies of workers on the nest (West-Eberhard 1969). This is workable only for marking small numbers of wasps; beyond a few tens of individuals, relying on randomly placed marks becomes too cumbersome and prone to errors of identification.

For larger colonies, a code that translates to numbers is essential. Karl von Frisch, working with honey bees, was apparently the first to do this, beginning in the 1920s. Using a color- and position-coded system of paint dots on the thorax and abdomen, he was able to uniquely mark up to several thousand individuals (von Frisch 1967). The numbered disks (Opalithplättchen), developed later for use on honey bees, are the wrong size for most wasp species and have rarely been used by wasp researchers.

19.6.3.1 Color Codes

For many years, we have used a coding system combining five colors of paint with positions on the thorax (Fig. 19.5). The colors are arranged alphabetically and code for numbers as follows: blue = 1 and 6, green = 2 and 7, red = 3 and 8, white = 4 and 9, yellow = 5. Position on the thorax determines the range: left side = 1–5, right side = 6–9. Positions from front to back represent ones, tens, hundreds, thousands. Thus, blue in the upper-left quadrant of the mesoscutum = 1, lower-left = 10, right half of the postscutum = 600, and so on (Fig. 19.6). Although von Frisch applied paint spots to the gaster as well as the thorax of his bees, we found that spots on the gaster are obscured by the folded wings when the wasp is on the nest. Hence, our system uses the thorax exclusively.

Others have developed different systems. Gadagkar uses ten colors (plus a blank) in four positions (thorax, abdomen, left wing, right wing) to achieve over 14,000 unique combinations; adding a fifth (unspecified) position increases the number to over 161,000 (Gadagkar 2001 and personal communication). Unlike the system described above, each wasp is identified by its unique combination of the color and position of its spots, rather than with a number. Another option is to use ten colors to represent digits 1–10, and assigning three different positions on the body to represent ones, tens, and hundreds (Brenner and Patterson 1988). Other marking systems are reviewed by Southwood (1978) and Walker and Wineriter (1981).

We have found that for field studies, using more than five colors is unworkable. Under low-light field conditions (early morning/late afternoon; forest understory) it becomes impossible to distinguish gray from silver, or dark blue from light blue, for example. We advise against marking the wings, as it risks fouling the wing-folding

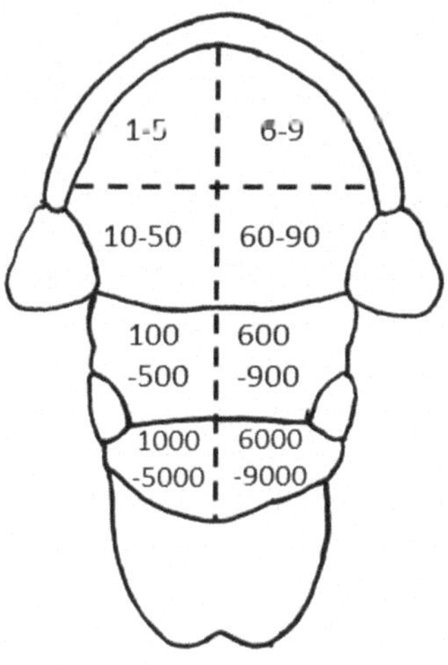

Fig. 19.5 Dorsal view of a wasp thorax showing the placement of spots. (Mesoscutum: top two quadrants = single digits; bottom two quadrants = tens. Scutellum = hundreds. Postscutellum = thousands)

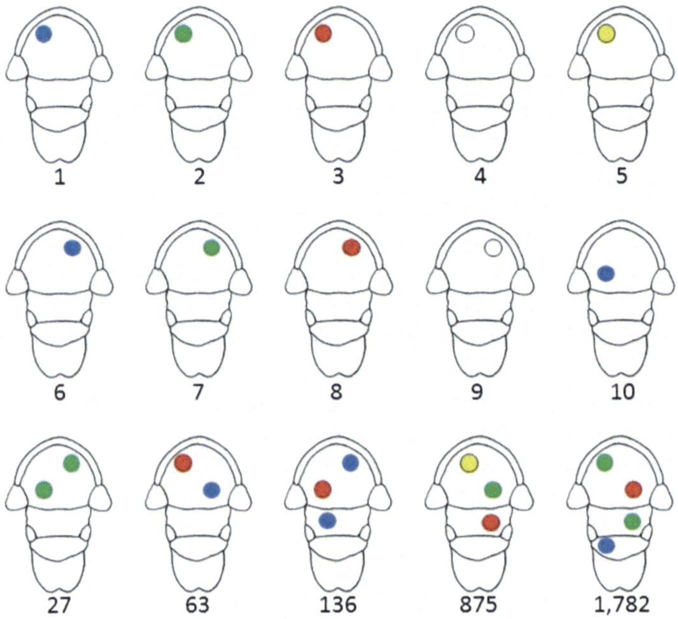

Fig. 19.6 Examples of the color-coded numbering system used by the authors

mechanism. Furthermore, especially for smaller wasp species, a paint spot is a substantial addition to the mass of the wing and could potentially hinder flight by altering wingbeat frequency.

19.6.3.2 Media

A wide variety of substances have been used to mark insects for individual or group recognition (reviewed by Hagler and Jackson 2001). Paints are an inexpensive solution. A number of wasp workers have used model airplane dope (e.g., Testors™ enamel). A droplet is applied to the wasp with a toothpick, insect pin, paper clip, or fine brush. In our experience, these paints are less than satisfactory. The dope comes in small screw-cap bottles, which are hard to handle under field conditions. If the bottles are left open for quick access, the solvent evaporates and the paint thickens and forms a skin; thick paint applied to cuticle will fail to make firm contact upon drying and will be at risk of flaking off. While this can be corrected by adding thinner, if too much is added, the paint risks running by capillarity into the wing articulations, potentially hindering flight or even proving fatal (Owen 1962). Opening and closing the bottles at each use requires two hands, risks spillage, and interrupts behavioral observation. Over time, the threads on cap and bottle become clogged with dried paint and have to be scraped clean.

All these difficulties are much reduced with the use of paint pens. These are available in both oil-based paint and water-based acrylic. Both dry quickly and are

Fig. 19.7 One brand of acrylic paint pens

Fig. 19.8 *P. occidentalis* worker held gently for marking in vise-like reverse-action bent-tip forceps. This individual's colors code for the number 64. (Photo: R. L. Jeanne)

opaque and waterproof, although oil-based paints were found to be more durable than water-based ones (Wineriter and Walker 1984). We have marked thousands of *Polybia occidentalis* individuals using a brand of oil-based paints (Fig. 19.7). The fine or extra-fine points make it quick and easy to apply spots precisely in the desired quadrant of the mesothorax (Fig. 19.8), despite the 1.6 mm width of this part of the body. Loss of spots due to flaking is rare. What works well for fieldwork is to drill five holes into the edge of a wood block, one for each color and sized to allow the paint pens to be inserted, cap down, tightly into them. With each pen resting lightly in its cap, it can quickly be picked up with one hand and used, with only a few seconds of exposure of the open pen before it is replaced loosely into its cap. RLJ fastens such a block inside the front wall of the plastic fishing tackle box he uses to carry forceps, mirrors, stopwatches, and other tools and supplies into the field (see Fig. 19.4).

Tests of the effects of these paints on the behavior and longevity of bees and wasps have shown mixed results. In a lab study, Testors PLA™ enamel applied to the heads of two species of halictine bee was shown to increase cooperative behavior in one and increase aggression and decrease cooperation in the other (Packer 2005). De Souza and coworkers tested Acrilex®, an acrylic paint, for its effect on behavior and survival of *Polistes* wasps (Souza et al. 2012). Wasps that had paint spots applied to them groomed at higher rates than controls for 5–7 min after marking; thereafter, the rate of grooming was indistinguishable from controls. There was no effect on longevity. The authors conclude that use of acrylics on social wasps is unlikely to influence social behavior in any measurable way. Although oil-based paints have not been tested on wasps, our use of them over many years gives us confidence that they have no effect on behavior or survival.

Another method is to paint the mesothorax with liquid-paper correction fluid and then write a number on it with indelible ink (McIntosh 1999). Different colors of correction fluid can be used as an additional variable. Our experience with this technique is that after a few days the correction fluid abrades and picks up stray dark streaks, making it increasingly hard to read the number as the wasp ages.

19.6.3.3 Handling Wasps for Marking

When it comes to paint marking, honey bees differ from many social vespids in at least two respects. Honey bee foragers can be marked without capture by applying the paint as they feed at a dish; pressure from the applicator, even if heavy enough to push their faces into their food solution, typically does not disturb them. *Ropalidia marginata* and *Polistes* spp., especially if newly eclosed, can be marked in this way on the nest (West-Eberhard 1969; Gadagkar 2001). In our experience this is not possible with vespines and many polistine wasps, which are extremely sensitive to movement nearby and will fly in response to contact with (or even to the approach of) a marking pen. Finally, smaller wasps, such as *Polybia occidentalis*, will not tolerate the kind of pressure from a paint applicator required to apply a spot.

The alternative is to capture the wasp on the nest and hold it for marking. Owen (1962) removed individual *P. fuscatus* from the nest with forceps, anesthetized them (with CO_2), marked them, and then returned them to the nest as they recovered. Naumann (1970), working with the much smaller *Protopolybia acutiscutis*, used the same method. In our work with *Polybia occidentalis*, capture enables the precise placement of the multiple dots required by the coding system described above (Figs. 19.5 and 19.6). Over many years of working with colonies of this wasp, we have found that the best method is to remove the selected individual directly from the nest surface using curved-tip reverse-action forceps (e.g., from SRA Soldering Products). These open when squeezed; when released they grip the wasp with a constant pressure that can be adjusted by bending the two arms at the base so that it is just sufficient to hold the wasp without injuring it; larger species require more pressure than smaller ones. The wasp is grasped from above around the thorax or the waist by releasing pressure on the forceps. The bent tips allow the wasp to be approached unnoticed from behind (rather than from above, as with straight tips). The opened bent tips can also be more precisely positioned around the wasp before closing, requiring only minute finger movements rather than whole wrist. To position the wasp for marking, it is transferred to a second pair of reverse-action bent-tip forceps, closed dorsoventrally over the posterior of the thorax (Fig. 19.8). This holds the wings out of the way while paint spots are applied to the thorax. Released after the paint dries (a few seconds), the wasp typically flies off and then returns to the nest after a few minutes. There is no evidence that this treatment affects the subsequent behavior of the marked wasps; rates of injury, fatal or otherwise, are negligible.

If a large random sample of a colony's adults must be marked, it may become expedient to capture them in bulk and chill them on ice to immobilize them before marking.

19.6.4 Uses of Video Recording

In the old days, 8- and 16-mm movie film was too expensive to be practical as a research tool, and was used primarily to document behavior for archiving, e.g., via *Encyclopaedia Cinematographica*. The advent of analog and now digital video has changed that. Digital camcorders have excellent macro-capabilities and are now cheap and compact enough to be used routinely in the field, and through repeated playbacks can enhance data collection in a variety of ways. The advantages are numerous: (1) obtaining precise counts and interarrival times of foragers in and out of a nest; (2) recording of interindividual interactions on the nest for detailed behavioral analysis; (3) analyzing the behavior of numerous simultaneously interacting individuals; (4) sharing of video with other scientists through the internet, as part of a publication, and/or through social media. In a recent study on gastral drumming in *Vespula germanica*, the use of recording made it possible to obtain rates of movement and trophallaxis for large samples of workers in a short amount of real time (Taylor and Jeanne 2018). Gathering such data would have been impossible without the use of video. Furthermore, by going backward through a video record, the behavior of an individual leading up to an event (e.g., its departure from the nest or from a swarm cluster) can be documented in detail (Sonnentag and Jeanne 2009) (see also Sect. 19.7.4)

On the downside, extracting data from videotapes can be very time-consuming. However, technology exists for streamlining the process. JWatcher, for example, is a free program that can be used as an event recorder to log the time at which keys are pressed while the observer follows the behavior of a focal insect as the video is played back (Blumstein and Daniel 2007). Its analysis routines can calculate time budgets, the duration of behavioral states, intervals between them, and run sequential analyses. Mignini and Lorenzi (2015) recently used the software in an analysis of the role of vibratory signals in *Polistes biglumis*. Image-based tracking technology can automate the collection of data on movement of individuals in video recordings (see Sect. 19.9).

19.6.5 Behavior-Sampling Techniques

Depending on the purpose of one's study, one or more behavioral sampling techniques will come into play, including ad libitum, focal-animal, and all-occurrences of a behavior pattern of interest. Altmann (1974) provides an excellent guide to these and other observational methods. Other useful resources include Martin and Bateson (1993) and, for analyzing sequential data, Bakeman and Gottman (1986).

19.6.6 Foraging

19.6.6.1 Collecting Foraged Loads

Prey loads brought to the nest by foragers can be collected and weighed and/or examined to determine the species. *Polistes* and *Mischocyttarus* foragers macerate the prey in the field before returning to the nest, so the identification of what was taken must rely on determining taxa based on fragments of cuticle. In contrast, *Polybia occidentalis*, *P. emaciata*, and others bring virtually intact prey to the nest, making it easy to identify large numbers of prey quickly (Hernandez et al. 2009; Yeison et al. 2013). Using reverse-action bent-tip forceps makes it easy to grasp returned prey foragers from the nest and make them drop their loads into vials of preservative. The same technique can be used to collect pulp loads from foragers (Jeanne 1986).

Water loads can be quantified by similarly restraining newly returned foragers and touching the tip of a microcapillary tube to their mouthparts. Lightly squeezing the gaster can encourage complete emptying of the crop (Roskens et al. 2010). If a regurgitated load fills a 10 μl capillary to 75% of its length, the load is 7.5 mg. Lightly tapping the tip of a capillary against the mouthparts of a larva will stimulate the release of larval saliva into the tube, where its volume can similarly be measured. Crop capacity of a wasp can be determined by letting a forager imbibe a sugar solution from a graduated capillary tube (Jeanne 1986). The same technique can be applied to nectar foragers (Fig. 19.9).

19.6.6.2 Foraging Distance

Especially for studies involving the use of social wasps in the biological control of agricultural pests, it is useful to know the distance to which wasps of a particular species will forage for prey. One approach is to determine the rates of return to the nest of workers released at a range of distances from the nest. This assumes that homing ability is based on knowledge of the territory surrounding the nest, which in turn is based on foraging experience. Clearly, older individuals, those likely to have gained such experience, will be able to home over greater distances than younger ones that have not yet begun to forage (Prezoto and Gobbi 2005) (for the use of harmonic radar tracking, see Sect. 19.9).

Fig. 19.9 *Vespula germanica* nectar forager imbibing sugar solution from a known-volume capillary tube. The ruler enables direct reading of amount imbibed (Photo: B. J. Taylor. Wisconsin, USA)

19.6.6.3 Recruitment to Food

A first step in determining whether a species has the ability to recruit nestmates to a food source is to train foragers to an artificial feeder. This can be difficult to accomplish for some species. Setting out feeders at some distance from the nest may or may not attract foragers from a nearby focal colony. It worked well for carrion-feeding species of *Agelaia* (Jeanne et al. 1995). In contrast, *Polybia occidentalis* foragers will rarely come to a sugar solution placed a few meters from a nest, and will readily abandon a feeder in the initial stages of training. Training this species to an artificial feeder was best accomplished by placing a dish of highly concentrated sucrose solution directly against the nest envelope below the entrance (Schueller et al. 2010; Taylor et al. 2010) (Fig. 19.10). Before long, several wasps will crawl onto the dish and feed from it, then return to the nest. We often found that as soon as the dish was moved away from the nest so that workers had to fly to it, they stopped coming (Schueller et al. 2010). Repeating the process several times, however, eventually resulted in some wasps learning to fly to and from the dish. Once trained to the dish, they typically continued to feed from it for several days (Taylor et al. 2011).

Wasps can learn to associate scents with food (Jandt and Jeanne 2005; Schueller et al. 2010; Taylor et al. 2011; Schueller 2012). Some studies have shown that wasps tend to be attracted to anise extract (Schueller et al. 2010; Taylor et al. 2010). Thus, if simple training to a single food source is required for an experiment, anise may be the best choice. However, if the aim is to test associative learning of a scent, anise should be avoided as its attractiveness could skew the results (Schueller et al. 2010). A feeder may consist simply of an open dish placed atop a tripod (Hrncir et al. 2007; Schueller et al. 2010; Taylor et al. 2010) (Fig. 19.10), or it may be more elaborate.

Fig. 19.10 The initial step in training wasps to a feeding dish. Also a method for catching a number of individuals at once for marking. (Photo: B. J. Taylor. Costa Rica)

Hrncir et al. (2007) inverted a glass cup filled with sugar solution onto a small acrylic plate cut with radial grooves. As the liquid in the grooves is drained by feeding wasps, they are kept filled from the reservoir in the glass cup.

Some wasp species are attracted to resources by "local enhancement," the visual cue of others feeding there (Raveret Richter 2000; Hrncir et al. 2007; Jeanne and Taylor 2009). To control for local enhancement, specialized feeders can eliminate its effect (Jandt and Jeanne 2005; Taylor et al. 2011). One type consists of a Syracuse dish covered with a metal tin with a rectangular hole cut into one side. A microscope slide inserted into the hole rests on the rim and slopes down into the sugar solution. To access the food, wasps have to walk down the slide and under the opaque cover. Thus, while feeding they are out of view of any wasps flying near the feeder.

In instances where nests are closely spaced, confirming the nest affiliation of wasps arriving at feeders can be accomplished in several ways. The most expedient is to mark all visitors to the feeder for individual recognition, then confirm their affiliation by observing each at the focal nest (Taylor et al. 2010). Another approach is to take advantage of the effect of colony odor. Schueller et al. (2010) marked foragers coming to the feeder, then presented each (held lightly in forceps) to wasps on the envelope of the focal colony or placed them inside the nest entrance. If the resident wasps attacked the introduced wasp, it was evidence that it did not belong to that colony. With this method, care must be taken to avoid disturbing the colony during the introduction. The greatest risk is that the handling of the wasp may cause it to release venom that could give a false negative with regard to membership. In addition, depending on the species, some colonies may accept non-nestmates (Sumner et al. 2007).

19.6.7 Testing Pheromones

19.6.7.1 Alarm

Although venom is the source of the alarm pheromone in all polistines so far tested, it is possible that in some species alarm substances are produced by other glands, either instead of the venom gland or in addition. *Vespula squamosa*, for example, produces an alarm pheromone in the head as well as in the venom (Landolt et al. 1999). The simplest approach to narrowing down the source(s) of alarm pheromone is to present crushed body parts—head, thorax, gaster, legs—of fresh-killed workers to other workers on the nest and scoring whether an alarm reaction is elicited. Once the tagma(ta) with activity are identified, the search can move to bioassaying specific glands in that body region by crushing and presenting them to nest workers in the same way. Because alarm pheromones are highly volatile and evaporate within seconds, each preparation must be presented to responders as soon after crushing as possible. Similarly, if the chemical components of a particular glandular secretion have been analyzed and likely active chemical components purchased or synthesized, they can be spotted on filter paper and tested in the same way.

For a more sophisticated approach to testing for alarm activity, it may be desired to quantify the responses of colony members to a candidate chemical. For example, Dani et al. (2000) showed that (2S,6R,8S)-2,8-dimethyl-1,7-dioxaspiro[5.5]undecane, a venom component of *Polybia occidentalis,* had alarm-releasing activity by spotting fixed amounts of it (in a solution of methanol) on filter paper and presenting it 1–3 cm upwind of an active nest. Behavioral responses were quantified and compared with controls (methanol only) in two ways. First, the number of wasps exiting the nest was determined from playbacks of video recordings of the nest. Second, the number of wasps attacking a nearby target (a plastic bottle wrapped in black paper) was determined from audible hits picked up by the camcorder's external microphone hung inside the bottle.

19.6.7.2 Emigration Trail

At least some Epiponini spot chemical trails that guide swarm members to a new nest site. An exception is *Apoica pallens*, which instead uses in-flight chemical calling (Howard et al. 2002). Circumstantial evidence for scent-marking can be obtained by observing the comings and goings of scouts a few meters from the swarm to see if they appear to be applying a glandular secretion to leaves. Gaster-dragging is a common scent-marking behavior seen in many epiponines (see Landolt et al. 1998 for a review), but it is possible that some may use glands other than sternal glands. To confirm that a chemical trail is produced and followed requires an experimental demonstration that swarm members can be led down an artificial trail (Jeanne 1981b). It would be worthwhile to investigate whether any independent-founding polistines engage in such marking, even if only around the new nest, as has been reported for *Mischocyttarus labiatus* by Litte (1981).

19.6.8 Nesting Behavior

A relatively neglected aspect of the biology of the Epiponini, in particular, concerns the details of nesting behavior. Much has been written about the final product, the architecture of the nest (Jeanne 1975a; Wenzel 1991), but relatively little about the behavioral sequences and stigmergic cues involved in building it (but see Downing and Jeanne 1988). Learning more about such details across species by mapping them onto a cladogram could very well provide new insights into how various nest types evolved. Moreover, there are tantalizing mysteries about how some species accomplish certain feats. For example, *Polybia occidentalis* constructs each new curved envelope so as to exactly parallel the curved surface of the comb being covered, but 2 cm out. How they maintain that distance without physically measuring the gap has not been answered. Careful observation at the nest can lead to hypotheses that can be tested with well-designed experiments.

19.7 Manipulating Colonies for Special Purposes

19.7.1 Age Polyethism

Studies of age polyethism require knowing the age of marked individuals. Various attempts have been made to estimate the ages of adult wasps based on morphological indicators. For example, the degree of pigmentation of the transverse apodeme of the first gastral sternite increases with age (Forsyth 1981). Unfortunately, this measure can at best indicate relative, not absolute, age and so is not of much use.

A better procedure is to introduce individually marked one-day-old adults into the observation colony and track their behavioral repertories as they age. Because newly eclosed adults take about a day to acquire their colony's odor from the nest, day-old individuals eclosing from colony A will be accepted and become part of the workforce of colony B. Jeanne and coworkers separated and incubated the combs of donor colonies of *P. occidentalis* (Jeanne et al. 1988). A source comb, kept dry and away from ants, will produce newly eclosing adults for a week or more. Each day the newly eclosed adult females were lightly anesthetized, marked for individual recognition (or merely for their day of eclosion), and dropped into the entrance of the observation nest. The tasks performed by these workers were then followed as they aged (Jeanne et al. 1988; O'Donnell and Jeanne 1992). In a slight variation, Simões and Zucchi (1980) used the lowermost of three combs of a nest of *Protopolybia exigua* as a daily source of newly eclosed individuals, which they marked and reintroduced to the same colony.

19.7.2 Colony Productivity

Michener (1964) argued that per-capita productivity decreases as colony size increases. Because of the wide range of size among founding swarms within a species, swarm-founders are ideal for testing the hypothesis. Including only founding swarms in the data set controls for the effect of stage of development on per-capita productivity. Colonies can be reset to the founding stage by dismantling the nest, then following the swarm to its new site (see next section). The colony is allowed to develop for *n* days, where *n* is just short of the egg-to-adult development time, and is then collected. All the adults will have been members of the founding swarm, and thus provide a measure of input, whereas the total brood biomass is a clean measure of output. Output/input gives a measure of per-capita productivity. Additional measures of output can include the number of brood cells and/or mass of the nest itself. Plotting per-capita productivity against colony size for a large sample of colonies provides a test of Michener's hypothesis. Only two species have been tested so far (*Polybia occidentalis*, *Parachartergus fraternus*); neither supports the hypothesis of decreasing per-capita output with increasing swarm size (Jeanne and Nordheim 1996; Bouwma et al. 2005; Bouwma et al. 2006).

19.7.3 Egg-to-Adult Development Time

Newly founded colonies can also be used to determine the egg-to-adult (E-A) development time, a fundamental life-history trait. For independent founders on their exposed combs, the data are easily obtained by keeping accurate daily records of the stage of brood in each cell, through cell-maps, tracking each newly laid egg until it gives rise to an adult. On the other hand, the multiple covered combs of most swarm founders make this difficult, if not impossible, so it is not surprising that brood-development times are known for less than a handful of species of epiponines. The scant data available suggest that E-A of swarm founders is significantly less than for independent-founding polistines (Jeanne in press), so data for more species would be welcome. For species whose nests are a single comb it is possible to keep track of brood development by periodically opening a flap of the envelope, as West-Eberhard (1978) did for *Metapolybia*. For species with multiple combs, this is impractical because the oldest brood will be in the upper comb, which is likely to be inaccessible without seriously damaging the nest. The alternative is to follow swarms from founding, then collecting them after *x* days, where the starting value of *x* is close to E-A for known species, around 30 days (Jeanne in press). The age of the oldest brood in the nest is determined and *x* is adjusted accordingly for subsequent trials until x = E-A by successive approximation. If the first colony in the series is collected a few days ahead of 30—25, for example—and the oldest comb is found to contain pupae, the comb can be incubated until the first adult ecloses. Adding the number of days of incubation to 25 will give a close approximation to the value of E-A, making it possible to close in on the value of *x* more quickly with subsequent colonies. It is important to keep in mind a complicating factor: as swarm size increases within a species, development time decreases (Howard and Jeanne 2004).

19.7.4 Working with Swarms

Although reproductive swarming characterizes all of the 200+ species of Epiponini, the details of behavior during this dispersal phase have been little studied. For some studies it is useful to know the population of the founding swarm. Working with *Polybia occidentalis*, Bouwma et al. (2003b) developed a method of obtaining precise counts. The swarm, either naturally occurring or induced, is captured at night by enclosing it and its twig in a plastic bag and refrigerating it overnight. Before daylight the next morning, the wasps are transferred to a capped 2 L plastic soda bottle via the removed bottom; after all are in the bottle, the bottom is sealed. The bottle is then clamped in an upright position to a tripod in the field and a 5 × 22-mm slit in the cap is untaped. With daybreak, the wasps become active and move up in the bottle. A camcorder zoomed in on the cap records the wasps as they exit through the slit. On video playback, a counter is used to tally the total number in the swarm, with repeated playbacks to confirm accuracy if necessary. Within an hour or two after leaving the bottle, all the wasps will again coalesce into a single cluster on a nearby leaf or twig.

For most studies involving swarming it is necessary to locate the new nesting site of a swarm. The first step is to mark some in the group so it can be recognized when found. A single spot of one color on the thorax for 3–5% of the population is sufficient (Bouwma et al. 2003a). If the swarm is lost during emigration, but a young nest is found a day or more later, its identity can be confirmed by the presence of marked individuals.

With some experience, it is possible to follow swarms as they emigrate to a new site. *P. occidentalis* takes several hours to a day to select a new site and move to it. Close observation of the swarm cluster can reveal when the activity level of swarm members increases, often a precursor to departure. Frequent inspections of vegetation a few meters in all directions from the swarm will often reveal where scouts are scent-marking vegetation on the way to a site they have found suitable. As time goes on, the concentration of scent marking will converge on a single direction from the swarm, and at that point, searching farther in that direction will reveal the scent-marked line leading to the selected new site. In the case of *P. occidentalis*, we have often been able in this way to find the chosen site—indicated by heavy scent marking by 10–20 scouts—long before the majority of the wasps arrive.

Our understanding of the behavior of the swarm up to the time it begins to emigrate is rudimentary. The typically dense packing of wasps in the swarm cluster makes it impossible to see any behavior except what little occurs on the surface facing the observer. Sonnentag and Jeanne (2009) got around this limitation by coaxing swarms of *P. occidentalis* onto a vertically mounted rectangle of plywood (Fig. 19.11). On this surface the wasps dispersed into numerous small clusters, exposing much of the interaction among swarm members to view and to videotaping, making it possible to work out much of the behavior leading up to the collective decision to make the move.

Fig. 19.11 Observation of behavior in a swarm. A swarm of *Polybia occidentalis* has been moved from its twig onto the board, where they disperse into smaller groups one or two wasps deep. Costa Rica. (Photo by Peter Sonnentag)

19.8 Studying Behavior in Artificially Housed Colonies

Housing colonies in nest boxes can be helpful if lab colonies need to be kept, or if predation on the colony is a concern. Also, while some groups of wasps build nests with open combs that allow observers to see colony activities, others nest in cavities and/or build an envelope that encloses the combs of the nest. This makes the use of an observation box necessary in order to study the inner workings of a colony. The use of such boxes to keep captive colonies became prominent in the 1970s for work on vespines (Greene 1991) and has enabled research in areas such as worker–queen interactions, worker–larva interactions, competition during colony founding, colony demographics, task partitioning, polyethism, communication, and the organization of foraging. The majority of studies using captive colonies have been conducted on two groups, *Polistes* and the vespine wasps.

Nest box design varies. *Polistes* have long been reared in captive or semi-captive conditions. Colonies can be kept in simple enclosures such as cardboard boxes (Reeve 1991) and provisioned with honey, water, and a source of protein (e.g., waxworms). If visual observations need to be performed, screen or plexiglass panels can be included in the nest box design (Post and Jeanne 1982; Jandt and Toth 2015). Boxes used in the field have a weather-tight roof and sides with a bottom of coarse wire mesh that allows wasps to pass through, but restricts access by predators (Judd and Carpenter 1996; Jandt and Toth 2015). Young colonies can be transplanted into such boxes. In areas with high densities of *Polistes*, foundresses often start nests in them (Rossi and Hunt 1988; Judd and Carpenter 1996; Hunt et al. 2007). These can then be moved to the lab, if needed.

Two main types of nest boxes have been used for studying vespine wasps. One style is an upright "vertical design" (Loope 2015). Here, the diameter of each comb is restricted to a width of approximately 3.75 cm by panes of glass on both sides. Because the nest is so narrow, it allows the observer to see what is happening between the nest combs. Furthermore, it allows the wasps to vertically expand the nest by adding additional combs below the existing ones as they would in a typical field nest (Loope 2015). The other type of nest box is a "horizontal design" originally described by Akre et al. (1976). In this type, the combs are separated and laid out next to one another in a single plane within the box. This design allows the colony to expand combs horizontally, but prevents vertical expansion. This type of box has undergone several improvements since the original design. Later iterations found that if the height of the box was reduced to 3 cm, wasps were discouraged from building envelope between the nest and the glass bottom, which would otherwise obstruct the view of the observer (Jandt and Jeanne 2005). Other improvements have included changes in the materials of the nest box itself. Originally, the box was constructed of wood, but Taylor et al. (2010) replaced it with nylon plastic because wooden nest boxes exposed to rain or high humidity warped. Further, they added slots for feeding the colony different types of food (Taylor et al. 2010; Taylor et al. 2012). Later versions of this same design replaced the Masonite ceiling with Plexiglas, which allowed the colony to be viewed from above and below the combs (Taylor and Jeanne 2018).

There are two options for providing access to food for artificially housed colonies. One is to allow the colony to forage outside via a short tube leading through a wall or window. This is all but mandatory for large colonies (i.e., the Vespinae and some Epiponini). The other is to limit the colony to foraging entirely indoors, by providing food either in the box housing the nest or via access to an adjacent foraging arena where food resources are provided by the experimenter. Completely enclosing colonies has been used with considerable success for *Polistes*, and both techniques have been successfully used with *Ropalidia marginata* (Gadagkar 2001).

There are some disadvantages to working with captive colonies. For one, most of the studies have been conducted with independent-founding species (*Polistes*, *Ropalidia*, *Vespula*), which do not readily adapt to changed nesting location unless they have already invested a considerable amount in the brood in the nest. Very few, if any, studies have been conducted on swarm-founding wasps held in enclosures, and it is unknown if colonies would find nest boxes desirable enough to remain within them. Secondly, if foraging is constrained to an enclosed arena in the lab, the artificial provision of an adequate diet may be an issue. Thirdly, a nest box restricts nest growth, since the colony cannot expand beyond the size of the box (Greene 1991). In addition, for subterranean-nesting species (e.g., *Vespula*) the entire task of cavity excavation is eliminated and therefore cannot be studied. It is unclear what effect this may have in an artificially housed colony, but it undoubtedly skews the allocation of worker effort away from what is experienced in the natural environment. Finally, rearing indoors may have effects on the colony that do not allow the results to be extended to natural colonies. For example, Jandt and coworkers found that for *Polistes* reared in a lab setting, some traits were skewed to be more gyne-like, possibly due to excess food. However, gene-expression analysis revealed that more worker-like genes were expressed. In addition, nest construction and male production were affected by rearing location (Jandt and Toth 2015). These results advise caution when extrapolating the results from lab-reared colonies to those in the field.

Use of nest boxes may provide some advantages for the study of behavior of Neotropical polistines. Cavity-nesters, such as many species of *Agelaia*, may adapt well to the "cavity" of a horizontal nest box, making possible close-up and detailed observation of behavior on the combs. For many of the species that construct envelopes, a narrow vertical design that constricts the width of the combs enough that in-nest behavior can be seen may be the best solution.

19.9 The Future

We foresee several ways that the development of new technology and methodologies will increase our understanding of Neotropical social wasp behavior and ecology. For one, the tools to study animal behavior and ecology will continue to increase in sophistication, allowing for ever-higher-throughput monitoring of individual behavior. Radio-frequency identification (RFID) tags, for example, attached to each individual are automatically read each time they pass by an RFID reader, such as when entering or leaving the nest. The tags are already small enough to attach to *Temnothorax* ants, and thus are likely suitable for most, if not all species of social wasps, provided the weight does not impede flight. This technology has already been used for several years on ants, bees, and a small number of social wasps to study behavior such as house-hunting during swarming, foraging, nest-drifting, shifts from nest work to foraging, adult life span, and reproductive parasitism (Sumner et al. 2007; Robinson et al. 2009; Van Oystaeyen et al. 2013; Nunes-Silva et al. 2018; Santoro et al. 2019). Future developments will likely include further miniaturization and longer reading distances.

Harmonic radar tracking has been utilized to track insects, including bees, over large spatial scales, and it was recently used to track the flight paths of bumble bee (*Bombus terrestris*) workers over their entire lifetimes (Woodgate et al. 2016). As transponders are further reduced in size, they will become available for use on smaller species of social wasps. The use of this technology, along with Geographic Information System (GIS) mapping techniques could allow us to understand how wasps explore their environment, how far they forage from the nest, the ontogeny of foraging behavior, and how foraging behavior changes with spatial heterogeneity of the environment.

Advances have also occurred in the area of image-based tracking technology (Mersch et al. 2013; Dell et al. 2014; Crall et al. 2015; Wario et al. 2015). The software tracks movement in video recordings and extracts information related to spatial use, activity patterns, and, increasingly, the characterization of specific behavioral acts. If individual identification is necessary, tags with unique barcodes or quick-response (QR) codes can be attached to each individual (Crall et al. 2015).

The advantages of these technologies are many. Perhaps most obvious is the amount of data that can be collected. In pre-tech days, a trained ethologist would be limited to the individuals they could readily observe or record with a camcorder. Current technologies allow for constant monitoring of multiple individuals, sometimes for months, and during a timeframe that can encompass millions of interactions (Mersch et al. 2013; Wario et al. 2015). Furthermore, the effort required to extract data from videotapes by hand is lengthy and labor-intensive. Now, the collection and compiling of data sets on behavior requires much less time, and in some cases is instantaneous, giving researchers access to the data in a fraction of the time. The data collected is also highly accurate, avoids observer bias, and precludes the need to correct for individual biases when multiple observers are working on the same study.

The methods are not without limitations. For example, RFID technology can only register particular individuals when they are within a few millimeters of the RFID reader. Similarly, image-based tracking technology is limited by what can be recorded via video. This is especially relevant to Neotropical social wasps, largely because the technology of housing colonies in observation nest boxes for most species is not developed. While it is certainly possible to video-record individuals on the outside of the nest or at a foraging site, it leaves out the richness of the behavioral interactions inside the nest for all species whose nests are covered by an envelope. If most Neotropical social wasps cannot be housed in observation boxes, the use of image-based tracking techniques to record internal interactions may be limited. All of the methods mentioned above are limited by the sophistication of the software that can be used to recognize individuals, track them, and categorize behavior into particular types. They are also limited by the statistical techniques that can be used to mine the extensive data sets that are produced. Fortunately, programs for tracking individuals, such as Track-a-Forager, BEEtag, and ToxTrac (programmed in Java, MATLAB ®, or R) are available for all platforms, have come a long way toward solving these issues, and will continue to improve (Crall et al. 2015; Van Geystelen et al. 2016; Rodriguez et al. 2018).

Acknowledgments The senior author acknowledges students, postdocs, and colleagues (including the junior author) who have worked with him in the tropics and, whether they know it or not, contributed in ways large and small to what is in this chapter: Roger Swain, David Post, Holly Downing, Eloy Castellon Bermudez, Michel Pratte, Monica Raveret Richter, Neal Williams, Sean O'Donnell, Larry Phelps, Jim Hunt, Malcolm Keeping, Karen London, Andy Bouwma, Kenneth Howard, Teresa Schueller, Peter Sonnentag, and Kevin Loope. We would also like to thank Kevin Mark for his assistance in the section on ether anesthesia. We also thank Priyantha Wijesinghe and Jim Carpenter for their assistance with the section on preservation for museum specimens and for molecular analysis.

References

Akre RD, Garnett WB, MacDonald JF, Greene A, Landolt PJ (1976) Behavior and colony development of *Vespula pensylvanica* and *V. atropilosa*. J Kansas Entomol Soc 49:63–84
Alloway TM (1972) Learning and memory in insects. Annu Rev Entomol 17:43–56
Altmann J (1974) Observational study of behavior: sampling methods. Behaviour 49:227–267
Aragão M, Andena SR (2016) The social wasps (Hymenoptera: Vespidae: Polistinae) of a fragment of Atlantic Forest in southern Bahia, Brazil. J Nat Hist 50:1411–1426
Arnett RH Jr (2000) American insects: a handbook of the insects of America north of Mexico. CRC Press, Boca Raton, Florida
Ashburner M, Thompson JN Jr (1978) The laboratory culture of *Drosophila*. In: Ashburner M, Wright TRF (eds) Genetics and biology of *Drosophila*. Academic Press, London, pp 1–109
Bakeman R, Gottman JM (1986) Observing interaction: an introduction to sequential analysis. Cambridge University Press, Cambridge, UK
Barbosa BC, Detoni M, Maciel TT, Prezoto F (2016) Studies of social wasp diversity in Brazil: over 30 years of research, advancements and priorities. Sociobiology 63:858–880
Blumstein DT, Daniel JC (2007) Quantifying behavior the JWatcher way. Sinauer Associates Incorporated, Sunderland, Massachusetts
Blüthgen N, Stork NE (2007) Ant mosaics in a tropical rainforest in Australia and elsewhere: a critical review. Austral Ecol 32:93–104
Bouwma AM, Bouwma PE, Nordheim EV, Jeanne RL (2003a) Adult mortality rates in young colonies of a swarm-founding social wasp. J Zool 260:11–16
Bouwma AM, Bouwma PE, Nordheim EV, Jeanne RL (2003b) Founding swarms in a tropical social wasp: adult mortality, emigration distance, and swarm size. J Insect Behav 16:439–452
Bouwma AM, Howard KJ, Jeanne RL (2005) Parasitism in a social wasp: effect of gregarines on foraging behavior, colony productivity, and adult mortality. Behav Ecol Sociobiol 59:222–233
Bouwma AM, Nordheim EV, Jeanne RL (2006) Per-capita productivity in a social wasp: no evidence for a negative effect of colony size. Insect Soc 53:412–419
Brenner RJ, Patterson RS (1988) Efficiency of a new trapping and marking technique for peridomestic cockroaches (Dictyoptera: Blattaria). J Med Entomol 25:489–492
Chao A (2019) Anne Chao's website. http://chao.stat.nthu.edu.tw/wordpress/software_download/softwarespader_online/
Chao A, Colwell RK (2017) Thirty years of progeny from Chao's inequality: estimating and comparing richness with incidence data and incomplete sampling. Sort-Statistics Operat Res Transact 41:3–54
Colwell RK (2013) EstimateS: statistical estimation of species richness and shared species from samples. Version 9 and earlier. User's Guide and application. http://purl.oclc.org/estimates
Corbara B, Carpenter JM, Céréghino R, Leponce M, Gibernau M, Dejean A (2009) Diversity and nest site selection of social wasps along Guianese forest edges: assessing the influence of arboreal ants. C R Biol 332:470–479
Crall JD, Gravish N, Mountcastle AM, Combes SA (2015) BEEtag: a low-cost, image-based tracking system for the study of animal behavior and locomotion. PLoS One 10:1–13
Dani FR, Jeanne RL, Clarke SR, Jones GR, Morgan ED, Francke W, Turillazzi S (2000) Chemical characterization of the alarm pheromone in the venom of *Polybia occidentalis* and of volatiles from the venom of *P. sericea*. Physiol Entomol 25:363–369
De Souza AR, Venancio DFA, Zanuncio JC, Prezoto F (2011) Sampling methods for assessing social wasps species diversity in a eucalyptus plantation. J Econ Entomol 104:1120–1123
Dell AI, Bender JA, Branson K, Couzin ID, de Polavieja GG, Noldus LP, Pérez-Escudero A, Perona P, Straw AD, Wikelski M, Brose U (2014) Automated image-based tracking and its application in ecology. Trends Ecol Evol 29:417–428
Desuo IC, Souza-Galheico CB, Shima SN, Santos GMM, Cruz JD, Bichara CC, Dias CTS (2011) An adaptive view of caste differentiation in the Neotropical wasp *Polybia (Trichothorax) sericea* Olivier (Hymenoptera: Vespidae). Neotrop Entomol 40:653–660

Downing HA, Jeanne RL (1988) Nest construction by the paper wasp, *Polistes*: a test of stigmergy theory. Anim Behav 36:1729–1739

Dvořák L, Landolt PJ (2006) Social wasps trapped in the Czech Republic with syrup and fermented fruit and comparison with similar studies (Hymenoptera Vespidae). Bulletin of Insectology 59:115–120

Edwards R (1980) Social wasps. Their biology and control. Rentokil Unlimited, East Grinstead, UK

Forsyth AB (1981) Swarming activity of polybiine social wasps (Hymenoptera: Vespidae: Polybiini). Biotropica 13:93–99

Gadagkar R (2001) The social biology of *Ropalidia marginata*. Toward understanding the evolution of eusociality. Harvard University Press, Cambridge, Massachusetts

Gomes B, Noll FB (2009) Diversity of social wasps (Hymenoptera, Vespidae, Polistinae) in three fragments of semideciduous seasonal forest in the northwest of São Paulo state, Brazil. Revista Brasileira de Entomologia 53:428–431

Greene A (1991) *Dolichovespula* and *Vespula*. In: Ross KG, Matthews RW (eds) The social biology of wasps. Cornell University Press, Ithaca, NY, pp 263–305

Hagler JR, Jackson CG (2001) Methods for marking insects: current techniques and future prospects. Annu Rev Entomol 46:511–543

Hastings MD, Queller DC, Eischen F, Strassmann JE (1998) Kin selection, relatedness, and worker control of reproduction in a large-colony epiponine wasp, *Brachygastra mellifica*. Behav Ecol 9:573–581

Hernandez J, Sarmiento CE, Fernandez C (2009) Foraging activity of *Polybia occidentalis venezuelana* (Hymenoptera, Vespidae). Revista Colombiana de Entomologia 35:230–234

Howard KJ, Jeanne RL (2004) Rates of brood development in a social wasp: effects of colony size and parasite infection. Insect Soc 51:179–185

Howard KJ, Smith AR, O'Donnell S, Jeanne RL (2002) Novel method of swarm emigration by the epiponine wasp, *Apoica pallens* (Hymenoptera Vespidae). Ethol Ecol Evolut 14:365–371

Hrncir M, Mateus S, Nascimento FS (2007) Exploitation of carbohydrate food sources in *Polybia occidentalis:* social cues influence foraging decisions in swarm-founding wasps. Behav Ecol Sociobiol 61:975–983

Hunt JH, Kensinger BJ, Kossuth JA, Henshaw MT, Norberg K, Wolschin F, Amdam GV (2007) A diapause pathway underlies the gyne phenotype in *Polistes* wasps, revealing an evolutionary route to caste-containing insect societies. Proc Natl Acad Sci U S A 104:14020–14025

Ishay JS, Pertsis V, Levtov E (1994) Duration of hornet sleep induced by ether anesthesia is curtailed by exposure to sun or UV irradiation. Experientia 50:737–741

Jacques GC, Pires E, Hermes MG, Faria LDB, Souza MM, Silveira LCP (2018) Evaluating the efficiency of different sampling methods to survey social wasps (Vespidae: Polistinae) in an anthropized environment. Sociobiology 65:515

Jandt JM, Jeanne RL (2005) German yellowjacket (*Vespula germanica*) foragers use odors inside the nest to find carbohydrate food sources. Ethology 111:641–651

Jandt JM, Toth AL (2015) Physiological and genomic mechanisms of social organization in wasps (family: Vespidae). Adv Insect Physiol 48:95–130

Jeanne RL (1972) Social biology of the Neotropical wasp *Mischocyttarus drewseni*. Bull Mus Comp Zool Harv Univ 144:63–150

Jeanne RL (1975a) The adaptiveness of social wasp nest architecture. Q Rev Biol 50:267–287

Jeanne RL (1975b) Social biology of *Stelopolybia areata* (Say) in Mexico (Hymenoptera: Vespidae). Insect Soc 22:27–34

Jeanne RL (1981a) Alarm recruitment, attack behavior, and the role of the alarm pheromone in *Polybia occidentalis* (Hymenoptera: Vespidae). Behav Ecol Sociobiol 9:143–148

Jeanne RL (1981b) Chemical communication during swarm emigration in the social wasp *Polybia sericea* (Olivier). Anim Behav 29:102–113

Jeanne RL (1986) The organization of work in *Polybia occidentalis*: costs and benefits of specialization in a social wasp. Behav Ecol Sociobiol 19:333–341

Jeanne RL (1991) The swarm-founding Polistinae. In: Ross KG, Matthews RW (eds) The social biology of wasps. Cornell University Press, Ithaca, New York, pp 191–231

Jeanne RL (2003) Social complexity in the Hymenoptera, with special attention to the wasps. In: Kikuchi T, Azuma N, Higashi S (eds) Genes, behaviors and evolution of social insects. Hokkaido University Press, Sapporo, Japan, pp 81–130

Jeanne RL (in press) Swarm-founding wasps. In: Starr CK (ed) Encyclopedia of social insects. Springer, Cham, Switzerland

Jeanne RL, Downing H, Post DC (1988) Age polyethism and individual variation in *Polybia occidentalis*, an advanced eusocial wasp. In: Jeanne RL (ed) Interindividual behavioral variability in social insects. Westview Press, Boulder, Colorado, pp 323–357

Jeanne RL, Hunt JH, Keeping MG (1995) Foraging in social wasps: *Agelaia* lacks recruitment to food (Hymenoptera: Vespidae). J Kansas Entomol Soc 68:279–289

Jeanne RL, Nordheim EV (1996) Productivity in a social wasp: per capita output increases with swarm size. Behav Ecol 7:43–48

Jeanne RL, Taylor B (2009) Individual and social foraging in social wasps. In: Jarau S, Hrncir M (eds) Food exploitation by social insects: ecological, behavioral, and theoretical approaches. CRC Press, Boca Raton, Florida, pp 53–79

Johnson NF, Triplehorn CA (2004) Introduction to the study of insects. Thompson Brooks Cole, Belmont, California

Judd TM, Carpenter JM (1996) *Polistes dominulus* (Hymenoptera: Vespidae) found in Michigan. Great Lakes Entomol 29:45–46

Landolt P, Reed H, Heath R (1999) An alarm pheromone from heads of worker *Vespula squamosa* (Hymenoptera: Vespidae). Fla Entomol 82:356–359

Landolt PJ, Jeanne RL, Reed HC (1998) Chemical communication in social wasps. In: Meer RKV, Breed MD, Winston ML, Espelie KE (eds) Pheromone communication in social insects. Westview Press, Boulder, Colorado, pp 216–235

Litte M (1981) Social biology of the polistine wasp *Mischocyttarus labiatus*: survival in a Colombian rain forest. Smithsonian Contribut Zool 327:1–27

Loope KJ (2015) Queen killing is linked to high worker-worker relatedness in a social wasp. Curr Biol 25:2976–2979

Mardan M, Rinderer TE (1980) Effects of carbon dioxide and cold anaesthesia on the hoarding behaviour of the honeybee. J Apic Res 19:149–153

Martin P, Bateson P (1993) Measuring behaviour. Cambridge University Press, Cambridge, UK

McIntosh RL (1999) Technique for labeling individual *Pissodes strobi* (Coleptera: Curculonidae) for mark-recapture studies. Canadian Entomol Suppl 131:131–136

Melo AM, Barbosa BC, Castro MM, Santos GMM, Prezoto F (2015) The social wasp community (Hymenoptera, Vespidae) and new distribution record of *Polybia ruficeps* in an area of Caatinga biome, northeastern Brazil. Check List 11:1–5

Mersch DP, Crespi A, Keller L (2013) Tracking individuals shows spatial fidelity is a key regulator of ant social organization. Science 340:1090–1093

Michener CD (1964) Reproductive efficiency in relation to colony size in hymenopterous societies. Insect Soc 11:317–342

Mignini M, Lorenzi CM (2015) Vibratory signals predict rank and offspring caste ratio in a social insect. Behav Ecol Sociobiol 69:1739–1748

Murakami ASN, Shima SN, Desuo IC (2009) More than one inseminated female in colonies of the independent-founding wasp *Mischocyttarus cassununga* von Ihering (Hymenoptera, Vespidae). Revista Brasileira de Entomologia 53:653–662

Naumann MG (1970) The nesting behavior of *Protopolybia pumila* in Panama (Hymenoptera: Vespidae). Ph.D. thesis. University of Kansas, Lawrence, Kansas

Nicolas G, Sillans D (1989) Immediate and latent effects of carbon dioxide on insects. Annu Rev Entomol 34:97–116

Noll FB, Gomes B (2009) An improved bait method for collecting Hymenoptera, especially social wasps (Vespidae: Polistinae). Neotrop Entomol 38:477–481

Noll FB, Zucchi R (2002) Castes and the influence of the colony cycle in swarm-founding polistine wasps (Hymenoptera, Vespidae, Epiponini). Insect Soc 49:62–74

Nunes-Silva P, Hrncir M, Guimarães JTF, Arruda H, Costa L, Pessin G, Siqueira JO, de Souza P, Imperatriz-Fonseca VL (2018) Applications of RFID technology on the study of bees. Insect Soc 66:15–24

O'Donnell S, Jeanne RL (1992) Lifelong patterns of forager behavior in a tropical swarm-founding wasp - effects of specialization and activity level on longevity. Anim Behav 44:1021–1027

Owen J (1962) The behavior of a social wasp *Polistes fuscatus* (Vespidae) at the nest, with special reference to differences between individuals. Ph.D. thesis. University of Michigan, Ann Arbor

Packer L (2005) The influence of marking upon bee behaviour in circle tube experiments with a methodological comparison among studies. Insect Soc 52:139–146

Peitsch D, Fietz A, Hertel H, de Souza J, Ventura DF, Menzel R (1992) The spectral input systems of hymenopteran insects and their receptor-based colour vision. J Comp Physiol A 170

Plath OE (1924) Do anesthetized bees lose their memory? Am Nat 58:162–166

Pomeroy N, Plowright RC (1979) Larval ejection following CO2 narcosis of bumble bees (Hymenoptera: Apidae). J Kansas Entomol Soc 52:215–217

Post DC, Jeanne RL (1982) Recognition of former nestmates during colony founding by the social wasp *Polistes fuscatus* (Hymenoptera: Vespidae). Behav Ecol Sociobiol 11:283–285

Prezoto F, Gobbi N (2005) Flight range extension in *Polistes simillimus* Zikán, 1951 (Hymenoptera, Vespidae). Braz Arch Biol Technol 48:947–950

Rau P (1933) The jungle bees and wasps of Barro Colorado Island. Phil Rau, Kirkwood, Missouri

Raveret Richter M (2000) Social wasp (Hymenoptera: Vespidae) foraging behavior. Annu Rev Entomol 45:121–150

Reed HC, Gallego J, Nelson J (1988) Morphological evidence for polygyny in post-emergence colonies of the red paper wasp, *Polistes perplexus* Cresson (Hymenoptera: Vespidae). J Kansas Entomol Soc 61:453–463

Reeve HK (1991) *Polistes*. In: Ross KG, Matthews RW (eds) The social biology of wasps. Cornell University Press, Ithaca, New York, pp 99–148

Ribbands CR (1950) Changes in the behaviour of honey-bees following their recovery from anaesthesia. J Exp Biol 27:302–310

Richards OW (1945) A revision of the genus *Mischocyttarus* de Saussure (Hymen., Vespidae). Transact Royal Entomol Soc London 95:295–462

Richards OW (1978) The social wasps of the Americas excluding the Vespinae. British Museum (Natural History), London

Richards OW, Richards MJ (1951) Observations on the social wasps of South America (Hymenoptera Vespidae). Transact Royal Entomol Soc London 102:1–170 + 174 plates

Robinson EJ, Richardson TO, Sendova-Franks AB, Feinerman O, Franks NR (2009) Radio tagging reveals the roles of corpulence, experience and social information in ant decision making. Behav Ecol Sociobiol 63:627–636

Rocha AA, Giannotti E (2016) External morphology of immatures during the post-embryonic development of *Mischocyttarus nomurae* Richards (Hymenoptera: Vespidae). Sociobiology 63:998–1004

Rodriguez A, Zhang H, Klaminder J, Brodin T, Andersson PL, Andersson M (2018) ToxTrac: a fast and robust software for tracking organisms. Methods Ecol Evol 9:460–464

Roskens VA, Carpenter JM, Pickett KM, Ballif BA (2010) Preservation of field samples for enzymatic and proteomic characterization: analysis of proteins from the trophallactic fluid of hornets and yellowjackets. J Proteome Res 9:5484–5491

Rossi AM, Hunt JH (1988) Honey supplementation and its developmental consequences: evidence for food limitation in a paper wasp, *Polistes metricus*. Ecol Entomol 13:437–442

Saga T (2019) Evaluation of the productivity of social wasp colonies (Vespinae) and an introduction to the traditional Japanese *Vespula* wasp hunting technique. J Vis Exp 151:e59044

Santoro D, Hartley S, Lester PJ (2019) Behaviourally specialized foragers are less efficient and live shorter lives than generalists in wasp colonies. Sci Rep 9:1–10

Schueller TI (2012) Aspects of integration in colonies of a highly eusocial wasp with emphasis on foraging, nest construction, and defense. Ph.D. thesis. The University of Wisconsin - Madison

Schueller TI, Nordheim EV, Taylor B, Jeanne RL (2010) The cues have it: nest-based, cue-mediated recruitment to carbohydrate resources in a swarm-founding social wasp. Naturwissenschaften 97:1017–1022

Silva-Pereira V d, Santos GMM (2006) Diversity in bee (Hymenopetra, Apoidea) and social wasp (Hymenoptera, Vespidae) community in Campos rupestres, Bahia, Brazil. Neotrop Entomol 35:165–174

Silveira OT (2002) Surveying neotropical social wasps: an evaluation of methods in the "Ferreira Penna" research station (ECFPn), in Caxiuanã, PA, Brazil (Hym., Vespidae, Polistinae). Papéis Avulsos de Zoologia 42:299–323

Silveira OT, Esposito MC, Santos JN d Jr, Gemaque FE Jr (2005) Social wasps and bees captured in carrion traps in a rainforest in Brazil. Entomol Sci 8:33–39

Simões D, Zucchi R (1980) Bionomics of *Protopolybia exigua exigua* (de Saussure) I – Age polyethism and life-table. (Hymenoptera, Vespidae, Polybiini). Naturalia 5:79–87

Somavilla A, de Oliveira ML, Silveira OT (2014) Diversity and aspects of the ecology of social wasps (Vespidae, Polistinae) in central Amazonian "terra firme" forest. Revista Brasileira de Entomologia 58:349–355

Sonnentag PJ, Jeanne RL (2009) Initiation of absconding-swarm emigration in the social wasp *Polybia occidentalis*. J Insect Sci 9:1–11

Sorensen AA, Busch TM, Vinson SB (1983) Factors affecting brood cannibalism in laboratory colonies of the imported fire ant, *Solenopsis invicta* Buren (Hymenoptera: Formicidae). J Kansas Entomol Soc 56:140–150

Southwood TRE (1978) Ecological methods. With particular reference to the study of insect populations. Chapman and Hall, New York

Souza AR d, Ribeiro B, Jose N, Prezoto F (2012) Paint marking social wasps: an evaluation of behavioral effects and toxicity. Entomol Exp Appl 144:244–247

Spradbery JP (1973) Wasps. An account of the biology and natural history of solitary and social wasps. University of Washington Press, Seattle, Washington

Sumner S, Lucas E, Barker J, Isaac N (2007) Radio-tagging technology reveals extreme nest-drifting behavior in a eusocial insect. Curr Biol 17:140–145

Taylor BJ, Jeanne RL (2018) Gastral drumming: a nest-based food-recruitment signal in a social wasp. Sci Nat 105

Taylor BJ, Nordheim EV, Jeanne RL (2012) Allocation of colony-level foraging effort in *Vespula germanica* in response to food resource quantity, quality, and associated olfactory cues. Ethology 118:594–605

Taylor BJ, Nordheim EV, Schueller TI, Jeanne RL (2011) Recruitment in swarm-founding wasps: *Polybia occidentalis* does not actively scent-mark carbohydrate food sources. Psyche 2011:7

Taylor BJ, Schalk DR, Jeanne RL (2010) Yellowjackets use nest-based cues to differentially exploit higher-quality resources. Naturwissenschaften 97:1041–1046

Van Geystelen A, Benaets K, de Graaf DC, Larmuseau MHD, Wenseleers T (2016) Track-a-forager: a program for the automated analysis of RFID tracking data to reconstruct foraging behaviour. Insect Soc 63:175–183

Van Oystaeyen A, Alves DA, Oliveira RC, do Nascimento DL, do Nascimento FS, Billen J, Wenseleers T (2013) Sneaky queens in *Melipona* bees selectively detect and infiltrate queenless colonies. Anim Behav 86:603–609

Visscher PK, Seeley TD (1989) Bee-lining as a research technique in ecological studies of honey bees. Am Bee J 129:536–539

von Frisch K (1967) The dance language and orientation of bees. Harvard University Press, Cambridge, Massachusetts

Walker TJ, Wineriter SA (1981) Marking techniques for recognizing individual insects. Fla Entomol 64:18–29

Wario F, Wild B, Couvillon MJ, Rojas R, Landgraf T (2015) Automatic methods for long-term tracking and the detection and decoding of communication dances in honeybees. Front Ecol Evol 3:103

Wenzel J (1991) Evolution of nest architecture. In: Ross KG, Matthews RW (eds) The social biology of wasps. Cornell University Press, Ithaca, New York, pp 480–519

West-Eberhard MJ (1969) The social biology of polistine wasps. Miscellaneous Publications of the Museum of Zoology, University of Michigan 140:1–101

West-Eberhard MJ (1978) Temporary queens in *Metapolybia* wasps: nonreproductive helpers without altruism? Science 200:441–443

Wineriter SA, Walker TJ (1984) Insect marking techniques: durability of materials. Entomol News 95:117–123

Woodgate JL, Makinson JC, Lim KS, Reynolds AM, Chittka L (2016) Life-long radar tracking of bumblebees. PLoS One 11

Yeison LG, John HD, Caraball P (2013) Foraging activity of the social wasp *Polybia emaciata* (Hymenoptera: Vespidae: Polistinae). Revista Colombiana de Entomologia 39:250–255

Chapter 20
Community Ecology of Social Wasps in Brazil: Forty Years of Studies

Gilberto M. de M. Santos, Emanuelle L. S. Brito, and Marcos Aragão

Abstract Social wasps present diverse functionalities in ecosystems, from predators, pollinators, detritivores to bioindicators; in addition, their dual participation in the food chain sometimes as carbohydrate consumers and sometimes as consumers of protein, made these organisms particularly attractive for study in community ecology. After 40 years from the first works about social wasps' communities of the Brazilian fauna, in which these primitives studies concentrated their efforts in describing the communities through faunistic analysis, and along the time passing to the niche analysis and more recently to the interaction networks. Thereby, the main product of these works was the information about the trophic and temporal niches and consequently, the contribution to aspects of the niche partitioning that are in consolidation for this group. In this chapter, with a historic review of researches with communities of social wasps in Brazil, we also attempt to provide brief insights on the assumptions used to access the structure of an ecological community, presenting community organization models described in the literature and examples of how these models are inserted in ecological studies of the Brazilian social wasps, and additionally, we indicate some aspects related to guilds and functional groups and how these communities and their ecological diversity are composed. Thus, this chapter poses some important questions about the study of social wasp communities, which we believe will be guidelines for the coming decades of studies on the ecological aspects of Polistinae fauna in Brazil.

Keywords Vespidae · Polistinae · Ecological communities · Neotropical wasps

---

G. M. de M. Santos (✉) · E. L. S. Brito · M. Aragão
Laboratório de Entomologia, Departamento de Ciências Biológicas, Programa de Pós-Graduação em Ecologia e Biodiversidade, Universidade Estadual de Feira de Santana, Bahia, Brazil

F. Prezoto et al. (eds.), *Neotropical Social Wasps*,
https://doi.org/10.1007/978-3-030-53510-0_20

20.1 The Ecological Importance of Social Wasps

The magnitude of biological diversity among social wasps is one indicator to the ecological importance of Polistinae in Brazil, where 369 species of social wasps are known, of which more than 100 are endemic (Carpenter and Marques 2001; Prezoto et al. 2007; Somavilla et al. 2017). In the last 20 years, there has been considerable advance in knowledge about Neotropical social wasps; the growing interest in these insects is justified by the diversity of ecological functions that they play in ecosystems services (Barbosa et al. 2016).

The ecological functions performed by social wasps are directly related to the trophic behavior of these insects, which can act as predators, pollinators or be necrophagous because depending on the nutritional demands of their colonies they have multiple positions in the food chain (Hunt et al. 1991; Carpenter and Marques 2001). The dichotomy in nutritional demands in the different phases of the life cycle of the social wasps, primarily herbivorous in adulthood (carbohydrate consumers) and primarily carnivorous when immature (protein consumers), allows polistines to occupy two prominent positions in food webs. Depending on the relationship between the number of adults and/or immature species in the colonies, these insects primarily collect carbohydrates present in flowers and fruit or sources of protein, either of animal origin through the predation of other insects (Prezoto et al. 2008a; Barbosa et al. 2014), or by collecting tissues in decayed animal carcasses (O'Donnell 1995; Silveira et al. 2005; Barbosa et al. 2015).

Besides, some studies have shown that social wasps can contribute effectively to pollination (Shuttleworth and Johnson 2009; Sühs et al. 2009) and participate in the complex of natural enemies, reducing herbivory in natural and agricultural systems (Hunt 2007; Prezoto et al. 2008a; Elisei et al. 2017). Urbini et al. (2006) and Souza et al. (2010) suggest that social wasps can be used as bioindicators of the degree of conservation in ecosystems, thanks to the great ecological plasticity of social wasps (Santos and Gobbi 1998). These insects also establish interspecific relationships, still poorly understood, with ants (Souza et al. 2013a, 2013b) and birds (Souza et al. 2017b).

20.2 Community Ecology of Social Wasps

The ecological concept of communities has been widely used since the beginning of the twentieth century, albeit with many divergences in its breadth of scope, especially in relation to the limits of use of some terms that are seldom employed, and arbitrarily neglected, such as communities, biocenoses, associations, groupings, congregations, assemblies, taxocenoses, and so on. The complexity of these levels makes them chaotic, and studies in this approach involve several distinct taxa and several distinct functional groups, making them difficult systems to describe and analyze.

Odum (1988) was one of the disseminators of the term "biological community" and defined it simply: a community as a set of populations. In general terms, communities are classically defined as a set of populations occurring in the same space and time (Begon et al. 2007), meaning populations as sets of individuals of the same species reproductively isolated, that is, distinct evolutionary entities. Several authors (e.g., Clements 1916; Whittaker 1975; Begon 1979; Gurevitch et al. 2009, and several works by the Odum brothers) have disseminated the idea that rather than a simple set of populations—the living part of an ecosystem—a community involves obligatory processes of interactions between the organisms that compose it and some level of redundancy and complementarity in the ecological functions of the different populations in a community. This concept should be assumed as a community *strictu* sensu.

Very complex concepts are closer to reality, but they are difficult to understand, while simpler concepts, even if they are not so close to factual reality, are more readily comprehensible. The academy opted for the comprehensibility of the concept of communities, using the classic "set of populations occurring in the same space and time"; however, this concept is applied in a broader way and allows the study of communities based on a taxon (e.g., ant community) and studies based on an ecological function (e.g.. community of seed dispersers), or studies that involve multiple criteria (e.g., a seed dispersal ant community).

In this text we define the term "community," *lato* sensu, as a set of populations (of social wasps) that interact with each other and with the environment in a given period, generating a system (hierarchical level) distinct from the simple sum of the characteristics of the populations involved. Moreover, from the interactions between the different component populations, new typical community properties, nonreducible properties, emerge, which cannot be understood only by studies of the component parts.

Richness, diversity, equitability, evenness, guild structure, structural stability, and composition are examples of emerging community-level properties. Several factors, biotic and abiotic, affect and are affected by the properties of local communities. In terms of biotic factors, interaction processes are among the main structuring agents of these communities, triggering the emergence of properties in ecosystems, at a hierarchically more complex level. The limits of similarity between populations thus depend on their emerging properties and how these properties are organized in a community (Begon et al. 2007).

Another issue involves a subtle and important difference, albeit neglected, between syntopic populations and co-occurring populations. Any population of the component set of a community is syntopic since by definition they occur in the same time and space; however, populations are considered co-occurring only if the presence of a population places selection pressure on another population component. That is, populations may be considered co-occurring if any interaction, agonist or non-agonist, affects the structure, composition, or dynamics of this community (Schoener 1986; Santos and Presley 2010). This premise leads to an essential questioning of the ecology of communities: how do certain species co-occur in the same community? Could it be driven by the process of competition or facilitation? A number of theories attempt to answer this question. For example, if we accept that competition among populations is an important factor with effects on community properties, there will be a little similarity expected in the use of resources or divergence in the way resources are used, so that social wasp species can co-occur. This hypothesis has led several researchers to measure the degree of resource sharing among populations that co-occur in a given community (e.g., Santos et al. 2010; Mello et al. 2011).

Classical ecology studies of communities provide information about the composition, given by the identification of the component populations; the structure, given by the way these populations are organized in the community; and ecological indexes, measurable data on population properties (e.g., abundance of individuals, age division) and measurable data on the properties of the communities themselves (e.g., richness, diversity).

The seminal studies of Heithaus (1979a, 1979b, 1979c) and the Brazilian tradition in studies of bees associated with flowers (Sakagami et al. 1967) were precursors of the Brazilian legacy of studies of floral wasp communities in Brazil. In fact, contemporaneously, research on wasp communities in Brazil is strongly associated with studies of flower wasp visitors. The first records of social wasps occurring in Brazil were consolidated from naturalistic expeditions conducted by Von Ihering (1904), Ducke (1904, 1905, 1907, 1918), Zikan (1949, 1951) and Araújo (1944, 1946, 1960). In the book *The Social Wasps of the Americas, Excluding the Vespinae*, Richards (1978) grouped all the records of social wasps from the American continent, including information on their biology and ecology. Although these studies touch on aspects of species distribution, composition, and richness, they are only studies of community, *lato* sensu.

The pioneering study on communities of Brazilian social wasps was conducted by Rodrigues and Machado (1982) and was basically a faunistic analysis. The large number of quantifiable variables obtained in flower visitor studies (Heithaus 1979a, 1979b, 1979c) was one of the justifications for the classic study of wasp communities by Mechi (1996) and others that followed (e.g., Diniz and Kitayama 1994; Santos 1996; Hermes and Köhler 2006; Aguiar and Santos 2007; Melo et al. 2015). More recently, studies on social wasp communities have focused not only on faunistic analysis and on the description of patterns found in communities (e.g., Santos et al. 2006), but on an elucidation of the processes that generate these patterns (e.g., Santos et al. 2010; Santos and Presley 2010; Mello et al. 2011; Clemente et al. 2012).

The methods initially used for studies of social wasp communities in Brazil involved faunistic analyses (evaluating parameters of richness, frequency, constancy, abundance, dominance, diversity, density, and distribution) (Mechi 1996; Santos 2000). In sequence, studies based on niche analysis (evaluating the width and overlap of trophic and temporal niches) and interaction networks (evaluating structural patterns, connectivity, stability, network diversity) were carried out (Santos et al. 2007a, 2007b). Much of the study of social wasp communities in Brazil addresses the use of food resources (floral resources and fruits), nesting habits (environmental requirements), and temporal patterns of resource use (Santos et al. 2007a; Santos and Presley 2010; Brugger et al. 2011; Souza et al. 2013a, 2013b; Barbosa et al. 2014).

20.3 Community Structure

There are different approaches to accessing information about biological communities:

1. Composition studies: propose in a simple way to describe the communities in terms of a citation of their component populations, and a checklist of the species of the community.
2. Faunistic analysis studies: evaluate descriptive metrics of the properties of community components (abundance, frequency, dominance, density) and community properties (richness, diversity, and equitability), sometimes adding elements of comparison between communities based on their composition (similarity indices and rarefaction curves).
3. Community dynamics studies: assess how the properties of community components and community properties vary by timescale (throughout the day, between seasons, long-term effects such as global changes).
4. Studies focused on the structure of communities: evaluate patterns presented by communities (nested, modular, gradient, and mixed) and seeking to explain the processes that generate these patterns (interactions among populations, physical factors of time).

In general, in Brazil, studies of social wasp communities use different approaches, combining different metrics and concepts in the same study. If, on the one hand, this mixed approach using different methods of analysis gives robustness to the conclusions, on the other hand, it allows the studies to speak superficially on several subjects without investigating any major question. For example, several studies about social wasps have clearly indicated interest in studying community patterns (Santos et al. 2007a, 2007b; Pereira-Bomfim and Antonialli-Junior 2012; Elisei et al. 2017) but in fact, did not name the structure found or the methods that would allow access to these structures.

Describing the patterns found in natural communities and proposing explanatory processes that generate these patterns is one of the historical objectives of ecology (Lewinsohn et al. 2006a, 2006b). What are the properties of communities, what properties emerge from the process of structuring communities? What properties are truly universal for communities? These are issues that permeate studies of communities and that also need to be investigated in studies on social wasp communities (Santos and Presley 2010).

Lewinsohn et al. (2006a), in their already classic work on pattern detection methods in natural communities, explain as part of their study objectives "the detection of patterns related to ecological and evolutionary processes that shape and maintain these communities" and "extend this approach to a range of possible non-random configurations in interactive communities." These authors also cite, describe, and discuss the theoretical framework of three methods for the detection of structure patterns in natural communities: matrices, network analysis, and community ordering. From the same point of view, Blüthgen et al. (2006) discuss the usefulness of niche analysis tools to access patterns of community structure. Both studies are focused on communities whose structure is driven by interactions between populations from the same community, usually in herbivory-based plant-guilds, pollination and/or dispersal services.

Among the methods summarized by Lewinsohn et al. (2006a, 2006b, Blüthgen et al. (2006), and Blüthgen and Klein (2011) to access the community structure, niche tools and network analysis are the most commonly used in Brazil to analyze the patterns of the social wasp communities and the processes that generate these patterns (e.g., Aguiar and Santos 2007; Santos et al. 2009a, 2010; Mello et al. 2011; Clemente et al. 2012).

All studies that used a niche approach (Santos and Presley 2010) or networks approach (Mello et al. 2011; Santos et al. 2010) or even matrices (Santos 2000) found a nested and modular structure. Although the amplitudes of the nesting degrees and the modularity varied the structure, the network topology did not change. These studies were carried out evaluating different communities of social wasps structured as a function of insect–plant mutual interactions, corroborating Olesen et al. (2007) and Donatti et al. (2011), who considered the nested and modular structures universal, common to all structured communities due to mutualistic interactions, especially in communities based on the populations of a single taxon.

20.4 Communities of Social Wasps and Analysis of Their Structuring Processes

It is important to note that the different approaches used in community studies seek to delimit more clearly the object of study at this level of organization, in order to better evaluate the processes that organize it. It is possible to identify the existence of interactions resulting from the coexistence of species in the same system, to evaluate how processes of interactions can help understand the functioning of the community. Among the tools commonly used to access the patterns of community structure are:

- Cluster analysis: Cluster analysis can reflect the characteristics of a community's composition versus the abiotic changes of an environment. Santos et al. (2009a, 2009b) used this type of metric for a semiarid environment, through the abundance of social wasps nests found in the environment, between different periods of the year and different types of vegetation, concluding that according to these environmental parameters the community suffered changes in the abundance of active nests in the environment, as well as species composition throughout the year and between different physiognomies. Hermes and Köhler (2006) also used this type of analysis to demonstrate the frequency of floral visitation by social wasps in two areas and to demonstrate how the communities of social wasps in two study sites grouped separately from each other.
- Network analysis: Networks are structures characterized by vertices, and edges that make the connections between these vertices. In community ecology studies,

vertices commonly represent populations, guilds, functional groups, entire communities, or even the abiotic factors structuring these networks. The connections commonly represent the interactions between these vertices and the strength of these interactions. The metrics of the network determine its structure and indicate the structure of the analyzed community, whether it is modular, nested, gradient, or combined (Lewinsohn et al. 2006a, 2006b). In this way, a network of mutualistic interactions, such as a wasp-plant, will have a nested and modular structure, since these are universal properties of these types of networks (Bascompte et al. 2003; Jordano et al. 2003; Dáttilo et al. 2016), and all studies of social wasp networks corroborate this assertion (e.g., Santos et al. 2010, 2014; Mello et al. 2011).

- Niche analysis: The analysis of niches allows many possibilities in an investigation, such as an evaluation of whether the use of a given resource or environmental condition contributes to the process of community structuring. Analysis of the amplitude and levels of niches overlap allows inferences about the degree of redundancy or complementarity of a community, proposals for explanatory models of stability, and even determine the community standard (Santos and Presley 2010), floral resources consumption (Aguiar and Santos 2007), temperature constrictions (Santos et al. 2007a, 2007b) activity schedule (Prezoto et al. 2008b; Santos et al. 2010; Sá 2013; Barbosa et al. 2014), hence, allowing the detection of structural patterns of the communities and the organizational models that they present.

Variations in the trophic niche amplitude and, consequently, in the degree of food specialization were recorded for bee and wasp species, both between habitats and in the same habitat at different stations in Costa Rica (Heithaus 1979b). The results of the analysis of social wasps have been interesting. For example, Aguiar and Santos (2007) observed that a group of wasps from the Caatinga area used a spectrum of sources of resources smaller than that of bees (Aguiar 2003), concentrating the collection of floral resources in a restricted group of plants, similar to that recorded in other studies (Mechi 2005). Another interesting observation involves the differences in the overlap of temporal and trophic niches in social wasps, communities when the study approach to the temporal niche observed a high temporal overlap (Santos and Presley 2010; Souza et al. 2013a, 2013b; Jacques et al. 2018), although a low overlap was found when studies explored the trophic niche of social wasps (Hermes and Köhler 2006; Aguiar and Santos 2007; Santos et al. 2007a, 2007b; Somavilla and Köhler 2012).

20.5 Main Models of Community Organization

The assumption that the more diverse the system, the more stable it is (Darwin 1859) has prompted researchers to confirm that increasing species richness increases the stability of a community. Several models have been proposed to explain this relationship. According to Peterson et al. (1998), the main models are: "species richness–diversity" (MacArthur 1955), "idiosyncratic" (Lawton 1994), "rivet" (Ehrlich and Ehrlich 1981), and "drivers and passengers" (Walker 1992).

20.5.1 Species Richness–Diversity Model

This model was formalized by MacArthur (1955) and is based on the premise that the greater the diversity of a system, the greater its stability. It proposes that the addition of species to an ecosystem increases the number of ecological functions present and that it is this increase that contributes to the stability of the ecosystem. However, no research has concluded that the relationship between the species richness and stability of a system has a constant rate of growth, which would reflect the simplicity of this model (Peterson et al. 1998).

However, some studies on social wasps in Brazil have applied this explanatory model. Most of these studies indicate that the species richness of social wasps is strongly associated with the structural complexity of vegetation. Santos (2000) showed that structurally complex environments have more niches available for the colonization and survival of a larger number of species. In subsequent studies, Santos et al. (2007a, 2007b), Clemente et al. (2012), and Locher (2012) also corroborated the idea that heterogeneous microhabitats may promote the coexistence of a larger number of species. So, the more heterogeneous/diverse the landscape, the greater the coexisting species richness, the greater the number of possible species interactions, and the greater the diversity of functional groups in these ecosystems, with positive effects on the properties that maintain the stability of the system (Clemente et al. 2012). Similar findings have also been reported in studies of agricultural ecosystems (Freitas et al. 2015).

20.5.2 Rivet Model

The rivet model, a metaphorical model suggested by Ehrlich and Ehrlich (1981), compares the ecological function of a species to the rivets attaching an airplane wing; a number of rivets may be lost before the complete loss of the wing. The rivet model proposes that the ecological function of the different species inhabiting a particular niche overlaps, so that even if a species is removed from the community, its ecological function may persist if the loss is compensated for by other species with similar functions. The rivet model builds on the richness–diversity model of MacArthur in terms of conceptual complexity, noting that if only richness influences stability, there would be no simple relationship between the removal/addition of a species and a real loss/gain in the stability of a community, simply because of the functional complementarity of the species in an ecosystem.

There is a certain number of rivets that an airplane could lose without there being any effect on stability, and once that threshold is reached, any further loss can lead to collapse; in the same way, there is a species loss threshold in an ecosystem. This model assumes that the ecological functional space is relatively small; therefore, as species are added to an ecosystem, their functions begin to overlap or complement, allowing ecological functions to persist despite the loss of a limited number of species since species with similar functions can compensate for the elimination or decline of these species. However, just as the loss of many fuselage rivets from an airplane is devastating, the loss of several species can lead to the complete collapse of the functional ecosystem, as the loss of several species with similar characteristics will result in degradation of ecological function due to the absence of species that could compensate for the functions performed by the lost species (Peterson et al. 1998; Pugnaire and Valladares 1999; Goswami et al. 2017).

20.6 Drivers and Passengers Model or Redundant Model

Walker's "drivers and passengers" model (1992) accepts the premise of species complementarity, whereby an increase or decrease in one species in a community has little effect on other species and proposes that ecological function resides in "drivers," i.e., functional groups of these species. The model proposed by Walker (1992, 1995) predicts that most species in a community occur redundantly and that there are distinct redundant groups, but with little association between the groups. This can be explained using the analogy of a bus with a driver and several passengers. In this system, several species are redundant, i.e., they share the same functions; these are the "passengers." The loss of any one of these "passenger" species would have little impact on the stability of the system. However, some species are considered key species since they present functional complementarity with the others; the loss of these "driver" species would strongly affect the stability of the system since no other species on this "bus" can perform the function of the "driver."

The drivers and passengers model supports the idea that there are species that are key to system maintenance, without which community stability collapses, and redundant species, which perform the same functions as each other and may be lost from the system without notable changes in structure or function as the loss can be compensated for by the community. In general, species operate independently, and Walker (1995) proposes that since most ecological functions reside in the strong influence of engine species, it is their presence or absence that determines the stability of a given ecological function of an ecosystem.

20.6.1 Idiosyncratic Model

In an ecological context, "idiosyncratic" means that each species occupying a given niche is ecologically different from the others. The idiosyncratic model, proposed by Lawton (1994), proposes that ecological function varies idiosyncratically as species richness increases. This model argues that the contribution of each species to ecological function is strongly influenced by interactions between species. Therefore, the effects of the addition or removal of a species depend entirely on the nature of the species introduced or removed, and also on the nature of the species with which it interacts. Thus, it can be said that the idiosyncratic model supports a relationship between ecological function and species richness (Peterson et al. 1998; Goswami et al. 2017). In this model, the stability of each community is driven by the emergence at the community level of unique interactions between its component populations.

All of these community organization model proposals address the relationship between ecosystem diversity and stability. Biological communities exhibit structural patterns that can be understood simply as an arrangement often found between comparable components (populations). The patterns found in communities are properties that emerge from the different possible arrangements of species at the population level. Community standards are not exclusive or unique; most communities exhibit some of the characteristics of each model. In addition, there may be communities that fit different models when evaluated through different processes or according to different factors. Thus, the same community could fit the drivers and passengers model when analyzed according to the time of use of a resource (temporal niche) or the rivet model when analyzed according to the type of resource used (trophic niche) (Lawton and Brown 1993; Peterson et al. 1998).

20.7 An Example of Models in Social Wasps' Studies

Wasp community studies based on activity schedules show high redundancy and low complementarity. Almost all such studies show that there are two modules of activity, with a peak in activity in the morning and another in the afternoon. These results lead us to believe that the wasp communities analyzed present a modular structure and the richness/stability relationship in these communities can be explained by the rivet model. The high redundancy in the activity schedules indicates that the loss of one species would not affect the structure of the system (e.g., Santos et al. 2010; Barbosa et al. 2014).

However, according to a study conducted by Santos and Presley (2010), one species, *Protonectarina sylveirae*, performs its foraging activities according to a totally different time schedule, complementing the schedule of other wasps in the system. The differentiation of the activity schedule of this species was not enough to change the modular pattern seen in other species, but the system happens to fit the drivers and passengers model since only *P. sylveirae* forages at certain times; with the withdrawal of this species, no other species would forage at that time. It is hypothesized that the same must occur in communities in which species of the genus *Apoica* are present, since they are nocturnal, making them unique among the Epiponini; other wasps are diurnal (Pickett and Wenzel 2007).

20.8 Guilds and Functional Groups of Social Wasp Communities

The plurality of biotic and abiotic factors that affect the structuring of a community is of such magnitude that it imparts characteristics of chaos to the system. The complexity of natural communities is such that it becomes difficult to define their boundaries, what exactly their component populations are, and the factors that affect their dynamics.

Some methodological proposals have emerged to support studies in community ecology. Contemporary ecologists have given increasing attention to the nonstrict phylogenetic classification of organisms to describe the structure and function of ecosystems (Gitay and Noble 1997). For example, groups of species that use similar resources in a similar way are called guilds, while taxonomically related species are known as assemblies or taxocenoses (Fig. 20.1), and species that have the same ecological function are called functional groups.

The term "guild" was initially proposed by Root (1967) and is used to refer to groups or sets of species that utilize the same environmental resources in a similar way and which overlap significantly in their niche requirements. Pianka (1980) modified the concept of guilds, adding the idea of competition arenas. According to

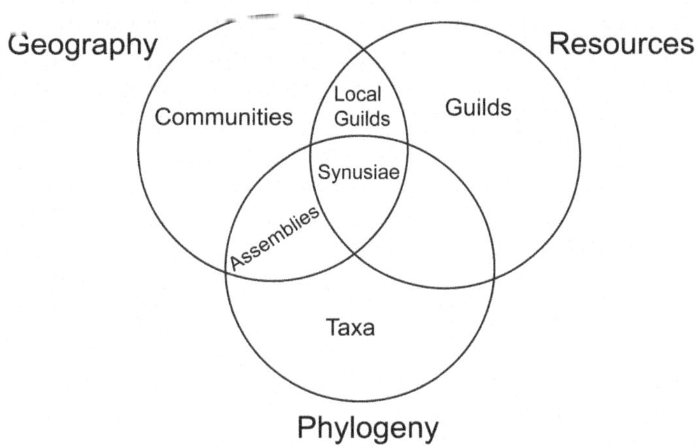

Fig. 20.1 Relationship between phylogeny, geography, and resources available in delimitation in community studies. (Source: Adapted from Fauth et al. (1996))

this new definition, groups of populations can only be considered guilds if the exploited resource is limiting, that is, if it does not meet the requirements of the set of populations that use it. There are also authors who prefer to use the term "functional group"; the concept of functional groups is focused on the ecological services that these populations provide in using these resources.

In general, there are several common ideas in the various definitions of guilds and functional groups. For Blondel (2003) there is no practical difference between the two terms since, metaphorically, they are two sides of the same coin. The main difference is that according to some definitions, species are grouped based on their use of the same resource (guilds), while in other definitions, species are grouped according to their responses to specific stimuli into categories such as pollinators, predators, and detritivores (functional groups) (Gitay and Noble 1997).

One of the main guilds of social wasps studied is that of floral visitors. The main purpose of floral visitation by social wasps is to collect nectar and fibers to build nests, or to predate small insects associated with flowers (Gess and Gess 1993). The guild of floral visitors can be analyzed through different aspects. For bees, this field of research has been extensively explored, mostly by analyzing the structure of mutualism networks (e.g., Olesen and Jordano 2002; Biesmeijer et al. 2005) and trophic niche amplitude and overlap (e.g., Wilms and Wiechers 1997; Aguiar 2003; Aguiar et al. 2013, 2017; Santos et al. 2013;), and by studying the role of social bees as pollinators (Heard 1999).

Social wasps represent a significant portion of the flower visitor guild, overlapping with bees in the exploitation of floral resources in various ecosystems (Heithaus 1979a, 1979b; Aguiar and Santos 2007; Santos et al. 2010), when both groups are compared (e.g., Heithaus 1979a, 1979b, 1979c; Mechi 1996, 2005; Santos et al. 1998; Silva-Pereira and Santos 2006; Hermes and Köhler 2006; Aguiar and Santos 2007; Mello et al. 2011; Clemente et al. 2013). Recent studies have focused on communities of wasps that visit decaying carcasses (e.g., Moretti et al. 2008; Barbosa et al. 2015), generally by analyzing the forensic importance of wasps, since in these cases the social wasps are part of another functional group, the detritivores.

By studying flower visitor guilds (of bees and social wasps), Aguiar and Santos (2007) and Santos et al. (2010) have shown that populations belonging to the same guild but different functional groups have the same nested structure, but with different degrees of redundancy and complementarity. Bee–plant interactions are a form of direct mutualism, in which bees provide pollination services, while wasp–plant interactions are a form of indirect mutualism in which wasps provide herbivory protection services (Santos et al. 2010; Mello et al. 2011). Although networks of mutualistic interactions have universal properties (nesting and modularity), the specific characteristics of different functional groups of the same guild (pollination or protection against herbivory) differentially affect community structure. Sühs et al. (2009), Shuttleworth and Johnson (2009), and Somavilla et al. (2016) have shown that wasps not only act as predators, providing protection services but also provide pollination services.

20.9 The Composition of Social Wasp Communities in Brazilian Ecosystems

Despite the considerable contribution of the great naturalists who studied the fauna of social wasps (opportunely cited), the number of Brazilian social wasps known to science continues to increase. The expansion of sampling efforts, in terms of sampling hours (Felizardo et al. 2017), size of geographical sampling areas (Raw 2016, Barroso et al. 2017; Souza et al. 2017a, 2017b; Raw 2018), and inclusion of samples from different landscapes within the same ecosystems (Santos et al. 2009a, 2009b; Brunismann et al. 2016; Somavilla et al. 2017; Melo et al. 2015; Souza et al. 2017a, 2017b) and even those from different ecosystems within the same geographic area (Santos et al. 2007a, 2007b; Souza et al. 2017a, 2017b), has added new species to the list of known Brazilian wasps. The number of recently discovered social wasps shows that the total diversity of these insects is still unknown and the known diversity is expected to increase further with increased collection efforts.

Studies comparing wasp communities in Brazil commonly use procedures proposed by Whittaker (1960) as indexes of similarity among the wasp species present in each locality to evaluate the extent of change in community composition, or degree of community differentiation, in relation to a given environmental gradient (such as complexity of vegetation structure or variations in temperature and humidity) (Santos et al. 2007a, 2007b). However, the absolute majority of studies lack discussions about alpha (local) and beta (regional) diversity and how they organize to determine total diversity. The analyses and discussions of these studies focus on the differences in the composition of social wasp communities among different habitats.

One of the central points of Whitaker's proposal for understanding biodiversity is the existence of a nucleus of common species commonly found in different habitats and the importance of rare species, with "turnover" between different habitats The alpha and beta diversity proposals of Whittaker (1960) are explanatory of the patterns of distribution of Brazilian social wasps.

All social wasp studies carried out in Brazil, from those carried out in ecosystems with high social wasp biodiversity, such as tropical rainforests, to those with simpler ecosystems and low biodiversity, have two features in common: they investigate (1) one species that can be found in different ecosystems under different environmental conditions and that occurs over a wide area of the national territory (*Apoica pallens, Polistes canadenses, P. versicolor, Polybia ignobilis, P. paulista, P. occidentalis, P. sericea, Protonectarina sylveirae, Protopolybia exigua*), with rare exceptions, and one or more other species within the same group; (2) the difference between the diversity of social wasps on a local and regional scale. The alpha diversity is relatively low, with a relatively small number of social wasp species per locality associated with relatively high beta diversity.

To simplify a complex fact, higher altitudes simulate the effects of latitude and result in lower relative diversity for several taxa. Perillo et al. (2017), who studied solitary bees and wasps, discuss the low diversity and richness that occurs at high altitudes due to physical factors of the flowering period and the relatively low availability of floral resources found at higher altitudes. According to these authors, along with the attitudinal gradients commonly found in mountainous systems, there is a continuous change in the landscape and changes in the boundaries of various environmental factors such as temperature, relative humidity, pressure, wind speed, and food availability.

The beta diversity of bees and wasps determined by Perillo et al. (2017) was mainly explained by the turnover of species along the altitudinal gradient. In support of this explanation, the authors found a greater similarity among communities in areas of similar altitude; the study demonstrated the occurrence of singleton and doubleton species at all altitudes.

A possible explanation for the high beta diversity of social wasps may be differences in the ecological valences of different species. While some eurythermal and ubiquitous species are widely distributed, stenotherms exhibit narrow ecological valence for a range of factors, requiring specific environmental conditions, which determines the rarity of these wasps. Studies of altitude carried out on social wasps in mountainous systems support this hypothesis. Albuquerque et al. (2015) and Souza et al. (2015) observed a reduction in the species richness of social wasps along an altitudinal gradient, with few very abundant species but a high abundance of rare species in the surrounding areas.

20.10 Final Considerations

There are still many gaps in our understanding of social wasp communities in Brazil. Knowledge of aspects of the behavior and ecology of populations of social wasps has been increasing, but research into the processes that generate patterns in wasp communities is still in its initial phase. We know of the flight capacity, level of aggressiveness, carrying capacity, and level of predation of various species of social wasps, but there are many questions still to be answered. For example:

- How does a set of wasp populations organize to form a community of natural enemies?
- What properties emerge when different species of wasps prey within the same system?
- Will patterns identified in communities of social wasps based on agonist interactions be the same as those predicted for communities based on mutualistic interactions?
- Is there dominance in the use of prey?
- What factors drive the formation of the dominance hierarchy?
- What are the limits of niche overlap according to the prey diet?
- How does the hourly activity factor permeate these interactions?
- How do body constraints affect interactions among populations?
- Is there a phylogenetic signal that drives community composition?

– Answering these questions is the next challenge in understanding the community ecology of social wasps.

20.11 Applied Social Wasp Community Ecology

Here, we provide an illustrative outline of community analysis based on two different approaches: niche analysis (Figs. 20.2, 20.3, 20.4, and 20.5) and network analysis (Fig. 20.6). In these fictitious examples, the niches of six species of social wasps (represented by different colors) according to flower visitation to 14 plants (P1 to P14 distributed in the x-axis) are presented. Other methodological approaches to analyzing the structure of communities can be found in reports by Lewinsohn et al. (2006a, 2006b).

1. Nested structure

Wasp community with a nested structure (Fig. 20.2), in which the most common wasps use resources from almost all plants. The lesser-used plants are used mainly by the most common wasps, generating asymmetry in the wasp–plant partnership. The characteristic asymmetry of nested communities is associated with high redundancy; the loss of a wasp species or visited plant does not significantly alter the community structure.

2. Modular structure

Wasp community with a modular structure (Fig. 20.3), in which different sets of wasps share the same resource (floral resource) or condition of resource use (visitation time); in this case, we have two groups of wasps, the niches of which overlap more within the two groups than between groups. Each group is complementary to the other in terms of flower visitation. The loss of a wasp species or visited plant would not significantly alter the community structure, although it would have greater effects in one of the modules than in the community as a whole.

3. Gradient structure

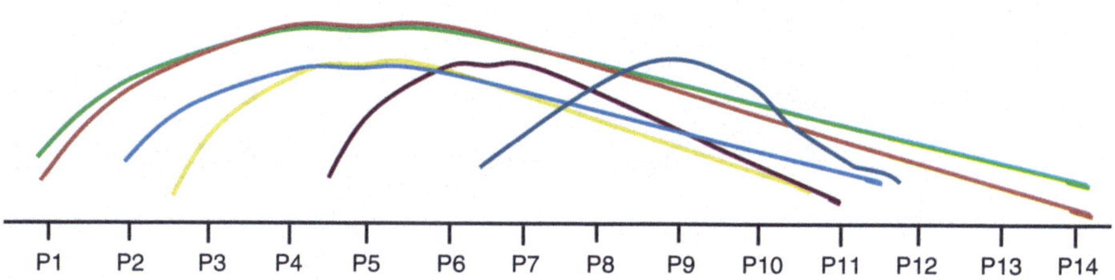

Fig. 20.2 Representation of nested structure. The color lines represent social wasps' species in a plants scale gradient

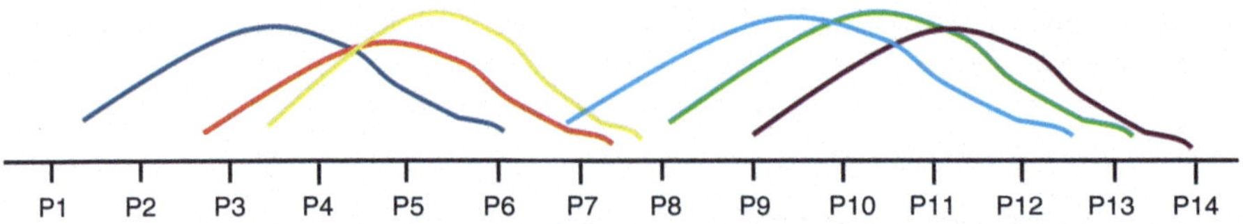

Fig. 20.3 Representation of modular structure. The color lines represent social wasps' species in a plants scale gradient

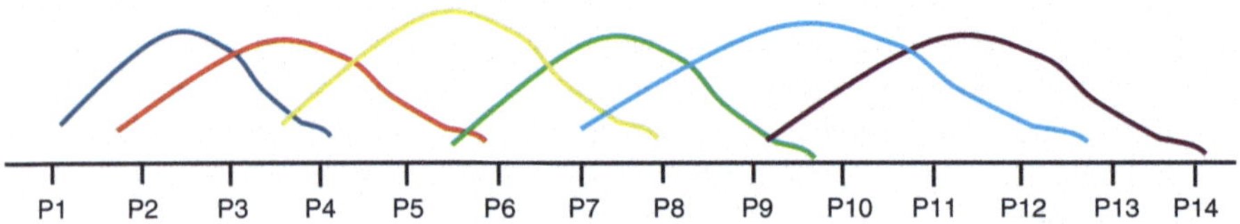

Fig. 20.4 Representation of gradient structure. The color lines represent social wasps' species in a plants scale gradient

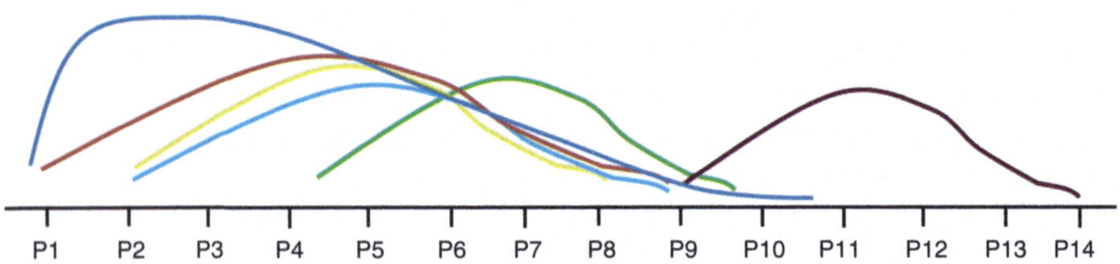

Fig. 20.5 Representation of mixed structure. The color lines represent social wasps' species in a plants scale gradient

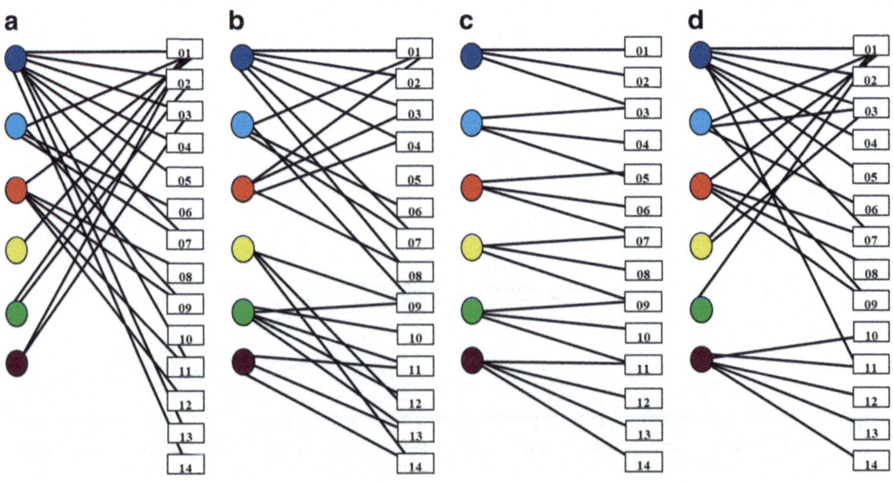

Fig. 20.6 Graphs representing social wasps' communities. (**a**) Nested structure, (**b**) modular structure, (**c**) gradient structure, and (**d**) mixed structure (with a modular and nested pattern)

Wasp community with a gradient structure (Fig. 20.4), in which there is a substitution in the wasp visitors and the visited plants. The gradient could be produced by changes in altitude or seasonality—conditions under which the partner populations fluctuate and are replaced in space or time. There is an obvious complementarity between visiting wasps; the absence of some wasp species may leave some plants unvisited. In this specific example, the loss of any wasp species would significantly alter the community structure.

4. Mixed structure

Wasp community represented by a mixed, nested, and modular model (Fig. 20.5). In this example, a single species of wasp form the module on the right and visits five plants, four of which are only visited by this species. The loss of any wasp species from the module on the left will not lead to drastic changes in the community, given the high overlap and redundancy in visitation. However, all visits in the right-hand module are made by a single wasp. The extinction of this species, in addition to changing the community structure, which would cease to be modular, would lead to the cessation of interactions with four plant species. The stability of this community is clearly dependent on this wasp (a key species); thus, it fits the drivers and passenger model.

References

Aguiar CML (2003) Utilização de recursos florais por abelhas (Hymenoptera, Apoidea) em uma área de Caatinga (Itatim, Bahia, Brasil). Rev Bras Zool 20:457–467. https://doi.org/10.1590/S0101-81752003000300015

Aguiar CML, Santos GMM (2007) Compartilhamento de recursos florais por vespas sociais (Hymenoptera: Vespidae) e abelhas (Hymenoptera: Apoidea) em uma área de Caatinga. Neotrop Entomol 36:836–842. https://doi.org/10.1590/S1519-566X2007000600003

Aguiar CML, Santos GMM, Martins CF, Presley SJ (2013) Trophic niche breadth and niche overlap in a guild of flower-visiting bees in a Brazilian dry forest. Apidologie 44:153–162. https://doi.org/10.1007/s13592-012-0167-4

Aguiar CML, Caramés J, França F, Melo E (2017) Exploitation of floral resources and niche overlap within an oil-collecting bee guild (Hymenoptera: Apidae) in a Neotropical savannah. Sociobiology. 64:78–84. https://doi.org/10.13102/sociobiology.v64i1.1250

Albuquerque CHB, Souza MM, Clemente MA (2015) Comunidade de vespas sociais (Hymenoptera, Vespidae) em diferentes gradientes altitudinais no sul do estado de Minas Gerais, Brasil. Biotemas 28(4):131–138. https://doi.org/10.5007/2175-7925.2015v28n4p131

Araújo RL (1944) Contribuição para o conhecimento do gênero "Synoecoides" Ducke, 1905 (Hym., Vespidae). Rev Bras Biol 4(3):399–348

Araújo RL (1946) Contribuição para o conhecimento do gênero *Metapolybia* Duck, 1905 (Hymenoptera, Vespidae). Arquivos Instituto Biológico 16:65–82

Araújo RL (1960) Insecta Amapaensia (Hymenoptera: Vespidae, Polybiinae). Studia Entomologica 3:251–253

Barbosa BC, Paschoalini MF, Prezoto F (2014) Temporal activity patterns and foraging behavior by social wasps (hymenoptera, polistinae) on fruits of *Mangifera indica* L. (Anacardiaceae). Sociobiology. 61(2):239–242. https://doi.org/10.13102/sociobiology.v61i2.239-242

Barbosa RR, Carriço C, Souto RNP et al (2015) Record of postmortem injuries caused by the Neotropical social wasp *Agelaia fulvofasciata* (Degeer) (Hymenoptera, Vespidae) on pig carcasses in the eastern Amazon region: implications in forensic taphonomy. Rev Bras Entomol 59(3):257–259. https://doi.org/10.1016/j.rbe.2015.07.004

Barbosa BC, Detoni M, Maciel TT, Prezoto F (2016) Studies of social wasp diversity in Brazil: over 30 years of research, advancements and priorities. Sociobiology 63(3):858. https://doi.org/10.13102/sociobiology.v63i3.1031

Barroso PCS, Somavilla A, Boldrini R (2017) Updating the geographic records of social wasps (Vespidae: Polistinae) in Roraima state. Sociobiology 64(2):339–346. https://doi.org/10.13102/sociobiology.v64i3.1789

Bascompte J, Jordano P, Melián CJ, Olesen JM (2003) The nested assembly of plant-animal mutualistic networks. Proc Natl Acad Sci U S A 100:9383–9387. https://doi.org/10.1073/pnas.1633576100

Begon M (1979) Investigating Animal Abundance: capture-recapture for biologists. Edward Arnold Limited. University Park Press, London, England

Begon M, Towsend CR, Harper JL (2007) Ecologia: de indivíduos a ecossistemas. Artmed, Porto Alegre, Brasil

Biesmeijer JC, Slaa EJ, Castro MS et al (2005) Connectance of Brazilian social bee: food plant networks is influenced by habitat, but not by latitude, altitude or network size. Biota Neotrop 5:85–93. https://doi.org/10.1590/S1676-06032005000100010

Blondel J (2003) Guilds or functional groups: does it matter? Oikos 100:223–231. https://doi.org/10.1034/j.1600-0706.2003.12152.x

Blüthgen N, Klein AM (2011) Functional complementarity and specialisation: the role of biodiversity in plant-pollinator interactions. Basic Appl Ecol 12:282–291. https://doi.org/10.1016/j.baae.2010.11.001

Blüthgen N, Menzel F, Blüthgen N (2006) Measuring specialization in species interaction networks. BMC Ecol 6:9. https://doi.org/10.1186/1472-6785-6-9

Brugger B, Souza LSA, Souza AR, Prezoto F (2011) Social wasps (*Synoeca cyanea*) damaging *Psidium* sp. (Myrtaceae) fruits in Minas Gerais state, Brazil. Sociobiology 6:533–536

Brunismann AG, Souza MM, Pires EP et al (2016) Social wasps (Hymenoptera: Vespidae) in deciduous seasonal forest in southeastern Brazil. J Entomol Zool Stud 4:447–452

Carpenter JM, Marques OM (2001) Contribuição ao estudo dos Vespídeos do Brasil (Insecta, Hymenoptera, Vespoidea, Vespidae). Publicações Digitais: Universidade Federal da Bahia 2:1–147

Clemente MA, Lange D, Del-Claro K et al (2012) Flower-visiting social wasps and plants interaction: network pattern and environmental complexity. Psyche 2012:1–10. https://doi.org/10.1155/2012/478431

Clemente MA, Lange D, Dáttilo W et al (2013) Social wasp-flower visiting guild interactions in less structurally complex habitats are more susceptible to local extinction. Sociobiology. 60:337–344. https://doi.org/10.13102/sociobiology.v60i3.337-344

Clements FE (1916) Plant sucession: analysis of development of vegetation. Publs Carnegie Instn, Washington, DC, USA

Darwin CR (1859) On the origin of species by means of natural selection or the preservation of favored races in the struggle for life. John Murray, London, England

Dáttilo W, Lara-Rodríguez N, Jordano P et al (2016) Unravelling Darwin's entangled bank: architecture and robustness of mutualistic networks with multiple interaction types. Proc R Soc B Biol Sci 283:20161564. https://doi.org/10.1098/rspb.2016.1564

Diniz IR, Kitayama K (1994) Colony densities and preferences for nest habitats of some social wasps in Mato Grosso state, Brazil (Hymenoptera, Vespidae). J Hym Res 3:133–143

Donatti CI, Guimarães PR, Galetti M et al (2011) Analysis of a hyper-diverse seed dispersal network: modularity and underlying mechanisms. Ecol Lett 14:773–781. https://doi.org/10.1111/j.1461-0248.2011.01639.x

Ducke A (1904) Sobre as Vespidas sociaes do Pará. Boletim do Museu Goeldi 4:317–374

Ducke A (1905) Sobre as Vespidas sociaes do Pará. Boletim do Museu Goeldi 4:652–698

Ducke A (1907) Connaissance de la faune Hyménoptérologique du nord-est du Brésil. Revue D'entomologique 23:73–96

Ducke A (1918) Catálogo de vespas sociaes do Brazil. Rev Mus Paulista 10:313–374

Ehrlich PR, Ehrlich AH (1981) Extinction: the causes and consequences of the disappearance of species. Random House, New York, USA

Elisei T, Valadares E, Albuquerque FA, Martins CF (2017) Diversity and structure of social wasps community (Hymenoptera: Vespidae, Polistinae) in neotropical dry forest. Sociobiology. 64(1):111–118. https://doi.org/10.13102/sociobiology.v64i1.1261

Fauth JE, Bernardo J, Camara M et al (1996) Simplifying the jargon of community ecology: a conceptual approach. Am Nat 147(2):282–286

Felizardo SPS, Borges RC, Santos JNA, Silveira OT (2017) Two more social wasp species (Vespidae, Polistinae) collect mullerian bodies from Cecropia trees (Urticaceae). Sociobiology. 64(2):442–450. https://doi.org/10.13102/sociobiology.v64i4.1684

Freitas JL, Pires EP, Oliveira TTC et al (2015) Vespas sociais (Hymenoptera: Vespidae) em lavouras de *Coffea arabica* L. (Rubiaceae) no Sul de Minas Gerais. Revista Agrogeoambiental 7(3):67–77

Gess FW, Gess SK (1993) Ethelogical studies of *Jugurtia confusa* Richards, *Ceramius capicula* Brauns, *C. linearis* Klug and *C. lichtensteinii* (Klug) (Hymenoptera: Masarinae) in the eastern Cape Province of South Africa. Ann Cape Prov Mus (Natural History) 13:63–83

Gitay H, Noble IR (1997) What are functional types and how should we seek them? In: Smith TM, Shugart HH, Woodward FI (eds) Plant functional types: their relevance to ecosystems properties and global change. Cambridge University Press. Cambridge, England

Goswami M, Bhattacharyya P, Mukherjee I, Tribedi P (2017) Functional diversity: an important measure of ecosystem functioning. Adv Microbiol 7:82–93. https://doi.org/10.4236/aim.2017.71007

Gurevitch J, Scheiner SM, Fox GA (2009) Ecologia vegetal. Artmed, Porto Alegre, Brasil

Heard TA (1999) The role of stingless bees in crop pollination. Annu Rev Entomol 44:183–206. https://doi.org/10.1016/0304-4238(87)90065-3

Heithaus ER (1979a) Community structure of Neotropical flower visiting bees and wasps: diversity and phenology. Ecology 60:190–202

Heithaus ER (1979b) Flower-feeding specialization in wild bee and wasp communities in seasonal Neotropical habitats. Oecologia 42:179–194

Heithaus ER (1979c) Flower visitation records and resource overlap of bees and wasps in northest Costa Rica. Brenesia 16:9–52

Hermes M, Köhler A (2006) The flower-visiting social wasps (Hymenoptera, Vespidae, Polistinae) in two areas of Rio Grande do Sul state, southern Brazil. Rev Bras Entomol 50:268–274. https://doi.org/10.1590/S0085-56262006000200008

Hunt JH (2007) The evolution of social wasps. Oxford University Press, New York, USA

Hunt JH, Brown PA, Sago KM, Kerker JA (1991) Vespid wasps eat pollen (Hymenoptera: Vespidae). J Kansas Entomol Soc 64:127–130. https://doi.org/10.2307/25085261

Jacques GC, Pikart TG, Santos VS, Vicente LO, Silveira LCP (2018) Niche overlap and daily activity pattern of social wasps (Vespidae: Polistinae) in kale crops. Sociobiology 65(2):312–319. https://doi.org/10.13102/sociobiology.v65i2.2670

Jordano P, Bascompte J, Olesen MJ (2003) Ivariant properties in coevolutionary networks of plant-animal interactions. Ecol Lett 6:69–81. https://doi.org/10.1046/j.1461-0248.2003.00403.x

Lawton JH (1994) What do species do in ecosystems? Oikos 71:367–374

Lawton JH, Brown VK (1993) Redundancy in ecosystems. In: Schulze ED, Mooney HA (eds) Biodiversity and ecosystem function. Springer. Washington, USA

Lewinsohn TM, Loyola RD, Prado PI (2006a) Matrizes, redes e ordenações: a detecção de estrutura em comunidades interativas. Oecologia Bras 10:90–104. https://doi.org/10.4257/oeco.2006.1001.06

Lewinsohn TM, Prado PI, Jordano P et al (2006b) Structure in plant-animal interaction assemblages. Oikos 113:174–184

Locher GA (2012) Estudo comparativo da diversidade de vespas sociais (Hymenoptera, Vespidae) em mata ciliar e cultura de cana-de- açúcar na região de Ipeúna, SP. Dissertation, Universidade Estadual Paulista

MacArthur R (1955) Fluctuation of animal populations and a measure of community stability. Ecology 36(3):533–536

Mechi MR (1996) Levantamento da fauna de vespas aculeata na vegetação de duas áreas de cerrado. PhD Thesis, Universidade Federal de São Carlos

Mechi MR (2005) Comunidade de vespas aculeata (Hymenoptera) e suas fontes florais. In: Pivello VR, Varanda EM (eds) O cerrado Pé-de-Gigante: Ecologia e conservação - Parque Estadual de Vassununga. SMA. São Paulo, Brasil

Mello MAR, Santos GMM, Mechi MR, Hermes MG (2011) High generalization in flower-visiting networks of social wasps. Acta Oecol 37:37–42. https://doi.org/10.1016/j.actao.2010.11.004

Melo AC, Barbosa BC, Castro MM et al (2015) The social wasp community (Hymenoptera, Vespidae) and new distribution record of *Polybia ruficeps* in an area of Caatinga biome, northeastern Brazil. Check List. 11(1):1530. https://doi.org/10.15560/11.1.1530

Moretti TC, Thyssen PJ, Godoy WAC, Solis DR (2008) Necrophagy by the social wasp *Agelaia pallipes* (Hymenoptera: Vespidae, Epiponini): possible forensic implications. Sociobiology 51:393–398

O'Donnell S (1995) Necrophagy by Neotropical swarm-founding wasps (Hymenoptera: Vespidae, Epiponini). Biotropica 27:133–136

Odum EP (1988) Ecologia. Guanabara, Rio de Janeiro, Brasil

Olesen JM, Jordano P (2002) Geographic patterns in plant-pollinator mutualistic networks. Ecology 83:2416–2424. https://doi.org/10.1890/0012-9658(2002)083[2416:GPIPPM]2.0.CO;2

Olesen JM, Bascompte J, Dupont YL, Jordano P (2007) The modularity of pollination networks. PNAS 104(50):19891–19896. https://doi.org/10.1073/pnas.0706375104

Pereira-Bomfim MGC, Antonialli-Junior WF (2012) Community structure of social wasps (Hymenoptera: Vespidae) in riparian forest in Batayporã, Mato Grosso do Sul, Brazil. Sociobiology. 59:755–765. https://doi.org/10.13102/sociobiology.V59I3.545

Perillo LN, de Neves FS, Antonini Y, Martins RP (2017) Compositional changes in bee and wasp communities along Neotropical mountain altitudinal gradient. PLoS One 12(7):e0182054. https://doi.org/10.1371/journal.pone.0182054

Peterson G, Allen CR, Holling CS (1998) Ecological resilience, biodiversity, and scale. Ecosystems 1:6–18. https://doi.org/10.1007/s100219900002

Pianka ER (1980) Guild structure in desert lizards. Oikos 35:194–201

Pickett KM, Wenzel JW (2007) Revision and cladistic analisys of the nocturnal social wasp genus, *Apoica* Lepeletier (Hymenoptera: Vespidae; Polistinae, Epiponini). Am Mus Novit 3562:1–30. https://doi.org/10.1206/0003-0082(2007)397%5B1:RACAOT%5D2.0.CO;2

Prezoto F, Ribeiro-Junior C, Oliveira-Cortes SA, Elisei T (2007) Manejo de vespas e marimbondos em ambiente urbano. In: Pinto AS, Rossi MM, Salmeron E (eds) Manejo de Pragas Urbanas. Editora CP2. Piracicaba, Brasil

Prezoto F, Cortes SAO, Melo AC (2008a) Vespas: de vilãs a parceiras. Ciência Hoje 43:70–72

Prezoto F, Ribeiro-Junior C, Guimarães DL, Elisei T (2008b) Vespas sociais e o controle biológico de pragas: atividade forrageadora e manejo das colônias. In: Vilela EF, Santos IA, Schoereder JH et al (eds) Insetos sociais: da biologia à aplicação. UFV, Viçosa, Brasil

Pugnaire FI, Valladares F (1999) Handbook of functional plant ecology. Marcel Dekker, New York, USA

Raw A (2016) New records of social wasps around Brasília (Hymenoptera: Vespidae: Polistnae). Sociobiology. 63:1073–1075. https://doi.org/10.13102/sociobiology.v63i4.1360

Raw A (2018) A founder-controlled, social wasp assemblage, and a recent severe fall in numbers. Sociobiology. 65:325–329. https://doi.org/10.13102/sociobiology.v65i2.1433

Richards OW (1978) The social wasps of the Americas excluding the Vespinae. British Museum, London, England

Rodrigues VM, Machado VLL (1982) Vespídeos sociais: espécies do Horto Florestal "Navarro de Andrade" de Rio Claro, SP. Naturalia 7:173–175

Root RB (1967) The niche exploitation of blue-grey gnatcatcher. Ecol Monogr 37:317–350

Sá CA (2013) Coleta de recursos glucídios por vespas sociais: mecanismos que regem seu padrão de atividade diário. Dissertation, Universidade Estadual de Feira de Santana

Sakagami SF, Laroca SF, Moure JSM (1967) Wild bee biocenotics in São José dos Pinhais (PR), South Brazil. J Fac Sci Hokkaido 16:253–291

Santos BB (1996) Ocorrência de vespídeos sociais (Hymenoptera, Vespidae) em pomar em Goiânia, Goiás, Brasil. Revista do Setor Ciências Agrárias 15(1):43–46

Santos GMM (2000) Comunidade de vespas sociais (Hymenoptera-Polistinae) em três ecossistemas do Estado da Bahia, com ênfase na estrutura da guilda de vespas visitantes de flores de Caatinga. PhD Thesis, Universidade de São Paulo

Santos GMM, Gobbi N (1998) Nesting habits and colonial productivity of *Polistes canadensis canadensis* (L.) (Hymenoptera-Vespidae) in a Caatinga area, Bahia state- Brazil. J Adv Zoo 19(2):63–69

Santos GMM, Presley S (2010) Niche overlap and temporal activity patterns of social wasps (Hymenoptera: Vespidae) in a Brazilian cashew orchard. Sociobiology 56:121–131

Santos GMM, Silva SOC, Bichara-Filho CC, Gobbi N (1998) Influencia del tamaño del cuerpo en el forrajeo de avispas sociales (Hymenoptera - Polistinae) visitantes de inflorescencias de Syagrus coronata (Martius) (Arecaceae). Revista Gayana de Zoologia 62(2):167–170

Santos GMM, Aguiar CML, Gobbi N (2006) Characterization of the social wasp guild (Hymenoptera: Vespidae) visiting flowers in the Caatinga (Itatim, Bahia, Brazil). Sociobiology 47:483–494

Santos GMM, Bichara-Filho CC, Resende JJ et al (2007a) Diversity and community structure of social wasps (Hymenoptera: Vespidae) in three ecosystems in Itaparica island, Bahia state, Brazil. Neotrop Entomol 36:180–185

Santos GMM, Cruz JD, Bichara-Filho CC et al (2007b) Utilização de frutos de cactos (Cactaceae) como recurso alimentar por vespas sociais (Hymenoptera, Vespidae, Polistinae) em uma área de caatinga (Ipirá, Bahia, Brasil). Rev Bras Zool 24:1052–1056. https://doi.org/10.1590/S0101-81752007000400023

Santos GMM, Bispo PC, Aguiar CML (2009a) Fluctuations in richness and abundance of social wasps during the dry and wet seasons in three phytophysiognomies at the tropical dry forest of Brazil. Environ Entomol 38:1613–1617. https://doi.org/10.1603/022.038.0613

Santos GMM, Cruz JD, Marques OM, Gobbi N (2009b) Social wasps (Hymenoptera: Vespidae) diversity in a Cerrado vegetation in Bahia state, Brazil. Neotrop Entomol 38:317–320. https://doi.org/10.1590/S1519-566X2009000300003

Santos GMM, Aguiar CML, Mello MAR (2010) Flower-visiting guild associated with the Caatinga flora: trophic interaction networks formed by social bees and social wasps with plants. Apidologie 41:466–475. https://doi.org/10.1051/apido/2009081

Santos GMM, Carvalho CAL, Aguiar CML et al (2013) Overlap in trophic and temporal niches in the flower-visiting bee guild (Hymenoptera, Apoidea) of a tropical dry forest. Apidologie 44:64–74. https://doi.org/10.1007/s13592-012-0155-8

Schoener TW (1986) Resource partitioning. In: Kikkawa J, Anderson DJ (eds) Community ecology – pattern and process. Blackwell Scientific Publications, Melbourne

Shuttleworth A, Johnson SD (2009) The importance of scent and nectar filters in a specialized wasp-pollination system. Funct Ecol 23:931–940. https://doi.org/10.1111/j.1365-2435.2009.01573.x

Silva-Pereira VS, Santos GMM (2006) Diversity in bee (Hymenoptera: Apoidea) and social wasp (Hymenoptera: Vespidae, Polistinae) community in "Campos Rupestres", Bahia, Brazil. Neotrop Entomol 35(2):165–174. https://doi.org/10.1590/S1519-566X2006000200003

Silveira OT, Esposito MC, Santos JN, Gemaque FE (2005) Social wasps and bees captured in carrion traps in a rainforest in Brazil. Entomol Sci 8:33–39. https://doi.org/10.1111/j.1479-8298.2005.00098.x

Somavilla A, Schoeninger K, Castro DGD et al (2016) Diversity of wasps (Hymenoptera: Vespidae) in conventonal and organic guarana (*Paullinia cupana* var. *sorbilis*) crops in the Brazilian Amazon. Sociobiology. 63(4):1051–1057. https://doi.org/10.13102/sociobiology.v63i4.1178
Somavilla A, Köhler A (2012) Preferência Floral de Vespas (Hymenoptera, Vespidae) no Rio Grande do Sul, Brasil. EntomoBrasilis 5(1):21–28. https://doi.org/10.12741/ebrasilis.v5i1.152
Somavilla A, Oliveira ML, Rafael JA (2017) Social wasps (Vespidae: Polistinae) from two National Parks of the Caatinga biome, in Brazil. Sociobiology. 64(3):334–338. https://doi.org/10.13102/sociobiology.v64i3.1593
Souza MM, Louzada J, Serrão JE, Zanuncio JC (2010) Social wasps (Hymenoptera: Vespidae) as indicators of conservation degree of riparian forests in Southeast Brazil. Sociobiology 56:387–396
Souza GK, Pikart TG, Jacques GC et al (2013a) Social wasps on *Eugenia uniflora* Linnaeus (Myrtaceae) plants in an urban area. 60(2):204–209. https://doi.org/10.13102/sociobiology.v60i2.204-209
Souza MM, Pires EP, Prezoto F (2013b) Nidification of *Polybia rejecta* (Hymenoptera: Vespidae) associated to *Azteca chartifex* (Hymenoptera: Formicidae) in a fragment of Atlantic Forest, in the state of Minas Gerais, southeastern Brazil. Biota Neotrop 13(2):390–392. https://doi.org/10.1590/s1676-06032013000300038
Souza MM, Pires EP, Elpino-Campos A, Louzada JNC (2014) Nesting of social wasps (Hymenoptera: Vespidae) in a riparianforest of rio das Mortes in southeastern Brazil. Acta Sci Biol Sci 36:189. https://doi.org/10.4025/actascibiolsci.v36i2.21460
Souza MM, Silva HNM, Dallo JB et al (2015) Biodiversity of social wasps (Hymenoptera: Vespidae) at altitudes above 1600 meters in the Parque Estadual da Serra do Papagaio, state of Minas Gerais, Brazil. EntomoBrasilis. 8:174–179. https://doi.org/10.12741/ebrasilis.v8i3.519
Souza MM, Brunismann AG, Pires EP (2017a) Nesting associations between *Chartergus globiventris* Saussure (Hymenoptera: Vespidae) and *Tolmomyias sulphurescens* Spix (Passeriformes: Tyrannidae) in southeastern Brazil. EntomoBrasilis. 10:51–53. https://doi.org/10.12741/ebrasilis.v10i1.638
Souza MM, Brunismann AG, Clemente MA (2017b) Species composition, relative abundance and distribution of social wasps fauna on different ecosystems. Sociobiology. 64(4):456–465. https://doi.org/10.13102/sociobiology.v64i4.1839
Sühs RB, Somavilla A, Köhler A, Putzke J (2009) Vespídeos (Hymenoptera, Vespidae) vetores de pólen de *Schinus terebinthifolius* Raddi (Anacardiaceae), Santa Cruz do Sul, RS, Brasil. Brazilian J Biosci 7:138–143
Urbini A, Sparvoli E, Turillazzi S (2006) Social paper wasps as bioindicators: a preliminary research with Polistes dominulus (Hymenoptera Vespidae) as a trace metal accumulator. Chemosphere 64:697–703. https://doi.org/10.1016/j.chemosphere.2005.11.009
Von Ihering R (1904) As vespas sociais do Brasil. Revista do Museu Paulista 6:9–309
Walker BH (1992) Biodiversity and ecological redundancy. Conserv Biol 6:18–23. https://doi.org/10.1046/j.1523-1739.1992.610018.x
Walker B (1995) Conserving biological diversity through ecosystem resilience. Conserv Biol 9:747–752. https://doi.org/10.1046/j.1523-1739.1995.09040747.x
Whittaker RH (1960) Vegetation of the Siskiyou Mountains, Oregon and California. Ecol Monogr 30:279–338
Whittaker RH (1975) Communities and ecosystems. Macmillan, Nova York
Wilms W, Wiechers B (1997) Floral resource partitioning between native Melipona bees and the introduced Africanized honey bee in the Brazilian Atlantic rain forest. Apidologie 28:339–355. https://doi.org/10.1051/apido:19970602
Zikán JF (1949) O gênero *Mischocyttarus* Saussure (Hymenoptera, Vespidae), com a descrição de 82 espécies novas. Parque Nacional do Itatiaia Boletim do Parque Nacional do Itatiaia 1:1–125
Zikán JF (1951) Beitrage zur Kenntnis der Arten der Gattung *Polistes* Latreille 1802, nebest Beschreibung. Arten Dusenia 2:225–236

Chapter 21
Interactions Between Wasps and Other Animals: Associations and Natural Enemies

Bruno Corrêa Barbosa, Tatiane Tagliatti Maciel, Alexandre Somavilla, and Fábio Prezoto

Abstract Some social wasp species are known for their interactions and associations with vertebrates, invertebrates, and other insects in general. In this chapter, we will therefore focus on the interspecific interactions performed by wasps and other animals in order to present the types of interactions, as well as some of the species involved. We list some harmonious relations, like the *Polybia rejecta* wasp and *Azteca* ants and *Cacicus cela* birds. We also list some disharmonious relations, like the parasitoids Hymenoptera in social wasps' nests and entomopathogenic fungi, like *Ophiocordyceps*, which can also attack social wasps.

Keywords Agonistic behavior · Behavioral ecology · Paper wasp · Polistinae · Trophic relationships

Species in nature interact with each other in a variety of ways, and the involved species may have their populations increased, decreased, or maintained at the same levels as a result of such interactions. According to these effects, interactions are respectively termed positive (+), negative (−), or neutral (0) (Del-Claro and Torezan-Silingardi 2012).

Positive relationships for the species involved are classified as mutualism, and may be mandatory or optional, and both species benefit from the interaction. Positive interaction for one and neutral for the other involved is categorized as commensalism, where one does not harm the other by benefiting. A positive relationship for

B. C. Barbosa (✉) · T. T. Maciel · F. Prezoto
Laboratório de Ecologia Comportamental e Bioacústica (LABEC), Programa de Pós-graduação em Biodiversidade e Conservação da Natureza, Universidade Federal de Juiz de Fora, Juiz de Fora, Minas Gerais, Brazil
e-mail: barbosa.bc@outlook.com; tatitagliatti@hotmail.com; fabio.prezoto@ufjf.edu.br

A. Somavilla
Instituto Nacional de Pesquisas da Amazônia, Manaus, Amazonas, Brazil
e-mail: alexandresomavilla@gmail.com

F. Prezoto et al. (eds.), *Neotropical Social Wasps*,
https://doi.org/10.1007/978-3-030-53510-0_21

one species and negative for the other indicates predation or parasitism, while negative interactions for both species signal competition (Towsend et al. 2006).

Social hymenopterans are known to play important roles in ecological interactions ranging from plants to animals, and these interactions have aroused great interest in science. We know that social wasps have interactions with mutualistic-like plants as pollinators (cf. Sühs et al. 2009; Clemente et al. 2012) and possible commensalism, as nesting brings protection to plants used as substrate.

Some social wasp species are known for their interactions and associations with vertebrates (e.g., Somavilla et al. 2013; Menezes et al. 2014), invertebrates, and other insects in general (e.g., Wilson 1975; Souza et al. 2013). In this chapter, we will therefore focus on the interspecific interactions performed by wasps and other animals in order to present the types of interactions, as well as some of the species involved.

21.1 Harmonious Relations in Social Events

Interspecific relations are more conspicuous and more frequent in ecosystems, occurring between individuals of different species, which can be harmonious when there is benefit for at least one species, or inharmonious when there is damage to one or both species. Many other animals associate with social wasps due to the fact that they often defend their nests by attacking with painful stings, thus obtaining some protection against predators, constituting a harmonious relationship.

Many bird species are commonly found in the Neotropical region associated with social wasps, nesting near breeding social wasp colonies (Fig. 21.1a), such as *Tolmomyias sulphurescens* Spix, 1825 which has records with 11 social wasp species (cf. Menezes et al. 2014). Members of the Tyrannidae family are also known for their aggressiveness (Sick 1997), and so it is possible that wasps could also benefit from the bird's territoriality, meaning that by defending its territory from potential competitors, it could also repel possible predators of wasp colonies.

Like the birds of the Tyrannidae family, the Trochilidae, popularly known as hummingbirds, are aggressive when defending their territory, and so their nests can be excellent nesting sites for less aggressive wasp species (Fig. 21.1b). Other bird species that can also be associated with social wasps are *Cacicus cela* Linnaeus 1758, for which several nests built near *Polybia rejecta* Fabricius, 1798 and ant (Fig. 21.1c) colonies can be observed due to its colonial habit, and *Galbula ruficauda* Cuvier, 1816, which were found associated with some *Apoica* and *Polybia* colonies (Fry 1972).

Invertebrates can also be recorded associated with social wasp colonies, such as *Azteca chartifex* Forel, 1896 ants, most commonly found with *P. rejecta* (Fig. 21.1d) (Somavilla et al. 2013; cf. Souza et al. 2013; Virgínio et al. 2015) and rarely with *Synoeca virginea* Fabricius, 1804 (Somavilla et al. 2013), where wasps seek association. Relations between wasps and *Azteca* spp. may also be of competition (disharmonious), as recorded for wasps of the *Polybia* and *Charterginus* genera, which were observed stealing food from *Azteca* ants in French Guiana and Costa Rica

Fig. 21.1 (**a**). Nest of the *Tolmomyias sulphurescens* bird associated with *Parachartergus fraternus* colony; (**b**). *Mischocyttarus cassununga* colony in *Chlorostilbon lucidus* nest; (**c**). *Cacicus cela* nests associated with *Polybia rejecta* and *Azteca chartifex* colonies; (**d**). *Polybia rejecta* colony associated with *Aztec chartifex* colony

(Lapierre et al. 2007). The *Camponotus* and *Pseudomyrmex* genera are also recorded in association with wasps of the *Parachartergus* and *Polybia* genera, respectively (cf. Richards 1978; Herre et al. 1986).

Social wasps may also associate with other wasps as well as other animals, benefitting from the aggressive behaviors exhibited by some species. The *Mischocyttarus* and *Polistes* genera are known to not exhibit aggressive behavior and can thus be associated with Epiponini tribe species, taking advantage of the aggressiveness and larger number of individuals that colonies of these species exhibit (e.g., Gorton 1978; London and Jeanne 1997). *Mischocyttarus immarginatus* Richards, 1940 has already been recorded in association with the *Polybia*, *Synoeca*, *Brachygastra*, and *Metapolybia* genera (Starr 1988; Gorton 1978; London and Jeanne 1997), and *Polistes carnifex* Fabricius, 1775 has nesting habits associated with *Polybia occidentalis* Olivier, 1791 (CORN 1972).

The coexistence of two species in the same nest is occasionally recorded occurring in the early stage of the colony, and one species may be expelled from the nest or subordinated when there is development and hierarchical stability in another species (i.e., Pinto et al. 2004; Montagna et al. 2012). Bees can also cohabit with social wasps, as was the case in the registration of the *Trigona cilipes* Fabricius, 1804 stingless bee in the same nest as the *Epipona tatua* (Cuvier, 1797) wasp (Rasmussen 2004).

Interspecific harmonious relationships that are not linked to nesting can also be relayed to social wasps, occurring on a smaller scale compared to ants where registration is most frequent. These relationships involve trophobionte hemiptera, which are obligatory parasites of vascular plants and pull food directly from the sap vessels through their specialized oral tract. Excess sap processed in the digestive tract is eliminated by the anus and known as "honeydew," mainly consisting of sugars plus a wide variety of chemical compounds like lipids, amino acids, vitamins, minerals and water (Hölldobler and Wilson 1990), serving as food for wasps and other animals.

Although the association is best documented for tropical wasps (e.g., Dehean & Turilazzi, 1992; Harris 1992; Markwell et al. 1993), Neotropical wasps are recorded performing the same behavior. There are reports of the *Agelaia*, *Mischocyttarus*, and *Polybia* genera visiting trophobionte hemiptera in Brazil, such as membracids. Sugden and McAllen (1994) recorded *Brachygastra mellifica* Say, 1837 collected honeydew from aphid and psyllids.

Occupation of social wasp nests (by other species) is poorly known and documented; however, it is also a harmonious form of interaction. Abandoned wasp nests, depending on where they were built and housed, may remain in good condition. Once abandoned, they may sometimes be reused with adaptations by subsequent generations of the same species or by other arthropods.

Tenants can be divided into two forms; (1) temporary are those that use the nests for a short period of time, which may be only for a night for shelter such as beetles, hemiptera and spiders, or where they will deposit their young to grow into adulthood like some other hymenopterans. Solitary wasps and bees with nesting habits in cavities opportunistically seek to occupy empty holes used by other insects, and may also occupy empty nest cells constructed by other wasp species (cf. Iwata 1976; Cowan 1991). The nests of genera such as *Mischocyttarus*, *Polistes*, and *Polybia* serve as shelters for these hymenoptera, as can be seen in Fig. 21.2. Pinto (2005) recorded cell occupation of *Polistes simillimus* Zikán, 1951 by solitary bees of the *Tetrapedia diversipes* Klug, 1810 species.

Permanent (2) occupants are those with perennial colonies such as *Camponotus* ants that use the protective shell of wasp nests as a protective structure for their colonies; *Polistes* wasps can also act as new occupants, while Prezoto and Nascimento (1999) registered *Polistes versicolor* occupying a *Mischocyttarus cassununga* nest, and Prezoto et al. (2002) also recorded other species of *Polistes ferreri* occupying and modifying a *Mischocyttarus cassununga* nest.

Fig. 21.2 Cells occupied by solitary bees from nesting habits in cavities. (**a**). *Polistes versicolor* nests with occupied cells; (**b**). *Polybia fastidiosuscula* nest with clay jars made by solitary hymenopterans

In addition to perennial colonies, other behaviors characterize occupants as permanent, as with other arthropods such as psocopters, collembolans, and mites that can live generations within active and abandoned nests. Spiders can reside in abandoned nests if feeding on these arthropods and benefit from the nest protection.

21.2 Disharmonious Relations in Social Wasps

Like other insects, social wasps are a food source for large numbers of animals, thus natural enemies such as predators, parasitoids, ectoparasites, fungi, and microorganisms will be approached in more detail in another chapter (see Chap. 23).

The main natural enemy of social wasps is ants, especially Ecitoninae (Fig. 21.3a). Ants' ability to detect social wasps is by sight and odor, as suggested by Chadab (1979). As a result of a coevolution, this is quite advantageous since when ants encounter a colony of wasps, they are unable to defend themselves from the attack, and their speed of escape is crucial as the ants attack suddenly and in large numbers, and particularly and especially pose a threat to immature forms (Maciel et al. 2016). Richards and Richards (1951) cite ant nesting as one of the few defenses found by tropical wasps against ant invasion, demonstrating the importance of ant association in protecting against aggressive birds and ants.

Other arthropods are also reported as predators of social wasps, however more opportunistically, such as Diptera of the Asilidae family, as well as Hemiptera and spiders of the Theridiidae, Thomisidae, and Salticidae families (Fig. 21.3b).

Social wasp colonies are resource-rich environments. Thus, many predators and parasitoids may be attracted to these desired resources (Soares et al. 2006) and destroy these colonies. Larvae and pupae are the target of many of these natural enemies, such as ants, birds, and parasitoid hymenopterans (Yamane 1996; Raw 1997; Clouse 2001). Despite being poorly studied, parasitoid attack may be the leading cause of mortality among social wasps in the early stages of development and be the main cause of colony abandonment.

Fig. 21.3 (**a**). Barring ants (*Labidus praedator*) attacking active *Polybia platycephala* nest; (**b**). Asilidae individual preying on social wasp of the *Polybia* genus

Until now, the occurrence of Chalcididae, Eulophidae, Trigonalidae, and Ichneumonidae parasitoids has mainly been reported in the social *Polistes* and *Mischocyttarus* genera, who find their nests by independent foundation and do not have protective shells for brood cells, and with lower frequencies for *Agelaia, Angiopolybia, Apoica, Brachygastra, Chartergus, Leipomeles, Metapolybia, Parachartergus, Polybia, Pseudopolybia*, and *Synoeca*, which are typically swarming wasps (Bertoni 1912; Weinstein and Austin 1991; Boucek 1992; Carmean and Kimsey 1998; Yu et al. 2005; Somavilla et al. 2015).

Although social insects are one of the best-studied insect groups and there is a large amount of literature on parasite fauna, host–parasite interactions, and immunological defense (Schmid-Hempel 1998; Schmid-Hempel and Ebert 2003); when we separate social insects, there is a limitation on studies on social wasp parasites, as pointed out by Hughes et al. (2003). The Strepsiptera order is described as ectoparasites of social wasps, having host genera such as *Apoica, Polybia*, and *Polistes* (Oliveira and Kogan 1962; Machado 1983; Hughes et al. 2003; Kudô et al. 2004).

Vertebrates such as birds are described as predators of social wasps (Windsor 1976; Strassmann, 1981); however, few bird species attack social wasp colonies to feed their offspring. The most attacked genera of social wasps in Brazil are: *Angiopolybia, Apoica, Leipomeles, Polybia, Protopolybia*, and *Pseudopolybia*. Attacking strategies vary by bird species, and can attack colonies by feeding on-site or breaking combs and flying to feed in another area to avoid wasp attacks (Thiollay 1994, Bhardwaj 2008).

Sazima (2014) recorded the foraging of *Melanerpes candidus* Otto, 1796, popularly known as the White Woodpecker, describing the attack as initially intense, then the wasp's defensive behavior begins and the bird leaves and waits for the wasps to disperse to forage for the colony. A similar strategy can be observed for *Frederickena viridis* Vieillot, 1816 (Mccann et al. 2014).

McCann et al. (2015) concluded in their work that the impact *Ibycter americanus* Boddaert, 1783 has on the colonies compares with the predatory impact of barring ants on some populations of Neotropical social wasps. Other vertebrates such as bats are also reported to be predators of social wasps (Jeanne 1970), however with a lower level of records.

In addition to these animals, entomopathogenic fungi can also attack social wasps, but unlike for ants that have many taxa reported in several studies on the influence of these manipulative fungi (Evans 1974, 1982; Evans et al. 2011; Barbosa et al. 2015), there are few records in Vespidae occurrences and little information about behavioral change caused by fungi. *Ophiocordyceps humbertii* was first recorded by Robin in 1853, infecting the *Vespa cincta* species in Senegal, but not formally described until Petch (1931). Tulasne and Tulasne (1865) and Cooke (1892) reported the attack of *Cordyceps*, a genus considered to be from the same group as *Ophiocordyceps*, in the *Polistes crinitus americanos* and *Polistes crinitus crinitus* species in the New World, but little information is available on how the host was found, the location where it was found, and especially about the signs of behavioral change caused by the fungus.

References

Barbosa BC, Halfeld VR, de Araújo JPM, Maciel TT, Prezoto F (2015) Record of Ophiocordyceps unilateralis sensu lato, the zombie-ant fungus, parasitizing Camponotus in an urban fragment of Atlantic rainforest in southeastern Brazil. Stud Neotropical Fauna Environ 50:01–03

Bertoni AW (1912) Contribucion a la biologia de las avispas y abejas del Paraguay (Hymenoptera). Anales del Museo Nacional de Historia Natural de Buenos Aires 22(1):97–146

Boucek Z (1992) The New World genera of Chalcididae of Hymenoptera. Mem Am Entomolog Instit 53:49–117

Bhardwaj GB (2008) Photo essay: a brief glimpse into the private world of the Oriental Honey-Buzzard Pernis ptilorhynchus. Indian Birds 4(5):154–156

Carmean D, Kimsey L (1998) Phylogenetic revision of the parasitoid wasp family Trigonalidae (Hymenoptera). Syst Entomol 23:35–76

Chadab R (1979) Early warning cues for social wasps attacked by army ants. Psyche 86:115–123

Clemente MA, Lange D, Del-Claro K, Prezoto F, Campos NR, Barbosa BC (2012) Flower-visiting social wasps and plants interaction: network pattern and environmental complexity. Psyche: J Entomol:2012

Clouse RM (2001) Some effects of group size on the output of beginning nests of Mischocyttarus mexicanus (Hymenoptera: Vespidae). Fla Entomol 84:418–425

Cooke MC (1892) Vegetable wasps and plant Worms. Society for Promoting Christian Knowledge, London, p 364

CORN ML (1972) Notes on the biology of Polistes carnifex (Hymenoptera, Vespidae) in Costa Rica and Colombia. Psyche 79:150–157

Cowan DP (1991) The solitary and presocial Vespidae. In: Ross KG, Matthews RW (eds) The social biology of wasps. Comell University Press, Ithaca, pp 33–73; Dejean A, Turillazzi S (1992) Territoriality during trophobiosis between wasps and homopterans. Trop Zool 5(2):237–247

Del-Claro K, Torezan-Silingardi HM (2012) Ecologia das interações plantas-animais: Uma abordagem ecológico-evolutiva. Technical Books Editora

Dejean A, Turillazzi S (1992) Territoriality during trophobiosis between wasps and homopterans. Tropical Zoology 5:237–247

Evans HC (1974) Natural control of arthropods, with special reference to ants (Formicidae), by fungi in tropical high forest of Ghana. J Appl Ecol 11:37–49

Evans HC (1982) Entomogenous fungi in the tropical forest ecosystems: an appraisal. Ecol Entomol 7:47–60

Evans HC, Elliot SL, Hughes DP (2011) Hidden diversity behind the zombie-ant fungus Ophiocordyceps unilateralis: four new species described from carpenter ants in Minas Gerais, Brazil. PLoS One 6:e17024

Fry CH (1972) Convergence between jacamars and bee-eaters. Ibis 112:257–259

Gorton RE (1978) Observations on the nesting behavior of Mischocyttarus immaginatus (rich) (Vespidae: Hymenoptera) in a dry forest in Costa Rica. Insect Soc 25:197–204

Harris RJ (1992) Competition between the introduced wasps Vespula germanica and V. vulgaris in honeydew beech forest, north-western South Island, New Zealand. Ph.D. thesis, University of Canterbury, Christchurch, N.Z., p 145

Herre EA, Windsor DM, Foster RB (1986) Nesting associations of wasps and ants on lowland Peruvian ant-plants. Psyche 93:321–330

Hölldobler B, Wilson EO (1990) The ants. Belknap Press of Harvard University Press, Cambridge: MA, p 732

Hughes DP, Beani L, Turillazzi S, Kathirithamby J (2003) Prevalence of the parasite Strepsiptera in Polistes as detected by dissection of immatures. Insect Soc 50(1):62–68

Iwata K (1976) Evolution of instinct: comparative ethology of Hymenoptera. New Delhi, Amerind, p 535

Jeanne RL (1970) Note on a bat (Phylloderma stenops) preying upon the brood of a social wasp. J Mamm 51:624–625

Kudô K, Yamane S, Mateus S, Tsuchida K, Itô Y, Miyano S, Zucchi R (2004) Parasitism affects worker size in the Neotropical swarm-founding social wasp, Polybia Paulista (Hymenoptera, Vespidae). Insect Soc 51(3):221–225

Lapierre L, Hespenheide H, Dejean A (2007) Wasps robbing food from ants: a frequent behavior? Naturwissenschaften 94(1):997–1001

London KB, Jeanne RL (1997) Site selection by a social wasp in a nesting association (Hymenoptera: Vespidae). J Insect Behav 10:279–288

Machado VLL (1983) Análise populacional de colônias de Polybia (Myrapetra) Paulista (Ihering, 1896) (Hymenoptera, Vespidae). Revista Brasileira de Zoologia 2(4):187–201

Maciel TT, Barbosa BC, Prezoto F (2016) Opportunistic predation of a colony of Polybia platycephala (Richards) (Hymenoptera, Vespidae) by Labidus praedator (smith) (Hymenoptera, Formicidae). Sociobiology 63(1):724–727

Markwell TJ, Kelly D, Duncan KW (1993) Competition between honey bees (Apis mellifera) and wasps (Vespula spp.) in honeydew beech (Nothofagus solandri var. solandri) forest. N Z J Ecol 17(2):85–93

Mccann S, Moeri O, Jones T, Gries GG (2014) Black-throated Antshrike preys on nests of social paper wasps in Central French Guiana. Revista Brasileira de Ornitologia-Brazilian Journal of Ornithology 22(3):300–302

McCann S, Scott C, Jones T, Moeri O, O'Donnell S, Gries G (2015) Red-throated Caracara, a falconid raptor, rivals predatory impact of army ants on social wasps. Insect Soc 62(1):101–108

Menezes JCT, Barbosa BC, Prezoto F (2014) Previously unreported nesting associations of the yellow-olive flycatcher (Tolmomyias sulphurescens) (Aves: Tyrannidae) with social wasps and bees. Ornitología Neotropical 25:363–368

Montagna TS, Neves ÉF, Antonialli-Junior WF (2012) First report of interspecific facultative social parasitism in the paper wasp genus Mischocyttarus Saussure (Hymenoptera, Vespidae). Revista Brasileira de Entomologia 56(2):263–265

Oliveira SJ, Kogan M (1962) Brazilian Strepsiptera (Insecta) parasitizing Vespidae, with the description of three new species of Xenos Rossius, 1793 (Stylopidae). Mem Inst Oswaldo Cruz 60(1):1–11

Petch T (1931) Notes on entomofenous fungi. Trans Br Mycol Soc 16:55–75

Pinto NPO (2005) Estudo de Caso: a reutilização de células de ninho abandonado de Polistes (Aphanilopterus) simillimus Zikán, 1951 (Hymenoptera: Vespidae, Polistinae) por Tetrapedia (Tetrapedia) diversipes Klug, 1810 (Hymenoptera: Apidae, Apinae). Revista de Etologia 7(2):67–74

Pinto NPO, Gobbi N, Noll FB, Penna MAH, Nazareth SL (2004) Coexistência interespecífica em Mischocyttarus Saussure (Hymenoptera: Vespidae, Mischocyttarini) durante a fase inicial de desenvolvimento de um ninho. Revista de Etologia 6(2):95–100

Prezoto F, Nascimento FS (1999) Nota sobre a ocorrência de usurpação de um ninho de Mischocyttarus cassununga por Polistes versicolor (Hymenoptera, Vespidae). Revista de Etologia 1(1):69–71

Prezoto F, Lima MAP, Andrade FR (2002) Nest of tropical paper wasp Mischocyttarus cassununga occupied and modified by Polistes ferreri (Hymenoptera: Vespidae) in South-Eastern Brazil. Sociobiology 39(3):407–410

Rasmussen C (2004) A stingless bee nesting with a paper wasp (Hymenoptera: Apidae, Vespidae). J Kansas Entomol Soc 77(4):593–601

Raw A (1997) Avian predation on individual neotropical social wasps (Hymenoptera, Vespidae) outside their nests. Ornitologia Neotropical 8:89–92

Richards OW, Richards MJ (1951) Observations on the social wasps of South America (Hymenoptera, Vespidae). Trans Entomol Soc London 102:1–170

Richards OW (1978) The social wasps of the Americas (excluding the Vespinae). British Museum of Natural History, London, p 580

Sazima I (2014) Tap patiently, hit safely: a preying tactic of the white woodpecker on social wasp nests. Revista Brasileira de Ornitologia-Brazilian Journal of Ornithology 22(3):292–296

Schmid-Hempel P, Ebert D (2003) On the evolutionary ecology of specific immune defence. Trends Ecol Evol 18:27–32

Schmid-Hempel P (1998) Parasites in social insects. Princeton University Press, Princeton, New Jersey, p 409

Sick H (1997) Ornitologia Brasileira. Ed. Nova Fronteira SA, Rio de Janeiro, Brazil

Soares MA, Gutierrez CT, Zanuncio JC, Bellini LL, Prezotto F, Serrão JE (2006) Pachysomoides sp. (Hymenoptera: Ichneumonidae: Cryptinae) and Megaselia scalaris (Diptera: Phoridae) parasitoids of Mischocyttarus cassununga (Hymenoptera: Vespidae) in Viçosa, Minas Gerais State, Brazil. Sociobiology 48(3):673–680

Somavilla A, Fernandes IO, Oliveira MLD, Silveira OT (2013) Association among wasps' colonies, ants and birds in central Amazonian. Biota Neotropica 13(2):308–313

Somavilla A, Schoeninger K, Carvalho AF, Menezes RS, Del Lama MA, Costa MA, Oliveira ML (2015) Record of Parasitoids in nests of social wasps (Hymenoptera: Vespidae: Polistinae). Sociobiology 62(1):92–98

Souza MMD, Pires EP, Prezoto F (2013) Nidification of Polybia rejecta (Hymenoptera: Vespidae) associated to Azteca chartifex (Hymenoptera: Formicidae) in a fragment of Atlantic Forest, in the state of Minas Gerais, southeastern Brazil. Biota Neotropica 13(3):390–392

Starr CK (1988) The nesting association of the wasps Mischocyttarus immarginatus and Polybia spp. in Costa Rica. Biotropica 20:171–173

Strassmann JE (1981) Parasitoids, predators, and group size in the paper wasp, Polistes exclamans. Ecology 62(5):1225–1233

Sugden EA, McAllen RL (1994) Observations on foraging, population and nest biology of the Mexican honey wasp, Brachygastra mellifica (say) in Texas [Vespidae: Polybiinae]. J Kansas Entomol Soc:141–155

Sühs RB, Somavilla A, Köhler A, Putzke J (2009) Vespídeos (Hymenoptera, Vespidae) vetores de pólen de Schinus terebinthifolius Raddi (Anacardiaceae), Santa Cruz do Sul, RS, Brasil. Revista Brasileira de Biociências 7(2)

Thiollay JM (1994) Family Accipitridae (Hawks and Eagles). In: del Hoyo J., Elliott A., Saragatal J. (eds). Hanbook of the birds of the World (Vol. 2, pp 52–205): New World Vultures to Guinea fowl./ Lynx Edicions, Barcelona

Towsend CR, Begon M, Harper JL (2006) Fundamentos em Ecologia, 2ª Edição edn. Editora Artmed, p 592

Tulasne LR, Tulasne C (1865) Selecta Fungorum Carpologia, vol 3. Imperial Press, Paris

Virgínio F, Maciel TT, Barbosa BC (2015) Nidificação de Polybia rejecta (Fabricius) (Hymenoptera: Vespidae). Associada à Azteca chartifex Forel (Hymenoptera: Formicidae) em Ecótono do Bioma Caatinga/Mata Atlântica, no Estado do Rio Grande do Norte EntomoBrasilis 8(3):242–245

Weinstein P, Austin AD (1991) The host-relationships of trigonalyid wasps (Hymenoptera: Trigonalyidae), with a review of their biology and catalogue to world species. J Nat Hist 18:209–214

Wilson EO (1975) Sociobiology: the new synthesis. Harvard University Press, Cambridge, p 697

Windsor DM (1976) Birds as predators on the brood of Polybia wasps (Hymenoptera: Vespidae: Polistinae) in a Costa Rican deciduous forest. Biotropica:111–116

Yamane S (1996) Ecological factors influencing the colony cycle in Polistes wasps. In: Turillazzi SE, West-Eberhard MJ (eds) Natural history and evolution of paper wasps. Oxford University Press, New York, pp 75–97

Yu DS, van Achterberg C, Horstmann K (2005) World Ichneumonoidea 2004. Taxonomy, biology, morphology and distribution. CD/DVD. British Columbia, Canada, Taxapad, Vancouver

Chapter 22
Interactions of Social Wasps with Microorganisms

D. Mayorga-Ch, C. Rodríguez-C, A. Ortíz-Reyes, M. Romero-Tabarez, and Carlos E. Sarmiento

Abstract Insects maintain many interactions with microorganisms, and these may become key factors for establishing new relationships, environments, and food resources. Such associations with microorganisms may play an important role in colony health, since sociability, especially in Hymenoptera, induce greater vulnerability to pathogens due to the genetic homogeneity and the greater physical proximity among the individuals of a colony. In addition, species-to-species interactions have great potential for the generation of systems with emergent properties. For example, symbiosis provides the host with the ability to use new metabolic pathways or products from the symbiont. Landmark examples of long-term and tightly developed interactions are photosynthetic species and other eukaryotes. The leaf cutter ant-microorganism relationship has generated antibiotics that protect the colony and open a significant new resource for the ants. These themes have been widely explored for many insects, including social groups such as termites, bees, and ants. However, in wasps, and especially in neotropical wasps, the topic deserves stronger research efforts.

Keywords Aspergillus · Pathogens · Defence strategies · Immune system · Microbial interactions

Insects maintain many interactions with microorganisms, and these may become key factors for establishing new relationships, environments, and food resources (Smith 1989). Such associations with microorganisms may play an important role

D. Mayorga-Ch · C. Rodríguez-C · C. E. Sarmiento (✉)
Instituto de Ciencias Naturales, Universidad Nacional de Colombia, Bogotá, Colombia
e-mail: dmayorgac@unal.edu.co; crodriguezco@unal.edu.co; cesarmientom@unal.edu.co

A. Ortíz-Reyes · M. Romero-Tabarez
Instituto de Ciencias Naturales, Universidad Nacional de Colombia, Bogotá, Colombia
Escuela Biociencias, Universidad Nacional de Colombia, Sede Medellín, Bogotá, Colombia
e-mail: adortizr@unal.edu.co; mromerota@unal.edu.co

F. Prezoto et al. (eds.), *Neotropical Social Wasps*,
https://doi.org/10.1007/978-3-030-53510-0_22

in colony health, since sociability, especially in Hymenoptera, induces greater vulnerability to pathogens due to the genetic homogeneity and the greater physical proximity among the individuals of a colony (Naug and Camazine 2002). In addition, species-to-species interactions have great potential for the generation of systems with emergent properties. For example, symbiosis provides the host with the ability to use new metabolic pathways or products from the symbiont (Douglas and Smith 1989). Landmark examples of long-term and tightly developed interactions are photosynthetic species and other eukaryotes. The leaf-cutter ant–microorganism relationship has generated antibiotics that protect the colony and open a significant new resource for the ants (Currie et al. 2006). These themes have been widely explored for many insects, including social groups such as termites, bees, and ants (Bourtzis and Miller 2009; Kaltenpoth and Engl 2014; Grubbs et al. 2015); however, in wasps, and especially in neotropical wasps, the topic deserves stronger research efforts.

In this chapter, we review research on the relationships between microorganisms and social wasps. Because the lack of work on this topic is even more noticeable for neotropical wasps, we present this chapter as a reference that may stimulate further studies in an area that could be very fruitful for understanding the evolution of wasp societies. The review starts with representative examples of symbiosis in Hymenoptera; later, it explores the immune and behavioral responses of social wasps, and finally reviews the different microorganisms that have been isolated from social wasps and their nests.

22.1 Microbial Interactions in Hymenoptera

The interaction between Hymenoptera and microorganisms is not a recent study area. In the last decade, the microbiota associated with different species of bees and ants has been widely studied, especially using molecular techniques for the identification of unculturable bacteria that have provided valuable information (Dittmer et al. 2016; Douglas 2011). Some groups such as *Apis mellifera*, *Bombus* spp. and leaf-cutting ants of the genera *Atta* and *Acromyrmex* have been better studied than others (Koch and Schmid-Hempel 2011; Engel et al. 2012; Aylward et al. 2012; Corby-Harris et al. 2014; Aykward et al. 2014; Meeus et al. 2015). In recent years, the study of wasp microbiomes has increased showing the reiterate presence of a few groups of bacteria, yeast, and fungi, suggesting that these could be among the main factors responsible for the evolutionary success of social insects (Douglas 2011).

Studies of *Apis mellifera* and the genus *Nasonia* (Pteromalidae) show that their microbiota is dominated by Proteobacteria, followed by Firmicutes, Bacteroidetes, and Actinobacteria. These phyla predominate in the digestive tracts of insects and are related to nutrition (Nechitaylo et al. 2014), while the microbiota associated with mammals is dominated by Bacteroidetes and Firmicutes (Douglas 2011; Nechitaylo et al. 2014; Dittmer et al. 2016). These differences are possibly a consequence of the wide separation in the evolutionary history of insects and mammals.

In insects, endosymbiotic microorganisms are hosted in special cells called bacteriocytes facilitating vertical transmission of microorganisms usually transovarically triggering phylogenetically congruent histories between the symbionts and their hosts. Since the digestive tracts of insects and mammals show wide differences not only in physiology but also in diet, they promote diverse microbiological associations (Douglas 2011).

Microorganisms play an important role in the life history of their hosts (Dittmer et al. 2016; Douglas 2011; Rodríguez-Hernández et al. 2019; Paludo et al. 2019). Nutrition and defense seem to be two essential aspects influenced by these taxa. For example, in *Sirex noctilio* Fabricius, 1793 (Siricidae) larvae feed on the monomers of glucose, mannose, galactose, and xylulose found in the liquid fraction of products of the metabolic action of the fungus *Amylostereum areolatum* (Chaillet ex Fries, 1828) that is dispersed in the feeding substrate during oviposition (Thompson et al. 2014). In this example, it is clear how the fungal symbiont enables its host to exploit a ubiquitous and hardly accessible resource such as cellulose during the larval stage. Adams et al. (2011) reported bacteria associated with the larvae and adult stages of *S. noctilio* that degrade cellulose and derived carbohydrates from the food, fulfilling a complementary role in the nutrition of this wasp species. Within these entities, species of the genera *Streptomyces* and *Pantoea* stand out as they have been widely reported by different researchers (Maccollom et al. 2009; Aykward et al. 2014). *Pantoea* has been described as a mutualist species of bacterium, benefiting the insects through nitrogen fixation, especially in nutrient-poor diets. For example, the fruit fly *Bactrocera tryoni* benefits from its associated *P. agglomerans* (Maccollom et al. 2009). These associations with bacteria, especially with species of *Pantoea*, can go beyond a two-way relationship. Leaf-cutting ants that grow their own fungal gardens for food are an example of multipartite interactions in which ants, the fungus symbiont *Leucoagaricus gongylophorus,* and a consortium of bacteria, defend the fungus gardens, eliminating toxic compounds from plants and participating in nitrogen fixation (Aylward et al. 2012). These interactions allow ants to use a wider variety of leaf substrates. *Megastigmus spermotrophus* (Torymidae) is a wasp that parasitizes plants, and a study revealed that interactions with microorganisms are fundamental in the exploitation of seeds, galls, or internal plant tissues (Paulson et al. 2014). Those wasps are also associated with a bacterial consortium, where species of the genus *Ralstonia* are abundant; it plays an important role in the nutrient cycling within the seeds, and, thus, supplements insect nutrition. In other groups such as the stingless bee *Scaptotrigona depilis* Moure, 1942 microorganisms play a deeper role in their life as these should be ingested to produce the hormones that stimulate pupation (Paludo et al. 2018, 2019).

The role of microorganisms in insect defense has been investigated, especially in those that exhibit social behavior. Actinobacteria are the most commonly reported, participating against pathogenic strains, parasitoids, and predators. Their ability to produce secondary metabolites with antimicrobial and antifungal activity might be the reason for their presence in ants (Hanshew et al. 2015). The Attini ants that are involved in a multipartite symbiosis with several microorganisms are among the best-studied examples. It is estimated that the association between the ants, the fungus they cultivate, and the pathogenic fungus *Escovopsis* was established 50 million years ago and is the result of a coevolutionary process with a high degree of specificity (Schultz 2008). The symbiosis of the leaf-cutter ants with filamentous bacteria of the genus *Pseudonocardia* (Actinomycetales) is added to this association. In this case, the bacteria that produce antibiotics are used by the ants to fight the pathogenic fungus *Escovopsis*. In this way, the Attini ants are involved in two mutualist associations. In one they obtain food and in the other they obtain antibiotics to defend their food source (Currie et al. 2006). To this quadripartite interaction is added a yeast of the genus *Phialophora* that, like the actinobacteria, are concentrated in the mesosoma of the ant, and would be the fifth member in this complex network of interactions with an apparently similar capacity to inhibit other microorganisms (Little and Currie 2007).

In the stingless bees *Melipona scutellaris* Latreille, 1811 the bacterium *Paenibacillus polymyxa* (Prazmowski, 1880) produces compounds that inhibits entomopathogenic fungi such as *Paenibacillus larvae* (White, 1906) and *Beauveria bassiana* (Bals-Criv, 1835) (Rodríguez-Hernandez et al. 2019). Microorganisms are also involved in communication within the colony, for example, the bacteria *Serratia marcescens* Bizio, 1823 produces compounds used by the ant *Atta sexdens* Linnaeus, 1758 to communicate food gathering and alarm within the colony (Silva-Junior et al. 2018).

Multiple structural modifications in insects reveal the long-term relationship that they have maintained with microorganisms. For example, the asymmetric mandible of *Sirex noctilio* larvae allows them to squeeze the xylem previously digested by the fungus, while the adult stage of the female wasp has specialized structures at the internal base of the ovipositor called mycangia, which allows the adult female to store the fungi conidia and inoculate it during the oviposition. In the Attini ants, the bacteria are found in foveae located in the propleural sclerite, or they can also be housed in crypts distributed throughout the exoskeleton. Both fovea and crypts are connected to exocrine glands that contribute to the nutrition of bacteria (Currie et al. 2006). In some groups of ants there are no crypts or foveae and bacteria are found directly on the cuticle around the channels of the exocrine glands (Currie et al. 2006).

Wasps of the genera *Philanthus*, *Philanthinus*, and *Trachypus* (Crabronidae) are closely associated with the bacterium symbiont *Streptomyces philanthi* (Kroiss et al. 2010; Kaltenpoth et al. 2012; Nechitaylo et al. 2014). The wasp females have a special cavity in their antennae where they store the microorganisms and a specialized gland that provides food to the bacteria. When the females build their breeding chambers, they deposit the bacteria on the wall and the larvae incorporate these organisms into their cocoons. At that moment, the bacterium produces two antibiotics, streptochlorin and derivatives of piericidin; both would protect the insect's pupa during this instar as they are very vulnerable to fungal attack (Kroiss et al. 2010; Kaltenpoth et al. 2012; Nechitaylo et al. 2014). Bacteria associated with the immune activity are not only located in the host body, the nests are also areas rich in microorganisms that protect the colonies from pathogens. An example of this interaction is present in nests of *Polistes* wasps. Bacteria of the genera *Streptomyces*, *Micromonospora*, and *Actinoplanes* have been isolated from nests of *Polistes dominulus* (Christ.) (Madden et al. 2013). These authors reveal a great diversity of actinomycetes isolated from the wasp nest and it has been observed that these groups are rich in antimicrobial activity providing the nests with a mixture of active compounds that keep the colonies protected from pathogenic microorganisms.

In addition to the role played by microorganisms in the nutrition and immunity of colonies, the participation of symbionts in speciation has been reported; Dittmer et al. (2016) for example, found that parasitoid wasps of the genus *Nasonia* harbor many species of *Wolbachia* and *Arsenophonus nasoniae* which may become an important reproductive barrier between the different species of this wasps genus. Paulson et al. (2014) also report the known influence of *Wolbachia* and *Rickettsia* in the induction of parthenogenesis and androcide in *Megastigmus spermotrophus* (Torymidae), respectively.

22.2 Immune Response to Pathogens in Social Wasps

It is first necessary to briefly characterize how infection occurs within a social insect species before describing the response of their defensive system. A colony may acquire a pathogen through several processes; first, the pathogen can actively search for a colony; second, one or several members of the colony may accidentally acquire the pathogen from the environment. Once the pathogen has infected an individual, horizontal transfer may occur by direct or indirect interaction between members of the same colony or between neighboring colonies. Third, vertical transference can also occur, where pathogens from a parent colony infect new colonies through founders (Cremer et al. 2007).

In general, the first defensive barrier of the insect is the cuticle; once this barrier is passed by, the immune responses take part in the defense (Nation 2008). A special case of the cuticle may occur with larvae of social wasps, as these lack a strong cuticle that protects them from infection by pathogens, so a stronger immune system or defense strategies might be required (Manfredini et al. 2008). Although insects have no adaptive immune response, it has been shown that once continuous attacks occur, the immune system becomes more sensitive (Schmid-Hempel 2005).

The immune reaction can be classified into humoral and cellular and they can act synergistically. The humoral reaction includes the synthesis of antimicrobial peptides, the coagulation of hemolymph, and melanization (Nation 2008). It begins with the secretion of pattern recognition proteins by the epithelial cells, fat cells, and hemocytes. These proteins initiate the signaling cascades that generate the synthesis of antimicrobial peptides through the activation of nuclear genes (Nation 2008). The antimicrobial peptides have a preponderant role in defense mechanisms in vespids. The secretion of cellular compounds such as phenol oxidase is performed mainly by granules synthetized in the Golgi apparatus, which contains large amounts of phagolysosomes (Manfredini et al. 2008). These compounds transform into melanin, which is essential for encapsulating pathogens, including parasitoids, parasites, viruses, bacteria and fungi (Wilson-Rich and Starks 2010).

The cellular mechanisms are performed by hemocytes and are named phagocytosis and encapsulation. Three main types of hemocytes have been identified from larvae of *Polistes dominulus* (Christ, 1791) (Manfredini et al. 2008). The first type has a large nucleus and basophilic cytoplasm; the second type has a smaller nucleus and a larger amount of cytoplasm whose shape may vary; the third type is described as a rounded cell that resembles a prohemocyte and that seems to be the precursor of the other two types (Manfredini et al. 2008). Each type of hemocyte differs in the morphology of the organelles according to their age. For example, the Golgi apparatus shows drops of lipids when the cell is old (Manfredini et al. 2008). Also, changes in these cells are evident according to the developmental stages of the organism. For example, in the younger instars of the larvae, the endoplasmic reticulum is a little fuller with flocculent material, while in the more developed stages it becomes thinner and is arranged in concentric structures. Also, phagolysosomes are produced towards the end of the larval development (Manfredini et al. 2008). The smaller hemocytes do not spread out; they tend to adhere to the substrate with small pseudopods and lamellipodia that are distributed around the cell. Larger hemocytes have the potential to move much faster, become very elongated cells, and develop numerous pseudopods and lamellipodia (Manfredini et al. 2008).

Venom and larval saliva have also been reported as active defense systems. For example, Baracchi et al. (2012) described antimicrobial peptides in the venom of *Parischnogaster mellyi* (Saussure, 1852) and *Parischnogaster alternata* Sakagami, 1969. These peptides are distributed in the cuticle of the wasp body by means of grooming, and there is evidence that they are not applied to the nest surface. In contrast, Turillazzi et al. (2006) identified defensive venom compounds in the nest of old world *Polistes* species. Such a difference in distribution is attributed to the low number of adults in Stenogastrine colonies or to their habit of removing meconia before larvae pupation, which strongly reduces microbial growth (Baracchi et al. 2012). Studies of the larval saliva of *Vespula germanica* (Fabricius, 1793) and *Polistes dominulus* have reported activity against gram-positive and gram-negative bacteria (Gambino 1993; Turillazzi et al. 2004). Turillazzi et al. (2004) suggested that such peptides are vital in the control of infections in social vespids that do not remove the meconia from the cells. As a result of trophallaxis between larvae and adults, the antimicrobial activity of larval saliva can provide defense for the entire colony (Turillazzi et al. 2004).

22.3 Behavioral Defense of Social Wasps Against Pathogens

The colony presents several strategies for protection against pathogens and parasitoids. These strategies range from individual behaviors that indirectly impact other members of the colony to interactions specifically aimed at the hygiene of the entire community. Even some authors have proposed that monogamy minimizes the possibility of acquiring pathogens, due to the lower interaction between individuals during mating (Cremer et al. 2007). However, it has also been suggested that polyandry introduces genetic variability that makes colonies less vulnerable to attack by pathogens (Liersch and Schmid-Hempel 1998).

Individual grooming has been understood as a strategy of hygiene that eliminates parasites and pathogens (Sumana and Starks 2004) since it removes many spores and particles that adhere to the surface of the body during foraging or while in contact with other individuals of the colony. A particularly important structure in social wasp grooming is the calcar, homologous in many other Hymenoptera (Nieves-Aldrey et al. 2006). The calcar consists of a row of setae arranged between the base of the tibia and the tibial spine and a concavity in the tibia itself. This structure makes antennae cleaning more efficient (Snodgrass 1956).

The behavioral pattern of individual grooming shows some trends; but variation is common, even in the individual. In *Polistes dominulus*, grooming behavior is carried out following several possible sequences: after grooming the head and especially the antennae, the next section to be cleaned may be the mesosoma or the metasoma. If the mesosoma is cleaned wasps are more likely to continue with the metasoma, and they almost never return to the head. In contrast, if after grooming the head the wasp grooms the abdomen, it is more likely to clean the head again rather than the metasoma (Sumana and Starks 2004). However, this pattern is not universal and in other groups, such as the Stenogastrinae, different sequences have been observed (Baracchi et al. 2012).

The frequency and duration of grooming vary between body regions. It is known, for example, that grooming the head is very frequent immediately after inspection in a cell. However, it has a shorter duration compared to the cleaning of other parts of the body. The cleaning of the mesosome has an intermediate duration and is less frequent than in the other parts of the body. Metasoma cleaning has a longer duration compared to that occurring in other body parts, and its frequency is close to that in other body regions' overall average. Grooming the metasoma occurs especially after the wasps stand motionless with the legs and wings close to the body and while the antennae are positioned low-down inferiorly (sitting). Finally, the head and the mesosome are groomed with the forelegs and the metasoma with the hind legs; this behavior eliminates the possibility of homogenizing the chemical signals in the body of the insect (Sumana and Starks 2004).

Social grooming is also a common protection strategy. In ants there is evidence of social grooming where some members of the colony groom others, and waste is stored in special buccal compartments for later disposal. Those buccal compartments can have glands that produce antiseptic substances. In bees, social grooming has been laboriously documented against the pathogen *Varroa destructor* (Boecking and Spivak 1999, Delfinado-Baker et al. 1992, Moore et al. 1995). There are studies on social grooming in wasps such as *Polistes versicolor* Olivier, 1791 (Carneiro Silva 1991), *Polistes ferreri* Saussure, 1853 (Sinzato et al. 2003), and some authors suggest that grooming is linked to social hierarchy or nest identity (Pietrobon 2005).

However, the topic has received less interest compared to other social insects (Quinet et al. 2008).

Other protection strategies are the avoidance or removal of infected or sick individuals. For example, in bees we can find individuals specialized in monitoring the entrance to prevent the entry of infected members (Cremer et al. 2007). In social wasps, individuals infected with *Steirnermema carpocapsae* have been reported in colonies of *Vespula vulgaris* (Linnaeus, 1758) and *Vespula pensylvanica* Saussure, 1857 (Gambino et al. 1992). No work on this subject was found by us for neotropical social wasps.

Food manipulation shows characteristics that are useful in the control of pathogens and parasites. *Polistes dominulus* remove the gut content of their prey before bringing them to the nest as pathogens and even toxic compounds may concentrate there (Richter 2000). As mentioned previously, larvae can also provide antimicrobial compounds for the adults and these in turn can pass them to the prey by means of the malaxation exercised by the workers who receive material from the foragers that arrive at the nest (Turillazzi et al. 2004).

Waste management is also a task that can involve specialized individuals or strategies. In ants there is a caste of senescent individuals that are completely dedicated to waste management within the nest, so that other nestmates minimize their contact with this hazardous material. Other interesting strategies that regulate the proliferation of pathogens in social wasps involve morphophysiological changes in their larvae. The absence of a connection between the intestine and the rectum during larval development allows accumulation of the by-products of metabolism in the ventricle, and only when the pupa is being formed is the connection of proctodeum with the ventricle established. All waste is expelled to the bottom of the cell in the form of a compressed material known as meconium (Turillazzi et al. 2004). It is worth mentioning that in *Vespula rufa* (Linnaeus, 1758) dead individuals are not removed and are found as part of the detritus (Greene 1991).

The disposition of the meconium varies between groups. In *Parischnogaster nigricans* (Cameron, 1902) and in *Anischnogaster laticeps* Vecht, 1972 (Stenogastrinae), which inhabit the eastern tropics, the adult closes the cell cover and, days later when the larva has expelled the meconium, the adults open an orifice at the base of the cell and extract the meconium (Turillazzi 1985; Hansell and Turillazzi 1995). In the genera of the Tribe Ropalidiini, *Belonogaster, Polybioides, Parapolybia*, and *Ropalidia*, which do not inhabit the neotropics, adults also open a small hole in the base of the cell and extract the meconium (Gadagkar 1991; Kojima 1996). In the *Belonogaster* species, a very particular phenomenon occurs because the larva is unable to expel the meconium on its own and the adults must pull the peritrophic sac from the larval body rear end to extract it or else the larvae die (Kojima 1983, 1996). In the genera *Ropalidia* and *Parapolybia*, adults close that hole with salivary secretions, while in the *Belonogaster* and *Polybioides* the larva turns around and closes the hole with silk while completing the cocoon (Kojima 1983, 1996).

Except for *Dolichovespula* and *Vespula vulgaris* that remove the meconia, in the Vespinae subfamily, and in the neotropical polistines the meconium is not removed from the cells but accumulates at the base of these cells as a series of compressed layers so that several depositions are seen in reused cells (Kojima 1983). The reuse of the cells in neotropical social wasps is widely known, and studies in *Mischocyttarus labiatus* (Fabricius, 1804) and *Mischocyttarus drewseni* (Saussure, 1954) show that setting up the cells for the new brood implies only the removal of silk remnants (Jeanne 1972; Litte 1981).

22.4 Design and Location of the Social Wasps' Nests Against Pathogens

The architecture of the nest determines its microclimatic characteristics and thus the potential for parasites and pathogenic microorganism proliferation (Hozumi and Inagaki 2010; Harris et al. 2000). In this context, water and humidity control is a crucial factor. Studies such as Harris et al. (2000) have documented that the fungus *Aspergillus flavus* Link, 1809 proliferate under experimental conditions of high relative humidity, affecting larval development (Harris et al. 2000). This species of fungus has been frequently recorded in social wasp colonies (Gambino and Thomas 1988; Fouillaud and Morel 1995; Glare et al. 1996).

Hozumi and Inagaki (2010) modeled the patterns of air currents in the nest of *Polybia spinifex* Richards, 1978 and showed that the vertical slit shape of the entrance and the communication tunnel stabilizes the temperature and relative humidity of the structure. The envelopes of *Polybia spinifex* and *Polybia occidentalis* (Olivier, 1791) nests have spines or projections that facilitate rainwater outflow taking advantage of water adherence to the surface and cohesion of water molecules. The combined effect of gravity and these characteristics carry the liquid out of the nest. Active removal of water from the nest is also performed by workers, which take liquid from the interior of the colony and regurgitate it outside.

22.5 Biota Associated with Social Wasps

Although studies on interactions between social wasps and microorganisms are less numerous than in other Hymenoptera, the results are interesting. The first records of microorganisms that affect social wasps date from the mid-eighteenth century with the story of the Franciscan father Torrubia in 1749, who recounts his encounters with skeletons of wasps in whose bellies he observed the germination of a "plant" (Edwards 1758). Today 156 species of microorganisms have been registered in association with social wasps, and these have been broken down into fungi that represent 69.9%, bacteria 24.4%, viruses 3.2%, nematodes 1.9%, and protozoa 0.6%. In general, the impact or role of these biota in the colonies is unknown and only a small fraction have been proven to be pathogens (Rose et al. 1999).

The following is a summary of the known groups and the relationships that have been described.

22.5.1 Fungi

Most of the microflora of nests or individuals of social wasps are made up of fungi, of around 109 species belonging to 35 genera in 23 families, particularly Aspergillaceae, Trichocomaceae, Clavicipitaceae, Cordycipitaceae, Ophiocordycipitaceae, and Thermoascaceae (Table 22.1).

As mentioned above, the study of the fungi associated with wasps and their colonies can be traced back to the story of the Franciscan father Torrubia in 1749. As Edwards (1758) pointed out, at that time it was thought that these "plants" grew to about three inches, formed buds that fell to the ground, giving rise to insect larvae. Today we know they are fungi probably of the genus *Ophiocordyceps* or *Cordyceps*.

Later, Cooke (1892) in his book, *Vegetable wasps and plant worms—a popular history of entomogenous fungi, or fungi parasitic upon insects*, makes a compendium of entomopathogenic fungi including several species catalogued as pathogens of vespids: *Cordyceps odyneri* Cuallet, 1886 was found by L. Quélet in a larva of the genus *Odynerus* (Eumeninae). As Cooke (1892) pointed out, probably *Cordyceps odyneri* has also been found in adult individuals of *Ancistrocerus parietum* (Linnaeus, 1758). However, due to the ambiguity of the description it is impossible to confirm that it is this fungus. Cooke (1892) also refers to adult wasps of the genera *Vespa* and *Polybia* which died after infection by *Ophiocordyceps sphecocephala* (Klotzsch ex Berkeley, 1843) (=*Cordyceps sphecocephala*). Other records of fungal pathogens in vespids reported in Cooke (1892) include *Ophiocordyceps humbertii* (CP Robin, 1865) (=*Cordyceps humberti*), *Hirsutella saussurei* (Cooke 1892) (=*Isaria saussurei*) found in *Polistes americana*, and *Ophiocordyceps ditmarii* (Quélet, 1877) (=*Cordyceps ditmarii*) and *Isaria sphecophila* Ditmar, 1817 found in *Vespa crabro* Linnaeus, 1758.

Toward the second half of the twentieth century, studies increased of pathogenic microorganisms, especially pathogenic fungi of vespids from temperate latitudes. The efforts focused on the genera *Vespula*, *Vespa*, and *Dolichovespula* due to their pests or invasive species characters (Rose et al. 1999). The most extensive publications relating to fungal microbiota in nests of social vespids are those by Fouillaud and Morel (1995) who found 52 species of fungi in *Polistes olivaceus* De Geer, 1773; later Rose et al. (1999) reported 50 species of fungi in species of the genera *Vespula*, *Vespa*, and *Dolichovespula*. More recently, Jayaprakash and Ebenezer (2010) isolated 24 fungal species in nests of the wasp *Ropalidia marginata*. Other more specific works are of Glare et al. (1996) who evaluated the pathogenicity of *Aspergillus flavus* Link, 1809 in adults and larvae of *Vespula vulgaris* and *Vespula germanica*; Hywel-Jones (1995) who studied infections of *Ophiocordyceps sphecocephala* (Klotzsch ex Berkeley, 1843) (=*Cordyceps sphecocephala*) in wasps and bees.

Table 22.1 Fungi registered in adults or nests of social wasps

Fungi	Host and substrate	Place	References
Apiosporaceae			
Arthrinium phaeospermum (Corda, 1837)	*Polistes olivaceus* (M)	France	Fouillaud and Morel (1995)
Arthrodermataceae			
Arthroderma sp.	*P. olivaceus* (M)	France	Fouillaud and Morel (1995)
Aspergillaceae			
Aspergillus aculeatus Iizuka, 1953	*P. olivaceus* (M,L)	France	Fouillaud and Morel (1995)
Aspergillus candidus Link, 1809	*P. olivaceus* (M)	France	Fouillaud and Morel (1995)
Aspergillus flavus Link, 1809	*P. olivaceus* (M,L)	France	Fouillaud and Morel (1995)
	R. marginata (N)	India	Jayaprakash and Ebenezer (2010)
Aspergillus flavus var. *Oryzae* (Ahlburg, 1878)	*V. vulgaris* (C,L)	California	Gambino and Thomas (1988)
Aspergillus flavus Link,1809*	*V. germanica* (L, A)	New Zeland	Glare et al. (1996)
	V. vulgaris (L,A)	New Zeland	Glare et al. (1996); Harris et al. (2000)
Aspergillus fumigatus Fresenius, 1863	*R. marginata* (N)	India	Jayaprakash and Ebenezer (2010)
Aspergillus japonicus Saito, 1906	*R. marginata* (N)	India	Jayaprakash and Ebenezer (2010)
Aspergillus niger Tieghem, 1867	*P. olivaceus* (M,L)	France	Fouillaud and Morel (1995)
	V. vulgaris (C,L)	California	Gambino and Thomas (1988)
	R. marginata (N)	India	Jayaprakash and Ebenezer (2010)
Aspergillus ochraceus K. Wilhelm, 1877	*P. olivaceus* (M,L)	France	Fouillaud and Morel (1995)
	R. marginata (N)	India	Jayaprakash and Ebenezer (2010)
Aspergillus sojae Sakaguchi & K. Yamada ex Murakami, 1971	*P. olivaceus* (M)	France	Fouillaud and Morel (1995
Aspergillus sparsus x *cremeus*	*P. olivaceus* (M)	France	Fouillaud and Morel (1995)
Aspergillus spp. (2 species)	*P. olivaceus* (M)	France	Fouillaud and Morel (1995)
Aspergillus sydowii (Bainier & Sartory, 1913)	*P. olivaceus* (L)	France	Fouillaud and Morel (1995)

(continued)

Table 22.1 (continued)

Fungi	Host and substrate	Place	References
Aspergillus tamarii Kita, 1913	*P. olivaceus* (M)	France	Fouillaud and Morel (1995)
Aspergillus ustus (Bainier, 1882)	*P. olivaceus* (L)	France	Fouillaud and Morel (1995)
Aspergillus wentii Wehmer, 1896	*P. olivaceus* (L)	France	Fouillaud and Morel (1995)
	V. pensylvanica (I)	California	Gambino and Thomas (1988)
Aspergillus neoniveus Samson, S.W. Peterson, Frisvad & Varga, 2011 (=*Emericella nivea*) 1973	*P. olivaceus* (M)	France	Fouillaud and Morel (1995)
Aspergillus nidulans (Eidam, 1883)	*R. marginata* (N)	India	Jayaprakash and Ebenezer (2010)
Aspergillus nidulans (Eidam, 1883) (=*Emericella nidulans*)	*P. olivaceus* (M,L)	France	Fouillaud and Morel (1995)
Aspergillus restrictus G. Smith, 1931	*V. vulgaris* (C,L)	California	Gambino and Thomas (1988)
Aspergillus niveus Bolchwitz, 1929	*V. vulgaris* (C,L)	California	Gambino and Thomas (1988)
Aspergillus terreus Thom 1918	*V. vulgaris* (C)	California	Gambino and Thomas (1988)
	R. marginata (N)	India	Jayaprakash and Ebenezer (2010)
Emericella sp.	*P. olivaceus* (M)	France	Fouillaud and Morel (1995)
Burenellidae			
Vairimorpha mesnili (Paillot, 1918)	*V. vulgaris* (L)	New Zeland	Rose et al. (1999)
Chaetomiaceae			
Chaetomium globosum Kunze ex Fries, 1829 (=*Chaetomium cochliodes*)	*P. olivaceus* (M,L)	France	Fouillaud and Morel (1995)
Chaetomium globosum Kunze ex Fries, 1829 (=*Chaetomium spirale*)	*R. marginata* (N)	India	Jayaprakash and Ebenezer (2010)
Chaetomium sp.	*P. olivaceus* (M,L)	France	Morel and Fouillaud (1992)
Cladosporiaceae			
Cladosporium cladosporioides (Fresenius, 1850)	*R. marginata* (N)	India	Jayaprakash and Ebenezer (2010)
Cladosporium sp.	*V. vulgaris* (C)	California	Gambino and Thomas (1988)
Clavicipitaceae			
Cordyceps odyneri Quélet, 1886 *	*Odynerus* sp. (L)	France	Cooke (1892)
Metarhizium anisopliae (Metschnikoff, 1879) *	*V. vulgaris* (L,A) *V. germánica* (L,A)	New Zeland	Harris et al. (2000)

(continued)

Table 22.1 (continued)

Fungi	Host and substrate	Place	References
	V. germanica (A)	Chile	Merino et al. (2007)
Cordycipitaceae			
Beauveria bassiana (Balsamo-Crivelli, 1835)*[a]	"On wasp"	United Kingdom	Petch (1948)
	Vespidae	Brazil	Humber et al. (2011)
	V. germanica (A)	Chile	Merino et al. (2007)
	Polistes sp.	Brazil	Humber et al. (2011)
	V. vulgaris (L,A) *V. germanica* (L,A)	New Zeland	Harris et al. (2000)
	Dolichovespula. maculata	USA	Thomas and Poinar (1973)
Beauveria pseudobassiana S.A. Rehner & R.A. Humber, 2011*	*V. vulgaris*	Alberta, Canada.	Humber et al. (2011)
Isaria sphecophila Ditmar, 1817*	*Vespa crabro*	–	Cooke (1892)
Isaria sphecophila Ditmar, 1817 (=*Hymenostilbe sphecophila*)*	"On wasps"	–	Petch (1948)
Isaria farinosa (Holmskjold, 1781) (=*Paecilomyces farinosus*)	*P. olivaceus* (M)	France	Fouillaud and Morel (1995)
	V. pensylvanica (A)	USA	Gambino and Thomas (1988)
Cunninghamellaceae			
Absidia cylindrospora Hagem, 1908	*P. olivaceus* (L)	France	Fouillaud and Morel (1995)
Dipodascaceae			
Geotrichum sp.	*P. olivaceus* (L)	France	Fouillaud and Morel (1995)
Dothioraceae			
Aureobasidium pullulans (de Bary, 1884)	*Vespula* sp. (N)	Colorado	Durrell (1965)
	R. marginata (N)	India	Jayaprakash and Ebenezer (2010)
	P. olivaceus (M)	France	Fouillaud and Morel (1995)
Aureobasidium sp.	*P. olivaceus* (L)	France	Fouillaud and Morel (1995)
Hypocreaceae			
Trichoderma viride Persoon, 1794	*R. marginata* (N)	India	Jayaprakash and Ebenezer (2010)
Trichoderma spp. (2 species)	*P. olivaceus* (L)	France	Fouillaud and Morel (1995)

(continued)

Table 22.1 (continued)

Fungi	Host and substrate	Place	References
Yeasts	*P. olivaceus* (M,I)	France	Morel and Fouillaud (1992)
	R. marginata (N)	India	Jayaprakash and Ebenezer (2010)
Mucoraceae			
Mucor hiemalis Wehmer, 1903	*P. olivaceus* (M,L)	France	Fouillaud and Morel (1995)
Mucor racemosus Fresenius, 1850	*Ropalidia marginata* (N)	India	Jayaprakash and Ebenezer (2010)
Mucor racemosus f. racemosus (=*Mucor varians*) (1917)	*Vespula* sp. (N)	Colorado	Durrell (1965)
Mucor nidicola (Madden et al. 2012)	*P. dominulus* (N)	USA	Madden et al. (2012)
Rhizopus microsporus var. *oligosporus* (Saito, 1905) (=*Rhizopus oligosporus*)	*P. olivaceus* (L)	France	Fouillaud and Morel (1995)
Rhizopus stolonifer (Ehrenberg, 1818)	*R. marginata* (N)	India	Jayaprakash and Ebenezer (2010)
Rhizopus spp. (2 species)	*P. olivaceus* (M)	France	Fouillaud and Morel (1995)
Mycosphaerellaceae			
Ramichloridium sp.	*P. olivaceus* (L)	France	Fouillaud and Morel (1995)
Nectriaceae			
Cylindrocarpon obtusiusculum (Saccardo, 1881) (=*Cylindrocarpon magnusianum*)	*P. olivaceus* (L)	France	Fouillaud and Morel (1995)
Fusarium culmorum x *equiseti*	*P. olivaceus* (M)	France	Fouillaud and Morel (1995)
Fusarium oxysporum Schlechtendal, 1824	*R. marginata* (N)	India	Jayaprakash and Ebenezer (2010)
Fusarium sambucinum var. sambucinum (=*Fusarium roseum*)	*Vespula* sp. (N)	Colorado	Durrell (1965)
	P. olivaceus (M)	France	Fouillaud and Morel (1995)
Fusarium spp. (2 species)	*P. olivaceus* (M,L)	France	Fouillaud and Morel (1995)
Ophiocordycipitaceae			
Ophiocordyceps sphecocephala (Klotzsch ex Berkeley, 1843) (=*Cordyceps sphecocephala*)*	"On wasps"	Khao Yai National Park, Thailand.	Hywel-Jones (1995)
	Polistes spp. (A) *Vespa* spp. (A)	Caribbean islands and South America	Cooke (1892)

(continued)

Table 22.1 (continued)

Fungi	Host and substrate	Place	References
O. ditmarii (Quélet, 1877) (=*Cordyceps ditmarii*)*	"On wasp"	Jura	Cooke (1892)
	V. vulgaris	Europe	Petch (1932)
Ophiocordyceps gentilis (Cesati, 1879)*	"On wasps"	Sarawak	Petch (1932)
Hirsutella saussurei (Cooke, 1892)*	*Polistes* spp.	USA, California, Fiyi, Caribean islands	Petch (1948)
Pleosporaceae			
Alternaria sp.	*Vespula vulgaris* (L)	England	Rose et al. (1999)
Alternaria alternata (Fries, 1832)	*P. olivaceus* (M)	France	Fouillaud and Morel (1995)
	R. marginata (N)	India	Jayaprakash and Ebenezer (2010)
Alternaria alternata (Fries, 1832) (=*Alternariatenuis*)	*Vespula* sp. (N)	Colorado	Durrell (1965)
Alternaria tenuissima (Nees, 1817)	*P. olivaceus* (M)	France	Fouillaud and Morel (1995)
Bipolaris hawaiiensis (M.B. Ellis, 1971) (=*Drechslera hawaiiensis*)	*R. marginata* (N)	India	Jayaprakash and Ebenezer (2010)
Cochliobolus lunatus R.R. Nelson & F.A. Haasis, 1964	*P. olivaceus* (M,L)	France	Fouillaud and Morel (1995)
Curvularia lunata (Wakker, 1898)	*R. marginata* (N)	India	Jayaprakash and Ebenezer (2010)
Curvularia ovoidea (Hiroë & N. Watanabe, 1934)	*R. marginata* (N)	India	Jayaprakash and Ebenezer (2010)
Ulocladium consortiale (Thümen, 1876) (=*Stemphylium ilicis*)	*Vespula* sp. (N)	Colorado	Durrell (1965)
Pleosporales			
Phoma sp.	*Vespula* sp. (N)	Colorado	Durrell (1965)
Saccharomycetaceae			
Candida glabrata (H.W. Anderson, 1917)	*Dolichovespula saxonica* (N)	Poland	Nadolski (2013)
Candida albicans (C.P. Robin, 1853)	*Vespa crabro* (N) *D. saxonica* (N)	Poland	Nadolski (2013)
Sclerotiniaceae			
Botrytis cinerea Persoon, 1794	*V. germanica*	France	Acolat (1953)
Strophariaceae			
Hebeloma radicosum (Bulliard, 1784)	*Vespula flaviceps lewisii* (M,P[b]) *V. vulgaris* (M,P[b])	Japan	Sagara et al. (1985)
Hebeloma spoliatum (Fries, 1838)	*Vespula flaviceps lewisii* (M,P[b]) *V. vulgaris* (M,P[b])	Japan	Sagara et al. (1985)

(continued)

Table 22.1 (continued)

Fungi	Host and substrate	Place	References
Thermoascaceae			
Paecilomyces amoeneroseus (Hennings, 1902)	*P. olivaceus* (L)	France	Fouillaud and Morel (1995)
Paecilomyces spp. (4 species)	*P. olivaceus* (M)	France	Fouillaud and Morel (1995)
Paecilomyces variotii Bainier, 1907	*P. olivaceus* (M,L)	France	Fouillaud and Morel (1995)
	R. marginata (N)	India	Jayaprakash and Ebenezer (2010)
Paecilomyces carneus (Duché & R. Heim, 1931)	*P. olivaceus* (L)	France	Fouillaud and Morel (1995)
Paecilomyces marquandii (Massee, 1898)	*P. olivaceus* (M)	France	Fouillaud and Morel (1995)
Paecilomyces lilacinus (Thom, 1910)	*P. olivaceus* (L)	France	Fouillaud and Morel (1995)
Trichocomaceae			
Penicillium canescens Sopp, 1912	*P. olivaceus* (L)	France	Fouillaud and Morel (1995)
Penicillium citrinum Thom, 1910	*P. olivaceus* (M,L)	France	Fouillaud and Morel (1995)
	R. marginata (N)	India	Jayaprakash and Ebenezer (2010)
Penicillium expansum Link, 1809	*P. olivaceus* (M)	France	Fouillaud and Morel (1995)
Penicillium frequentans Westling, 1911	*R. marginata* (N)	India	Jayaprakash and Ebenezer (2010)
Penicillium glabrum (Wehmer, 1893)	*P. olivaceus* (L)	France	Fouillaud and Morel (1995)
Penicillium jensenii x *canensces*	*P. olivaceus* (M)	France	Fouillaud and Morel (1995)
Penicillium oxalicum Currie & Thom, 1915	*R. marginata* (N)	India	Jayaprakash and Ebenezer (2010)
Penicillium simplicissimum (Oudemans, 1903)	*P. olivaceus* (M,L)	France	Fouillaud and Morel (1995)
Penicillium spp. (9 species)	*P. olivaceus* (M,L)	France	Fouillaud and Morel (1995)
Penicillium steckii K.M. Zalessky, 1927	*P. olivaceus* (M,L)	France	Fouillaud and Morel (1995)
Penicillium variabile Sopp, 1912	*R. marginata* (N)	India	Jayaprakash and Ebenezer (2010)
Penicillium verrucosum Dierckx, 1901	*P. olivaceus* (M)	France	Fouillaud and Morel (1995)
Penicillium diversum Raper & Fennell, 1948	*P. olivaceus* (M)	France	Fouillaud and Morel (1995)

(continued)

Table 22.1 (continued)

Fungi	Host and substrate	Place	References
Eupenicillium shearii Stolk & D.B. Scott, 1967	*P. olivaceus* (M)	France	Fouillaud and Morel (1995)
Talaromyces trachyspermus (Shear, 1902)	*P. olivaceus* (M)	France	Fouillaud and Morel (1995)
Thermomyces lanuginosus Tsiklinsky, 1899	*P. olivaceus* (L)	France	Fouillaud and Morel (1995)
Penicillium sp.*	*V. germanica*	Australia	Humber et al. (2011)
Penicillium sp.	*Vespa cracro* (N)	Poland	Nadolski (2013)
Trichosphaeriaceae			
Khuskia oryzae H.J. Hudson, 1963 *H.J. Hudson*	*P. olivaceus* (L)	France	Fouillaud and Morel (1995)

The letters indicate where the fungus was found: L = larva, I = larval intestine, M = meconia, N = nest, A = Adult, C = nest cells . * = species recorded as vespid pathogens. The names of the microorganisms were updated using the Mycobank database (www.mycobank.org)
[a]Pathogens confirmed by Koch's postulates
[b]Found in abandoned nests

Both Fouillaud and Morel (1995) and Jayaprakash and Ebenezer (2010) point out that *Aspergillus* and *Penicillium* are the genera with the highest number of isolated strains in the flora of *Polistes olivaceus* and *Ropalidia marginata.* The genera *Aspergillus* and *Penicillium* are characterized by containing many species with variable characteristics (Geiser et al. 2007) that colonize an equally large number of environments. Some of their species reproduce by the direct production of mycelium, others by conidiophores where conidiospores are produced, and even some of them reproduce sexually forming fruiting bodies (Krijgsheld et al. 2013; Calvo and Cary 2015). Their taxonomy is still a matter of discussion and the difficulties for identification require multidisciplinary tools (Leitão 2009).

There is limited understanding of the ecological role played by different species of fungi in wasps, as well as the importance of wasps for fungi and few studies can be cited. Rose et al. (1999) found that most of the fungi in nests of wasps are saprophytes that frequently thrive in nest materials such as paper, food waste, and even on dead individuals. Durrell (1965) reports hyphae intertwined with the fibers of *Vespula* sp. nest, perhaps providing mechanical resistance to the nest. Stefanini et al. (2016) reported how *Polistes dominulus* provides an auspicious environment in its digestive tract for successive periods of sporulation and germination of *Saccharomyces cerevisiae* Meyen, 1838 and *Saccharomyces paradoxus* Bachinskaya, 1914. These take place thanks to changes in pH in the wasp's guts. As a result, there are high rates of genetic crosses between yeasts that account for their high heterozygosity and hybridization. In this way, *Polistes dominulus* maintains genetic variability of the species. However, it is not known how much the yeasts benefit their host. On the other hand, it has also been argued that, since wasps are generalist predators, their nests can be reservoirs of entomopathogenic fungi that, although they do not directly affect wasps, act as pathogens for other insect taxa (Morel and Fouillaud 1992; Fouillaud and Morel 1995; Lester et al. 2015).

Under normal conditions many fungal species are found in wasps and their colonies; however, it is very rare to find nests that have failed as a consequence of a fungal attack. The report of *A. flavus* as a vespid pathogen commonly found in healthy nests (Jayaprakash and Ebenezer 2010; Fouillaud and Morel 1995) suggests that colonies have mechanisms to control fungi that under experimental conditions can be pathogenic (Harris et al. 2000). Equally active nests must deal with saprophytic and cellulitic fungi to maintain their viability. It is also known that some fungi may increase the damage caused by parasites in nests of social vespids. In nests of *Vespa crabro* parasitized by larvae of the beetle *Quedius brevicornis* (Thomson, 1860), the wasps cease maintenance of the affected area allowing the proliferation of fungi and, thus, increasing the damage caused by the parasite (Jerzy Nadolski 2012).

Little is known about the fungi associated with vespids and their nests in the tropics since most fungal records have been noted in temperate latitudes. A few reports from Brazil and Chile have been published recording the presence of fungi in wasp colonies (Table 22.1). We need to make efforts to evaluate this topic, making special emphasis on clarifying the nature of these interactions.

22.5.2 Bacteria

There are very few studies focused on evaluating the bacterial microbiota in social vespids and their nests. The most representative are the ones carried out by Morel and Fouillaud (1992) in *Polistes olivaceus*, Madden et al. (2013) in *Polistes dominulus*, and Nadolski (2013) in *Vespa crabro* and *Dolichovespula saxonica* (Fabricius, 1793). The diversity of bacteria isolated from wasps is considerably lower than that of fungi, with approximately 38 species of bacteria belonging to 20 genera and 10 families (Table 22.2).

Among the bacteria that have been isolated from wasps are those belonging to the Actinobacteria phylum, characterized by their great ability to interact with other organisms (Seipke et al. 2012). Bacteria of the *Streptomyces* genus constitute 77% of the sequenced strains (Madden et al. 2013). This genus belongs to the Streptomycetaceae. *Streptomyces* is the most important and extensive genus of Actinobacteria and establishes symbiotic interactions with invertebrates with a primordially protective function thanks to the secretion of secondary metabolites that the host uses as antifungal, antibacterial, and antiviral, and other agents (Chater 2006; de Lima Procópio et al. 2012; Seipke et al. 2012). Some of the strains studied by Madden et al. (2013) show great similarity with *Streptomyces griseus* and with *Streptomyces* sp., the first known for the production of streptomycin and the first antibiotic with an effect on *Mycobacterium tuberculosis* (Schatz and Waksman 1944). The second *Streptomyces* sp. has been isolated from leaf-cutter ants and is

Table 22.2 Bacterial micro-flora associated with the vespids

Bacteria	Host and Substrate	Place	References
Moraxellaceae			
Acinetobacter sp.	*P. olivaceus* (I)	France	Morel and Fouillaud (1992)
Paenibacillaceae			
Paenibacillus alvei (Cheshire & Cheyene, 1885) *Bacillus alvei* (Bas.)	*P. olivaceus* (M,I)	France	Morel and Fouillaud (1992)
Brevibacillus brevis (Migula, 1900) *Bacillus brevisk* (Bas.)	*P. olivaceus* (M)	France	Morel and Fouillaud (1992)
Bacillaceae			
Bacillus cereus Frankland & Frankland, 1887	*P. olivaceus* (M,I)	France	Morel and Fouillaud (1992)
Bacillus circulans Jordan, 1890	*P. olivaceus* (M)	France	Morel and Fouillaud (1992)
Bacillus coagulans Hammer, 1915 emend. De Clerck et al. 2004	*P. olivaceus* (M)	France	Morel and Fouillaud (1992)
Bacillus firmus Bredemann & Werner, 1933	*P. olivaceus* (M,I)	France	Morel and Fouillaud (1992)
Bacillus lentus Gibson, 1935	*P. olivaceus* (M)	France	Morel and Fouillaud (1992)
Bacillus licheniformis (Weigmann, 1898) Chester, 1901	*P. olivaceus* (M)	France	Morel and Fouillaud (1992)
Bacillus megaterium de Bary, 1884	*P. olivaceus* (M,I)	France	Morel and Fouillaud (1992)
Bacillus pumilus Meyer & Gottheil, 1901	*P. olivaceus* (M,I)	France	Morel and Fouillaud (1992)
Bacillus subtilis (Ehrenberg, 1835) Cohn, 1872	*P. olivaceus*(M)	France	Morel and Fouillaud (1992)
Lysinibacillus sphaericus (Meyer & Neide, 1904) Ahmed et al. 2007 *Bacillus sphaericus* (Bas.)	*P. olivaceus* (M,I)	France	Morel and Fouillaud (1992)
Enterobacteriaceae			
Citrobacter freundii (Braak, 1928) Werkman & Gillen, 1932	*P. olivaceus*(I)	France	Morel and Fouillaud (1992)
Citrobacter koseri Frederiksen, 1970 *Citrobacter diversus* (NA)	*V. crabro* (N)	Poland	Nadolski (2013)
Cronobacter sakazakii (Farmer et al. 1980) Iversen et al. 2008 *Enterobacter sakazakii* (Bas.)	*P. olivaceus* (M)	France	Morel and Fouillaud (1992)
Enterobacter cloacae (Braak, 1928) Werkman & Gillen, 1932	*P. olivaceus* (M,I)	France	Morel and Fouillaud (1992)
	V. crabro (N)	Poland	Nadolski (2013)
Enterobacter cancerogenus (Urosevic, 1966) Dickey & Zumoff, 1988 *Enterobacter taylorae* (syn. heterotipico)	*P. olivaceus* (M)	France	Morel and Fouillaud (1992)
Enterococcus faecalis (Andrewes & Horder, 1906) Schleifer & Kilpper-Bälz, 1984	*V. crabro* (N) *D. saxonica* (N)	Poland	Nadolski (2013)

(continued)

Table 22.2 (continued)

Bacteria	Host and Substrate	Place	References
Klebsiella pneumoniae (Schroeter, 1886) Trevisan, 1887	*V. crabro* (N)	Poland	Nadolski (2013)
Klebsiella oxytoca (Flügge, 1886) Lautrop, 1956	*P. olivaceus* (M)	France	Morel and Fouillaud (1992)
Klebsiella pneumoniae subsp. *ozaenae* (Abel, 1893)	*P. olivaceus*(I)	France	Morel and Fouillaud (1992)
Serratia marcescens Bizio, 1823	*P. olivaceus* (I)	France	Morel and Fouillaud (1992)
Escherichia coli (Migula, 1895) Castellani & Chalmers, 1919	*Vespa crabro* (N) *D. saxonica* (N)	Poland	Nadolski (2013)
Shimwellia blattae (Burgess et al. 1973) Priest & Barker, 2010 *Escherichia blattae* (Bas.)	*V. crabro* (N)	Poland	Nadolski (2013)
Enterococcus sp.	*V. crabro* (N)	Poland	Nadolski (2013)
Micrococcaceae			
Kocuria varians (Migula, 1900) Stackebrandt et al. 1995 *Micrococcus varians* (Bas.)	*P. olivaceus* (M,I)	France	Morel and Fouillaud (1992)
Pseudomonadaceae			
Pseudomonas alcaligenes Monias, 1928	*P. olivaceus* (I)	France	Morel and Fouillaud (1992)
Pseudomonas sp.	*P. olivaceus* (M)	France	Morel and Fouillaud (1992)
Burkholderiaceae			
Burkholderia pseudomallei (Whitmore, 1913) Yabuuchi et al. 1993 *Pseudomonas pseudomallei* (Bas.)	*P. olivaceus* (I)	France	Morel and Fouillaud (1992)
Staphylococcaceae			
Staphylococcus aureus Rosenbach, 1884	*V. crabro*	Poland	Nadolski (2013)
Staphylococcus epidermidis (Winslow & Winslow, 1908) Evans, 1916	*V. crabro*	Poland	Nadolski (2013)
Staphylococcus hominis Kloos & Schleifer, 1975	*P. olivaceus* (M,I)	France	Morel and Fouillaud (1992)
Staphylococcus saprophyticus (Fairbrother, 1940) Shaw et al. 1951	*V. Crabro* (N) *D. saxonica* (N)	Poland	Nadolski (2013)
Staphylococcus sciuri Kloos et al. 1976	*P. olivaceus* (M,I)	France	Morel and Fouillaud (1992)
Streptomycetaceae			
Streptomyces spp. (23 strains)	*P. dominulus* (N)	USA	Madden et al. (2013)
Micromonosporaceae			
Micromonospora spp. (6 strains)	*P. dominulus* (N)	USA	Madden et al. (2013)

(continued)

Table 22.2 (continued)

Bacteria	Host and Substrate	Place	References
Actinoplanes spp. (1 strains)	*P. dominulus* (N)	USA	Madden et al. (2013)

Letters indicate: L = outer larval bacteria have been isolated, I = larval intestine, M = meconia, N = nest, A = Adult. Nomenclature follows the Leibniz Institute database DSMZ—German Collection of Microorganisms and Cell Cultures, Germany, Prokaryotic Nomenclature Up-to-date (www.dsmz.de/bacterial diversity/prokaryotic-nomenclature-up-to-date)
Bas basonim, *NA* not accepted name

capable of inhibiting the growth of the pathogenic fungus *Escovopsis* by secretion of antifungal substances (Haeder et al. 2009).

It is important to highlight that strains with great similarity to *Streptomyces flavovirens* (Waksman, 1923) were found in many localities, suggesting that their presence in *P. dominulus* nests is independent of geographic locations (Madden et al. 2013). The frequent in vitro antibacterial activity of many of these strains and their similarity between these isolates and others found in Attini ants lead to Madden et al. (2013) suggesting that there may be a long-term association between wasps and *Streptomyces*.

An interesting case where the presence of bacterial strains may play a role in the success of wasp colonization has been documented for the foreign *P. dominulus* in North America (Cervo et al. 2000). Even though this species shows a lower immune response compared to that of the native *Polistes fuscatus* (Fabricius, 1793) according to the analysis of production of phenol oxidases (Wilson-Rich and Starks 2010), the presence in *P. dominulus* of *Streptomyces* bacteria with antibiotic capacities may explain the success of this wasp in its dispersal capacity by decreasing the pressure of pathogens (Cervo et al. 2000).

22.5.3 Protozoa

The few protozoa reported in social wasps belong to the genus *Crithidia* (Urdaneta-Morales 1983). Previous reports indicating the genera *Nosema* and *Vairimorpha* (Rose et al. 1999; Morel and Fouillaud 1992) as Protozoa are no longer valid as the Microsporidia are currently grouped with fungi (Capella-Gutiérrez et al. 2012) (Table 22.3).

The genus *Crithidia* is known to be an intestinal flagellate parasite of invertebrates, especially of insects (Wallace 1979); once an insect is infected, the immune system triggers both cellular and humoral responses, silencing some routes and culminating in the production of highly active defense products such as antimicrobial peptides (AMPs), lysozymes, proteins containing thioesters (TEPs), lectins, profenol oxidase (PPO), and finally the melanization response (Sadd and Barribeau

Table 22.3 Protozoa, nematodes, and viruses that have been reported associated with Vespidae or their nests

Pathogen	Host	Place	References
Protozoa			
Trypanosomatidae			
Crithidia sp.	*V. squamosa* (Drury, 1773) (A)	USA	Urdaneta-Morales (1983)
Nematodes			
Steinernematidae			
Seteinernema carpocapsae (Weiser, 1955)*	*V. pensylvanica* (L, A)	USA	Gambino et al. (1992)
	V. vulgaris (L, A)		
	V. rufa atropilasa (A)	USA	Poinar and Ennik (1972)
Mermithidae			
Pheromermis pachysoma (V. Linstow)*	*V. vulgaris*	United Kingdom	Poinar et al. (1976)
Sphaerulariidae			
*Sphaerularia bombi**	*V. rufa*	–	Bedding (1984)
	V. vulgaris (R)		
Sphaerularia sp.*	*Vespa simillima* (R)	Japan	Sayama et al. (2007)
Virus			
Dicistroviridae			
Genus *Cripavirus*			
*Cricket paralysis virus**	*V. germanica* (L)	New Zeland	Rose et al. (1999)
	V. vulgaris	–	Martignoni and Iwai (1986)
	V. crabro	–	
Genus *Aparavirus*			
Kashmir bee virus strain 1y 2	*Vespula germanica* (P)	New Zeland	Rose et al. (1999)
Genus *Triatovirus*			
Black queen cell virus	*V. germanica* (P)	Injected, not pathogenic	Rose et al. (1999)
Iflaviridae			
Genus *Iflavirus*			
Bee slow paralisis virus[a]	*V. vulgaris*	–	Martignoni and Iwai (1986)
	V. crabro	–	

(continued)

Table 22.3 (continued)

Pathogen	Host	Place	References
	Dolichovespula media (Retzius, 1783)	–	
Sacbrood virus	*V. germanica* (P)	Inyectado, no patogénico.	Rose et al. (1999)
Poxiviridae			
Genus: "Entomopoxvirus 1990"[b]	*P. olivaceus* (M)	Isla la reunion	Morel and Fouillaud (1992)
Reoviridae			
Genus: *Cypovirus* "*cytoplasmic polyhedrosis virus group 1987*"	*P. olivaceus* (M)	Isla la reunion	Morel and Fouillaud (1992)
Baculoviridae			
Genus: "*Nuclear polyhedrosis viruses 1991*"[a]	*P. olivaceus* (M)	Isla la reunion	Morel and Fouillaud (1992)
Genus: *Betabaculovirus* "*granulosis viruses*"	*P. olivaceus* (M)	Isla la reunion	Morel and Fouillaud (1992)

The letters indicate where the microorganism was found L = larva, A = adult, R = queen, P = pupa. * = species recorded as vespid pathogens. The virus nomenclature followed the updates available in the International Committee on Taxonomy of Viruses. (ictvonline.org)
[a]Subsequently this genus was separated into *Alphabaculovirus*, *Deltabaculovirus*, and *Gammabaculoviru*. It is impossible to know the species since it was not reported in the study
[b]Later, this genus was separated into *Alphaentomopoxvirus, Betaentomopoxvirus, and Gammaentomopoxvirus*; however, it is impossible to know the species since it was not reported in the study

2013). The induction of the production of specific antimicrobial peptides has been reported in *Drosophila melanogaster* Meigen, 1830 and it has been established that they are different from those generated under a bacterial or fungal infection (Boulanger et al. 2001). In Vespidae, very little is known about the interactions with this group, to the point that the species that act as parasites are unknown and consequently so are the mechanisms they use to counteract them.

22.5.4 Nematodes

The first records of nematode infections in social wasps are described in Baird (1853), who reports *Gordius vespae vulgaris* Baird 1853 in *Vespula vulgaris*. This nematode was considered later by Poinar et al. (1976) as *Pheromermis pachysoma* (V. Linstow). Currently, four species of nematodes have been reported, all of them pathogenic for vespids (Table 22.3).

Experimentally it has been demonstrated that *Steinernema carpocapsae* (Weiser, 1955) is a pathogen of *Vespula vulgaris* and *Vespula pensylvanica*. Nevertheless, grooming behaviors in these species are effective against that nematode allowing some colonies to recover after infection (Gambino et al. 1992). In an interesting study, Poinar and Ennik (1972) showed that the death of adult individuals of *Vespula* is not caused by the action of the nematode *S. carpocapsae* but due to the septicemia produced by the nematod symbiont bacterium of the genus *Xenorhabdus*, which is released by the nematode in the hemocoel of the wasp soon after infection. The nematodes proliferate in the corpse of the wasp, and the juvenile infective form is produced. Therefore, the removal of sick or dead individuals from the colony is an efficient strategy to combat infection by *S. carpocapsae* (Gambino et al. 1992).

22.5.5 *Virus*

To date, approximately nine genera distributed among five families of viruses have been registered in social wasps (Table 22.3). Fouillaud and Morel (1994) and Morel and Fouillaud (1992) conducted several studies that demonstrated the presence of viruses in the meconia and intestines of *Polistes olivaceus* larvae. These studies show that these viral bodies maintain their virulence after being digested but do not affect the larvae. However, there are other types of viruses that are capable of infecting vespids, such as the Cricket paralysis virus (CrPV) and the Kashmir bee virus found in *Vespula germanica* larvae and pupae (Rose et al. 1999).

Different viruses generate different symptoms and consequences in the host insects. That is the case of the Dicistroviridae, that are able to decrease fertility, decrease life expectancy, generate paralysis, and damage intestinal tissues. The virus of the genera *Cripavirus* and *Aparavirus* induce damage of the tracheae, the fatty bodies, and the nervous system, causing paralysis (Bonning 2009). Entities of the genus *Ifavirus* (Ifaviridae) may be associated with greater aggressiveness in bees (Terio et al. 2008). The *Entomopoxvirus* (Poxiviridae) interrupt the signaling cascade of the immune system generated by the host cell (Kristen 2010). It is worth mention that these viruses also are pathogenic to humans. Finally, entities of the Baculoviridae are capable of inhibiting molt and apoptosis (Clem et al. 1991; O'Reilly and Miller 1989).

22.6 Conclusions

This review reveals an immense field of research that may have an important role for understanding the success of various groups of social wasps, as well as clarifying the role of microorganisms in the social organization of the group, either by generating health problems or as allies in the various activities of the colony. Given the climatic and diversity characteristics that occur in the tropics, it will be very interesting to compare the results of studies in temperate zones with those coming from tropical regions, especially if they focus on the impact that microorganisms may have on the ecology and social interaction of the colonies.

Bibliography

Acolat L (1953) Les materiaux des nids de guepes. Annals Scientifiques Universite de Besancon 8:39–43

Adams AS, Jordan MS, Adams SM, Suen G, Goodwin LA, Davenport KW, Currie CR, Raffa K (2011) Cellulose-degrading bacteria associated with the invasive woodwasp *Sirex noctilio*. ISME J 5:1323–1331

Aylward FO, Burnum KE, Scott JJ, Suen G, Tringe SG, Adams SM, Barry KW, NIcora CD, Piehowski PD, Purvine SO, Starrett GJ, Goodwin LA, Smith RD, Lipton MS, Currie CR (2012) Metagenomic and metaproteomic insight into bacterial communities in leaf-cutter ant fungus gardens. ISME J 6:1688–1701

Aykward FO, Suen G, Biedmann PHW, Adams AS, Scott JJ, del Malfatti SA, Rio TG, Tringe SG, Poulsen M, Raffa KF, Klepzig KD, Currie CR (2014) Convergent bacterial microbiotas in the fungal agricultural systems of insects. MBio 5(6):e02077–e02014

Baird SF (1853) Catalogue of the species of entozoa, or intestinal worms, contained in the collection of the British Museum. Order of the Trustees, London, p 132

Baracchi D, Mazza G, Turillazzi S (2012) From individual to collective immunity: the role of the venom as antimicrobial agent in the Stenogastrinae wasp societies. J Insect Physiol 58(1):188–193

Bedding RA (1984) Nematode parasites of hymenoptera. In: Nickle WR (ed) Plant and insect nematodes. Marcel Dekker, New York, pp 755–795

Bonning BC (2009) The Dicistroviridae: an emerging family of invertebrate viruses. Virol Sin 24(5):415–427

Boulanger N, Ehret-Sabatier L, Brun R, Zachary D, Bulet P, Imler JL (2001) Immune response of *Drosophila melanogaster* to infection with the flagellate parasite *Crithidia* spp. Insect Biochem Mol Biol 31(2):129–137

Bourtzis K, Miller TA (eds) (2009) Insect symbiosis, vol 3. CRC Press, Boca Raton

Calvo AM, Cary JW (2015) Association of fungal secondary metabolism and sclerotial biology. Front Microbiol 6:62. https://doi.org/10.3389/fmicb.2015.00062

Capella-Gutiérrez S, Marcet-Houben M, Gabaldón T (2012) Phylogenomics supports microsporidia as the earliest diverging clade of sequenced fungi. BMC Biol 10(1):47

Chater KF (2006) *Streptomyces* inside-out: a new perspective on the bacteria that provide us with antibiotics. Philos Trans R Soc B Biol Sci 361(1469):761–768

Cervo R, Zacchi F, Turillazzi S (2000) *Polistes dominulus* (Hymenoptera, Vespidae) invading North America: some hypotheses for its rapid spread. Insect Soc 47(2):155–157

Clem RJ, Fechheimer M, Miller LK (1991) Prevention of apoptosis by a Baculovirus gene during infection of insect cells. Science 254(5036):1388–1390

Cooke MC (1892) Vegetable wasps and plant worms. A popular history of entomogenous fungi, or fungi parasitic upon insects. Committee of General Literature and Education, London, pp 30–72

Corby-Harris V, Maes P, Anderson KE (2014) The bacterial communities associated with Honey Bee (*Apis mellifera*) Foragers. PLoS One 9(4):1–13. https://doi.org/10.1371/journal.pone.0095056

Cremer S, Armitage SAO, Schmid-Hempel P (2007) Social immunity. Curr Biol 17(16):693–702

Currie CR, Poulsen M, Mendenhall J, Boomsma JJ, Billen J (2006) Coevolved crypts and exocrine glands support mutualistic bacteria in fungus-growing ants. Science 311(5757):81–83

Dittmer J., van Opstal EJ, Shropshire JD, Bordenstein SR, Hurst GDD, Brucker RM (2016) Disentangling a Holobiont –recent advances and perspectives in *Nasonia* wasps. Front Microbiol 7:1478, 1–1478,18. https://doi.org/10.3389/fmicb.2016.01478
Douglas AE, Smith DC (1989) Are endosymbioses mutualistic? Trends Ecol Evol 4(11):350–352
Douglas AE (2011) Lessons from studying insect symbioses. Cell Host Microbe 10(4):359–367
Durrell LW (1965) Fungi in nests of paper wasps. Am Midl Nat 73(2):501–503
Edwards G (1758) Gleanings of natural history. Royal College of Physicians, London
Engel F, Martinson VG, Moran NA (2012) Functional diversity within the simple gut microbiota of the honey bee. Proc Natl Acad Sci 109(27):11002–11007
Fouillaud M, Morel G (1994) Characterization of cytoplasmic and nuclear polyhedrosis viruses recovered from the nest of *Polistes hebraeus* F. (Hymenoptera; Vespidae). J Invertebr Pathol 64(2):89–95
Fouillaud M, Morel G (1995) Fungi associated with nests of the paper wasp *Polistes hebraeus* (Hymenoptera: Vespidae) on La Reunion Island. Environ Entomol 24(2):298–305
Frisvad JC (2015) Taxonomy, chemodiversity, and chemoconsistency of *Aspergillus*, *Penicillium*, and *Talaromyces* species. Front Microbiol 5:773
Frisvad JC, Larsen TO (2015) Chemodiversity in the genus *Aspergillus*. Appl Microbiol Biotechnol 99(19):7859–7877
Gadagkar R (1991) *Belonogaster, Mischocyttarus, Parapolybia*, and independent-founding *Ropalidia*. In: Ross KG, Matthews RW (eds) The social biology of wasps. Cornell University Press, Ithaca, pp 149–190
Gambino P (1993) Antibiotic activity of larval saliva of *Vespula* wasps. J Invertebr Pathol 61:110
Gambino P, Pierluisi GJ, Poinar GO (1992) Field test of the nematode *Steinernema feltiae* (Nematoda: Steinernematidae) against yellowjacket colonies (Hym.: Vespidae). Entomophaga 37(1):107–114
Gambino P, Thomas GM (1988) Fungi associated with two *Vespula* (Hymenoptera: Vespidae) species in the eastern San Francisco Bay area. Pan Pac Entomol 64(2):107–113
Geiser DM, Klich MA, Frisvad JC, Peterson SW, Varga J, Samson RA (2007) The current status of species recognition and identification in *Aspergillus*. Stud Mycol 59:1–10
Glare TR, Harris RJ, Donovan BJ (1996) *Aspergillus flavus* as a pathogen of wasps, *Vespula* spp. in New Zealand. N Z J Zool 23(4):339–344
Goodrich-Blair H (2007) They've got a ticket to ride: *Xenorhabdus nematophila–Steinernema carpocapsae* symbiosis. Curr Opin Microbiol 10(3):225–230
Greene A (1991) Dolichovespula and Vespula. In: Ross KG, Matthews RW (eds) The social biology of wasps. Cornell University Press, Ithaca, pp 263–308
Grubbs KJ, Scott JJ, Budsberg KJ, Read H, Balser TC, Currie CR (2015) Unique honey bee (*Apis mellifera*) hive component-based communities as detected by a hybrid of phospholipid fatty-acid and fatty-acid methyl ester analyses. Plos One 10(4):1–17. https://doi.org/10.1371/journal.pone.0121697
Haeder S, Wirth R, Herz H, Spiteller D (2009) Candicidin-producing *Streptomyces* support leaf-cutting ants to protect their fungus garden against the pathogenic fungus *Escovopsis*. Proc Natl Acad Sci 106(12):4742–4746
Hansell MH, Turillazzi S (1995) Nest structure and building material of three species of *Anischnogaster* (Vespidae Stenogastrinae) from Papua New Guinea. Trop Zool 8(1):203–219
Hanshew AS, McDonald BR, Diaz CD, Djiéto-Lordon C, Blatrix R, Currie CR (2015) Characterization of Actinobacteria associated with three ant-plant mutualisms. Microb Ecol 69(1):192–203
Harris RJ, Harcourt SJ, Glare TR, Rose EAF, Nelson TJ (2000) Susceptibility of *Vespula vulgaris* (Hymenoptera: Vespidae) to generalist entomopathogenic fungi and their potential for wasp control. J Invertebr Pathol 75(4):251–258
Hill CL, Booth TF, Stuart DI, Mertens PPC (1999) Lipofectin increases the specific activity of cypovirus particles for cultured insect cells. J Virol Methods 78(1–2):177–189

Hozumi S, Inagaki T (2010) A long slit-like entrance promotes ventilation in the mud nesting social wasp, *Polybia spinifex*: visualization of nest microclimates using computational fluid dynamics. J Insect Sci 10(175):1–12. https://doi.org/10.1673/031.010.14135
Humber RA, Hansen KS, Wheeler MM (2011) Catalog of species. USDA-ARS Biological Integrated Pest Management Research, Ithaca, 505 p
Hywel-Jones N (1995) *Cordyceps sphecocephala* and a *Hymenostilbe* sp. infecting wasps and bees in Thailand. Mycol Res 99(2):154–158
Jayaprakash A, Ebenezer P (2010) A new report on mycobiota associated with *Ropalidia marginata* paper nests. Indian J Sci Technol 3(1):6–8
Jeanne RL (1972) Social biology of the Neotropical wasp *Mischocyttarus drewseni*. Bull Museum Comparative Zool 144(3):150
Kaltenpoth M, Engl T (2014) Defensive microbial symbionts in Hymenoptera. Funct Ecol 28(2):315–327
Kaltenpoth M, Yildirim E, Gürbüz MF, Herzner G, Strohmd E (2012) Refining the roots of the Beewolf-Streptomyces symbiosis: antennal symbionts in the rare genus *Philanthinus* (Hymenoptera, Crabronidae). Appl Environ Microbiol 78(3):822–827
Koch H, Schmid-Hempel P (2011) Socially transmitted gut microbiota protectet bumble bees against an intestinal parasite. Proc Natl Acad Sci 108(48):19288–19292
Kojima J (1996) Meconium egestion by larvae of *Ropalidia* Guérin (Hymenoptera: Vespidae) without adult aid, with a note on the evolution of meconium extraction behaviour in the tribe Ropalidiini. Aust J Entomol 35(1):73–75
Kojima J (1983) Peritrophic sac extraction in *Ropalidia fasciata* (Hymenoptera, Vespidae). Entomol Soc Jpn 51(4):502–508
Krijgsheld P, Bleichrodt RV, Van Veluw GJ, Wang F, Müller WH, Dijksterhuis J, Wösten HAB (2013) Development in *Aspergillus*. Stud Mycol 74:1–29
Kristen WL (2010) Characterization of the anti-viral effects inhibited by the Vaccinia virus K1 protein. Tesis de doctorado en Microbiologia – University of Illinois at Urbana-Champaign, Urbana
Kroiss J, Kaltenpoth M, Scheneider B, Schwinger M, Hertweck C, Maddula RK, Strhom E, Svatoš A (2010) Symbiotic streptomycetes provide antibioticcombination prophylaxis for wasp offspring. Nat Chem Biol 6:261–263
Leitão AL (2009) Potential of *Penicillium* species in the bioremediation field. Int J Environ Res Public Health 6(4):1393–1417
Lester PJ, Bosch PJ, Gruber MAM, Kapp EA, Peng L, Brenton-Rule EC, Buchanan J, Stanislawek WL, Archer M, Corley JC, Masciocchi M, Van Oystaeyen A, Wenseleers T (2015) No evidence of enemy release in pathogen and microbial communities of common wasps (*Vespula vulgaris*) in their native and introduced range. PLoS One 10(3):e0121358
Liersch S, Schmid-Hempel P (1998) Genetic variation within social insect colonies reduces parasite load. Proc R Soc B 265(1392):221–225
de Lima Procópio RE, da Silva IR, Martins MK, de Azevedo JL, de Araújo JM (2012) Antibiotics produced by *Streptomyces*. Braz J Infect Dis 16(5):466–471
Litte M (1981) Social biology of the Polistine wasp *Mischocyttarus labiatus*: survival in a Colombian rain forest. Smithsonian Contribution to Zoology, 327
Little AE, Currie CR (2007) Symbiotic complexity: discovery of a fifth symbiont in the attine ant–microbe symbiosis. Biol Lett 3(5):501–504
MacCollom GB, Lauzon CR, Sjogren RE, Meyer WL, Olday F (2009) Association and attraction of Blueberry Maggot fly Curran (Diptera: Tephritidae) to *Pantoea* (*Enterobacter*) *agglomerans*. Environ Entomol 38(1):116–120
Madden AA, Grassetti A, Soriano JN, Starks PT (2013) Actinomycetes with antimicrobial activity isolated from paper wasp (Hymenoptera: Vespidae: Polistinae) nests. Environ Entomol 42(4):703–710
Madden AA, Stchigel AM, Guarro J, Sutton D, Starks PT (2012) *Mucor nidicola* sp. nov. a fungal species isolated from an invasive paper wasp nest. Int J Syst Evol Microbiol 62:1710–1714

Manfredini F, Dallai R, Ottaviani E (2008) Circulating hemocytes from larvae of the paper wasp *Polistes dominulus* (Hymenoptera, Vespidae). Tissue Cell 40(2):103–112

Martignoni ME, Iwai PJ (1986) A catalog of viral diseases of insects, mites and ticks. Department of Agriculture, Forest Service, Pacific Northwest Research Station, Portland, 57 p

Meeus I, Parmentier L, Billiet A, Maebe K, Van Nieuwerburgh F, Deforce D, Wäckers F, Vandamme P, Smagghe G (2015) 16S rRNA amplicon sequencing demonstrates that indoor-reared bumblebees (*Bonbus terrestris*) harbor a core subset of bacteria normally associated with the wild host. PLoS One 10(4):1–15. https://doi.org/10.1371/journal.pone.0125152

Merino L, France A, Gerding M (2007) Selection of native fungi strains pathogenic to *Vespula germanica* (Hymenoptera: Vespidae). Agricultura Técnica 67(4):335–342

Meyer V, Wu B, Ram AFJ (2011) *Aspergillus* as a multi-purpose cell factory: current status and perspectives. Biotechnol Lett 33(3):469–476

Miller LK, Ball LA (eds) (1998) The insect viruses. Springer Science & Business Media, New York, 414 p

Morel G, Fouillaud M (1992) Presence of microorganisms and viral inclusion bodies in the nests of the paper wasp *Polistes hebraeus* Fabricius (Hymenoptera, Vespidae). J Invertebr Pathol 60(2):210–212

Nadolski J (2013) Factors restricting the abundance of wasp colonies of the European hornet *Vespa crabro* and Saxon wasp *Dolichovespula saxonica* (Hymenoptera: Vespidae) in an urban area in Poland. Entomol Fenn 24(4):204–215

Nation JL (2008) Insect physiology and biochemeistry, 2nd edn. CRC Press, Boca Raton, pp 367–384

Naug D, Camazine S (2002) The role of colony organization on pathogen transmission in social insects. J Theor Biol 215(4):427–439

Nechitaylo TY, Westermann M, Kaltenpoth M (2014) Cultivation reveal physiological diversity among defensive '*Streptomyces philanthi*'symbionts of beewolf digger wasps (Hymenoptera, Crabronidae). BMC Microbiol 14:202, 1–202,15. https://doi.org/10.1186/s12866-014-0202-x

Nieves-Aldrey JL, Fontanal-Cazalla FM, Fernández F (2006) Filogenia and Evolución en Hymenoptera. In: Fernández F, Sharkey MJ (eds) Introdución a los Hymenoptera de la Región Neotropical. Sociedad Colombiana de Entomologia and Universidad Nacional de Colombia, Bogotá, pp 37–55

O'Reilly DR, Miller LK (1989) A baculovirus blocks insect molting by producing Ecdysteroid UDP-glucosyl transferase. Science 245(4922):1110–1112

Paulson AR, Aderkas P, Perlman SJ (2014) Bacterial associates of seed-parasitic wasps (Hymenoptera: *Megastigmus*). BMC Microbiol 14:224, 1–224,15. https://doi.org/10.1186/s12866-014-0224-4

Paludo CR, Menezes C, Silva-Junior EA, Vollet-Neto A, Andrade-Dominguez A, Pishchany G, Khadempour L, Nascimento FS, Currie CR, Kolter R, Clardy J, Pupo M (2018) Stingless bee larvae require fungal steroid to pupate. Nature Sci rep 8(1):1122–1133

Paludo CR, Pishchany G, Andrade-Dominguez A, Silva-Junior EA, Menezes C, Nascimento FS, Currie R, Kolter R, Clardy J, Pupo MT (2019) The microbial community modulates the growth of the symbiotic fungus necessary for the metamorphosis of stingless bees. PloS one 14(7):e0219696

Petch T (1932) Cordyceps on Hymenoptera. Trans Br Mycol Soc 16:219–221

Petch, T. (1948). A revised list of British entomogenous fungi. Trans Br Mycol Soc, 31(3–4), 286–304

Pietrobon TAO (2005) Glândula ectomandibular e comportamento de *Polistes versicolor* (Olivier) (Hymenoptera, Vespidae). Tesis de doctorado en Ciencias Biológicas – Universidade Estadual Paulista "Júlio de Mesquita Filho", Rio Claro

Poinar GO, Ennik F (1972) The use of *Neoaplectana carpocapsae* (Steinernematidae: Rhabditoidea) against adult yellowjackets (Vespula spp. Vespidae: Hymenoptera). J Invertebr Pathol 19(3):331–334

Poinar GO, Lane RS, Thomas GM (1976) Biology and redescription of *Pheromermis pachysoma* (v. Linstow) n. gen. n. comb. (Nematoda: Mermithidae), a parasite of yellowjackets (Hymenoptera: Vespidae). Nematologica 22(3):360–370

Quinet Y, Heredia P, Marlier J, Biseau J (2008) Defesa Quimica em formigas do gênero *Crematogaster* Lund 1831 (Myrmicinae). In: Viela EF, dos Santos IA, Schoereder JH, Serrão JE, Campos LA, Lino-Neto J (eds) Insectos Sociais. Universidade Federal de Viçosa, Minas Gerais, pp 68–86

Richter MR (2000) Social wasp (Hymenoptera: Vespidae) foraging behavior. Annu Rev Entomol 45(1):121–150

Rodriguez-Hernandez D, Melo WG, Menegatti C, Lourenzon VB, Nascimento FS, Pupo MT (2019). Actinobacteria associated with stingless bees biosynthesize bioactive polyketides against bacterial pathogens. NJ Chem 43:10109–10117

Rose EAF, Harris RJ, Glare TR (1999) Possible pathogens of social wasps (Hymenoptera: Vespidae) and their potential as biological control agents. N Z J Zool 26(3):179–190

Sagara N, Kitamoto Y, Nishio R, Yoshimi S (1985) Association of two *Hebeloma* species with decomposed nests of vespine wasps. Trans Br Mycol Soc 84(2):349–352

Sadd BM, Barribeau SM (2013) Heterogeneity in infection outcome: lessons from a bumblebee-trypanosome system. Parasite Immunol 35(11):339–349

Sayama K, Kosaka H, Makino S (2007) The first record of infection and sterilization by the nematode *Sphaerularia* in hornets (Hymenoptera, Vespidae, *Vespa*). Insect Soc 54(1):53–55

Schatz A, Waksman SA (1944) Effect of streptomycin and other antibiotic substances upon *Mycobacterium tuberculosis* and related organisms. Soc Exp Biol Med 57(2):244–248

Schmid-Hempel P (2005) Evolutionary ecology of insect immune defenses. Annu Rev Entomol 50:529–551

Schultz TR (2008) Major evolutionary transitions in ant agriculture. Proc Natl Acad Sci USA 105(14):5435–5440

Segato F, de Damásio ARL, Lucas RC, Squina FM, Prade RA (2014) Genomics review of holocellulose deconstruction by aspergilli. Microbiol Mol Biol Rev 78(4):588–613

Seipke RF, Kaltenpoth M, Hutchings MI (2012) *Streptomyces* as symbionts: an emerging and widespread theme? FEMS Microbiol Rev 36(4):862–876

Stefanini I, Dapporto L, Berná L, Polsinelli M, Turillazzi S, Cavalieri D (2016) Social wasps are a *Saccharomyces* mating nest. Proc Natl Acad Sci 113(8):2247–2251

Smith JM (1989) Generating novelty by symbiosis. Nature 341(6240):284–285

Snodgrass RE (1956) Anatomy of the honey bee, vol 107. Cornell University Press, Londres, p 108

Sumana A, Starks PT (2004) Grooming patterns in the primitively eusocial wasp *Polistes dominulus*. Ethology 110(10):825–833

Terio V, Martella V, Camero M, Decaro N, Testini G, Bonerba E, Tantillo G, Buonavoglia C (2008) Detection of a honeybee iflavirus with intermediate characteristics between kakugo virus and deformed wing virus. New Microbiol 31(4):439–444

Thomas GM, Poinar GOJ (1973) Report of diagnoses of diseased insects 1962-1972. Hilgardia 42(8):261–360

Thompson BM, Bodart J, McEwen C, Gruner DS (2014) Adaptations for symbiont-mediated external digestion in *Sirex noctilio* (Hymenoptera: Siricidae). Ann Entomol Soc Am 107(2):453–460

Turillazzi S, Perito B, Pazzagli L, Pantera B, Gorfer S, Tancredi M (2004) Antibacterial activity of larval saliva of the European paper wasp *Polistes dominulus* (Hymenoptera, Vespidae). Insect Soc 51(4):339–341

Turillazzi S (1985) Brood rearing behaviour and larval development in *Parischnogaster nigricans* serrei (Du Buysson) (Hymenoptera Stenogastrinae). Insect Soc 32(2):117–127

Turillazzi S, Mastrobuoni G, Dani FR, Moneti G, Pieraccini G, la Marca G, Bartoluccib G, Peritoc B, Dapporto L (2006) Dominulin A and B: two new antibacterial peptides identified on the cuticle and in the venom of the social paper wasp *Polistes dominulus* using MALDI-TOF, MALDI-TOF/TOF, and ESI-Ion trap. J Am Soc Mass Spectro 17(3):376–383

Urdaneta-Morales S (1983) Some biological characteristics of a *Crithidia* species from the yellow-jacket wasp, *Vespula squamosa* (Hymenoptera, Vespidae). Rev Braz Biol 43(1):409–412
Van den Berg MA (2011) Impact of the *Penicillium chrysogenum* genome on industrial production of metabolites. Appl Microbiol Biotechnol 92:45–53
Wallace FG (1979) Biology of the Kinetoplastida of arthropods. In: Lumsden WHR, Evans DA (eds) Biology of kinetoplastida. Academic Press, Lodres, pp 213–235
Wilson-Rich N, Starks PT (2010) The *Polistes* war: weak immune function in the invasive *P. dominulus* relative to the native *P. fuscatus*. Insect Soc 57(1):47–52

Chapter 23
Artificial Environments for Studying Eusocial Wasps

André Rodrigues de Souza, Camila Folly Baptista, Gabrazane Venâncio Marques Teixeira, and Maria Augusta Pereira Lima

Abstract Animal research is often performed in a laboratory, semi-field, or other artificial environments. This approach has the potential to better control variables during experimentation. Paper wasps often build their nests in different human-made structures (metal, plastic), but also in nonhuman-made ones (leaf, tree), suggesting considerable flexibility in the requirements for nesting. Additionally, nests of many species are composed by an uncovered comb in which all adults and immatures can be easily observed. Therefore, keeping the wasp colonies in artificial environments seems a promising way to observe their biology in detail. In this chapter, we describe how artificial environments can be used to study paper wasps.

Keywords Lab rearing · Rational rearing · Nest building · Nutrition

23.1 Introduction

Animal research may involve the collection of biological material, behavioral observation, and experimental manipulation. To minimize the influence of random variables, animal research is often performed in a laboratory, semi-field, or other artificial environments. Eusocial wasps that build stelocyttarous and gymnodomous nests, such as the independent-founding *Polistes* and *Mischocyttarus*, are good biological models for behavioral studies, because their colony mates and social interactions can be promptly observed without the need for much manipulation. Therefore,

A. R. de Souza (✉)
Departamento de Biologia, Faculdade de Filosofia Ciências e Letras de Ribeirão Preto, Universidade de São Paulo, Ribeirão Preto, São Paulo, Brazil

C. F. Baptista
Departamento de Biologia Geral, Universidade Federal de Viçosa, Viçosa, Minas Gerais, Brazil

G. V. M. Teixeira · M. A. P. Lima
Curso de Pós-Graduação em Biologia Animal - Departamento de Biologia Animal, Universidade Federal de Viçosa, Viçosa, Minas Gerais, Brazil

F. Prezoto et al. (eds.), *Neotropical Social Wasps*,
https://doi.org/10.1007/978-3-030-53510-0_23

keeping the wasp colonies in artificial environments can prove to be useful for studying their biology in detail; although, the development of methods for wasp keeping should be improved. Colonies of *Polistes* and *Mischocyttarus* are easy to find, collect, and transport, as they are quite sinanthropic—presumably, a desirable trait for species intended to be reared in artificial environments. The challenge, however, is to provide the basic resources for colony development and a proper stimulus to build an artificial environment that simulates the important natural conditions.

In this chapter, we describe how artificial environments can be used to study paper wasps such as the independent-founding ones. To this, we have used examples from literature as well as information on primitively eusocial paper wasps that we have obtained during our laboratory investigations.

23.2 Artificial Environments: What they Are and why Use them in Eusocial Wasp Research

In this chapter, we define artificial environment as a set of structures humans use to maintain the colony, nest, adult and/or immature wasps for a certain period of time. Basically, we discuss rearing boxes, shelters, and observation arenas. Wasps can be either totally confined to the artificial environment (e.g., laboratory rearing boxes) or they can be partially confined, as in the semi-field artificial environment that allows wasps to forage in the field, while keeping the colony partially protected in a convenient place (a vespiary, a plantation, or near the laboratory). Artificial environments have been used to investigate wasp physiology (Strambi 1990), demography (Nadeau and Stamp 2003), foraging behavior (Daugherty et al. 2011; Sheehan and Tibbetts 2008), vibrational communication (Suryanarayanan et al. 2011), and sexual behavior (de Souza et al. 2014), among others.

The use of an artificial environment has advantages such as allowing a more detailed observation of insects for longer periods of time and the possibility to control environmental conditions, providing more accurate experimental manipulation.

23.3 The Use of Artificial Environments to Study Eusocial Wasps

A variety of artificial environments has been used to investigate many questions about wasps. In Table 23.1, we grouped the research by the type of biological material to be maintained/observed (colony, nest, adult, immature). This illustrates which artificial environments allow to investigate, together or separately, different phases of a wasp's life cycle. We also describe the purpose of each study, to demonstrate different contexts in which artificial environments may be useful. Finally, we provide a brief description of the artificial environment, illustrating the variety of mate-

Table 23.1 Some examples of artificial environments used in wasp research

Aim	Material	Reference
Colony		
To produce males for immunocompetence assays	Acrylic box (40 × 40 × 20 cm)	Cappa et al. (2015)
To produce females for longevity assay	Acrylic box (30 × 33 × 34 cm) with two meshed windows (perimeter 52.5 cm)	Southon et al. (2015)
To produce females and test the effect of diet on the expression of visual signals	Wood box with one acrylic removable side (14 × 15.5 × 20 cm)	Tibbetts and Curtis (2007)
To produce females for behavioral assays on nestmate recognition	Glass box (15 × 15 × 15 cm)	Signorotti et al. (2014)
To keep colonies in the laboratory and test the effect of this environment on colony development and wasp physiology	Acrylic box (30 × 30 × 30 cm)	Jandt et al. (2015)
To attract wasps to build their nest	Wood box (14 × 14 × 14 cm)	Jandt et al. (2015)
To keep colonies in a corn plantation and promote biological pest control	Wood box with a roof shape (30 × 40 × 1 cm) attached to a batten (1.8 m high)	Prezoto and Machado (1999)
To keep colonies in a *Eucalyptus* plantation and promote biological pest control	Plastic box (13 × 17 × 11 cm) open at the bottom, that fixes in a tree around 1.5 m above the ground	Elisei et al. (2012)
Nest with immature brood		
To test the effect of temperature on the expression of visual signals	Plastic box (4 × 4 × 6 cm) with a meshed lead	Green et al. (2012)
Adult		
To test male longevity	Wood box (45 × 45 × 45 cm) with some meshed sides	Sen and Gadagkar (2010)
To investigate male–male and male–female interactions	Plastic arena (8 × 8 × 2 cm)	Izzo and Tibbetts (2012)
To promote dominance trials between females	Plastic arena (7.5 × 12.5 × 4 cm)	Tibbetts and Dale (2004)
To observe male–male competition in a body size-controlled population	Meshed box (4 × 2 × 2 m)	Beani and Zaccaroni (2015)
Immature		
To manipulate olfactory experiences in newly emerged wasps	Plastic microtube (2 × 1.5 cm)	Signorotti et al. (2014)

rials, sizes, and shapes. This table provides only a few representative examples of how artificial environments have helped researchers to answer their questions. Depending on the purpose of the research, it may be necessary to provide the artificial environment with water, carbohydrate (e.g., 1:1 sugar and water solution), protein (e.g., immature beetles or fly larvae), and material for nest building (e.g., cardboard). Details on how to transfer wasps to artificial shelters will not be described here, but can be found in Prezoto and Machado (1999) and Prezoto et al. (2007).

23.4 Designs of Artificial Environments Used to Study Wasps

Despite the widespread use of artificial environments to study wasps, detailed information that allows their replication is often lacking (for exceptions, see Gadagkar and Gadagkar 2001; Southon et al. 2015; Jandt et al. 2015). To help those interested in this tool, designs of some artificial environments that we have successfully used (Figs. 23.1, 23.2, and 23.3) are reported here.

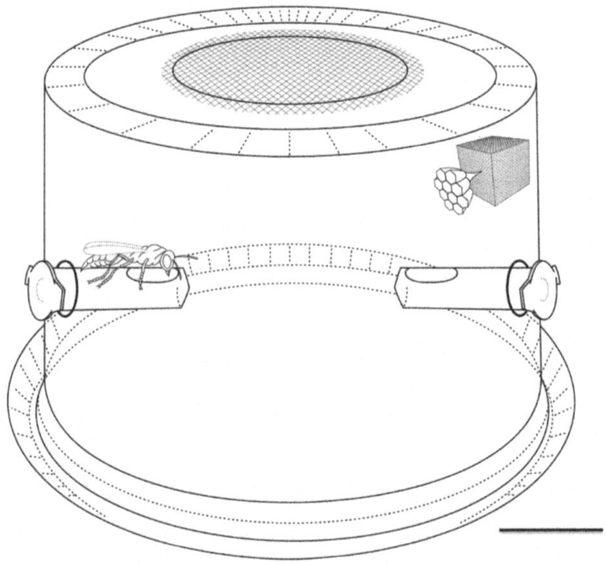

Fig. 23.1 Artificial environment for rearing and transporting a small colony of *Polistes versicolor* in early development (preemergence). It is composed of a plastic box (volume 1 L) meshed at the top. Note that drilled plastic microtubes are used as feeders. Foam is glued to the inner side of the cage and used for nest attachment (by means of pins). Scale bar = 3 cm

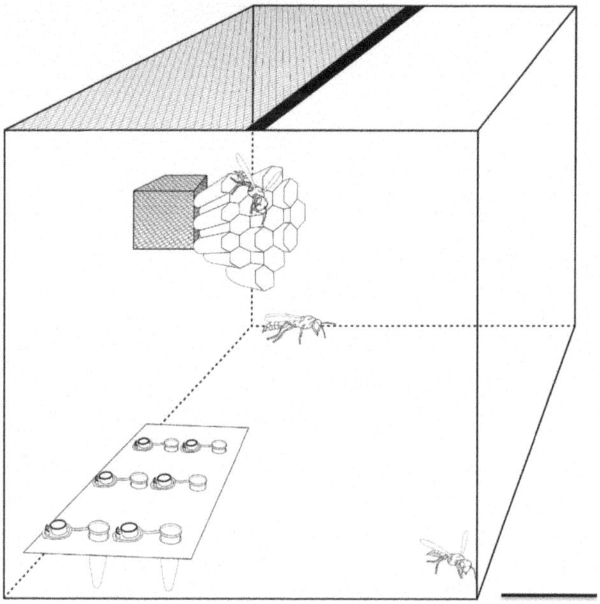

Fig. 23.2 An artificial environment that is comprised of a glass box for rearing and observing a colony of *Polistes versicolor*. It is composed of four sides with dimensions of 25 × 25 cm each, an upper side of 25 × 25 cm partially screened and the bottom with a dimension of 40 × 40 cm, which functions as a "tray" that supports the cage. Feeders are made with several microtubes inserted into a cardboard platform. Scale bar = 5 cm. (Of note, these boxes were formerly developed for European *Polistes* by the Italian research group at the Dipartimento di Biologia, Universitá degli Studi di Firenze-Italy, and then successfully applied to the Neotropical species *P. versicolor*, described in this chapter)

23.5 Limitations for the Study of Wasps in Artificial Environments

Depending on the research question, artificial environments may not be suitable for wasp rearing. This is because some aspects of the wasps' biology develop better (or only) in the natural environment. For example, when transferring small colonies of *Polistes versicolor* (Olivier, 1791) to plastic boxes housed in the laboratory (Fig. 23.1), we noticed, in some colonies, a delay in nest construction. Specifically, the wasps did not increase the height of the nest cells even though the boxes were filled with cardboard and water for nest building. As a result, the larvae became too big and fell from the nest.

An artificial environment can also modify the wasps' biology, as described in the comparison between *Polistes fuscatus* (Fabricius, 1793) reared in the field and in laboratory boxes (Jandt et al. 2015). These authors found that laboratory rearing (characterized by excess nutrition) changes the wasps' physiology, increasing the production of queen-like individuals (greater body size and more fatty tissue) relatively to field rearing. In addition, the same study demonstrated that the expression

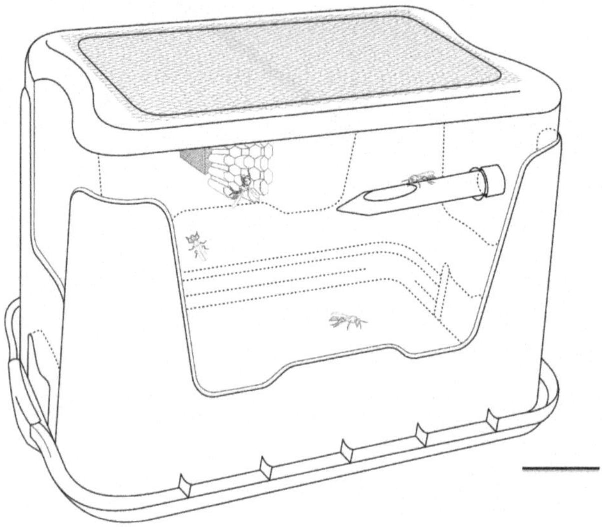

Fig. 23.3 An artificial environment for rearing and transporting a colony of *Polistes versicolor*. It is composed of a plastic box, with dimensions around 60 × 30 × 45 cm. The upper side is meshed and the bottom is removable. Note the presence of a Falcon tube on the right side, used as feeder. Scale bar = 10 cm

of caste-related genes was biased in the opposite direction. That is, wasps that developed in the laboratory expressed more worker-like genes. Finally, laboratory-reared colonies had a lower nest building rate (as we observed for *P. versicolor*) and produced males a little earlier than in colonies that had developed under natural conditions. Thus, experiments carried out in artificial environments should have adequate hypotheses, and when possible, results should be validated with complementary observations in the natural environment (e.g., Cappa et al. 2014). Moreover, such experiments should have proper controls, with similar conditions and resources used among the different treatments. Finally, depending on the research question, researchers should perform many different treatments trying to isolate the effects of each variable on the tested parameters.

The above recommendations refer to wasps reared in the laboratory (captivity), but what about the use of artificial environments to study wasps in the field? An example would be the case of artificial shelters for transfer of wasps to agricultural crops (e.g., Prezoto and Machado 1999; Elisei et al. 2012). We still do not know if the biology of the wasps is altered under these conditions. At least, it is reassuring to know that wasps are highly sinanthropic insects (Lima et al. 2000; Prezoto et al. 2007; Alvarenga et al. 2010), often nesting on different human-made substrates. In conclusion, we stress on the importance of adequate research questions, the value of the comparison between colonies kept in natural and artificial environments, and the necessity to use adequate controls to validate the experiments using wasps kept in artificial environments.

23.6 Tips for Keeping and Observing Wasps in Artificial Environments

Here are some tips to help those who want to use artificial environments to study wasps:

1. When keeping colonies in laboratory boxes, be sure to protect them from ants. We always put the boxes above trays with water and soap.
2. When observing wasps in arenas, make sure they are well lit with artificial illuminators or natural light. This is especially important in visual communication studies (de Souza et al. 2014). Also, be sure to clean the arena at each trial (e.g., Liebert et al. 2010) in order to remove odors that may potentially affect the subsequent trials. We line the arena with a new filter paper at each test.
3. When transferring colonies from one environment to another (in the field), make sure that the new environment has the minimal conditions needed for establishment of wasps, such as, availability of water and food and protection against bad weather, to maximize the success of the transfer.
4. Finally, when trying to establish wasps in artificial environments, be aware that not all wasps behave in the same manner. Even phylogenetically closely related species may have considerable differences in adaptation to these environments (Gadagkar and Gadagkar 2001).

23.7 Final Considerations

In this chapter, we have described how artificial environments have been used to investigate eusocial wasps. We hope this information will inspire more basic and applied research on these insects. We also hope that the methods discussed here will help in the development and improvement of the artificial rearing of eusocial wasps, drawing attention to the limitations of such approaches.

Ideally, these methods should be standardized, to allow comparisons between studies developed by different research groups. In addition, the methods and environments used to maintain colonies should affect the biology and natural behavior of wasps minimally, increasing the reliability of the results obtained in the laboratory, semi-field, or field, making possible extrapolations about what will occur in the natural environment.

References

Alvarenga RB, de Castro MM, Santos-Prezoto HH, Prezoto F (2010) Nesting of social wasps (Hymenoptera, Vespidae) in urban gardens in Southeastern Brazil. Sociobiology 55(2):445–452

Beani L, Zaccaroni M (2015) Experimental male size manipulation in *Polistes dominula* paper wasps: being the right size. Ethol Ecol Evolut 27(2):185–199

Cappa F, Beani L, Cervo R, Grozinger C, Manfredini F (2015) Testing male immunocompetence in two hymenopterans with different levels of social organization: live hard, die young? Biol J Linn Soc 114(2):274–278

Cappa F, Manfredini F, Dallai R, Gottardo M, Beani L (2014) Parasitic castration by Xenos vesparum depends on host gender. Parasitology 141(8):1080

Daugherty THF, Toth AL, Robinson GE (2011) Nutrition and division of labor: effects on foraging and brain gene expression in the paper wasp *Polistes metricus*. Mol Ecol 20(24):5337–5347

Elisei T, Ribeiro Junior C, Fernandez AJ, Nunes JV, Souza AR, Prezoto F (2012) Management of social wasp colonies in eucalyptus plantations (Hymenoptera: Vespidae). Sociobiology 59(3):1167–1174

Gadagkar R, Gadagkar R (2001) The social biology of Ropalidia marginata: toward understanding the evolution of eusociality. Harvard University Press

Green JP, Rose C, Field J (2012) The role of climatic factors in the expression of an intrasexual signal in the paper wasp *Polistes dominulus*. Ethology 118(8):766–774

Izzo AS, Tibbetts EA (2012) Spotting the top male: sexually selected signals in male *Polistes dominulus* wasps. Anim Behav 83(3):839–845

Jandt JM, Thomson JL, Geffre AC, Toth AL (2015) Lab rearing environment perturbs social traits: a case study with *Polistes* wasps. Behav Ecol 26(5):1274–1284

Liebert AE, Wilson-Rich N, Johnson CE, Starks PT (2010) Sexual interactions and nestmate recognition in invasive populations of *Polistes dominulus* wasps. Insect Soc 57(4):457–463

Lima MAP, Lima JR, Prezoto F (2000) Levantamento dos gêneros, flutuação das colônias e hábitos de nidificação de vespas sociais (Hymenoptera, Vespidae) no Campus da UFJF, Juiz de Fora, MG. Revista Brasileira de Zoociências 2(1):69–80

Nadeau H, Stamp N (2003) Effect of prey quantity and temperature on nest demography of social wasps. Ecol Entomol 28(3):328–339

Prezoto F, Machado VLL (1999) Transferência de colônias de vespas (*Polistes simillimus* Zikán, 1951) (Hymenoptera, Vespidae) para abrigos artificiais e sua manutenção em uma cultura de *Zea mays*. Revista Brasileira de Entomologia 43(3–4):239–241

Prezoto F, Ribeiro Júnior C, Oliveira SA, Elisei T (2007) Manejo de vespas e marimbondos em ambiente urbano. In: Pinto AS, Rossi MM, Salmeron E (eds) Manejo de pragas urbanas. CP2, Belo Horizonte, pp 125–130

Sen R, Gadagkar R (2010) Natural history and behaviour of the primitively eusocial wasp *Ropalidia marginata* (Hymenoptera: Vespidae): a comparison of the two sexes. J Nat Hist 44(15–16):959–968

Sheehan MJ, Tibbetts EA (2008) Robust long-term social memories in a paper wasp. Curr Biol 18(18):R851–R852

Signorotti L, Cappa F, d'Ettorre P, Cervo R (2014) Novel insights into the ontogeny of nestmate recognition in *Polistes* social wasps. PLoS One 9(5):e97024

Southon RJ, Bell EF, Graystock P, Sumner S (2015) Long live the wasp: adult longevity in captive colonies of the eusocial paper wasp *Polistes canadensis* (L.). PeerJ 3:e848

de Souza AR, Júnior CAM, do Nascimento FS, Lino-Neto J (2014) Sexy faces in a male paper wasp. PLoS One 9(5):e98172

Strambi A (1990) Physiology and reproduction in social wasps. In: Engels W (ed) Social insects: an evolutionary approach to castes and reproduction. Springer, Berlin, pp 59–75

Suryanarayanan S, Hermanson JC, Jeanne RL (2011) A mechanical signal biases caste development in a social wasp. Curr Biol 21(3):231–235

Tibbetts EA, Curtis TR (2007) Rearing conditions influence quality signals but not individual identity signals in *Polistes* wasps. Behav Ecol 18(3):602–607

Tibbetts EA, Dale J (2004) A socially enforced signal of quality in a paper wasp. Nature 432(7014):218–222

Chapter 24
Economic Importance of Neotropical Social Wasps

Thiago Elisei, Cleber Ribeiro Junior, and Fábio Prezoto

Abstract The study of the foraging activity of social wasps sheds light on their ecological interactions with the environment in which they inhabit. These interactions are the focus of several studies. Research has shown the importance of some species as potential agents of pest control, preying caterpillars of different species. Other studies have demonstrated the action of wasps in pollination processes, since they use floral nectaries in the search for resources with carbohydrate. Some studies with wasps do not focus on forage behavior, but on their biology. There are researches that have revealed the medicinal action of poisons of some groups of Polistinae. Social wasps can also be used as indicators of environmental quality, and their nesting habit is one of the indication forms. However, Polistinae is not always supportive of the human economy. Some research has shown that they cause damage to some crops, destroying fruits or intimidating farmers. This study aims to present ways in which social wasps can be applied in the human economy.

Keywords Polistinae · Biological control · Bioindicators · Polinizators · Plague

24.1 Why Social Wasps Are Important to Humans?

The Vespidae family is divided into subfamilies, being Polistinae, Vespinae, and Stenogastrinae those that present eusocial behavior (Bell and Sumner 2013). Theses subfamilies exhibit the behaviors of division of labor into castes, overlap of generations in the colony, and offspring care, evidencing characteristics of eusocial groups

T. Elisei (✉)
Universidade Federal da Paraíba, Programa de Pós-Graduação em Ciências Biológicas
Departamento de Zoologia, João Pessoa, PB, Brazil

C. R. Junior
Universidade Federal de Viçosa, Programa de Pós-graduação em Biologia Celular e Estrutural, Departamento de Biologia Geral, Viçosa, MG, Brazil

F. Prezoto
Universidade Federal de Juiz de Fora, Programa de Pós-graduação em Comportamento e Biologia Animal, Departamento de Zoologia, Juiz de Fora, MG, Brazil

F. Prezoto et al. (eds.), *Neotropical Social Wasps*,
https://doi.org/10.1007/978-3-030-53510-0_24

(Wilson 1971). This social behavior, according to Hamilton (1964a, b), is built by altruist behaviors based on genetic success.

However, new studies have revealed that eusociality can also be the result of three types of interactions: (1) Coercion – aggressive social behavior, which punishes and polices selfish individual behavior. (2) Parental manipulation – behavior exhibited by the parents, which aims at the persuasion of the children to take care of the brothers. (3) Policing – inhibition behavior of the reproduction activity of certain workers that can happen by the destruction of eggs or physical aggression (Ratnieks & Wenseleers 2008).

This eusocial trait needs a complex organization for maintenance of the nest and individuals. For this, the wasp's social colonies utilize resources acquired from the ambient environment. They search for carbohydrates from different sources, such as nectar from flowers, which gives this group an important role in pollination (Hunt et al. 1991; Quirino and Machado 2001; Nádia et al. 2007; Santos et al. 2007). Other fodder material is animal protein, the main food for immature brood in the colony, supplied by carcasses of dead animals (Greene 1991; O'Donnell 1995) and mainly through the predation of invertebrates, caterpillars being the main group slaughtered (Richter 2000; Carpenter and Marques 2001; Prezoto et al. 2006; Bichara et al. 2009; Elisei et al. 2010; Saraiva et al. 2017) (Fig. 24.1). Due to its predation on the damaging groups in agriculture, Polistinae emerges as agent of biocontrol on populations of pest species.

The study of the foraging activity and behavior of social wasps revels importants ecological interactions with the environment. These studies could be tools in the human lives. In this chapter, we introduce some methods to apply social wasp in the economic development, with the aim to turn them into friends of humans and not a problem, as they are treated today.

24.2 Social Wasps as Biological Controller on Agricultural Pests

Integrated Pest Management (IPM) is a low-cost and low environmental impact tool. It is applied to reduce the economic losses caused to plantations by harmful plague species. In this system, natural methods are used to control populations that are causing damage to plantations, and the use of the term "Ecological Pest Management" is also correct (Gravena 1992). In Parra et al. (2002), the authors cite the use of predators and parasites in integrated pest management studies, such as bedbugs (*Podisus nigrispinus, Geocoris* spp., *Nabis* spp., *Orius* spp. and *Zellus* spp.,); green lacewings (*Chrysoperla* spp.,); ladybugs (*Cycloneda sanguinea* and *Scymnus* spp.); beetles (*Calosoma sp.* and *Lebia concinna*); treasures (*Dorus* sp.); and the microhymenopterans *Trichogramma pretiosum, Cotesia flavipes* and *Cerastomicra intmaculata* (Hymenoptera: Calcidae). Among the entomopathogens

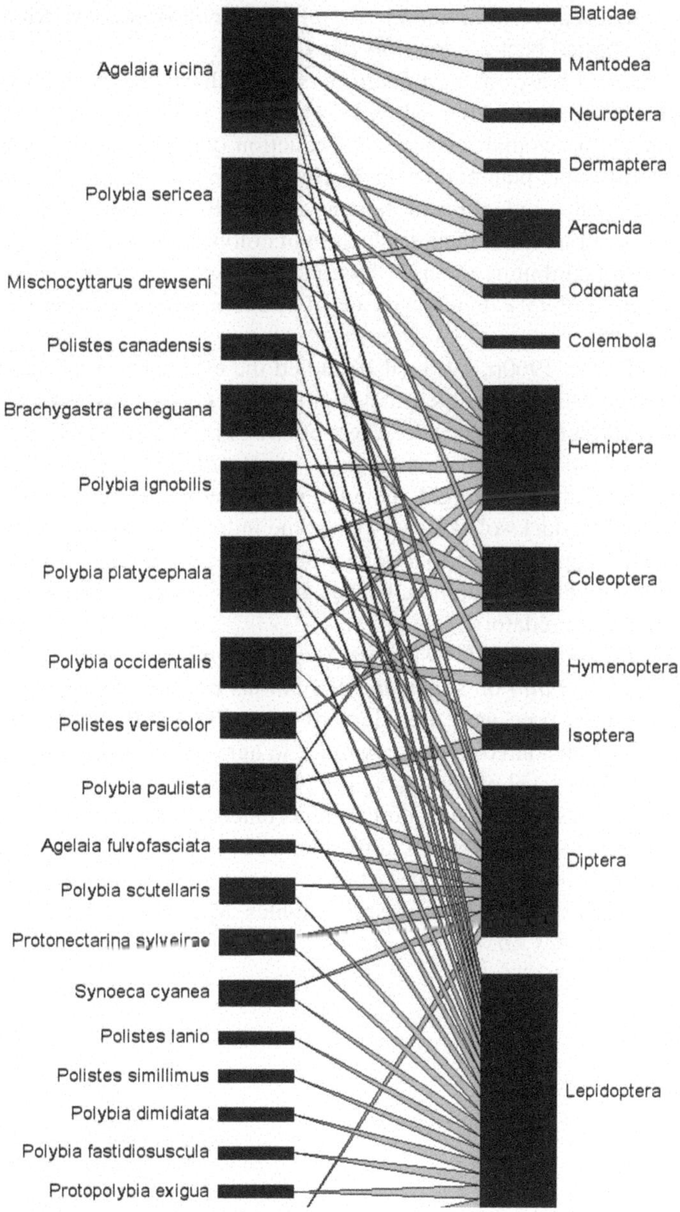

Fig. 24.1 Ecological network between social wasps and their preys. The left columns on each side represent social wasps, while the right columns represent order of captured prey. The height of each rectangle is directly proportional to the connections between wasps and their prey

used are the fungi *Beauveria bassiana, Nomuraea rileyi* (Moniliales: Moniliaceae), and the widely applied bacteria *Bacillus thuringiensis.*

The use of social wasps as pest controllers is an alternative scarcely tested and consequently also applied infrequently. However, the predatory behavior of social wasps generated studies that revealed the reduction of damage caused by agricultural pests in different plantations. In North Carolina, USA, Rabb and Lawson (1957) verified a 68% reduction in damage caused by the caterpillar *Protoparce sexta* in tobacco culture, after the introduction of colonies of *Polistes exclamans* and *Polistes fuscatus.* Combining the use of wasps and chemicals, Lawson et al. (1961) verified a reduction of 95% in damage caused by *Pieris rapae* caterpillar in tobacco culture.

Morimoto (1960a, 1960b, 1961) also studied the efficiency of wasps as natural enemies of crop pests and verified that a colony of *Polistes* sp. uses about 2000 *P. rapae caterpillars* during its cycle. The work of Cornell et al. (1987) reported the importance of wasp predation on groups of *Hemileuca lucina* (Lepidoptera, Saturniidae). The authors highlighted the indirect influence of predation on the caterpillars; due to the attacks of *Polistes* spp., some individuals would leave the situation of aggregation and migrate to less favorable areas of feeding. This disturbance ended up damaging the development of the caterpillars, making them more susceptible to the attack of predators.

Studies by Prezoto et al. (1994) with *Polistes simillimus* and Giannotti et al. (1995) with *Polistes lanio* on sugarcane plantations revealed that between 90 and 95% of the captured prey were Lepidoptera caterpillars (Fig. 24.2). In this way, these caterpillars, considered to cause damage to agriculture, were reduced to rates considered below the level of damage to the plantation.

Lolato and Moraes (1997) verified the importance of the predation by Vespidae wasps on herbivorous caterpillars of citrus plants. In cotton plantation, Shang-Chiu (1976) registered a control of 70–80% of *Pieris rapae* and *Etiella zinckenella*, seven days after the introduction of *Polistes* spp. colonies. Marques et al. (2005) recorded the predation of *Polistes versicolor* on *Heraclides anchysiades* caterpillars in a crop

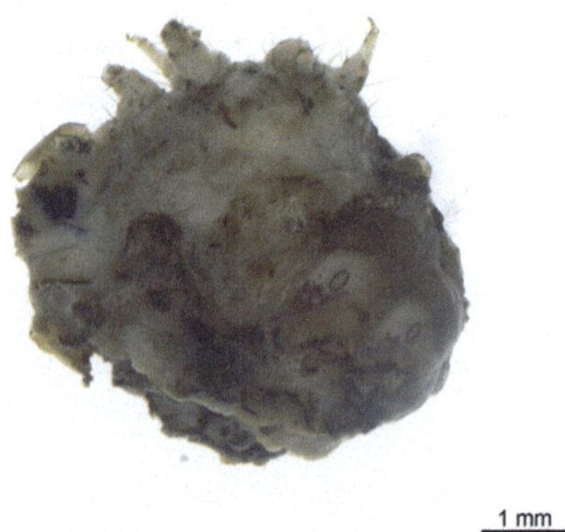

Fig. 24.2 Caterpillar crushed captured by *Polistes canadensis* in urban area

Fig. 24.3 Workers of *Polistes versicolor* sharing one caterpillar crushed in the colony

of *Citrus senensis*. Already, Prezoto et al. (2006) carried out the identification of prey captured by *P. versicolor* colonies and evidenced the potential of the species as agent for biological control of pests, mainly of deforming caterpillars. The same species also utilized caterpillars captured in eucalyptus plantation as protein resource (Elisei et al. 2010) (Fig. 24.3). Santana-Júnior et al. (2012) reported the important action of *Polybia* wasps in the control of caterpillars in cucumber plantation. Jacques et al. (2018) reported the survey of the species of social wasps that forage in kale (*Brassica oleraceae* var. *acephala*), verifying that *Polybia ignobilis, Protonectarina sylveirae*, and *Protopolybia sedula* were the most common wasp species.

Several other studies of forage activity analysis have revealed the predation of social wasps occurring mainly on caterpillars (Butignol 1992; Prezoto et al. 1994; Giannotti et al. 1995; Prezoto and Machado 1999a, b; Richter 2000 Andrade and Prezoto 2001; Prezoto et al. 2006; Bichara et al. 2009). However, the use of social wasps in IPM is still very incipient. In an economy in which the use of organic products, free of pesticides, grows in an exponential curve, the use of social wasps in management programs is a tool that should be stimulated and disseminated among agroecological producers. The preservation and conservation of Polistinae community in areas of agricultures is the easy way for their controller action.

24.3 Social Wasp Colony Transfers: Tool in the Integrated Pest Management

In an Integrated Pest Program with the use of biocontroller, an important step is the introduction of the natural enemy. This controller species must be manageable and can be manipulated in order to increase the population or simply introduced in application areas. In this context, studies that intend to use social wasps should verify the feasibility of colonies management, analyzing the possibility of transferring their nests to artificial shelters. For these purposes, wood, metal, or plastic shelters have been used (Kirkton 1970; Gillaspy 1979; Shang-Chiu 1976; Turillazzi 1980; Prezoto and Machado 1999a, b; Elisei et al. 2012).

After transfers, the colonies should be analyzed as to their longevity and forage behavior. The success in the method and viability of the species must be assessed by the first returns with prey. Another important fact is the knowledge of the action range extension of the species that will be used in the management. This reduces the number of colonies managed, preventing overlapping of foraging areas, reducing the chance of competition among workers, which can harm the success of transfers.

Polistes is a genus that stands out in transfer studies. This is because their nests do not have a protective envelope and are fixed to the substrate by only a single peduncle. These characteristics make the group easier to manipulate and to observe their behavior. Prezoto and Machado (Prezoto and Machado 1999a, b) obtained success when they transferred colonies of *P. simillimus* to shelters manufactured with wood and metal for corn planting area. Butignol (1992) carried out transfers of *P. versicolor* colonies to wooden shelters in an urban area, and verified the viability of this process with this species. Elisei et al. (2012), also with *P. versicolor*, relocated colonies in plastic shelters, fixed on eucalyptus plants, in order to verify the viability of applying the species in management (Fig. 24.4). The authors registered success of 85% and the average longevity of 80 days.

24.3.1 Selective Insecticides on Social Wasp

According to Zeck (1985), it is possible to adopt the joint use of alternative control methods to integrate management of certain pests, thus allowing the rational use of phytosanitary products. However, what we observe is the use of these products

Fig. 24.4 Colony of *Polistes versicolor* translocated to artificial shelter in eucalyptus plantation

without control. The indiscriminate use of insecticides can promote the occurrence of a phenomenon known as the resurgence of pests (Gerson and Cohen 1989; Soares et al. 1995), increasing the number of harmful insects in plantations after apparent reduction. An alternative to integrated pest management is the use of selective chemical plant protection products, which are compounds that have potential to control the pest with the least impact on natural enemies (Soares & Busoli 2000, Degrande et al. 2002). This is because the maintenance of natural enemies is important to avoid resurgence of the harmful species.

Studies of selective insecticides have already been developed with some species of social wasps. Picanço et al. (1988), evaluated the selectivity of the following active components of the insecticides: carbaril, deltamethrin, methyl parathion, permethrin, and trichlorophon, used in the control of *Ascia monuste orseis (*Godart) (Lepidoptera: Pieridae) in relation to its main predator *P. ignobilis* (Haliday). According to the data, it was observed that methyl parathion and carbaril were not selective to *P. ignobilis*. Deltamethrin presented intermediate degree of selectivity. The most selective active components were trichlorophenol and permethrin.

Santos et al. (2003) studied the selectivity of the active principles present in different insecticides traditionally recommended for the control of *Spodoptera frugiperda* (Lepidoptera, Noctuidae) against the natural enemy *Polybia sericea*. The active ingredients of the insecticides tested were: trichlorfon, carbaril, deltamethrin, lufenuron. The lufenuron-containing insecticide was the most selective. Trichlorophon was classified as an intermediate insecticide, with carbaril and deltamethrin not being selective.

Galvan et al. (2002), testing the selectivity of eight insecticides on social wasps, found that *Brachygastra lecheguana* is the most acephate-tolerant species as *P. sylveirae* is the most susceptible, and *P. exigua* remains in intermediate situation. *P. exigua* is more tolerant to deltametrine than *B. lecheguana* and *P. sylveirae*. Already, *B. lecheguana* and *P. exigua* are more tolerant to abamectin and carbaril than *P. sylveirae*.

For the appropriate use of various insecticides it is important to observe the proper dosage of the insecticides in their preparation. The type of selectivity will depend not only on the active ingredient of each insecticide, but also on the characteristics of the natural enemies. It is likely that the variation in the degree of selectivity may be related to the metabolic velocity of these insecticides in the wasp species or even by the molecular weight of the active substances (Santos et al. 2003).

Another important factor is analysis of the forage times of the natural enemies. Social wasps present high activity in the warmest period of the day, between 10:00 h and 15:00 h (Ribeiro Jr et al. 2008; Elisei et al. 2010, Elisei et al. 2013). Thus, avoiding the application of insecticides in this period would reduce the chances of killing wasps.

24.4 Polistinae in Bioindication and Environmental Damage Reduction

The relationship between individuals and the environment can highlight the importance of preserving and conserving ecosystems. The study of these relationships produces information that could be applied in the human economy. Bioindication is one of these applications, constituting a very widespread and useful tool in studies with different animal groups. Social wasps are cited by authors, who verified species' preferences to certain types of environments. Social wasps' species can be found living only in preserved environments; there are species that support anthropization; and there are those who nest almost exclusively in human constructions (Santos et al. 2007; Souza et al. 2010; de Souza and Zanuncio 2012). Thus, finding some species could be a way to analyze the ecological conditions of the area.

Social wasps feed their brood with caterpillars. This fact allows the analysis of accumulated substances in the immatures' bodies in the colony. An example was the diagnosis of the presence of heavy metals in the environment with the examination of individuals in colonies of *Dolichovespula saxonica* and *Polistes dominulus* (Kowalczyk and Watala 1989; Urbini et al. 2006). There is not researchs about accumulation of inorganic substances in social wasps, in Neotropical regions. This could be a important focus of study, applied in the ambiental indicator programs.

Another relevance of social wasps is that their presence in integrated management can generate a reduction in the use of agricultural defensives. This is because the chemical control of agricultural pests presents numerous problems to the environment. One of them is that phytosanitaries do not present high selectivity, as has already been mentioned in this chapter. This fact causes elimination of natural enemies, favoring the resurgence of the pest, as well as the emergence of secondary pests (Gallo et al. 2002; Pedigo 1999; Reis et al. 1998). Another is the environmental damage resulting from soil and water contamination, which can result in the elimination of entire ecosystems (Veiga et al. 2006). In cotton plantations, in some cases, 12–14 pesticide sprays are needed to control insect pests and mites (Richetti et al. 2004). Soares et al. (2002), in a study on the impact of chemical insecticides on natural enemies and production costs, revealed that in the areas where IPM was adopted, there was a reduction in the need for chemical applications in the agrosystem.

Brazil is the world champion in pesticide application and in the last decade consumption has increased by 190% (Rigotto et al. 2014). As an alternative to this scenario, there is use of insects in controlling agricultural pests, which reduces the need of phytosanitary products in plantations. As already mentioned, social wasps have an effective action of predation on numerous groups of lepidoptera. In this way, the simple preservation and conservation of the wasp colonies can be a way of decreasing plagues' injuries.

24.5 Social Wasps as Floral Visitors

The nectar resource is the main food of adults of Polistinae, being offered to immatures as an alternative source of nutrition. It is collected mainly in flowers, social wasps being an important optional pollinator. Several studies report the pollinating behavior of Polistinae, such as Mechi (1996), in which it was verified that up to 25 species visited plants in the Brazilian savanna. Already Vitali-Veiga and Machado (2001) have registered six species visiting nectars of *Gleditsis triacanthos* in the urban environment of Rio Claro, São Paulo. Silva-Pereira and Santos (2006) recorded 13 species of social wasps visiting 39 species of plants, in Caatinga. Aguiar and Santos (2007) recorded 13 species visiting 22 species of plants, in work done in the state of Bahia. Suhs et al. (2009) verified a Polistinae richness of 25 collecting nectar in red Aroeira flowers, in the state of Rio Grande do Sul, being more frequent than bees. Clemente et al. (2013) verified that in 27 plant morphotypes, 15 species of social wasps made visits to flowers nectaries, in a study conducted in Ibitipoca State Park, Minas Gerais state.

Although the pollination of social wasps is considered less than compared to that performed by bees, they also play an important role in the ecosystems. They could be an alternative for the disappearance of bees. Thus, the reduction in the population of Apidae and the increase of the interaction network studies highlight Polistinae as an important group in the maintenance of natural environments and environmental preservation (Santos et al. 2010; Mello et al. 2011; Clemente et al. 2013).

24.6 Social Wasps in Medicine

The venom produced by social wasps presents a variety of chemical compounds and therefore has been studied for its applicability in medicine to verify its pharmacological effects. A study with *Polybia paulista* verified the anticancer property of its poison (Leite et al. 2015).

Vinhote (2015), also studying *P. paulista,* verified the action of tripanocida as an effective treatment against Chagas' disease. Wang et al. (2012, 2013), for the same social wasp, verified that peptides found in its poison can eliminate *Escherichia coli*, *Staphylococcus aureus*, *Staphylococcus epidermidis*, and *Bacillus subtilis*. Considering antibiotic activity, Mendes et al. (2004) verified the bactericidal action of the *Agelaia pallipes* poison against gram-positive and gram-negative bacteria. Neves et al. (2016) verified the antimycobacterial activity of peptides of the social wasp *Polybia dimorpha.* Other researchers revealed that there is potential in the development of painkillers from the poisons of social wasps, such as *Parachartegus fraternus*, *Apoica* sp., *Agelaia pallipes* (Baptista-Saidemberg et al. 2010; Gonçalves et al. 2016).

Although wasp venom can be used for medicinal purposes, it can also cause injury to humans. Hymenopterans (ants, bees, wasps,) are often related to accident with stings. Cases with bees are the most frequent; however, wasp accidents are quite relevant (Martins and Becil Junior 2018). Accidents occur due to aggressive behavior for nest protection; often it is installed in human buildings, due to the synanthropic habit of many species. In severe cases, anaphylactic shock may result in serious risk to the life of the person being stung.

24.7 When the Polistinae Are Plague

In this chapter, we focused on the importance of wasps against pest species. However, in some crops, the wasps can be the agents of plant damage. In Brazil, *Synoeca cyanea* was described as the cause of damage on fruits of *Myrciaria* sp., Brazilian Grape Tree (De Souza et al. 2010) and *Psidium* sp., Guava (Brugger et al. 2011) (Fig. 24.5). Barbosa et al. (2014) verified *S. cyanea* been the only social wasp specie which was drilling the skin of mango fruits. They reported too some social wasps' species explored fruits with preexistent orifices, caused by *S. cyanea* and other insects, as *Atta* ants and *Trigona spinipes*.

Hickel and Schuck (1995) described six genera of *Polistinae* that attacked grape fruits (Hickel and Schuck 1995). Brugger et al. (2017) registered *Polybia scutellaris* damaging fruit of *Syzygium jambos* (L.) (Myrtaceae), visiting fruit on the ground and fruit on the branches. A study by Santos and Presley (2010) described the action of social wasp using cashew fruits as a glucidic resource. The authors registered that wasps do not damage intact fruits but their presence on fruits or nests associated with the orchard can decrease harvest efficiency by intimidating farm workers. This intimidation by social wasp can be a factor present in a lot of orchards, which would result in accidents.

Fig. 24.5 *Synoeca cyanea* obtaining carbohydrates from *Psidium* spp. (Guava) fallen on the ground

This chapter brings forward Polistinae species as relevants economics factors, in different sectors of the human economy. Like that, the preservation of this group should always be considered in the process of occupation urban and rural. At last, social wasps can be the hero or the villain, everything depends on point of view.

References

Aguiar CML, Santos GM d M (2007) Compartilhamento de Recursos Florais por Vespas Sociais (Hymenoptera: Vespidae) e Abelhas (Hymenoptera: Apoidea) em uma Área de Caatinga. Neotrop Entomol 36(6):836–842

Andrade FR, Prezoto F (2001) Horários de atividade forrageadora e material coletado por *Polistes ferreri* Saussure, 1853 (Hymenoptera, Vespidae), nas diferentes fases de seu ciclo biológico. Revta Bras Zoociências 3(1):117–128

Baptista-Saidemberg NB, Saidemberg DM, Souza BM, César-Tognoli LMM, Ferreira VMR, Mendes MA, Cabrera MPS, Neto JR, Palma MS (2010) Protonectin(1-6): a novel chemotactic peptide from the venom of the social wasp *Agelaia pallipes pallipes*. Toxicon 56:880–889

Barbosa BC, Paschoalini MF, Prezoto F (2014) Temporal activity patterns and foraging behavior by social wasps (Hymenoptera, Polistinae) on fruits of *Mangifera indica* L. (Anacardiaceae). Sociobiology 61(2):239–242

Bell E, Sumner S (2013) In: eLS (ed) Ecology and social organisation of wasps. Wiley, Chichester

Bichara CC, Santos GMD, Resende JJ, Da Cruz JD, Gobbi N, Machado VLL (2009) Foraging behavior of the swarm-founding wasp, Polybia (Trichothorax) sericea (Hymenoptera, Vespidae): prey capture and load capacity. Sociobiology 53(1):61–69

Brugger BP, Souza L, Souza AR, Prezoto F (2011) Social Wasps (*Synoeca cyanea*) damaging *Psidium* sp (Myrtaceae) Fruits in Minas Gerais State, Brazil. Sociobiology 57(3):533–535

Brugger BP, Castro BMC, Prezoto F, Serrão JE, Zanuncio JC (2017) Feeding by the social wasp *Polybia scutellaris* (Hymenoptera: Vespidae) on *Syzygium jambos* (Myrtaceae) fruits in Minas Gerais, Brazil. Fla Entomol 100(1):172–173

Butignol CA (1992) Observações sobre a bionomia da vespa predadora *Polistes versicolor* (Olivier, 1791) (Hymenoptera, Vespidae) em Florianópolis/SC. Anais da Sociedade Entomológica do Brasil 19:201–206

Carpenter JM, Marques OM (2001) Contribuição ao estudo dos vespídeos do Brasil (Insecta, Hymenoptera, Vespoidea, Vespidae). Publicações digitais, 2. Universidade Federal da Bahia

Clemente MA, Lange D, Dáttilo W, Del-Claro K, Prezoto F (2013) Social wasp-flower visiting guild interactions in less structurally complex habitats are more susceptible to local extinction. Sociobiology 60(3):337–344

Cornell JC, Stamp NE, Bowers MD (1987) Developmental change in aggregation, defense and escape behavior of buckmoth caterpillars, *Hemileuca lucina* (Saturniidae). Behav Ecol Sociobiol 20(6):383–388

de Souza MM, Zanuncio JC (2012) Marimbondos: vespas sociais (Hymenoptera: Vespidae). Editora UFV, 79 p

De Souza AR, Venâncio DFA, Prezoto F (2010) Social wasps (Hymenoptera: Vespidae: Polistinae) damaging fruits of *Myrciaria* sp. (Myrtaceae) Sociobiology 55(2):297–299

Degrande PE, Reis PR, Carvalho GA, Belarmino LC (2002) Metodologia para avaliar o impacto de pesticidas sobre inimigos naturais. p. 75–81. In: Parra JRP, Botelho PSM, Corrêa-Ferreira BS, Bento JMS (eds) Controle Biológico no Brasil: Parasitóides e predadores, Manole, São Paulo, 635 p

Elisei T, Nunes JV, Ribeiro Junior C, Fernandes Junior AJ, Prezoto F (2010) Uso da vespa social *Polistes versicolor* no controle de desfolhadores do eucalipto. Pesq Agrop Brasileira 45(9):958–964

Elisei T, Nunes JV, Ribeiro Junior C, Fernandes JAJ, Prezoto F (2012) Management of social wasp colonies in Eucalyptus plantations (Hymenoptera:Vespidae). Sociobiology 59(4):1167–1174

Elisei T, Nunes J, Junior CR, Junior AF, Prezoto F (2013) What is the Ideal Weather for Social Wasp Polistes versicolor (Olivier) go to Forage?. EntomoBrasilis 6(3):214–216

Jacques GC, Pires E, Hermes MG, Faria LDB, Souza MM, Silveira LCP (2018) Evaluating the efficiency of different sampling methods to survey social wasps (Vespidae: Polistinae) in an anthropized environment. Sociobiology 65(3):515

Gallo D, Nakano O, Silveira-Neto S, Carvalho RPL, Baptista GC, Berti-Filho E, Parra JRP, Alves SB, Vendramin JD, Marchini LC, Lopes JRS, Omoto C (2002) Manual de Entomologia Agrícola, Piracicaba, Fealq, 649 p

Galvan TL, Picanço MC, Bacci L, Pereira EJG, Crespo ALB (2002) Seletividade de oito inseticidas a predadores de lagartas em citros. Pesq Agrop Brasileira 37(2):117–122

Gerson V, Cohen E (1989) Ressurgences of spider mites (Acari: Tetranychidae) induced by synthetic pyrethroids. Exp Appl Acarol 6(1):29–46

Giannotti E, Prezoto F, Machado VLL (1995) Foraging activity of *Polistes lanio lanio* (Fabr.) (Hymenoptera, Vespidae). Anais da Sociedade Entomológica do Brasil 24(3):455–463

Gillaspy JE (1979) Management of *Polistes* wasps for caterpillar predation. Southwestern Entomol 4:334–352

Gonçalves JC, Rangel M, Mayer A, Mortari MR (2016) Antinociceptive properties of the mastoparan peptide Agelaia-MPI isolated from social wasps. Toxicon 120:15–21

Gravena S (1992) Controle biológico no Manejo Integrado de Pragas. Pesq Agrop Brasileira 27:281–299

Greene A (1991) Dolichovespula and Vespula, p. 263–304. In: Ross KG, Matthews RW (eds) The social biology of wasps. Ithaca and Londom, Comstock Publishing Associates, 678p

Hamilton WD (1964a) Genetical evolution of social behaviour I. J Theor Biol 7:1–16

Hamilton WD (1964b) Genetical evolution of social behaviour 2. J Theor Biol 7:17–52

Hickel ER, Schuck E (1995) Vespas e abelhas atacando a uva no Alto Vale do Rio do Peixe. Agropecuaria Catarinense 8:38–40

Hunt JH, Brown PA, Sago KM, Kerker JA (1991) Vespid wasps eat pollen (Hymenoptera: Vespidae). J Kansas Entomol Soc 64(2):127–130

Kirkton RM (1970) Habitat management and its effects on populations of *Polistes* and *Iridomyrmex*. Proc Tall Timbers Conf 2:243–246

Kowalczyk JK, Watala C (1989) Content of some heavy metal ions in various developmental stages of the social wasp *Dolichovespula Saxonica* (Fabr.) (Hymenoptera, Vespidae). Bull Environ Contam Toxicol 43(3):415–420

Lawson FR, Rabb RL, Guthrie FE, Bowery TG (1961) Studies of an integrated control system for hornworms on tobacco. J Econ Entomol 54:93–97

Leite NB, Aufderhorst-Roberts A, Palma MS, Connell SD, Neto JR, Beales PA (2015) PE and PS lipids synergistically enhance membrane Poration by a peptide with anticancer properties. Biophys J 109:936–947

Lolato A, Moraes JC (1997) Ocorrência da "lagarta minadora das folhas de citros" *Phyllocnistis citrella* Stainton, 1856 (Lepidoptera: Gracilariidae) em Lavras, sul de Minas Gerais. Ciência e Agrotecnologia 21(4):521–522

Marques OM, Carvalho CAL, Santos GMM, Bichara-Filho CC (2005) Defensive behavior of caterpillars of *Heraclides anchysiadescapys* (Lepidoptera: Papilionidae) against the social wasp *Polistes versicolor versicolor* (Hymenoptera: Vespidae). Magistra 17(1):28–32

Martins A, Becil Junior MR (2018) Acidentes com animais peçonhentos da ordem Hymenoptera (abelhas e vespas): principais complicações em países da América Latina e Caribe accidents with venomous animals of the order hymenoptera (bees and wasps): main complications in Latin American and Caribbean countries. Braz J Health Rev 1(1):220–232

Mechi MR (1996) Levantamento da fauna de vespas Aculeata na vegetação de duas áreas de cerrado, Ph.D. Thesis, Universidade Federal de São Carlos, São Carlos

Mello MAR, Santos GMM, Mechi MR, Hermes MG (2011) High generalization in flower-visiting networks of social wasps. Acta Oecol 37:37–42

Mendes MA, Souza BM, Marques MR, Palma MS (2004) Structural and biological characterization of two novel peptides from the venom of the neotropical social wasp *Agelaia pallipes pallipes*. Toxicon 44:67–74. 46

Morimoto R (1960a) Polistes wasps as natural enemies of agricultural and forest pests. I. Studies on the social Hymenoptera of Japan. X Sci Bull Fac Agric Kyushu Univ 18:109–116

Morimoto R (1960b) Polistes wasps as natural enemies of agricultural and forest pests. II. Studies on the social Hymenoptera of Japan. XI Sci Bull Fac Agric Kyushu Univ 18:117–132

Morimoto R (1961) Polistes wasps as natural enemies of agricultural and forest pests. III. Studies on the social Hymenoptera of Japan. XII Sci Bull Fac Agric Kyushu Univ 18:243–252

Nádia TL, Machado IC, Lopes AV (2007) Polinização de *Spondias tuberosa* Arruda (Anacardiaceae) e análise dapartilha de polinizadores com *Ziziphus joazeiro* Mart. (Rhamnaceae),espécies frutíferas e endêmicas da caatinga. Revista Brasil Bot 30(1):89–100

Neves RC, Trentini MM, Castro e Silva J, Simon KS, Bocca AL, Silva LP (2016) Antimycobacterial activity of a new peptide Polydim-I isolated from Neotropical social wasp *Polybia dimorpha*. PLoS One 11(3):e0149729

O'Donnell S (1995) Necrophagy by Neotropical swarm-founding wasps (Hymenoptera: Vespidae, Epiponini). Biotropica 27:133–136

Parra JRP, Botelho PSM, Corrêa-Ferreira BS, BENTO JMS (2002) Controle Biológico no Brasil: parasitóides e predadores. São Paulo, Ed. Manole, 609p

Pedigo LP (1999) Entomology and pest management, 3rd edn. Prentice Hall, Englewood, 691 p

Picanço M, Ribeiro LJ, Leite GLD, Gusmão MR (1988) Seletividade de inseticidas a *Polybia ignobilis* (Halyday) (Hymenoptera, Vespidae) predador de *Ascia monuste orseis* (Godart) (Lepidoptera: Pieridae). An Soc Entomol Bras 27:85–90

Prezoto F, Machado VLL (1999a) Ação de *Polistes* (Aphanilopterus) simillimus Zikán (Hymenoptera, Vespidae) no controle de Spodoptera frugiperda (Smith) (Lepidoptera, Noctuidae). Rev Bras Zool 16:841–851

Prezoto F, Machado VLL (1999b) Transferência de colônias de vespas (*Polistes* simillimus Zikán, 1951) (Hymenoptera, Vespidae) para abrigos artificiais e sua manutenção em uma cultura de Zea mays L. Rev Bras Entomol 43:239–241

Prezoto F, Giannotti E, Machado VLL (1994) Atividade forrageadora e material coletado pela vespa social *Polistes simillimus* Zikán, 1951 (Hymenoptera, Vespidae). Insecta 3:11–19

Prezoto F, Santos-Prezoto HH, Machado VLL, Zanuncio JC (2006) Prey captured and used in *Polistes versicolor* (Olivier) (Hymenoptera: Vespidae) nourishment. Neotrop Entomol 35:707–709

Quirino ZGM, Machado IC (2001) Biologia da polinização e da reprodução de três espécies de *Combretum loefl.* (Combretaceae). Rev Bras Bot 24:181–193

Rabb RL, Lawson FR (1957) Some factors influencing the predation of *Polistes* wasps on tobacco hornworm. J Econ Entomol 50:778–784

Ratnieks F, Wenseleers T (2008) Altruism in insect societies and beyond: voluntary or enforced? Trends Ecol Evol 23(1):45–52

Reis PR, Chiavegato LG, Moraes GJ, Alves EB, Sousa EO (1998) Seletividade de agroquímicos ao ácaro predador *Iphiseiodes zuluagai* Denmark & Muma (Acari: Phytoseiidae). Anais da Sociedade Entomológica do Brasil, Londrina 27(2):265–274

Richetti A, Melo Filho GA, Lamas FM, Staut LA, Fabrício AC (2004) Estimativa do custo de produção de algodão, safra 2004/05, para Mato Grosso do Sul e Mato Grosso. Embrapa Pecuária Oeste. Embrapa, ComunicadoTécnico, Dourados, p 9116

Richter MR (2000) Social wasp (Hymenoptera: Vespidae) foraging behavior. Annu Rev Entomol 45:121–150

Rigotto RM, Vasconcelos DP, Rocha MM (2014) Uso de agrotóxicos no Brasil e problemas para a saúde pública. Cad Saúde Pública 30(7):1–3

Santos GMM, Presley SJ (2010) Niche overlap and temporal activity patterns of social wasps (Hymenoptera: Vespidae) in a Brazilian Cashew Orchard. Sociobiology 56(1):121–131

Santos LP, Resende JJ, Santos GMM, Bichara-Filho CC, Reis VPGS (2003) Seletividade de inseticidas a *Polybia* (Trichothorax) sericea (Olivier, 1791) (Hymenoptera, Vespidae) em condições de laboratório. Rev Bras Zoociências 5:1). 33–1). 44

Santos GMM, Filho CCB, Resende JJ, Cruz JD, Marques OM (2007) Diversity and community structures of social wasps (Hymenoptera: Vespidae) in three ecosystems in Itaparica Island, Bahia State, Brazil. Neotrop Entomol 36:180–185

Santos GMM, Aguiar CML, Mello MAR (2010) Flower-visiting guild associated with the Caatinga flora: trophic interaction networks formed by social bees and social wasps with plants. Apidologie 41:466–475

Saraiva NB, Prezoto F, Fonseca MG, Blassioli-Moraes MC, Borges M, Laumann RA, Auad AM (2017) The social wasp *Polybia fastidiosuscula* Saussure (Hymenoptera: Vespidae) uses herbivore induced maize plant volatiles to locate its prey. J Appl Entomol 141:620–629

Shang-Chiu (1976) A preliminary study on the bionomics of hunting wasps and their utilization in cotton insect control. Acta Entomol Sinica 19:313–318. (Inst. Agr.For.Sci. SHANG-CHIU Dist. Honan Prov., Inst. Agr.For.Sci. Yu-Cheng Co., Sci. Res. Sta. Degng, Jiao Prod.Brig., Ku-Chen Peop.Comm. Shang-Chiu Co.)

Silva-Pereira V, Santos GMM (2006) Diversity in bee (Hymenoptera: Apoidea) and social wasp (Hymenoptera: Vespidae, Polistinae) community in "campos rupestres", Bahia, Brazil. Neotrop Entomol 35:165–174

Soares JJ, Busoli AC (2000) Efeito de inseticidas em insetos predadores em culturas de algodão. Pesq Agrop Brasileira 35(9):1889–1894

Soares JJ, Busoli BA, Braz AC (1995) Impacto de herbicidas sobre artrópodos benéficos associados ao algodoeiro. J Econ Entomol 30(9):1135–1140

Soares JJ, Freire EC, dos Santos IF, Moulin MC, de Alencar AR, Silva MS, Silva EB (2002) Resultados da pesquisa com algodão herbáceo em Barreiras, BA: safra 2000/2001. EMBRAPA Algodão, Campina Grande. 3p. EMBRAPA Algodão. Comunicado Técnico, 144

Souza MM, Ladeira TE, Elpino-Campos A, Campos PC, Louzada J (2010) Ecologia de vespas sociais (Hymenoptera, Vespidae) no Campo Rupestre na Área de Proteção Ambiental, APA, São José, Tiradentes, MG. MGBiota 3:01–30

Suhs RB, Somavilla A, Köhler A, Putzke J (2009) Vespídeos (Hymenoptera, Vespidae) vetores de pólen de *Schinus terebinthifolius* Raddi (Anacardiaceae), Santa Cruz do Sul, RS, Brasil. Revista Brasileira de Biociências 7:2

Santana-Júnior PA, Gonring AHR, Picanço MC, Ramos RS, Martins JC, Ferreira DO (2012) Natural biological control of Diaphania spp. (Lepidoptera: Crambidae) by Social Wasps. Sociobiology, 59(2): 561-571

Turillazzi S (1980) Use of artificial nests for rearing and studying *Polistes* wasps. Psyche 87(1/2):131–140

Urbini A, Sparvoli E, Turillazzi S (2006) Social paper wasps as bioindicators: a preliminary research with *Polistes* dominulus (Hymenoptera Vespidae) as a trace metal accumulator. Chemosphere 64(5):697–703

Veiga MM, Silva DM, Veiga LBE, Faria MVC (2006) Análise da contaminação dos sistemas hídricos por agrotóxicos numa pequena comunidade rural do Sudeste do Brasil. Caderno de Saúde Pública 22(11):2391–2399

Vinhote JFC (2015) Ação tripanocida de um mastoparano de *Polybia paulista* e seu possível mecanismo de ação. Tese de doutorado, 109p

Vitali-Veiga MJ, Machado VLL (2001) Entomofauna visitante de *Gleiditsia triacanthos* L.–Leguminosae durante o seu período de floração. Revista Bioikos 15(1):29–38

Wang K, Yan J, Chen R, Dang W, Zhang B, Zhang W, Wang R (2012) Membrane-active action mode of *Polybia*-CP, a novel antimicrobial peptide isolated from the venom of *Polybia paulista*. Antimicrob Agents Chemother 56(6):3318–3323

Wang K, Yan J, Dang W, Liu X, Chen R, Zhang J, Yang Z (2013) Membrane active antimicrobial activity and molecular dynamics study of a novel cationic antimicrobial peptide *Polybia*-MPI, from the venom of *Polybia paulista*. Peptides 39:80–88

Wilson EO (1971) The Insect Societies. Cambridge, The Belknap Press Harvard University. 548p

Zeck WM (1985) The future of narrow versus broad – spectrum inseticides. In: Hilton JL (ed) Agricultural chemical on the future. Rowman & Allanheld, New Jersey, pp 293–309, Beltsville symposia in agricultural research, 8: 464P

Index

F. Prezoto et al. (eds.), *Neotropical Social Wasps*,
https://doi.org/10.1007/978-3-030-53510-0

Q

The manufacturer's authorised representative in the EU is Springer Nature Customer Service Centre GmbH, Europaplatz 3, 69115 Heidelberg, Germany. If you have any concerns regarding our products, please contact ProductSafety@springernature.com

Printed and bound by CPI Group (UK) Ltd, Croydon, CR0 4YY
30/09/2026
02225506-0013